Application of Electroencephalography (EEG) Signal Analysis in Disease Diagnosis

Application of Electroencephalography (EEG) Signal Analysis in Disease Diagnosis

Guest Editors

Serena Dattola
Fabio La Foresta

Basel • Beijing • Wuhan • Barcelona • Belgrade • Novi Sad • Cluj • Manchester

Guest Editors

Serena Dattola
IRCCS Centro Neurolesi
Bonino-Pulejo
Messina
Italy

Fabio La Foresta
Mediterranea University of
Reggio Calabria
Reggio Calabria
Italy

Editorial Office
MDPI AG
Grosspeteranlage 5
4052 Basel, Switzerland

This is a reprint of the Special Issue, published open access by the journal *Applied Sciences* (ISSN 2076-3417), freely accessible at: https://www.mdpi.com/journal/applsci/special_issues/EEG_signal_analysis.

For citation purposes, cite each article independently as indicated on the article page online and as indicated below:

Lastname, A.A.; Lastname, B.B. Article Title. *Journal Name* **Year**, *Volume Number*, Page Range.

ISBN 978-3-7258-2911-8 (Hbk)
ISBN 978-3-7258-2912-5 (PDF)
https://doi.org/10.3390/books978-3-7258-2912-5

© 2024 by the authors. Articles in this book are Open Access and distributed under the Creative Commons Attribution (CC BY) license. The book as a whole is distributed by MDPI under the terms and conditions of the Creative Commons Attribution-NonCommercial-NoDerivs (CC BY-NC-ND) license (https://creativecommons.org/licenses/by-nc-nd/4.0/).

Contents

About the Editors vii

Jhosmary Cuadros, Lucía Z-Rivera, Christian Castro, Grace Whitaker, Mónica Otero, Alejandro Weinstein, et al.
DIVA Meets EEG: Model Validation Using Formant-Shift Reflex
Reprinted from: *Appl. Sci.* 2023, 13, 7512, https://doi.org/10.3390/app13137512 1

Héctor Juan Pelayo-González, Verónica Reyes-Meza, Ignacio Méndez-Balbuena, Oscar Méndez-Díaz, Carlos Trenado, Diane Ruge, et al.
Quantitative Electroencephalographic Analysis in Women with Migraine during the Luteal Phase
Reprinted from: *Appl. Sci.* 2023, 13, 7443, https://doi.org/10.3390/app13137443 21

Koun-Tem Sun, Kai-Lung Hsieh and Shih-Yun Lee
Using Mental Shadowing Tasks to Improve the Sound-Evoked Potential of EEG in the Design of an Auditory Brain–Computer Interface
Reprinted from: *Appl. Sci.* 2023, 13, 856, https://doi.org/10.3390/app13020856 33

Alka Rachel John, Zehong Cao, Hsiang-Ting Chen, Kaylena Ehgoetz Martens, Matthew Georgiades, Moran Gilat, et al.
Predicting the Onset of Freezing of Gait Using EEG Dynamics
Reprinted from: *Appl. Sci.* 2023, 13, 302, https://doi.org/10.3390/app13010302 57

Irina Tarasova, Olga Trubnikova, Darya Kupriyanova, Irina Kukhareva, Irina Syrova, Anastasia Sosnina, et al.
Effect of Carotid Stenosis Severity on Patterns of Brain Activity in Patients after Cardiac Surgery
Reprinted from: *Appl. Sci.* 2023, 13, 20, https://doi.org/10.3390/app13010020 66

Soo-Bin Lee, Ji-Won Kwon, Sahyun Sung, Seong-Hwan Moon and Byung Ho Lee
Delirium after Spinal Surgery: A Pilot Study of Electroencephalography Signals from a Wearable Device
Reprinted from: *Appl. Sci.* 2022, 12, 9899, https://doi.org/10.3390/app12199899 79

Mariia Chernykh, Bohdan Vodianyk, Ivan Seleznov, Dmytro Harmatiuk, Ihor Zyma, Anton Popov and Ken Kiyono
Detrending Moving Average, Power Spectral Density and Coherence: Three EEG-Based Methods to Assess Emotion Irradiation during Facial Perception
Reprinted from: *Appl. Sci.* 2022, 12, 7849, https://doi.org/10.3390/app12157849 89

Amedeo D'Angiulli, Guillaume Lockman-Dufour and Derrick Matthew Buchanan
Promise for Personalized Diagnosis? Assessing the Precision of Wireless Consumer-Grade Electroencephalography across Mental States
Reprinted from: *Appl. Sci.* 2022, 12, 6430, https://doi.org/10.3390/app12136430 103

Serena Dattola and Fabio La Foresta
Effect of Rehabilitation on Brain Functional Connectivity in a Stroke Patient Affected by Conduction Aphasia
Reprinted from: *Appl. Sci.* 2022, 12, 5991, https://doi.org/10.3390/app12125991 122

Hezam Albaqami, Ghulam Mubashar Hassan and Amitava Datta
Wavelet-Based Multi-Class Seizure Type Classification System
Reprinted from: *Appl. Sci.* 2022, 12, 5702, https://doi.org/10.3390/app12115702 132

Daniele Pirrone, Emanuel Weitschek, Primiano Di Paolo, Simona De Salvo and Maria Cristina De Cola
EEG Signal Processing and Supervised Machine Learning to Early Diagnose Alzheimer's Disease
Reprinted from: *Appl. Sci.* **2022**, *12*, 5413, https://doi.org/10.3390/app12115413 . . . . **149**

Maria Semeli Frangopoulou and Maryam Alimardani
qEEG Analysis in the Diagnosis of Alzheimer's Disease: A Comparison of Functional Connectivity and Spectral Analysis
Reprinted from: *Appl. Sci.* **2022**, *12*, 5162, https://doi.org/10.3390/app12105162 . . . . **162**

Yun Lo, Yi-Tse Hsiao and Fang-Chia Chang
Use Electroencephalogram Entropy as an Indicator to Detect Stress-Induced Sleep Alteration
Reprinted from: *Appl. Sci.* **2022**, *12*, 4812, https://doi.org/10.3390/app12104812 . . . . **176**

Lina Abou-Abbas, Imene Jemal, Khadidja Henni, Youssef Ouakrim, Amar Mitiche and Neila Mezghani
EEG Oscillatory Power and Complexity for Epileptic Seizure Detection
Reprinted from: *Appl. Sci.* **2022**, *12*, 4181, https://doi.org/10.3390/app12094181 . . . . **191**

Alexander A. Fingelkurts and Andrew A. Fingelkurts
Quantitative Electroencephalogram (qEEG) as a Natural and Non-Invasive Window into Living Brain and Mind in the Functional Continuum of Healthy and Pathological Conditions
Reprinted from: *Appl. Sci.* **2022**, *12*, 9560, https://doi.org/10.3390/app12199560 . . . . **205**

Cornelia Herbert
Decoding of Processing Preferences from Language Paradigms by Means of EEG-ERP Methodology: Risk Markers of Cognitive Vulnerability for Depression and Protective Indicators of Well-Being? Cerebral Correlates and Mechanisms
Reprinted from: *Appl. Sci.* **2022**, *12*, 7740, https://doi.org/10.3390/app12157740 . . . . **266**

About the Editors

Serena Dattola

Serena Dattola received her M.Sc. in Electronic Engineering from the University of Messina in 2013. In the following years, she won research grants at the University of Messina and the Mediterranea University of Reggio Calabria. She received her PhD from the Mediterranea University of Reggio Calabria in 2021 and was a research fellow in biomedical engineering. Her research activity mainly focuses on EEG signal processing, brain active source reconstruction, and brain connectivity analysis. Currently, she is a researcher at IRCCS Centro Neurolesi Bonino-Pulejo in Messina. She is a peer reviewer for international scientific journals and a speaker at international conferences.

Fabio La Foresta

Fabio La Foresta received his MS degree summa cum laude in Electronic Engineering at the University of Messina in 1998. Since then, he has been a Visiting Researcher at the French Research Institute ISEN. From 2000 to 2003, he was a PhD fellow in "Advanced Technologies for Engineering" at the University of Messina. From 2003 to 2005 he was a Researcher in biomedical engineering at the Mediterranea University of Reggio Calabria. In 2004 he received his PhD from the University of Messina. From 2005 to 2016 he was an Assistant Professor in electrical engineering and since 2016 he has been an Associate Professor in electrical engineering at the Mediterranea University of Reggio Calabria. Since 2006 he has carried out teaching in BS and MS courses in engineering. He conducts research activity in advanced systems for the processing of biomedical data. He is the author/coauthor of about 100 papers and he is the editor and reviewer of international journals and conferences.

Article

DIVA Meets EEG: Model Validation Using Formant-Shift Reflex

Jhosmary Cuadros [1,2,3], Lucía Z-Rivera [2,4], Christian Castro [2,4], Grace Whitaker [2], Mónica Otero [5,6], Alejandro Weinstein [2,4], Eduardo Martínez-Montes [7], Pavel Prado [8] and Matías Zañartu [1,2,*]

1. Department of Electronic Engineering, Universidad Técnica Federico Santa María, Valparaíso 2390123, Chile; jhosmary.cuadros@sansano.usm.cl
2. Advanced Center for Electrical and Electronic Engineering, Universidad Técnica Federico Santa María, Valparaíso 2390123, Chile; lucia.zepeda@postgrado.uv.cl (L.Z.-R.); christian.castro@uv.cl (C.C.); grace.whitaker@usm.cl (G.W.); alejandro.weinstein@uv.cl (A.W.)
3. Grupo de Bioingeniería, Decanato de Investigación, Universidad Nacional Experimental del Táchira, San Cristóbal 5001, Venezuela
4. Escuela de Ingeniería Civil Biomédica, Facultad de Ingeniería, Universidad de Valparaíso, Valparaíso 2350026, Chile
5. Facultad de Ingeniería, Arquitectura y Diseño, Universidad San Sebastián, Santiago 8420524, Chile; monica.otero@uss.cl
6. Centro Basal Ciencia & Vida, Universidad San Sebastián, Santiago 8580000, Chile
7. Brain Mapping Division, Cuban Neuroscience Center, Habana 11300, Cuba; eduardo@cneuro.edu.cu
8. Escuela de Fonoaudiología, Facultad de Odontología y Ciencias de la Rehabilitación, Universidad San Sebastián, Santiago 7510602, Chile
* Correspondence: pavel.prado@uss.cl (P.P.); matias.zanartu@usm.cl (M.Z.)

Featured Application: An extension of the DIVA model to include EEG is presented and initially validated using group-level statistics. The DIVA_EEG expands the number of scenarios in which vocal and speech behaviors can be assessed and has potential applications for personalized model-driven interventions.

Abstract: The neurocomputational model 'Directions into Velocities of Articulators' (DIVA) was developed to account for various aspects of normal and disordered speech production and acquisition. The neural substrates of DIVA were established through functional magnetic resonance imaging (fMRI), providing physiological validation of the model. This study introduces DIVA_EEG an extension of DIVA that utilizes electroencephalography (EEG) to leverage the high temporal resolution and broad availability of EEG over fMRI. For the development of DIVA_EEG, EEG-like signals were derived from original equations describing the activity of the different DIVA maps. Synthetic EEG associated with the utterance of syllables was generated when both unperturbed and perturbed auditory feedback (first formant perturbations) were simulated. The cortical activation maps derived from synthetic EEG closely resembled those of the original DIVA model. To validate DIVA_EEG, the EEG of individuals with typical voices (N = 30) was acquired during an altered auditory feedback paradigm. The resulting empirical brain activity maps significantly overlapped with those predicted by DIVA_EEG. In conjunction with other recent model extensions, DIVA_EEG lays the foundations for constructing a complete neurocomputational framework to tackle vocal and speech disorders, which can guide model-driven personalized interventions.

Keywords: auditory feedback; DIVA model; EEG; feedback perturbation; vocal compensation

1. Introduction

Effective oral communication is a basic and valued human daily activity [1,2]. A key aspect of this function is the sensory-motor integration for the control of speech production, which has been shown to be critical for speech acquisition [3] and that is affected in speech and voice disorders including vocal hyperfunction [4,5], stuttering and other disfluencies [6,7], as well as in neurodegenerative diseases (Parkinson's disease) [8,9].

Studies on sensory-motor integration have traditionally used the altered auditory feedback paradigm [3], i.e., vocal compensations elicited by perturbations in the intensity, frequency, and temporality of the auditory feedback of one's own voice.

Auditory perturbations have been studied via two approaches: (1) some trials are perturbed randomly, generating a reflexive compensatory response on the part of the participant, and (2) the perturbation is gradual, inducing the adaptation to the perturbation response. Both methods consist of recording the participant's voice through a microphone, artificially altering speech formants or fundamental frequency, and playing back the altered vocalization to the participant in near real time through headphones [10]. Only a few studies (e.g., [11–19]) have been carried out regarding compensation in response to formant perturbation.

Research on speech production and acquisition has proposed several models of speech motor control [20]. For example, the Directions into Velocities of Articulators (DIVA) model has been developed using control theory concepts and anatomo-physiological information of brain networks. This model represents a unified neurocomputational framework that accounts for different aspects of speech production, including compensatory behaviors due to sensory feedback perturbations [21,22]. Following predictive coding [3], the DIVA model uses sensory feedback information to track and correct transient deviations from the desired vocalization. This is achieved by generating error signals that modify previously learned speech-motor programs and reconfiguring the set of motor commands associated with the activation of the articulatory and laryngeal musculature. Therefore, the DIVA model has laid the foundation for a great deal of research regarding the role of auditory feedback on speech production and acquisition in both normal-hearing and hearing-impaired populations [23–28]. Furthermore, it has become a valuable tool for assessing the etiology of stuttering, apraxia, and other speech pathologies [3,29].

The theoretical bases of the DIVA model are supported by empirical work demonstrating increased activity of the prefrontal, Rolandic and superior temporal cortices in response to auditory feedback perturbations, which has been observed using different functional modalities [30–35]. Nevertheless, the match between DIVA model predictions and experimentally acquired brain activity has been exclusively tested using functional magnetic resonance imaging (fMRI) [3,18,36]. It remains to be seen if a similar match is observed when brain activity is assessed through the electroencephalogram (EEG). It may be advantageous to the field of speech production to verify the DIVA model with EEG, as this neuroimaging modality is a direct measure of the electrical activity of the brain and allows for the representation of whole-brain oscillatory dynamics with high temporal resolution [37,38]. Furthermore, EEG is a portable, low-cost technology with relatively broad availability. Considering the large number of EEG studies assessing vocal and speech behaviors in disturbed acoustic environments [39–41], an extension of the DIVA model to EEG may contribute to disentangling key neural mechanisms of sensorimotor integration for speech-motor control.

Therefore, this study aims to investigate whether the brain activations intrinsic to DIVA match the brain activity maps estimated from EEG. To achieve this goal, the dynamics of the different DIVA maps (i.e., sets of brain nodes that collectively represent a particular type of information) [3] were obtained in three simulated conditions: (1) undisturbed auditory feedback; (2) auditory feedback with up-shifted first formant (F1); and (3) auditory feedback with down-shifted F1. The DIVA map activations corresponding to each condition were the input of a generative EEG model, which allowed for the construction of EEG scalp distributions. This extension of the DIVA model will be referred to as DIVA_EEG. Using models for solving the inverse problem in EEG, the brain cortical generators of the simulated EEG were estimated. These brain activation maps were used as a template in the experimental phase of the study, in which the event-related potentials (ERPs) elicited by each of the conditions were obtained. The cortical generators of the ERPs were estimated using source localization methods, and empirical cortical activation maps were compared with the EEG theoretical templates.

1.1. DIVA Model

DIVA is a neurocomputational model used to simulate speech production and acquisition and it is initially designed for the English language. Each module of DIVA corresponds to a brain region activated during speech programing and production (e.g., premotor cortex, motor cortex, auditory and somatosensory cortex, cerebellum). The DIVA model is constructed as an adaptive neural network that allows for the simulation of the movement of the vocal articulators (lips, tongue, larynx, palate, and mandible) to generate speech. It also contains both a feedforward and a feedback control mechanism [3]. Figure 1 shows the structure of the model.

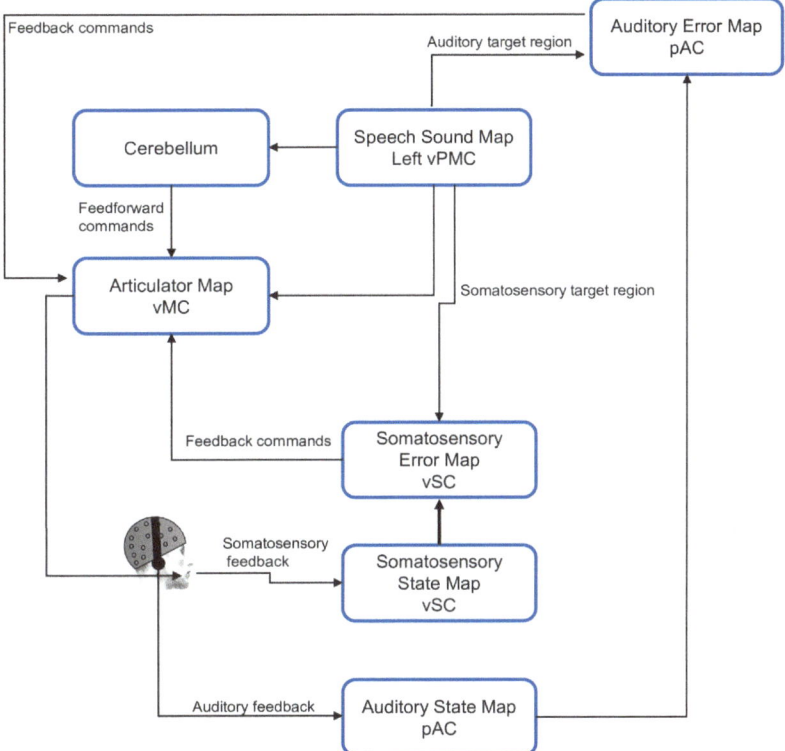

Figure 1. DIVA model scheme. vMC, ventral motor cortex; vPMC, ventral premotor cortex; vSC, ventral somatosensory cortex; pAC, posterior auditory cortex.

In the model, the production of a phoneme or syllable starts with the activation of the Speech Sound Map. Then, this information is sent to the Articulatory Velocity and Positions Maps located in the motor cortex, which control the movement of the speech articulators (vocal tract). The Auditory State Map and the Somatosensory State Map provide auditory and sensory information about how phonemes or syllables are produced. When a mismatch between the desired and actual speech production is detected, both the Auditory Error map and the Somatosensory Error Map are activated and generate a signal to correct the vocalization [3,18,36].

1.2. Electroencephalography (EEG)

EEG is a useful tool in clinical and research for assessing neurodevelopmental and behavioral disorders, state of consciousness, as well as in neurofeedback applications, brain–computer interfaces, among others [42–44]. The main advantage of EEG lies in its non-invasive approach for measuring the electrical activity collectively produced by large groups of neurons in the brain during information processing, with resolution in the order of milliseconds. Due to the macroscopic character of this activity and the variety of possible neural configurations responsible for a particular EEG scalp topography, it is impossible to univocally determine the EEG brain generators [45]. There are physical-mathematical algorithms that attempt to find a reasonable solution to this issue, termed the EEG inverse problem. These methods aim to estimate the brain areas responsible of the electrical potential distributions measured on the scalp [46–48].

Considering that measurements (potentials on the scalp) are only possible on a finite set of sensors and the geometric and electromagnetic characteristics of the conductive volume (head) in a discrete set of points, this relationship can be written as Equation (1), [47,48]:

$$\Phi = K \cdot J \tag{1}$$

in which K is the matrix that expresses the linear relationship between the electric potentials on the scalp (Φ) and the average primary current density (J) at the intracerebral points.

2. Materials and Methods

The construction and the subsequent validation of DIVA_EEG consisted of two phases: DIVA model Simulation and Experimental Phase, which are illustrated in Figure 2 and described in the following subsections.

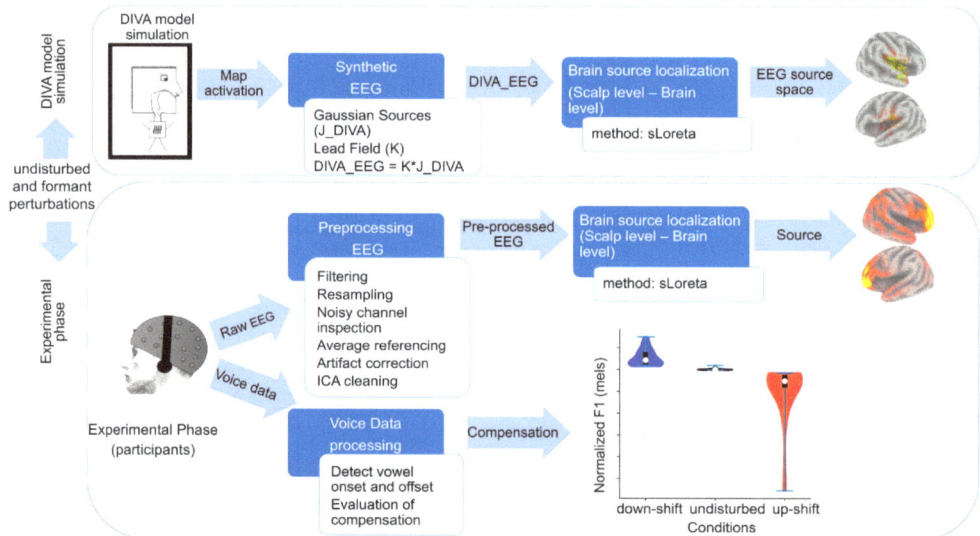

Figure 2. Block diagram illustrating the methodology proposed for the construction of DIVA_EEG. Both the DIVA model Simulation and the Experimental Phase of the study are presented.

2.1. DIVA Model Simulation

In the present study, the main objective was to model the spatio-temporal dynamics of DIVA to obtain a template of the cortical activation associated with the DIVA observed via EEG. The outcome is the generation of EEG topographical maps that represent the activation of the different DIVA maps in each experimental condition (undisturbed, up-disturbed and down-disturbed auditory feedback).

2.1.1. Simulated Speech

We chose the phoneme /e/ (defined in the model) as this vowel can readily be transformed in sounds to resemble the phoneme /æ/ (by increasing the F1 frequency) or the phoneme /ɪ/ (by decreasing the F1 frequency). The perturbation size (F1 change in Hz) was 350 Hz. Three simulations were carried out: undisturbed, down-shift, and up-shift, under experiment type: 'Reflexive responses'. The duration of the simulation was 550 ms, and the disturbance was applied throughout the simulation.

2.1.2. Generation and Source Localization of Synthetic EEG

During simulation, the output of each DIVA node is associated with the computational load (denoted L in [3]), a term that represents the instantaneous neural activity of the node. These neural activities served as input for the EEG generative model. Therefore, point sources for the DIVA-EGG generation were seeded in brain locations that match the different nodes in the original DIVA model [3]. Table S1 shows the brain coordinates for the centroids of the seeds Traces of the synthetic EEG are displayed in Figure S1. A full-brain activity pattern was then constructed by treating the electrical activity of the seeds as Gaussian activity sources (J_DIVA) that added-up together at each brain location. The standard deviation of the normal distribution was 2. Voxels with amplitudes lower than 0.01 times the maximum amplitude were deemed inactive. The synthetic EEG (DIVA_EEG) was obtained by multiplying the simulated brain activity (J_DIVA) and the lead field K. The lead field K was computed by using a head model of three concentric, piece-wise homogeneous, and isotropic spheres [49]. Voltages (DIVA_EEG) were obtained in 64-scalp locations (a 64-electrode layout that followed the 10/20 international system for electrode placement). The DIVA-EEG is expressed by the following equation:

$$DIVA\_EEG = K \cdot J\_DIVA, \qquad (2)$$

where the matrix $DIVA\_EEG$ has one row for each EEG sensor and one column for each time (size $Nsen \times Nt$), K has the number of DIVA model components as columns and is of size $Nsen \times Nc$, and J_DIVA contains the time series of the different seeds of the model and is of (size $Nc \times Nt$).

Brain source localizations were estimated using the standardized Low-Resolution Electromagnetic Tomography method (sLORETA, [50]; for a review, see [51]). sLORETA is based on an appropriately standardized version of the minimum norm current density estimation which overcomes problems intrinsic to the estimation of deep sources of EEG.

2.2. Experimental Phase

2.2.1. Participants

Thirty individuals with typical voices were enrolled in this study (mean age 24 ± 3.8 years). This sample size is larger than the minimum sample necessary to conduct F-tests (repeated measure ANOVA) sensitive to large effect sizes with a statistical power of 0.8. Furthermore, the sample is sufficiently large to conduct two-tailed t-tests, able to sense large effect sizes with a statistical power of 0.8. Participants were recruited if they (1) were right-handed, (2) had no history of psychological, neurological, or speech-language disorders, (3) did not have prior training in singing, and (4) had normal binaural hearing (hearing threshold ≤ 20 dB HL at all octave frequencies between 250 and 8000 Hz). Before the experimental session, participants signed a written consent form, which was approved by the Research and Ethics Committee of the Faculty of Medicine, Universidad de Valparaíso, Chile (assessment code 52015), in compliance with the national guidelines for research with human subjects and the Declaration of Helsinki.

2.2.2. Experimental Setup

This work reports reflexive responses in controls tested in an altered auditory feedback paradigm such as that utilized in [18].

Participants were seated in a comfortable chair inside a double-walled, sound-attenuating booth meeting the ANSI S3.1-1999 standard. A microphone (B&K 4961) was positioned approximately 10 cm from the participants' mouth at a 45-degree offset in the axial direction.

The acoustic signal was calibrated to physical units of dB SPL (dB re 20 µPa) using a Larson Davis calibrator (model CAL200, Depew, NY, USA).

Speech was sampled at 48 kHz using a MOTU Microbook IIc sound card and the CueMix FX software. Participants' voices were played back to them over closed-back, over-the-ear AKG K240 Studio Headphones, with a mean latency of ~18 ms. This latency is lower than that at which feedback delays are perceived (50 ms) [52]. The speech level of the participant determined the amplitude of the speech playback.

Participants were instructed to read a series of texts presented on a screen (white font on a black background) positioned 70 cm away and adjusted in the vertical axes to the eye level of the participants at a comfortable conversational pitch and loudness. The text series comprised repetitions of the Spanish monosyllabic words: /mes/, /pep/, and /ten/. Words were presented for 2.5 s, at a presentation rate of 0.25 Hz (one word every 4 s to prevent the participants from developing a constant rhythm and the automatic character of their production). A total of 648 stimuli were presented, distributed in 6 blocks of 108 trials. In each block, stimuli were distributed in a random order. Participants were asked to sustain the vocalization of the vowel until the end of each word's presentation. No additional instructions were provided.

A 10-trial training session was conducted prior to the start of the experiment to ensure that participants were familiar with the experimental setup, familiar with stimulus timing, and comfortable with sustaining vocalizations.

2.2.3. Feedback Perturbation

To apply the auditory perturbations, we used Audapter [29,53], a publicly available software for tracking and shifting the frequency of F1 in near real time. Both stimulus presentation and data collection were controlled by a custom MATLAB (R2022b) script (Mathworks, Natick, MA, USA) (Figure 3).

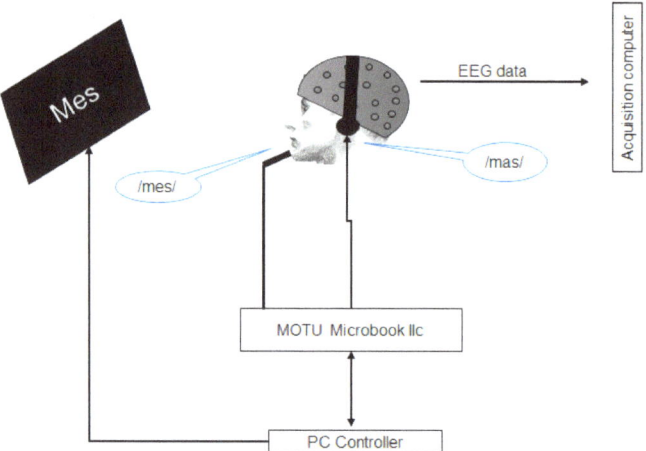

Figure 3. Schematic of the apparatus for applying formant perturbations. Participants produced monosyllabic words containing the vowel /e/ while their auditory feedback was perturbed toward the participant-specific vowel /a/ (e.g., participants produced /mes/ but heard a word that sounded like /mas/).

Following previous studies [18], the frequency of F1 for the auditory feedback was increased 30 percentage points relative to the produced speech signal on 1/6 of the trials (up-shift conditions: 108 trials), decreased 30 percentage points on another 1/6 of the trials (down-shift condition: 108 trial), and unaltered on the remaining 2/3 of the trials (432). After the transformation, the pronunciation of the phoneme /e/ approached either the pronunciation of the phoneme /a/ in the words /mas/, /pap/, and /tan/ (up-shifted F1), or the pronunciation of the phoneme /i/ in the words /mis/, /pip/, and /tin/ (down-shifted F1) [54]. The perturbation values were different from that used in the DIVA model because the vowel triangle of the vowels in Spanish differ from that of the triangle of vowels in English)

2.2.4. Processing of Acoustic Signals

Vowel onset and offset were first automatically identified with a Linear Predictive Coding model to find the frequency of F1 [55]. The compensation was evaluated in the time window between 120 and 500 ms after the vowel onset. This time window corresponds with the time at which the beginning of vocal compensations occurs [9,17,18,56,57]. Previous studies have shown that corrective responses begin between 100 and 200 milliseconds (usually 150 ms) after the onset of the perturbations and increase at least for the following 400 ms [11,15,19].

The compensatory response for each subject was calculated as follows: First, for each stimulus word the average F1 trajectory is calculated for all undisturbed trials (baseline trials). Second, the trajectory of F1 from each perturbed trial was normalized to the control condition, by subtracting the baseline from the perturbed trials. Compensatory response magnitude was calculated for each subject as the average F1 value within 120–500 ms after vowel onset [17,57].

2.2.5. EEG Acquisition and Analysis

EEG was recorded using the ActiveTwo BioSemi system (BioSemi, Amsterdam, Netherlands) with ActiView acquisition software (BioSemi) with 64 scalp electrodes (10–20 electrode placement). External electrodes were placed in periocular locations to record blinks and eye movements. Analog filters were set at 0.03 and 100 Hz. During the analog/digital conversion, signals were sampled at 4096 Hz, with 24 bits of resolution. The EEG signal was pre-processed offline using standard procedures implemented in Brain Vision Analyzer 2.0® (Brain Products GmbH, Munich, Germany). Recordings were re-referenced to the average of all channels and band-pass filtered between 0.1 and 40 Hz using a zero-phase shift Butterworth filter of order 8. Data were downsampled to 512 Hz. Independent Component Analysis (ICA) was used for correcting EEG artifacts induced by blinking and eye movements (following [54]). Data were segmented from -200 to 500 ms around the onset of vocalization. Semiautomatic criteria implemented in Brain Vision Analyzer were used for rejecting noisy epochs. ERPs were obtained by averaging baseline-corrected epochs. N1 and P2 peaks were identified using semiautomatic procedures. Electrodes in occipital, parietal locations and in the midline were pooled (Iz, O1, O2, Oz, P10, P7, P8, P9, PO7, PO8), and N1 and P2 amplitudes were computed as the average voltage in a two-point window around the corresponding peak amplitude. The amplitude of the N1-P2 complex was obtained and compared between conditions (unperturbed feedback, up-shifted pitch, and down-shifted pitch) using a repeated measure ANOVA ($p \leq 0.05$).

2.2.6. ERP Source Localization

Brain generators of the N1-P2 complex were estimated using the standardized Low-Resolution Electromagnetic Tomography Analysis (sLORETA). For this, the 10–20 electrode layout was registered onto the scalp MNI152 coordinates. A signal-to-noise ratio of 1 was chosen for the regularization method used to compute the EEG transformation matrix (forward operator for the inverse solution problem). The standardized current density maps were obtained using a head model of three concentric spheres in a predefined source space of 6242 voxels (voxel size of $5 \times 5 \times 5$ mm) of a reference brain (MNI 305, Brain Imaging Centre, Montreal Neurologic Institute) [58,59]. A brain segmentation of 82 anatomic compartments (cortical areas) was implemented using the automated anatomical labeling (AAL90) atlas [60].

The cortical activations (standardized current density) maps were estimated for each scalp voltage distribution in the time windows between −5 ms relative to the peak N1 amplitude and +5 ms relative to the peak P2 amplitude. Cortical activations maps obtained for the different scalp distributions were averaged. Brain cortical activity (voxel-wise activity) of the different conditions were paired-wise compared (undisturbed feedback vs. up-shifted formant, undisturbed feedback vs. down-shifted formant, and up-shifted formant vs. down-shifted formant) using two tailed t-test (α = 0.05). Results were corrected for multiple comparisons using non-parametric permutation tests (5000 randomizations) as implemented in Loreta_Key [61,62].

2.2.7. Match between DIVA Related (Simulated) and ERP (Real) Cortical Activation Maps

Binarized representations of the cortical activation maps associated with feedback perturbations (maps that resulted from the statistical analyses) were obtained for both the model-driven synthetic EEG and the N1-P2 complex of the ERP (real EEG). The binarized maps were overlapped. The match between the theoretical (predicted by the model) and real (obtained from the experimental data) cortical maps was computed as a function of the number of voxels belonging to a particular AAL region that were active during the vocalization.

3. Results

3.1. DIVA Model Simulation

The activation of the cortical maps of the DIVA model during the vocalization of the phoneme /e/ with undisturbed auditory feedback is illustrated in Figure 4. DIVA maps provided by the model activated at different times with respect to the onset of the simulated vocalization. The first maps were activated at t = 0 (onset of the vocalization) and were the motivation, initiation, speech, somatosensory target (somato-t) and auditory target (auditory-t) maps (Figure 4A). While the activity of the motivation map reduced to 0 directly following the vocalization onset, the activity of the initiation map remained constant (value of 1) throughout the vocalization. The articulator map (articulator) activated 10 ms after the onset of the vocalization. This was followed by the activation of the somatosensory state map (somato-s) (25 ms), the somatosensory error (somato-e) (30 ms), the feedback map (35 ms), and the auditory state (auditory-s) (55 ms after the vocalization onset). As the auditory feedback was not disturbed, the auditory error map was not activated.

Cortical activations feed into the EEG generative model, which resulted in EEG scalp distributions that characterized the different phases (stages) of the cortical dynamics (Figure 4B). Current density maps in the cerebral cortex were estimated from the EEG scalp distributions using sLORETA (Figure 4B). The EEG sources estimated with the inverse solution method closely resembled the brain distribution of DIVA maps (cortical seeds used for the EEG generation). Auditory feedback perturbations (both down- and up shift in F1) were reflected in the activity profile of the DIVA model (Figure 5A). While the activity changes of the Auditory state map clearly followed the direction of the perturbations, Somatosensory state maps changed minimally. Evident increases in the activity of the Feedback map were obtained in the presence of auditory feedback perturbation. Noteworthy, the feedback perturbation triggered the activation of both the Auditory error map and the Somatosensory error map, which are typically suppressed in undisturbed conditions.

Due to the auditory feedback perturbation, differences were observed in both the EEG scalp distributions and the activity of the EEG generators estimated with sLORETA (Figure 5B). The shifts in F1 resulted in increased bilateral activation of frontal, temporal and parietal cortical areas (Figure 5C, left and middle panels), including the orbital, opercular and triangular parts of the inferior frontal gyrus, the middle and superior frontal gyri, the Rolandic operculum, the Heschl gyrus, the temporal pole, as well as the middle and superior temporal gyri (Table S2, Supplementary Materials). The downward and upward shifts in F1, although equal in magnitude, resulted in different EEG source-space maps (Figure 5C, right panel). This asymmetry was reflected as an increase in the cortical activity elicited by down-shifted feedback perturbations in comparison with that induced by up-shifted perturbations. The differences in activity were mainly observed in frontal and parietal brain areas (bilaterally), including the primary somatosensory and motor cortices (Table S2, Supplementary Materials).

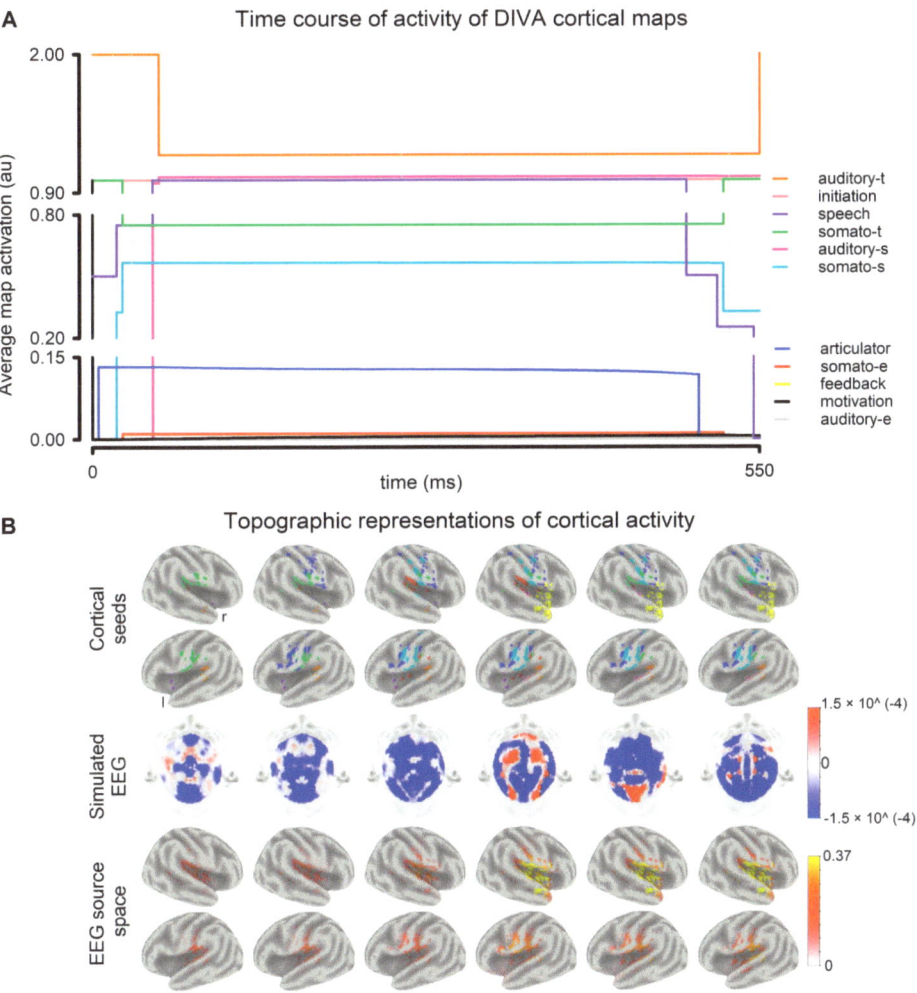

Figure 4. Simulations of the brain cortical activity associated with the different DIVA maps during the vocalization of the phoneme /e/ with undisturbed feedback: (**A**) Time course of activity of DIVA cortical maps. t: target, s: state, e: error (**B**) Topographic representations of cortical activity for time t = 0, 10, 25, 250, 510, 550 ms relative to the onset of the vocalization. top panel: cortical seeds. middle panel: simulated EEG. bottom panel: source space representation of the synthetic EEG.

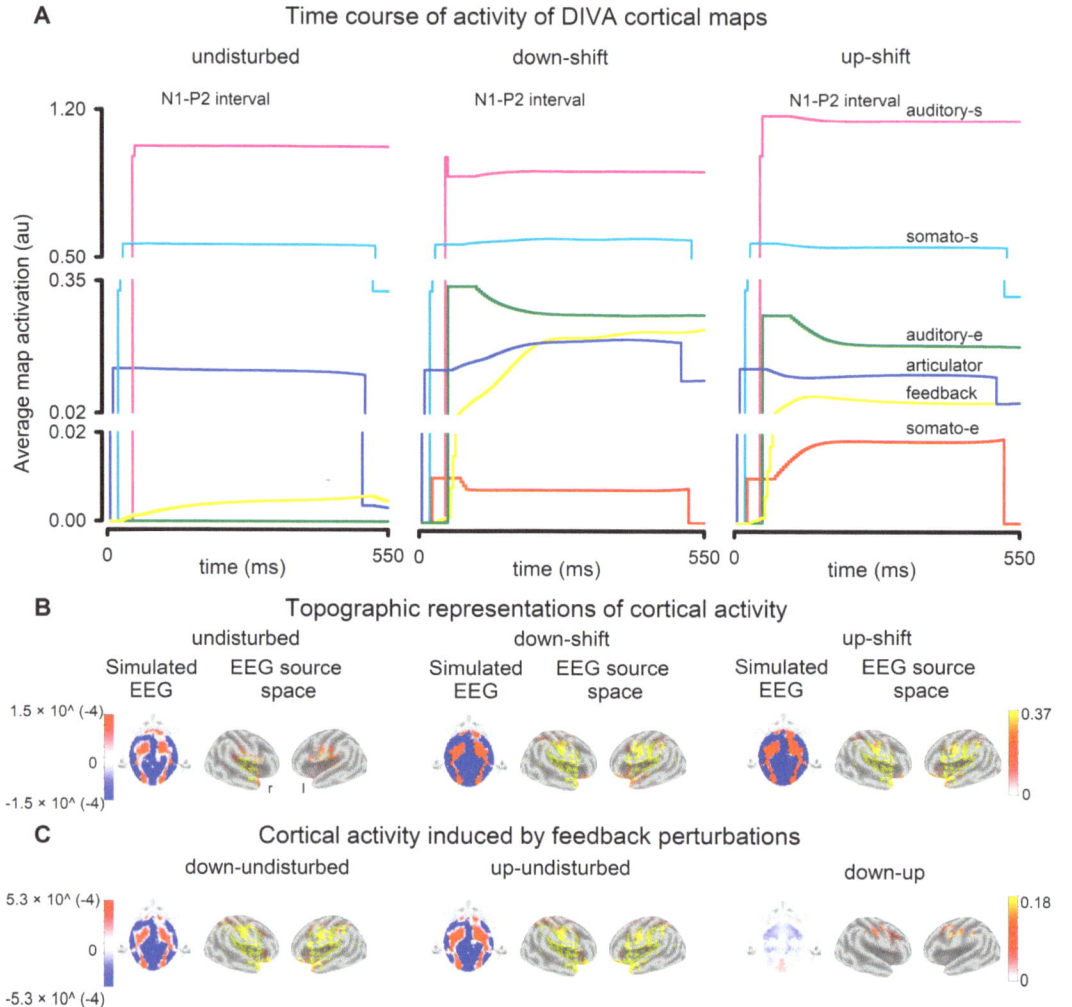

Figure 5. Simulations of the brain cortical activity associated with the different DIVA maps elicited by auditory feedback perturbations (F1 shifts) during the vocalization of the phoneme /e/. (**A**) Time course of activity of the DIVA cortical maps whose activity varied in response to feedback perturbations. Activities in undisturbed, downshifted, and upshifted conditions are presented. The shaded area represents the N1-P2 interval of the ERP. t: target, s: state, e: error (**B**) Scalp topography and source space representation of the synthetic EEG estimated in the time interval that corresponds to the generation of the N1-P2 complex. (**C**) Synthetic EEG (N1-P2 interval) contrasted across conditions.

3.2. Behavioral and Physiological Data

During the formant-shift experiment, F1 varied between conditions ($F_{(29,2)} = 23.052$, $p < 0.001$), as participants compensated for auditory feedback perturbations (Figure 6A, right panel). The F1 deviations counteracted the perturbational formant-shifts, such that F1 compensations were in the opposite direction to the perturbations (Figure 6A, left panel). The F1 of both types of compensations significantly differed from that of vocalizations elicited during unperturbed feedback (Holm post hoc test, $p < 0.0.5$).

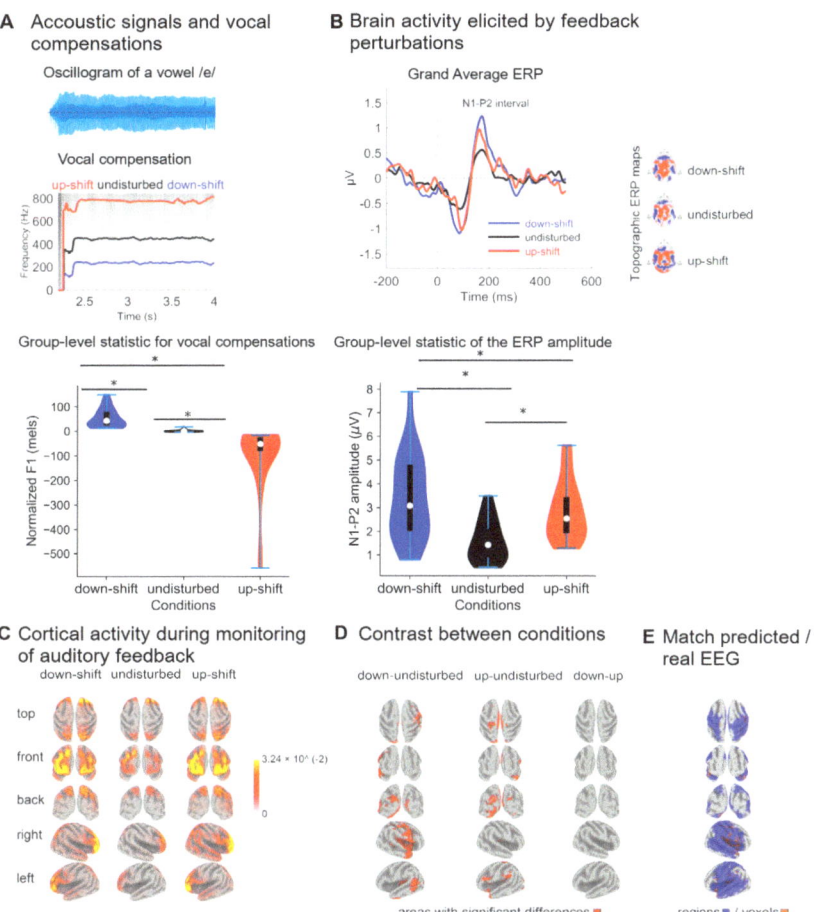

Figure 6. Acoustic and electrophysiological parameters describing the monitoring of one's own vocalization. (**A**) Examples of vocal compensations elicited by F1 perturbations in the auditory feedback. In the left panel, an oscillogram representative of the phoneme /mes/ is illustrated. Likewise, the direction of the perturbation is indicated at the top of each chart. The mean F1 values of vocalizations produced in unperturbed acoustic conditions and those of vocal compensations to perturbed auditory feedback are presented in the right panel, along with the corresponding sample distributions. (**B**) Event-related potential (ERP) elicited by actively monitoring the auditory feedback of one's own vocalizations. In the left panel, the grand average of the ERP elicited by both unperturbed and F1-shifted auditory feedback are presented. The shaded area indicates the N1-P2 complex. Scalp topography of the N1-P2 complex is illustrated in the middle panel. The mean amplitude of the N1-P2 complex elicited by unperturbed and perturbed auditory feedback are presented in the right panel, along with the corresponding sample distribution. (**C**) Current density maps illustrating the brain generators of the N1-P2 complex in the different conditions (unperturbed and perturbated auditory feedback). (**D**) Differences in the cortical activity obtained in response to unperturbed and perturbated auditory feedbacks. The difference between the current density maps elicited by F1 perturbations of equal magnitude and opposite directions is presented in the right panel. (**E**) Cortical sources of the N1-P2 complex elicited in response to F1 perturbations in the auditory feedback of one's own vocalizations that are predicted by the DIVA model. They are illustrated both areas and voxels for which the activity predicted by the model overlapped that estimated from the real EEG. Statistically significant differences between groups are represented by *.

F1 perturbation induced changes in the cortical activity associated with monitoring the sensory feedback of one's own voice, which was reflected in the N1-P2 amplitude of the ERP obtained across conditions ($F_{(29,2)}$ = 29.047, $p < 0.001$) and the changes in ERP scalp topography (Figure 6B). The N1-P2 amplitude elicited in response to both upward and downward perturbations was higher than that obtained when auditory feedback was unperturbed (Holm post hoc test, $p < 0.001$). The N1-P2 amplitude did not differ when F1 was upward and downward perturbed (Holm post hoc test, $p = 0.36$).

The cortical source of the ERP associated with monitoring of one's own voices were estimated in large portions of the frontal, temporal, and parietal lobes (Figure 6C). It is worth noting that the activity of the N1-P2 generators significantly varied in response to F1 perturbations (t-test, 5000 randomizations) (Figure 6D). Downshifted F1 perturbations induced right lateralized activation of areas including the opercular, triangular and orbital parts of the inferior frontal gyrus, the Heschl gyrus (primary auditory cortex), the temporal pole, the middle and inferior temporal gyri, the Rolandic Operculum (including the primary somatosensory and motor cortices), the lingual gyrus (Figure 6D, left panel) and several sensory association cortical regions (Table S3, Supplementary Materials). Upshifted F1 perturbations resulted in a more diffuse cortical activation (Figure 6D, middle panel). Nevertheless, the cortical activations elicited by downward and upward shifts in F1 were not statistically significantly different (t-test, 5000 randomizations) (Figure 6D, right panel). Results for uncorrected comparisons are presented in Table S5, Supplementary Materials.

3.3. Match between DIVA Simulations and Real EEG

As upshifted and downshifted F1 perturbations did not result in statistically different cortical activations, current density maps elicited by both types of auditory feedback perturbations were merged into a single representation. This was carried out separately for activations derived from DIVA simulations (Figure 5C) and real EEG (Figure 6D), respectively. Both representations of cortical activations were binarized and contrasted to assess if cortical activity derived from DIVA simulations predicted the EEG source space of the ERP elicited by auditory feedback perturbations.

A match between the predicted and real cortical activations was obtained. This was reflected at the level of brain areas (Figure 6E left panel). Overlapping regions included the opercular part of the right inferior frontal gyrus, the Rolandic operculum (bilaterally), the temporal pole (bilaterally), the Heshl gyrus (bilaterally), the superior temporal gyrus (bilaterally), the left middle temporal gyrus, the supramarginal gyrus (bilaterally), the parietal superior gyrus (bilaterally), as well as limbic areas such as the hippocampus (bilaterally) and the insula (bilaterally) (Table 1). Overlapping was also obtained at the voxel level (Figure 6E, right panel) in frontal, temporal, parietal and limbic areas mentioned above (Table S4, Supplementary Materials).

Table 1. Areas for which DIVA model predictions matches experimentally acquired EEG.

Brain Lobe	AAL Region	Hemisphere
Frontal	Precentral	(bilateral)
	Frontal_Inf_Oper	(right)
	Rolandic_Oper	(bilateral)
Limbic	Insula	(bilateral)
	Cingulum_Mid	(bilateral)
	Cingulum_Post	(right)
	Hippocampus	(left)
	ParaHippocampal	(bilateral)
Temporal	Heschl	(bilateral)
	Temporal_Sup	(bilateral)
	Temporal_Pole_Sup	(bilateral)
	Temporal_Mid	(bilateral)
	Temporal_Pole_Mid	(left)

Table 1. *Cont.*

Brain Lobe	AAL Region	Hemisphere
Parietal	Postcentral	(bilateral)
	Parietal_Sup	(bilateral)
	Parietal_Inf	(right)
	SupraMarginal	(bilateral)
	Paracentral	(right)
Occipital	Lingual	(bilateral)
	Fusiform	(bilateral)

4. Discussion

In this study, an extension of the DIVA model to EEG, referred to as DIVA_EEG, is presented. Neural activity of the DIVA maps associated with the vocal production and the monitoring of one's own voice were fed into generative models of EEG. The scalp topographies of the EEG obtained in response to auditory feedback perturbations were simulated (Figures 4 and 5). Brain sources of the synthetic EEG were estimated and compared with those of the ERP (real EEG) obtained when conducting the altered auditory feedback paradigm in healthy participants (Figure 6). At the region level, a 91.5% overlapping was obtained between the model-predicted cortical activity for the control of speech production and that estimated from the experimentally acquired EEG. The overlapping between the real and predicted representations of brain activity was of 57.6% at the voxel level. Noteworthy, all the seed regions used for the EEG generative model were represented in the brain activity maps estimated from real EEG.

4.1. DIVA_EEG

Other modifications of the DIVA model preceded the development of DIVA_EEG. For instance, DIVA has been extended to incorporate physiologically based laryngeal motor control [63] or simplified for assessing the relative contribution of feedback and feedforward control mechanisms to sensorimotor adaptation [64]. Furthermore, DIVA has been translated to open-source codes, thereby facilitating their integration with freely available machine learning tools [65]. The DIVA environment, which also comprises the gradient order DIVA (GODIVA) for the analysis of speech sequencing [66], is now enriched with a new neuroimaging modality (EEG).

Several aspects need to be considered when interpreting the synthetic EEG that resulted from the activations of the different DIVA maps. First, DIVA_EEG comprise anatomical priors since the locations of seeds for the EEG generation are the same as for the nodes in the original DIVA model [3], which in turn were obtained from fMRI feedback perturbations protocols [18,21]. Noteworthy, since brain activity reflected in the EEG is mainly restricted to the cerebral cortex [18,36], DIVA_EEG does not include subcortical regions, which are already considered in DIVA. Second, the brain activity of DIVA_EEG seeds are simulated as Gaussian functions that extend 2mm from the centroid. Therefore, seeds in the model can be considered as a point source for the EEG generation since the seed size is lower than the voxel size of the head model used in this study for solving the EEG inverse problem [50]. Third, the main outcome of the study is presenting the first version of DIVA_EEG. The scalp topography and the cortical source of the synthetic EEG obtained with DIVA_EEG (Figures 4 and 5) are highly dependent on the head model and the theoretical considerations selected for constructing the generative EEG model and solving the EEG inverse problem. Further refinement of the DIVA_EEG can result from including individual head models [67,68], generating brain activity maps that combine the EEG obtained from DIVA_EEG and the BOLD signal obtained with DIVA [3], and testing the replicability of the results as a function of the EEG generative model [69] and the source estimation method [70]. Noteworthy, future developments can use the computational load of the nodes (the instantaneous neural activity) as input of mean field models (e.g., neural mass models) to generate oscillatory EEG-like signals for assessing the EEG oscillatory dynamic [71], including cross-frequency coupling. This aspect is relevant since accurate speech encoding has been associated with the coupling of theta oscillations that tracks slow speech fluctuations and gamma-spiking activity related to phoneme-level responses [72].

4.2. Vocal Compensations

Unlike the DIVA simulations, where feedback perturbations are generated by modifying the F1 of a close vowel (the English vowel /e/), the behavioral compensations of the participants were assessed by modifying an open vowel (the Spanish vowel /e/). Nevertheless, in both simulated and real perturbations, upshifts in F1 transformed the target vowel in an open vowel (/æ/ and /a/ for English and Spanish, respectively). Likewise, downshifts in F1 transformed the target vowel in a close vowel (/ɪ/ and /i/ for English and Spanish, respectively). The vocal compensations elicited by these feedback perturbations, which typically opposes to the F1 shift (Figure 6A), replicate previous studies in which the compensatory behaviors of speakers of the target language have been assessed (e.g., Spanish [73], English [15,17,19,22] and Mandarin [74]).

Noteworthy, while compensatory behaviors typically opposed to F1 perturbations, compensations in the same direction to the F1 shift occasionally occurred (Figure S2, Supplementary Materials). This is in line with previous studies and supports the idea that, although compensations are primarily a reflex, their magnitude is modulated by several factors including attention [75], the predictability of the perturbation [1,76] and the vocal training of the participants [39,77]. Furthermore, the F1 during the compensations (Figure 6A) were closer to the F1 of the unperturbed auditory feedback than to that of the disturbed feedback, a result that has been previously reported [78,79]. Considering the interaction between different DIVA cortical maps, this has been explained by a counteracting effect of the activation of the somatosensory feedback controller on the activation of the auditory feedback controller [80].

4.3. ERP Elicited by Perturbations

The increased amplitude of the N1-P2 complex of the ERP elicited by auditory feedback perturbations (Figure 6B) can be considered the electrophysiological hallmark of the sensorimotor integration processes underlying the speech production [40,81,82]. The N1 component has been associated with the activation of the primary and secondary auditory cortices [83–85] and reflects the auditory processing of basic properties of acoustic stimuli. In addition, it has been suggested that P2 represents the coordinated activity of neural generators located in sensory, motor and frontal cortical regions, which might include auditory and speech-related motor areas involved in sensorimotor integration [83,86,87]. The changes in the ERP elicited by auditory feedback perturbations can be partially explained by the predictive coding models, which posits that processing of sensory information is facilitated when the sensory input is predictable [88–90]. This idea was initially proposed to explain the decreased amplitudes of N1 during active speech as compared with that obtained during the passive listening of own voices [34,83,91]. This attenuation was supposed to reflect filtering processes in which redundant information in the sensory feedback is cancelled by neural codes generated in motor-related cortical areas [92]. The hypothesis of predictability has been subsequently refined using feedback perturbations protocols [34,83,93]. Evidence shows that, the larger the differences between the expected and the incoming auditory feedback, the greater the ERP amplitude [34,83,93]. This is likely mediated by learning and reinforcing mechanisms in which predicted perturbations are segregated from the auditory re-afference, such that the disparity between the ongoing auditory feedback and the predicted feedback is reduced [1,77,83,85,94].

4.4. EEG Source Localization

Several methodological approaches have been used to assess the neural correlates of vocal production and control. They include, for example, the analysis of local field potentials with cortical electrodes [83] and the use of transcranial magnetic stimulation [85,95].

While these procedures enable the role of anatomically restricted brain regions to be investigated, the analysis of the whole brain activity is facilitated by methods to solve the EEG inverse problem [96,97]. The latter approach was used in this study to estimate the neural generators of the ERP elicited by self-produced speech (Figure 6C). Feedback perturbations resulted in increased activity of frontal, temporal and regions that have been traditionally associated with speech production and speech motor control (Table S3, Supplementary Materials). This group of regions include the precentral gyrus, the supplementary motor area, and the Rolandic operculum (frontal lobe), the insula (limbic lobe), the Heschl gyrus as well the inferior and superior temporal gyri (temporal lobe), and the postcentral gyrus (parietal lobe) [86].

Furthermore, differences in activity were also obtained in the occipital lobe and other limbic areas. Although this result needs to be validated, evidence suggest that speech-driven spectrotemporal receptive fields that are sensitive to pitch are located in the calcarine area, an occipital cortical region that display strong functional connections with early auditory areas [98]. Likewise, the medial and the posterior cingulate cortices have been proposed as hubs of the syllable and speech production network, respectively [99]. These networks also comprise the hippocampus, the amygdala and the insula (limbic areas), as well as the cuneus, the lingual gyrus and the inferior, middle and superior occipital gyri (occipital areas) [99].

4.5. Comparing Simulated and Experimentally Acquired Brain Cortical Map for Speech Motor Control

The cortical activation maps in DIVA_EEG, instead of being represented as the set of nodes obtained from DIVA, were constructed by implementing an EEG generative model to simulate EEG scalp topographies, from which current density maps in the brain were estimated. This allowed for a fair comparison between the model-based brain activity maps and those estimated from experimentally acquired EEG. An appropriate match between the predicted and the EEG-driven cortical maps was obtained, at the level of both cortical regions and voxels (Figure 6E). Differences between these cortical representations may be due to different factors, including the use of point sources for generating the synthetic EEG. Therefore, tunning the size and shape of the brain areas used as seeds for the EEG generation shall be considered for further developments of DIVA_EEG. Noteworthy, all the cortical regions selected as seeds in DIVA_EEG were present in the cortical activation maps estimated from real EEG (Table S5, Supplementary Materials). The fact that brain activation maps estimated from both synthetic and experimentally acquired EEG extends beyond the seed regions of DIVA_EEG primarily relies on the following aspects. First, the spatial resolution and precision of the EEG source estimation methods in lower than that of the fMRI. In the case of LORETA, the cortical activity is represented in a grid of 6239 voxels, each of $5 \times 5 \times 5$ mm [50], which is much larger than the typical $1 \times 1 \times 1$ mm voxel size of the fMRI data. Second, one of the assumptions made for solving the EEG inverse problem using LORETA is that the electrical activity of neighboring voxels has maximal similarity [100], which leads to smooth cortical activations. Third, different statistical approaches have been used for estimating speech-related cortical activation maps from fMRI [3,18,86] and EEG [95]. Finally, fMRI and EEG reflect the hemodynamic and electrical activity of the brain, respectively. In other words, these neuroimaging modalities are different in nature and have largely different dynamics. Therefore, complementary but different results are expected when assessing brain activity from EEG and fMRI. A less restricted set of cortical regions resulted from the EEG feedback perturbation paradigm (Table S4, Supplementary Materials) when compared with its analogue fMRI paradigm [3,18,86]. This indicates that speech production, rather than relying on a discrete and reduced set of brain areas, is controlled by a broadly distributed network in which information is interchanged between primary nodes (seeds in DIVA_EEG) and between them and occipital, frontal and limbic areas.

5. Conclusions

The extension of DIVA to include a new neuroimaging modality (EEG) will expand the use of this neurocomputational tool for assessing different aspects of speech motor control, including sensorimotor integration and predictive coding. DIVA_EEG was validated using group-level statistics of the behavior and the EEG acquired from volunteers with typical voices. Further research is needed to ascertain if the configuration parameters of DIVA_EEG can predict vocal compensatory behaviors and brain activation at individual level. Subject-specific simulations can be fostered by incorporating vocal fold control models, as carried out in LaDIVA [63], which provide a complete set of biomechanical parameters for vocal function assessment. In fact, vocal fold models associated with LaDIVA have been successfully used for subject-specific modeling of vocal hyperfunction [101]. Likewise, further extension of DIVA_EEG may consider neurophysiological muscle activation schemes for controlling vocal fold models [102] for assessing reflective and adaptive vocal behaviors at the laryngeal level. The latter may incorporate the parametrization of the sensory adaptation elicited by continuous and repetitive stimulation [103,104]. These developments are the foundations for constructing a complete and comprehensive neurocomputational framework to tackle vocal and speech disorder, which can guide model-driven personalized interventions.

Supplementary Materials: The following supporting information can be downloaded at: https://www.mdpi.com/article/10.3390/app13137512/s1, Table S1: Cortical regions selected as seeds in DIVA_EEG, Figure S1: Synthetic EEG (DIVA_EEG) traces, and electrode layout, Table S2: Numbers of active voxels in AAL90 areas in DIVA model simulations, Table S3: Numbers of active voxels in AAL90 areas reflecting the cortical activity in feedback reflexive paradigms, Numbers of active voxels in AAL90 areas reflecting the by cortical activity in auditory feedback reflexive paradigms, Table S4: Numbers of active voxels in AAL90 areas by cortical activity when comparing DIVA-EEG model and real EEG, Figure S2: Example of compensations in the same direction of the down- shift F1, Table S5: Cohen Effect Size of the cortical activations.

Author Contributions: Conceptualization: J.C., E.M.-M., P.P. and M.Z.; methodology, J.C., P.P. and E.M.-M.; software, J.C., A.W., M.O., E.M.-M. and C.C.; validation, J.C., L.Z.-R., A.W. and C.C.; formal analysis, J.C., L.Z.-R., M.O. and P.P.; investigation, J.C. and P.P.; resources, M.Z.; data curation, J.C. and L.Z.-R.; writing—original draft preparation, J.C., G.W. and P.P.; writing—review and editing, J.C., G.W., P.P. and M.Z.; visualization, P.P. and J.C.; supervision, P.P. and M.Z.; project administration, M.Z.; funding acquisition, M.Z. All authors have read and agreed to the published version of the manuscript.

Funding: This research was funded in part by the U.S. National Institutes of Health (NIH), National Institute on Deafness and Other Communication Disorders Grant P50 DC015446 (M.Z.), Agencia Nacional de Investigación y Desarrollo de Chile (ANID) grants BASAL FB0008 (M.Z.), FONDECYT 1230828 (M.Z.), FONDECYT 1231132 (A.W.), FONDECYT Postdoctorado 3210508 (M.O.), BASAL FB210008 (M.O.). The content is solely the responsibility of the authors and does not necessarily represent the official views of the National Institutes of Health.

Institutional Review Board Statement: The study was conducted in accordance with the Declaration of Helsinki, and approved by the Institutional Review Board of Universidad de Valparaíso (protocol code 52015, approved on April 2019).

Informed Consent Statement: Informed consent was obtained from all subjects involved in the study.

Data Availability Statement: The data presented in this study are available on request from the corresponding author to protect the personal information of participants.

Conflicts of Interest: M.Z. and A.W. have a financial interest in Lanek SPA, a company focused on developing and commercializing biomedical devices and technologies. Their interests were reviewed and are managed by Universidad Técnica Federico Santa María and Universidad de Valparaíso, respectively, in accordance with their conflict-of-interest policies.

References

1. Scheerer, N.E.; Jones, J.A. The Predictability of Frequency-Altered Auditory Feedback Changes the Weighting of Feedback and Feedforward Input for Speech Motor Control. *Eur. J. Neurosci.* **2014**, *40*, 3793–3806. [CrossRef]
2. Parrell, B.; Lammert, A.C.; Ciccarelli, G.; Quatieri, T.F. Current Models of Speech Motor Control: A Control-Theoretic Overview of Architectures and Properties. *J. Acoust. Soc. Am.* **2019**, *145*, 1456–1481. [CrossRef]
3. Guenther, F.H. *Neural Control of Speech*; The MIT Press: Cambridge, MA, USA, 2016; ISBN 978-0-262-33698-7.
4. Aaron, A.S.; Abur, D.; Volk, K.P.; Noordzij, J.P.; Tracy, L.F.; Stepp, C.E. The Relationship Between Pitch Discrimination and Fundamental Frequency Variation: Effects of Singing Status and Vocal Hyperfunction. *J. Voice* **2023**, S0892199723000103. [CrossRef]
5. Abur, D.; Subaciute, A.; Kapsner-Smith, M.; Segina, R.K.; Tracy, L.F.; Noordzij, J.P.; Stepp, C.E. Impaired Auditory Discrimination and Auditory-Motor Integration in Hyperfunctional Voice Disorders. *Sci. Rep.* **2021**, *11*, 13123. [CrossRef] [PubMed]
6. Max, L.; Guenther, F.H.; Gracco, V.L.; Ghosh, S.S.; Wallace, M.E. Unstable or Insufficiently Activated Internal Models and Feedback-Biased Motor Control as Sources of Dysfluency: A Theoretical Model of Stuttering. *Contemp. Issues Commun. Sci. Disord.* **2004**, *31*, 105–122. [CrossRef]
7. Civier, O.; Bullock, D.; Max, L.; Guenther, F.H. Computational Modeling of Stuttering Caused by Impairments in a Basal Ganglia Thalamo-Cortical Circuit Involved in Syllable Selection and Initiation. *Brain Lang.* **2013**, *126*, 263–278. [CrossRef]
8. Vojtech, J.M.; Stepp, C.E. Effects of Age and Parkinson's Disease on the Relationship between Vocal Fold Abductory Kinematics and Relative Fundamental Frequency. *J. Voice* **2022**, S0892199722000704. [CrossRef]
9. Abur, D.; Subaciute, A.; Daliri, A.; Lester-Smith, R.A.; Lupiani, A.A.; Cilento, D.; Enos, N.M.; Weerathunge, H.R.; Tardif, M.C.; Stepp, C.E. Feedback and Feedforward Auditory-Motor Processes for Voice and Articulation in Parkinson's Disease. *J. Speech Lang. Hear. Res.* **2021**, *64*, 4682–4694. [CrossRef]
10. *Speech Production and Perception: Learning and Memory*; Fuchs, S.; Cleland, J.; Rochet-Capellan, A. (Eds.) Peter Lang D: New York, NY, USA, 2019; ISBN 978-3631726914.
11. Parrell, B.; Agnew, Z.; Nagarajan, S.; Houde, J.; Ivry, R.B. Impaired Feedforward Control and Enhanced Feedback Control of Speech in Patients with Cerebellar Degeneration. *J. Neurosci.* **2017**, *37*, 9249–9258. [CrossRef]
12. Cai, S.; Beal, D.S.; Ghosh, S.S.; Guenther, F.H.; Perkell, J.S. Impaired Timing Adjustments in Response to Time-Varying Auditory Perturbation during Connected Speech Production in Persons Who Stutter. *Brain Lang.* **2014**, *129*, 24–29. [CrossRef]
13. Cai, S.; Beal, D.S.; Ghosh, S.S.; Tiede, M.K.; Guenther, F.H.; Perkell, J.S. Weak Responses to Auditory Feedback Perturbation during Articulation in Persons Who Stutter: Evidence for Abnormal Auditory-Motor Transformation. *PLoS ONE* **2012**, *7*, e41830. [CrossRef]
14. Cai, S.; Ghosh, S.S.; Guenther, F.H.; Perkell, J.S. Focal Manipulations of Formant Trajectories Reveal a Role of Auditory Feedback in the Online Control of Both Within-Syllable and Between-Syllable Speech Timing. *J. Neurosci.* **2011**, *31*, 16483–16490. [CrossRef]
15. Niziolek, C.A.; Guenther, F.H. Vowel Category Boundaries Enhance Cortical and Behavioral Responses to Speech Feedback Alterations. *J. Neurosci.* **2013**, *33*, 12090–12098. [CrossRef]
16. Purcell, D.W.; Munhall, K.G. Compensation Following Real-Time Manipulation of Formants in Isolated Vowels. *J. Acoust. Soc. Am.* **2006**, *119*, 2288–2297. [CrossRef]
17. Reilly, K.J.; Dougherty, K.E. The Role of Vowel Perceptual Cues in Compensatory Responses to Perturbations of Speech Auditory Feedback. *J. Acoust. Soc. Am.* **2013**, *134*, 1314–1323. [CrossRef]
18. Tourville, J.A.; Reilly, K.J.; Guenther, F.H. Neural Mechanisms Underlying Auditory Feedback Control of Speech. *NeuroImage* **2008**, *39*, 1429–1443. [CrossRef]
19. Daliri, A.; Chao, S.-C.; Fitzgerald, L.C. Compensatory Responses to Formant Perturbations Proportionally Decrease as Perturbations Increase. *J. Speech Lang. Hear. Res.* **2020**, *63*, 3392–3407. [CrossRef]
20. Franken, M.K.; Acheson, D.J.; McQueen, J.M.; Hagoort, P.; Eisner, F. Consistency Influences Altered Auditory Feedback Processing. *Q. J. Exp. Psychol.* **2019**, *72*, 2371–2379. [CrossRef]
21. Kearney, E.; Guenther, F.H. Articulating: The Neural Mechanisms of Speech Production. *Lang. Cogn. Neurosci.* **2019**, *34*, 1214–1229. [CrossRef]
22. Tourville, J.A.; Guenther, F.H. The DIVA Model: A Neural Theory of Speech Acquisition and Production. *Lang. Cogn. Process.* **2011**, *26*, 952–981. [CrossRef]
23. Lane, H.; Denny, M.; Guenther, F.H.; Hanson, H.M.; Marrone, N.; Matthies, M.L.; Perkell, J.S.; Stockmann, E.; Tiede, M.; Vick, J.; et al. On the Structure of Phoneme Categories in Listeners With Cochlear Implants. *J. Speech Lang. Hear. Res.* **2007**, *50*, 2–14. [CrossRef]
24. Kearney, E.; Nieto-Castañón, A.; Falsini, R.; Daliri, A.; Heller Murray, E.S.; Smith, D.J.; Guenther, F.H. Quantitatively Characterizing Reflexive Responses to Pitch Perturbations. *Front. Hum. Neurosci.* **2022**, *16*, 929687. [CrossRef]
25. Castro, C.; Prado, P.; Espinoza, V.M.; Testart, A.; Marfull, D.; Manriquez, R.; Stepp, C.E.; Mehta, D.D.; Hillman, R.E.; Zañartu, M. Lombard Effect in Individuals With Nonphonotraumatic Vocal Hyperfunction: Impact on Acoustic, Aerodynamic, and Vocal Fold Vibratory Parameters. *J. Speech Lang. Hear. Res.* **2022**, *65*, 2881–2895. [CrossRef]
26. Perkell, J.S.; Denny, M.; Lane, H.; Guenther, F.; Matthies, M.L.; Tiede, M.; Vick, J.; Zandipour, M.; Burton, E. Effects of Masking Noise on Vowel and Sibilant Contrasts in Normal-Hearing Speakers and Postlingually Deafened Cochlear Implant Users. *J. Acoust. Soc. Am.* **2007**, *121*, 505–518. [CrossRef]

27. Frankford, S.A.; Heller Murray, E.S.; Masapollo, M.; Cai, S.; Tourville, J.A.; Nieto-Castañón, A.; Guenther, F.H. The Neural Circuitry Underlying the "Rhythm Effect" in Stuttering. *J. Speech Lang. Hear. Res.* **2020**, *64*, 2325–2346. [CrossRef]
28. Perkell, J.S.; Guenther, F.H.; Lane, H.; Matthies, M.L.; Perrier, P.; Vick, J.; Wilhelms-Tricarico, R.; Zandipour, M. A Theory of Speech Motor Control and Supporting Data from Speakers with Normal Hearing and with Profound Hearing Loss. *J. Phon.* **2000**, *28*, 233–272. [CrossRef]
29. Tourville, J.T.; Cai, S.; Guenther, H. Frank Exploring Auditory-Motor Interactions in Normal and Disordered Speech. *J. Acoust. Soc. Am.* **2013**, *133*, 3564. [CrossRef]
30. Zarate, J.M. Neural Substrates Governing Audiovocal Integration for Vocal Pitch Regulation in Singing. *Ann. N. Y. Acad. Sci.* **2005**, *1060*, 404–408. [CrossRef]
31. Toyomura, A.; Koyama, S.; Miyamaoto, T.; Terao, A.; Omori, T.; Murohashi, H.; Kuriki, S. Neural Correlates of Auditory Feedback Control in Human. *Neuroscience* **2007**, *146*, 499–503. [CrossRef]
32. Takaso, H.; Eisner, F.; Wise, R.J.; Scott, S.K. The Effect of Delayed Auditory Feedback on Activity in the Temporal Lobe While Speaking: A Positron Emission Tomography Study. *J. Speech Lang. Hear. Res. JSLHR* **2010**, *53*, 226–236. [CrossRef]
33. Fu, C.H.Y.; Vythelingum, G.N.; Brammer, M.J.; Williams, S.C.R.; Amaro, E.; Andrew, C.M.; Yágüez, L.; Van Haren, N.E.M.; Matsumoto, K.; McGuire, P.K. An FMRI Study of Verbal Self-Monitoring: Neural Correlates of Auditory Verbal Feedback. *Cereb. Cortex* **2006**, *16*, 969–977. [CrossRef]
34. Heinks-Maldonado, T.H.; Nagarajan, S.S.; Houde, J.F. Magnetoencephalographic Evidence for a Precise Forward Model in Speech Production. *NeuroReport* **2006**, *17*, 1375–1379. [CrossRef]
35. Niziolek, C.A.; Nagarajan, S.S.; Houde, J.F. What Does Motor Efference Copy Represent? Evidence from Speech Production. *J. Neurosci.* **2013**, *33*, 16110–16116. [CrossRef]
36. Golfinopoulos, E.; Tourville, J.A.; Bohland, J.W.; Ghosh, S.S.; Nieto-Castanon, A.; Guenther, F.H. FMRI Investigation of Unexpected Somatosensory Feedback Perturbation during Speech. *NeuroImage* **2011**, *55*, 1324–1338. [CrossRef]
37. Goriely, A.; Kuhl, E.; Bick, C. Neuronal Oscillations on Evolving Networks: Dynamics, Damage, Degradation, Decline, Dementia, and Death. *Phys. Rev. Lett.* **2020**, *125*, 128102. [CrossRef]
38. Rossini, P.M.; Di Iorio, R.; Vecchio, F.; Anfossi, M.; Babiloni, C.; Bozzali, M.; Bruni, A.C.; Cappa, S.F.; Escudero, J.; Fraga, F.J.; et al. Early Diagnosis of Alzheimer's Disease: The Role of Biomarkers Including Advanced EEG Signal Analysis. Report from the IFCN-Sponsored Panel of Experts. *Clin. Neurophysiol.* **2020**, *131*, 1287–1310. [CrossRef]
39. Patel, S.; Hebert, K.; Korzyukov, O.; Larson, C.R. Effects of Sensorimotor Voice Training on Event-Related Potentials to Pitch-Shifted Auditory Feedback. *PLoS ONE* **2023**, *18*, e0269326. [CrossRef] [PubMed]
40. Chen, Z.; Liu, P.; Wang, E.Q.; Larson, C.R.; Huang, D.; Liu, H. ERP Correlates of Language-Specific Processing of Auditory Pitch Feedback during Self-Vocalization. *Brain Lang.* **2012**, *121*, 25–34. [CrossRef]
41. Korzyukov, O.; Karvelis, L.; Behroozmand, R.; Larson, C.R. ERP Correlates of Auditory Processing during Automatic Correction of Unexpected Perturbations in Voice Auditory Feedback. *Int. J. Psychophysiol.* **2012**, *83*, 71–78. [CrossRef]
42. *Niedermeyer's Electroencephalography*; Schomer, D.L.; Lopes da Silva, F.H. (Eds.) Oxford University Press: Oxford, UK, 2017; Volume 1, ISBN 978-0-19-022848-4.
43. Blenkmann, A. Localización de Fuentes de Actividad Cerebral. Ph.D. Thesis, Universidad Nacional de La Plata, Buenos Aires, Argentina, 2012.
44. Sörnmo, L.; Laguna, P. *Bioelectrical Signal Processing in Cardiac and Neurological Applications*; Elsevier: Amsterdam, The Netherlands, 2005; ISBN 978-0-12-437552-9.
45. Grave de Peralta, R.; González Andino, S.; Gómez González, C.M. Bases biofísicas de la localización de los generadores cerebrales del electroencefalograma. Aplicación de un modelo de tipo distribuido a la localización de focos epilépticos. *Rev. Neurol.* **2004**, *39*, 748. [CrossRef]
46. Liu, F.; Qin, J.; Wang, S.; Rosenberger, J.; Su, J. Supervised EEG Source Imaging with Graph Regularization in Transformed Domain. In *Brain Informatics*; Zeng, Y., He, Y., Kotaleski, J.H., Martone, M., Xu, B., Peng, H., Luo, Q., Eds.; Lecture Notes in Computer Science; Springer International Publishing: Cham, Switzerland, 2017; Volume 10654, pp. 59–71. ISBN 978-3-319-70771-6.
47. Sanei, S.; Chambers, J.A. *EEG Signal Processing: Sanei/EEG Signal Processing*; John Wiley & Sons Ltd.: West Sussex, UK, 2007; ISBN 978-0-470-51192-3.
48. Tong, S.; Thakor, N.V. *Quantitative EEG Analysis Methods and Clinical Applications*; Artech House: Boston, MA, USA, 2009; ISBN 978-1-59693-205-0.
49. Hallez, H.; Vanrumste, B.; Grech, R.; Muscat, J.; De Clercq, W.; Vergult, A.; D'Asseler, Y.; Camilleri, K.P.; Fabri, S.G.; Van Huffel, S.; et al. Review on Solving the Forward Problem in EEG Source Analysis. *J. NeuroEng. Rehabil.* **2007**, *4*, 46. [CrossRef]
50. Pascual-Marqui, R.D. Standardized Low-Resolution Brain Electromagnetic Tomography (SLORETA): Technical Details. *Methods Find. Exp. Clin. Pharmacol.* **2002**, *24*, 5–12.
51. Grech, R.; Cassar, T.; Muscat, J.; Camilleri, K.P.; Fabri, S.G.; Zervakis, M.; Xanthopoulos, P.; Sakkalis, V.; Vanrumste, B. Review on Solving the Inverse Problem in EEG Source Analysis. *J. NeuroEng. Rehabil.* **2008**, *5*, 25. [CrossRef]
52. Kalinowski, J.; Stuart, A.; Sark, S.; Armson, J. Stuttering Amelioration at Various Auditory Feedback Delays and Speech Rates. *Int. J. Lang. Commun. Disord.* **1996**, *31*, 259–269. [CrossRef]

53. Cai, S.; Boucek, M.M.; Ghosh, S.S.; Guenther, F.H.; Perkell, J.S. A System for Online Dynamic Perturbation of Formant Trajectories and Results from Perturbations of the Mandarin Triphthong /Iau/. In Proceedings of the 8th International Seminar on Speech Production, Strasbourg, France, 8–12 December 2008; pp. 65–68.
54. Chaumon, M.; Bishop, D.V.M.; Busch, N.A. A Practical Guide to the Selection of Independent Components of the Electroencephalogram for Artifact Correction. *J. Neurosci. Methods* **2015**, *250*, 47–63. [CrossRef]
55. Formant-Analyzer 2023. Available online: https://github.com/fulldecent/formant-analyzer (accessed on 1 March 2023).
56. Niziolek, C.A.; Parrell, B. Responses to Auditory Feedback Manipulations in Speech May Be Affected by Previous Exposure to Auditory Errors. *J. Speech Lang. Hear. Res.* **2021**, *64*, 2169–2181. [CrossRef]
57. Daliri, A. A Computational Model for Estimating the Speech Motor System's Sensitivity to Auditory Prediction Errors. *J. Speech Lang. Hear. Res.* **2021**, *64*, 1841–1854. [CrossRef]
58. Collins, D.L.; Neelin, P.; Peters, T.M.; Evans, A.C. Automatic 3D Intersubject Registration of MR Volumetric Data in Standardized Talairach Space. *J. Comput. Assist. Tomogr.* **1994**, *18*, 192–205. [CrossRef]
59. Evans, A.C.; Collins, D.L.; Mills, S.R.; Brown, E.D.; Kelly, R.L.; Peters, T.M. 3D Statistical Neuroanatomical Models from 305 MRI Volumes. In Proceedings of the 1993 IEEE Conference Record Nuclear Science Symposium and Medical Imaging Conference, San Francisco, CA, USA, 30 October–6 November 1993; IEEE: San Francisco, CA, USA, 1993; pp. 1813–1817.
60. Rolls, E.T.; Joliot, M.; Tzourio-Mazoyer, N. Implementation of a New Parcellation of the Orbitofrontal Cortex in the Automated Anatomical Labeling Atlas. *NeuroImage* **2015**, *122*, 1–5. [CrossRef]
61. Pascual-Marqui, R.D.; Michel, C.M.; Lehmann, D. Low Resolution Electromagnetic Tomography: A New Method for Localizing Electrical Activity in the Brain. *Int. J. Psychophysiol.* **1994**, *18*, 49–65. [CrossRef]
62. Nichols, T.E.; Holmes, A.P. Nonparametric Permutation Tests for Functional Neuroimaging: A Primer with Examples. *Hum. Brain Mapp.* **2002**, *15*, 1–25. [CrossRef]
63. Weerathunge, H.R.; Alzamendi, G.A.; Cler, G.J.; Guenther, F.H.; Stepp, C.E.; Zañartu, M. LaDIVA: A Neurocomputational Model Providing Laryngeal Motor Control for Speech Acquisition and Production. *PLoS Comput. Biol.* **2022**, *18*, e1010159. [CrossRef]
64. Kearney, E.; Nieto-Castañón, A.; Weerathunge, H.R.; Falsini, R.; Daliri, A.; Abur, D.; Ballard, K.J.; Chang, S.-E.; Chao, S.-C.; Heller Murray, E.S.; et al. A Simple 3-Parameter Model for Examining Adaptation in Speech and Voice Production. *Front. Psychol.* **2020**, *10*, 2995. [CrossRef]
65. Kinahan, S.P.; Liss, J.M.; Berisha, V. TorchDIVA: An Extensible Computational Model of Speech Production Built on an Open-Source Machine Learning Library. *PLoS ONE* **2023**, *18*, e0281306. [CrossRef]
66. Bohland, J.W.; Bullock, D.; Guenther, F.H. Neural Representations and Mechanisms for the Performance of Simple Speech Sequences. *J. Cogn. Neurosci.* **2010**, *22*, 1504–1529. [CrossRef]
67. Valdés-Hernández, P.A.; von Ellenrieder, N.; Ojeda-Gonzalez, A.; Kochen, S.; Alemán-Gómez, Y.; Muravchik, C.; Valdés-Sosa, P.A. Approximate Average Head Models for EEG Source Imaging. *J. Neurosci. Methods* **2009**, *185*, 125–132. [CrossRef]
68. Barzegaran, E.; Bosse, S.; Kohler, P.J.; Norcia, A.M. EEGSourceSim: A Framework for Realistic Simulation of EEG Scalp Data Using MRI-Based Forward Models and Biologically Plausible Signals and Noise. *J. Neurosci. Methods* **2019**, *328*, 108377. [CrossRef]
69. Wang, H.E.; Bénar, C.G.; Quilichini, P.P.; Friston, K.J.; Jirsa, V.K.; Bernard, C. A Systematic Framework for Functional Connectivity Measures. *Front. Neurosci.* **2014**, *8*, 405. [CrossRef]
70. Prado, P.; Birba, A.; Cruzat, J.; Santamaría-García, H.; Parra, M.; Moguilner, S.; Tagliazucchi, E.; Ibáñez, A. Dementia ConnEEGtome: Towards Multicentric Harmonization of EEG Connectivity in Neurodegeneration. *Int. J. Psychophysiol.* **2022**, *172*, 24–38. [CrossRef]
71. Otero, M.; Lea-Carnall, C.; Prado, P.; Escobar, M.-J.; El-Deredy, W. Modelling Neural Entrainment and Its Persistence: Influence of Frequency of Stimulation and Phase at the Stimulus Offset. *Biomed. Phys. Eng. Express* **2022**, *8*, 045014. [CrossRef]
72. Hyafil, A.; Fontolan, L.; Kabdebon, C.; Gutkin, B.; Giraud, A.-L. Speech Encoding by Coupled Cortical Theta and Gamma Oscillations. *eLife* **2015**, *4*, e06213. [CrossRef]
73. Martin, C.D.; Niziolek, C.A.; Duñabeitia, J.A.; Perez, A.; Hernandez, D.; Carreiras, M.; Houde, J.F. Online Adaptation to Altered Auditory Feedback Is Predicted by Auditory Acuity and Not by Domain-General Executive Control Resources. *Front. Hum. Neurosci.* **2018**, *12*, 91. [CrossRef]
74. Cai, S.; Ghosh, S.S.; Guenther, F.H.; Perkell, J.S. Adaptive Auditory Feedback Control of the Production of Formant Trajectories in the Mandarin Triphthong /Iau/ and Its Pattern of Generalization. *J. Acoust. Soc. Am.* **2010**, *128*, 2033–2048. [CrossRef]
75. Hu, H.; Liu, Y.; Guo, Z.; Li, W.; Liu, P.; Chen, S.; Liu, H. Attention Modulates Cortical Processing of Pitch Feedback Errors in Voice Control. *Sci. Rep.* **2015**, *5*, 7812. [CrossRef]
76. Behroozmand, R.; Sangtian, S.; Korzyukov, O.; Larson, C.R. A Temporal Predictive Code for Voice Motor Control: Evidence from ERP and Behavioral Responses to Pitch-Shifted Auditory Feedback. *Brain Res.* **2016**, *1636*, 1–12. [CrossRef]
77. Chen, Z.; Chen, X.; Liu, P.; Huang, D.; Liu, H. Effect of Temporal Predictability on the Neural Processing of Self-Triggered Auditory Stimulation during Vocalization. *BMC Neurosci.* **2012**, *13*, 55. [CrossRef]
78. Larson, C.R.; Burnett, T.A.; Kiran, S.; Hain, T.C. Effects of Pitch-Shift Velocity on Voice F0 Responses. *J. Acoust. Soc. Am.* **2000**, *107*, 559–564. [CrossRef]
79. Liu, H.; Larson, C.R. Effects of Perturbation Magnitude and Voice F0 Level on the Pitch-Shift Reflex. *J. Acoust. Soc. Am.* **2007**, *122*, 3671–3677. [CrossRef]

80. Smith, D.J.; Stepp, C.; Guenther, F.H.; Kearney, E. Contributions of Auditory and Somatosensory Feedback to Vocal Motor Control. *J. Speech Lang. Hear. Res.* **2020**, *63*, 2039–2053. [CrossRef]
81. Toyomura, A.; Miyashiro, D.; Kuriki, S.; Sowman, P.F. Speech-Induced Suppression for Delayed Auditory Feedback in Adults Who Do and Do Not Stutter. *Front. Hum. Neurosci.* **2020**, *14*, 150. [CrossRef]
82. Behroozmand, R.; Larson, C.R. Error-Dependent Modulation of Speech-Induced Auditory Suppression for Pitch-Shifted Voice Feedback. *BMC Neurosci.* **2011**, *12*, 54. [CrossRef]
83. Behroozmand, R.; Oya, H.; Nourski, K.V.; Kawasaki, H.; Larson, C.R.; Brugge, J.F.; Howard, M.A.; Greenlee, J.D.W. Neural Correlates of Vocal Production and Motor Control in Human Heschl's Gyrus. *J. Neurosci. Off. J. Soc. Neurosci.* **2016**, *36*, 2302–2315. [CrossRef]
84. Butler, B.E.; Trainor, L.J. Sequencing the Cortical Processing of Pitch-Evoking Stimuli Using EEG Analysis and Source Estimation. *Front. Psychol.* **2012**, *3*, 180. [CrossRef]
85. Wang, J.; Mathalon, D.H.; Roach, B.J.; Reilly, J.; Keedy, S.K.; Sweeney, J.A.; Ford, J.M. Action Planning and Predictive Coding When Speaking. *NeuroImage* **2014**, *91*, 91–98. [CrossRef]
86. Behroozmand, R.; Shebek, R.; Hansen, D.R.; Oya, H.; Robin, D.A.; Howard, M.A.; Greenlee, J.D.W. Sensory–Motor Networks Involved in Speech Production and Motor Control: An FMRI Study. *NeuroImage* **2015**, *109*, 418–428. [CrossRef]
87. Parkinson, A.L.; Flagmeier, S.G.; Manes, J.L.; Larson, C.R.; Rogers, B.; Robin, D.A. Understanding the Neural Mechanisms Involved in Sensory Control of Voice Production. *NeuroImage* **2012**, *61*, 314–322. [CrossRef]
88. Guenther, F.H.; Ghosh, S.S.; Tourville, J.A. Neural Modeling and Imaging of the Cortical Interactions Underlying Syllable Production. *Brain Lang.* **2006**, *96*, 280–301. [CrossRef]
89. Houde, J.F.; Nagarajan, S.S. Speech Production as State Feedback Control. *Front. Hum. Neurosci.* **2011**, *5*, 82. [CrossRef]
90. Houde, J.F.; Chang, E.F. The Cortical Computations Underlying Feedback Control in Vocal Production. *Curr. Opin. Neurobiol.* **2015**, *33*, 174–181. [CrossRef]
91. Ford, J.M.; Gray, M.; Faustman, W.O.; Roach, B.J.; Mathalon, D.H. Dissecting Corollary Discharge Dysfunction in Schizophrenia. *Psychophysiology* **2007**, *44*, 522–529. [CrossRef]
92. Bendixen, A.; SanMiguel, I.; Schröger, E. Early Electrophysiological Indicators for Predictive Processing in Audition: A Review. *Int. J. Psychophysiol.* **2012**, *83*, 120–131. [CrossRef]
93. Liu, H.; Meshman, M.; Behroozmand, R.; Larson, C.R. Differential Effects of Perturbation Direction and Magnitude on the Neural Processing of Voice Pitch Feedback. *Clin. Neurophysiol.* **2011**, *122*, 951–957. [CrossRef]
94. Korzyukov, O.; Sattler, L.; Behroozmand, R.; Larson, C.R. Neuronal Mechanisms of Voice Control Are Affected by Implicit Expectancy of Externally Triggered Perturbations in Auditory Feedback. *PLoS ONE* **2012**, *7*, e41216. [CrossRef]
95. Behroozmand, R.; Sangtian, S. Neural Bases of Sensorimotor Adaptation in the Vocal Motor System. *Exp. Brain Res.* **2018**, *236*, 1881–1895. [CrossRef]
96. Dai, G.; Chen, M.; Chen, X.; Guo, Z.; Li, T.; Jones, J.A.; Wu, X.; Li, J.; Liu, P.; Liu, H.; et al. A Causal Link between Left Supplementary Motor Area and Auditory-Motor Control of Vocal Production: Evidence by Continuous Theta Burst Stimulation. *NeuroImage* **2022**, *264*, 119767. [CrossRef]
97. Shum, M.; Shiller, D.M.; Baum, S.R.; Gracco, V.L. Sensorimotor Integration for Speech Motor Learning Involves the Inferior Parietal Cortex: Speech Motor Adaptation. *Eur. J. Neurosci.* **2011**, *34*, 1817–1822. [CrossRef]
98. Venezia, J.H.; Richards, V.M.; Hickok, G. Speech-Driven Spectrotemporal Receptive Fields Beyond the Auditory Cortex. *Hear. Res.* **2021**, *408*, 108307. [CrossRef]
99. Valeriani, D.; Simonyan, K. The Dynamic Connectome of Speech Control. *Philos. Trans. R. Soc. B Biol. Sci.* **2021**, *376*, 20200256. [CrossRef]
100. Fallgatter, A.J.; Bartsch, A.J.; Zielasek, J.; Herrmann, M.J. Brain Electrical Dysfunction of the Anterior Cingulate in Schizophrenic Patients. *Psychiatry Res. Neuroimaging* **2003**, *124*, 37–48. [CrossRef]
101. Galindo, G.E.; Peterson, S.D.; Erath, B.D.; Castro, C.; Hillman, R.E.; Zañartu, M. Modeling the Pathophysiology of Phonotraumatic Vocal Hyperfunction With a Triangular Glottal Model of the Vocal Folds. *J. Speech Lang. Hear. Res.* **2017**, *60*, 2452–2471. [CrossRef]
102. Manriquez, R.; Peterson, S.D.; Prado, P.; Orio, P.; Galindo, G.E.; Zañartu, M. Neurophysiological Muscle Activation Scheme for Controlling Vocal Fold Models. *IEEE Trans. Neural Syst. Rehabil. Eng.* **2019**, *27*, 1043–1052. [CrossRef]
103. Prado-Gutierrez, P.; Martínez-Montes, E.; Weinstein, A.; Zañartu, M. Estimation of Auditory Steady-State Responses Based on the Averaging of Independent EEG Epochs. *PLoS ONE* **2019**, *14*, e0206018. [CrossRef]
104. Prado-Gutierrez, P.; Castro-Fariñas, A.; Morgado-Rodriguez, L.; Velarde-Reyes, E.; Martínez, A.D.; Martínez-Montes, E. Habituation of Auditory Steady State Responses Evoked by Amplitudemodulated Acoustic Signals in Rats. *Audiol. Res.* **2015**, *5*, 113. [CrossRef]

Disclaimer/Publisher's Note: The statements, opinions and data contained in all publications are solely those of the individual author(s) and contributor(s) and not of MDPI and/or the editor(s). MDPI and/or the editor(s) disclaim responsibility for any injury to people or property resulting from any ideas, methods, instructions or products referred to in the content.

Article

Quantitative Electroencephalographic Analysis in Women with Migraine during the Luteal Phase

Héctor Juan Pelayo-González [1,*], Verónica Reyes-Meza [2,*], Ignacio Méndez-Balbuena [1], Oscar Méndez-Díaz [1], Carlos Trenado [3,4], Diane Ruge [3], Gregorio García-Aguilar [1] and Vicente Arturo López-Cortés [1]

1. Facultad de Psicología, Benemérita Universidad Autónoma de Puebla, Puebla 72960, Mexico; ignacio.mendez@correo.buap.mx (I.M.-B.); huevesillo@hotmail.com (O.M.-D.); gregorio.garcia@correo.buap.mx (G.G.-A.); vicente.lopez@correo.buap.mx (V.A.L.-C.)
2. Centro Tlaxcala de Biología de la Conducta, Universidad Autónoma de Tlaxcala, Tlaxcala de Xicohténcatl 90000, Mexico
3. Laboratoire de Recherche en Neurosciences Cliniques (LRENC), 34000 Montpellier, France; carlos.trenadoc@gmail.com (C.T.); diane.ruge@gmail.com (D.R.)
4. Institute of Clinical Neuroscience and Medical Psychology, Heinrich Heine University Düsseldorf, 40225 Düsseldorf, Germany
5. Institute of Neurology, University College London (UCL), Queen Square, London WC1N 3BG, UK
* Correspondence: hector.pelayo@correo.buap.mx (H.J.P.-G.); veronica.reyes@uatx.mx (V.R.-M.)

Abstract: Migraine is a common, headache disorder characterized by recurrent episodes of headache often associated with nausea, vomiting, photophobia, and phonophobia. Prior to puberty, boys and girls are equally affected. Female preponderance emerges after puberty. Migraine pathophysiology is not fully understood, and although the hormonal effect of estrogen is significant, it is not clear how hormonal phases affect brain excitability and EEG patterns in women with migraine. The objective of this research was to study the effect of migraine on the resting-state EEG activity of women during the luteal phase. This work compares electroencephalographic (EEG) absolute power in different frequency bands and scalp areas between young women who suffer from migraine and had a migraine attack within 24 h prior to EEG recording (experimental) and ten age-matched young healthy women (controls), all with normal menstrual cycles. For women with migraine, we found a significant decrease/increase in alpha power in the occipitoparietal/frontocentral area, significant decrease in beta power for all areas, significant decrease in delta power in the temporal area, and significant decrease in theta power in the frontocentral and occipitoparietal area. We concluded that women with migraine have a distinct electroencephalographic pattern during the luteal phase in comparison with control women. A possible explanation might be an intermittent rhythmic activity linked to pain.

Keywords: EEG; migraine; luteal phase; absolute power

1. Introduction

Migraine is a form of neurovascular headache [1], with a high incidence (>12%). Predominantly women (3:1 compared with men) are affected and intensity is variable [2]. Migraine is a public health problem, and it is also one of the main causes of incapacity for work because fifty percent of migraine patients interrupt their daily activities due to attacks and most of them require rest at home in dark places [3].

Migraine symptoms negatively affect quality of life as well as academic and work performance and limit the realization of daily activities [4]. In general, it is defined as an episodic attack of intense, pulsating, and unilateral headache that may last from 4 to 72 h ranging from once a week to once a year [5]. The age group with the highest prevalence of migraine is between 25 and 55 years old [6], however, there are reports that indicate that the number of episodes decreases as the age of the patients increases, or at least the prevalence of unilateral and pulsating pain [7]. With regard to triggers that induce migraine, it has been reported that drinking alcohol, smoking, living with stress, neck pain, and hormonal changes are the main stimuli that play a role [8].

Migraine's etiology involves a neurovascular mechanism and is characterized by cortical hypersensitivity [9]. However, for some authors, it is mainly a neuronal pathology with secondary vascular effects [10].

Migraine has been associated with hormonal changes; several authors have reported that there is an increase in the reproductive years, and menstruation specifically is considered one of the most common triggers for migraine, also affecting the level of pain, the duration of symptoms, and the response to treatment [11–13].

The menstrual cycle is a series of natural changes in hormone production and the structures of the uterus and ovaries of the female reproductive system that makes pregnancy possible. The four phases of the menstrual cycle are menstruation, the follicular phase, ovulation, and the luteal phase. Common menstrual problems include heavy or painful periods and premenstrual syndrome. Migraines that occur during the menstrual period tend to be more disabling than those that happen at different times of the month [14].

Moreover, hormonal contraceptive use during the reproductive years and hormone replacement in menopause alter the levels of sex hormones, and these events and interventions are associated with a change in the prevalence and intensity of headaches [14].

Regarding the mechanism for menstrual migraine, it was reported that an estrogen level drop during the late luteal phase may be a marker of vulnerability to migraine symptoms in women that experience migraine [15]. In line with this, others reported a drop in estrogen levels prior to menstruation, which was explained by the effects of estrogens and progestogens on central serotonergic and opioidergic neurons, modulating their neuronal activity and the density of the 5-HT1 type receptor [16,17].

Although there are some techniques for studying the central mechanisms involved in chronic pain like migraine, electroencephalography (EEG) stands out as a valuable, non-invasive tool because it provides reliable and relevant information about brain functioning during rest, sensory stimulation, and cognitive tasks. In addition, this technique is safe, low-cost, and employs a straightforward methodology, thus making it an appropriate tool for use in clinical practice.

EEG has been applied to assess brain function in several chronic pain syndromes. The American Neurology Academy has suggested that EEG can be used in people with symptoms associated with headaches and migraine [18]. One of the first studies in which EEG alterations were associated with headache dates back to 1959, when Golla and Winter [19], described two main types of EEG frequency response during flickering light: (a) a peak in the alpha band and a rapid decline with an increase in stimulus frequency above 14 f/s, and (b) a flat-top showing a response maintained up to or above 20 f/s. These authors concluded that the spatial distribution of the cerebral mechanisms involved in the flicker response resulted from the disturbance in the cardio-vascular barostatic mechanisms.

Several years later, Wasler et al. [20], reported frontal intermittent rhythmic delta activity (FIRDA) in EEG during and shortly after migraine episodes in migraine patients with episodes of impairment of consciousness and neurological deficit, indicating dysfunction of the upper brainstem and occipital and medial temporal lobes. Schoenen et al. [21], identified markedly reduced alpha activity in one occipital area on the side of the headache in 19 of 22 patients with migraine. Sixteen of these patients had a concomitant reduction in theta activity in the same location. In all patients except one, when they were re-examined seven days after a migraine attack, the EEG asymmetries had disappeared. According to the authors, unilateral EEG changes can thus be detected during migraine attacks and could be associated with unilateral disturbances of cortical electrogenesis. Later, Nyrke et al. [22], found an increase in higher alpha rhythm variability within 72 h following a migraine attack. Bjork et al. [23], found increased relative theta activity and attenuated medium-frequency photic responses in migraineurs without aura compared to controls. On the other hand, O'Hare et al. [24] found the lower alpha band (8 to 10 Hz) power was increased in the migraine group compared with the control group, which may provide a mechanism for increased multiplicative noise.

Silberstein [25], described that migraine was more present in women than in men, suggesting that changes in estrogen levels at menarche, menstruation, pregnancy, and menopause may trigger or change the prevalence of migraine. For example, the fall in the levels of estrogen that occurs during menstruation triggers menstrual migraine, whereas the sustained high estrogen levels during pregnancy frequently result in headache relief. The same author argued that estrogen produces changes in prostaglandins, hypothalamic opioids, and prolactin secretion, which may, in part, account for the genesis of headache. For this reason, one might assume that EEG patterns could be influenced by the menstrual cycle and the pain experienced. Becker et al. [26], reported mean alpha frequency cyclic changes in EEG activity, i.e., slower alpha waves during the follicular phase and faster alpha waves during luteal phase, as well as theta and beta small cyclic changes in women during both spontaneous and oral contraceptive-controlled menstrual cycles. Solís-Ortiz et al. [27], studied the effect of the menstrual cycle on EEG power during rest (eyes open and closed) in healthy women with no oral contraceptive effects. They reported lower EEG absolute power during the follicular phase; high power in delta, theta, and alpha 1 (7.5–9.5 Hz) during the luteal phase; high alpha 2 (9.5–12.5 Hz), beta 1 (12.5–17.5 Hz), and beta 2 (17.5–30 Hz) during the menstrual phase; and lower relative power in low alpha and higher in high alpha during the luteal phase. In addition, there was higher interhemispheric correlation between frontal regions during ovulation and between occipitals during the luteal phase, with no significant asymmetries. Thus, the authors concluded with the observation of a lower activation of frontal regions during the luteal phase and higher activation of central–parietal regions during the menstrual phase. Furthermore, Baehr et al. [28], found EEG frontal alpha asymmetry in a group of women suffering from premenstrual dysphoric disorder in comparison to a control group during the luteal period. Haraguchi et al. [29], reported lower alpha, theta, and gamma MEG power during the menstrual phase in comparison with outside this phase in healthy women.

Platzer et al. [30], investigated the effect of the menstrual cycle on brain activation and connectivity patterns by using fMRI in naturally cycling women performing cognitive tasks (spatial navigation and verbal fluency). The authors found no significant difference in task performance throughout the menstrual cycle, and changes in brain activation patterns were similar during both tasks. They also reported a hippocampal activation during the follicular phase and a boosting effect of progesterone in fronto-striatal activation during the luteal phase. Moreover, right-hemispheric frontal activation was suggested to result from inter-hemispheric decoupling and to be involved in the down-regulation of hippocampal activation. Hidalgo and Pletzer [31], assessed brain activation during an N-back verbal memory task in women with a regular menstrual cycle. They were able to corroborate a hormone-mediated inter-hemispheric decoupling that enhanced frontal activity and the disinhibition of the salience brain network and striatum during the luteal phase. The authors interpreted these results in relation to a top-down differential regulation in higher hormone level phases and a hyperactive bottom-up network during the luteal phase, which could explain the vulnerability of this phase to menstrual cycle-associated disorders.

To the best of our knowledge, this is the first study that addresses the effect of migraine on the resting-state EEG of women during the luteal phase. We focused on the luteal phase as it is characterized by a change in estrogen levels, which may trigger migraine symptoms. Based on the mentioned migraine studies that reported EEG power changes in different frequency bands (delta, theta, alpha), we expected a difference in EEG power between the with and without migraine conditions in women during the luteal phase.

2. Material and Methods

2.1. Subjects

Twenty female right-handed subjects participated in our study. There were ten women with migraine: mean age 25.4 ± SD 1.9 years, mean years of condition 4.9 (±SD 2.6), mean schooling 14.5 ± SD 2.1 years. In the case of healthy women: mean age 26.7 ± SD 1.9, mean schooling 15.5 ± SD 3.5.

We used the Oldfield questionnaire [32], to test the handedness.

A local Ethics Committee of Maestría en Diagnóstico y Rehabilitación Neuropsicológica from the Faculty of Psychology of Benemérita Universidad Autónoma de Puebla (DCECEN) approved the experimental protocol. All women participated in accordance with the Declaration of Helsinki as it was established by the World Medical Association in 1964 [33]. Subjects participated with understanding and informed consent.

We collected data through a semi-structured interview (Table 1). The experimental group described pain originating from all over the frontal and central areas.

Table 1. Questionnaire results.

Patient	Pain Place L (Left), R (Right)	Symptoms
1	Frontal L	Dizziness
2	Fronto-Parietal R	Blurred vision
3	Frontal R	None
4	Frontal L–R	None
5	Frontal L	Dizziness
6	Frontal L-R	Dizziness
7	Frontal L-R	None
8	Frontal L	Dizziness
9	Frontal R	Dizziness
10	Frontal L	Dizziness, blurred vision

Nausea was the most frequent symptom associated with pain during a migraine attack. Vomiting and dizziness were also present, as well as blurred vision.

For convenience, our study was cross-correlational with non-probability statistical sampling. The migraine of participants was diagnosed by a neurologist, according to 'the International Classification of Headache Disorders, 3rd edition [34]. Patients were recruited through a request from the neurology service to complete the neurological assessment. We also used semi-structured interviews to collect demographic data, as well as information on how migraine episodes occur: triggering (beginning), episode duration, pain location, frequency, and associated symptoms.

The control group was formed based on the voluntary participation of women awaiting gynecological consultation (control group).

Inclusion Criteria for All Participants

Patients with migraine (experimental group) had the following signs: headache attacks lasting 4–72 h, pulsating quality, moderate or severe pain intensity, aggravation by or causing avoidance of routine physical activity (e.g., walking or climbing stairs), nausea, vomiting, photophobia, or phonophobia. We selected patients with migraine who reported a migraine attack in the last 24 h prior to the EEG session.

Volunteer women had no signs of migraine or headache.

All women with regular spontaneous menstrual cycles were screened in a standardized interview.

All women in the luteal phase were selected with the use of a calendar to establish the first day of menstrual bleeding, the average cycle length, and the length of the luteal phase.

Exclusion criteria:

Anxiety or depression;

Neuroleptic or drug use for chronic pain, depression, or epilepsy;

Alcohol or drug use;

Report of irregular menstrual cycles;

Pregnant or lactating during the last 12 months;

Taking oral contraceptives during the last four months.

2.2. EEG Session

Subjects sat comfortably in an electrically shielded, dimly lit room. We recorded EEGs in rest conditions with closed eyes for three minutes. Subjects did not report having anxiety during the experimental session, but women with migraine (experimental group), reported migraine attack in the last 24 h prior to the EEG session.

2.3. Recordings

We collected EEGs (bandpass DC-200 Hz, sampling rate 250 Hz, NicVue System, Nicolet Biomedical Inc., Middleton, Wisconsin, USA) from 20 scalp positions (according to 10–20 System) referenced to the ear lobes, with the ground electrode at the forehead. We set the Notch filter at 60 Hz, and we kept electrode impedances under 5 kOhm. For further analysis, we recorded electrooculograms (using the same bandpass and sampling rate as for the EEG) to exclude trials contaminated by eye movements. We stored data and analyzed them offline.

3. Data Analysis

3.1. EEG Spectral Power Analysis

We performed offline visual artifact rejection to exclude contaminated segments. After that, we concatenated segments of 60 s (15,000 points) from each participant. We computed spectral power (SP) for the considered frequency bands by using customized software scripts programed in MATLAB (Mathworks, 2019), using the following formula:

$$SP_C(f) = \frac{1}{n}\sum_{i=1}^{n} C_i(f)C_i^*(f)$$

where C_i represents the Fourier-transformed channel c for a given segment number ($i = 1, n$) and "*" indicates the complex conjugate.

We calculated the SP in the following frequencies: delta (0.5–3.5 Hz), theta (4–7.5 Hz), alpha (8–12 Hz), and beta (13–30 Hz). The frequency resolution we selected was 0.1 Hz.

3.2. Statistical Analysis

We calculated the areas under the curve for each SP frequency band. To test for any statistical difference in the SP, we normalized data values between 0 and 1. Because our data were not normally distributed (Shapiro–Wilk normality test $p < 0.05$), we used a nonparametric statistical analysis for two independent groups (U Mann–Whitney test) for the absolute power comparisons in each band per topographic area. In each band, we grouped the electrodes' signals to the average by cortical areas. In the frontocentral topographical area, we averaged the following electrodes: Fp1, Fp2, F7, F3, Fz, F4, F8, C3, Cz, and C4. In the occipitoparietal area, we averaged the following electrodes: P3, Pz, P4, O1, Oz, and O2. In the temporal area, we averaged the following electrodes: T3, T4, T5, and T6.

The null hypothesis was that the dependent variables were the same across the factors. We report effects as significant (two-tailed) if $p \leq 0.05$. The statistical analysis was performed using the software SPSS 25.

4. Results

The task was performed for all subjects according to the instructions. None of the participants reported fatigue during the experiment, but some showed anxiety and signs of irritability, such as sweating, agitation, and claustrophobic sensation.

4.1. Topographical Analysis

We obtained the following results for the absolute power analysis in different cortical areas and frequency bands.

4.2. Alpha

We found a significantly higher normalized power in the experimental group for the alpha band in the frontocentral area (U = 23, z = −2.04, r = 0.45, $p < 0.05$) in comparison with the control group. In contrast, we found a significantly higher normalized power in the occipitoparietal area in the control group (U = 6, z = −1.92, r = 0.43, $p < 0.05$) in comparison with the experimental group (Figure 1).

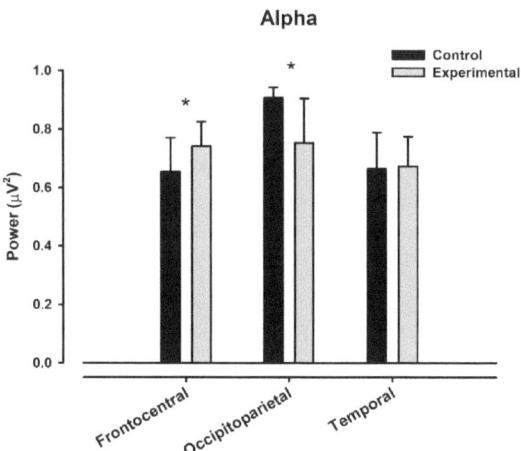

Figure 1. Topographic comparisons in alpha band. The * means a statistically significant difference ($p < 0.05$).

4.3. Beta

We found a significantly higher normalized power in the control group for the beta band in the frontocentral area (U = 19, z = −2.34, r = 0.52, $p < 0.05$), occipitoparietal area (U = 2, z = −2.56, r = 0.57, $p < 0.05$), and temporal area (U = 0, z = −2.31, r = 0.52, $p < 0.05$) in comparison with the experimental group (Figure 2).

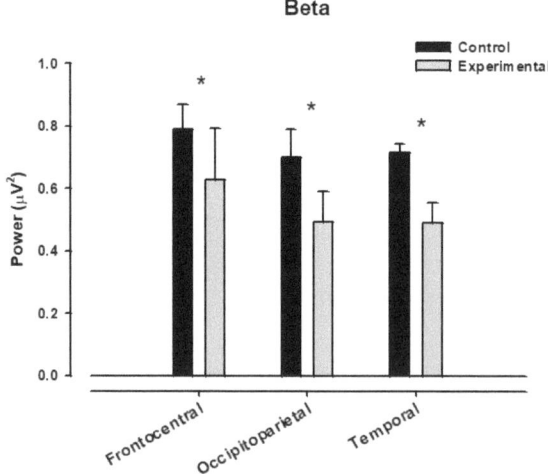

Figure 2. Topographic comparisons in beta band. The * means a statistically significant difference ($p < 0.05$).

4.4. Delta

We found a significantly higher normalized power in the control group for the delta band in the temporal area (U = 0, z = −2.31, r = 0.52, $p < 0.05$) in comparison with the experimental group (Figure 3).

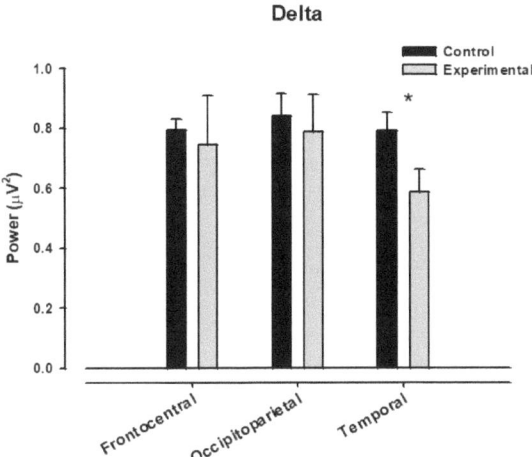

Figure 3. Topographic comparisons in delta band. The * means a statistically significant difference ($p < 0.05$).

4.5. Theta

We found a significantly higher normalized power in the control group for the theta band in the frontocentral area (U = 24, z = −1.97, r = 0.44, $p < 0.05$) and occipitoparietal area (U = 5, z = −2.10, r = 0.47, $p < 0.05$) in comparison with the experimental group (Figure 4).

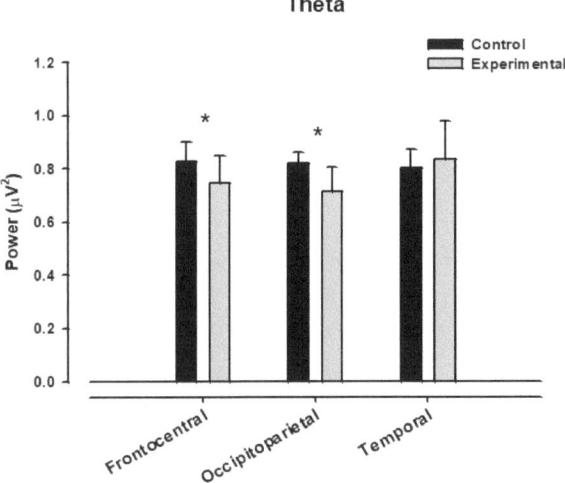

Figure 4. Topographic comparisons in theta band. The * means a statistically significant difference ($p < 0.05$).

4.6. Grand Average Band Analysis

For each band, we averaged the normalized absolute power for all electrodes (Figure 5). We found a significantly higher amplitude in the theta (U = 5, Z= −0.32, $p < 0.05$, r = 0.07) and beta (U = 2, Z= −2.56, $p < 0.05$, r = 0.57) bands in the control group in comparison with the experimental group.

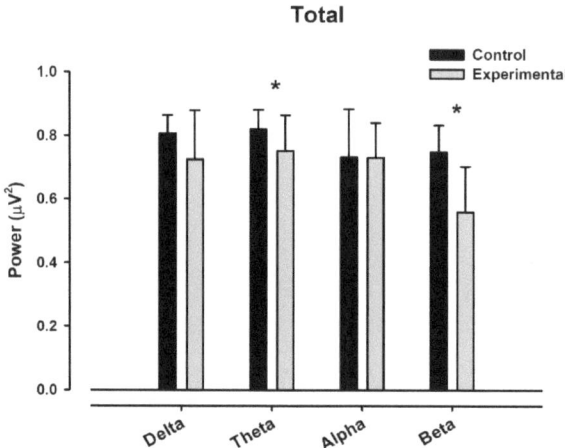

Figure 5. Total averaged absolute power for delta, theta, alpha, and beta bands. The * means a statistically significant difference ($p < 0.05$).

4.7. Independent Channel Analysis

We performed the same analysis for each electrode as for cortical areas. Because the statistics to compare all electrodes with all the bands in the two groups would require a multidimensional analysis, we do not present comparisons of these results. For this reason, we simplified the analysis by areas. We only describe qualitatively what we see in the corresponding averaged areas The mean absolute power figure, for individual electrodes in the 10–20 system (both groups) is presented in the Supplementary Material because we believe it might provide new data on the distribution of frequency bands in the scalp.

We found more specific differences in the frontal areas: the theta band was larger in Fp1 and Fp2 in the control group than in the experimental group. In F7 and F8, beta was larger in the control group than in the experimental group. Meanwhile, the alpha band in the same areas was larger in the experimental group than in the control group. In F3 and F4, beta was larger in the control group than in the experimental group (see Figure S1).

We also found that in the somatosensory-related leads C3 and C4, the beta band was larger in the control group than the experimental group. We also found that the theta band over C3 was larger in the control group than in the experimental group. In contrast, the theta band over C4 was larger in the experimental group than in the control group (see Figure S1).

5. Discussion

It has been reported that migraine is three times more frequent in women than in men of reproductive age. This fact is associated with the changes of hormonal mechanisms that occur in females throughout the menstrual cycle [35].

A specific phase of the menstrual cycle that is related to psychological, cognitive, and physical changes is the luteal phase. The symptoms and signs of this period are distinct and have been named pre-menstrual syndrome. During this period, a persistent headache of the migraine type that decreases after three days of menstruation has been reported [36].

The present study showed that women with no migraine presented the following relevant encephalographic characteristics during the luteal phase: alpha occurrence over the parieto-occipital area, a predominance of beta rhythm over the whole scalp, theta increase over the frontocentral and parietooccipital area, and an increase in delta for the temporal areas.

In the case of women with migraine, we observed an increase in alpha in the frontocentral area and a reduction in alpha in occipitoparietal area. Few studies have been carried out in patients with migraine, and the results have not been very consistent. However, the first studies reported an increase in alpha rhythm variability in the headache-free phase [21,22,37–40].

In our study, we observed a reduction in beta in all the considered cortical areas except the frontopolar leads. This result is in contrast with studies that reported an excess of beta activity during attacks [41], in migraine patients. Walker [42], reported an excess of high-frequency beta activity (21–30 Hz) in all four cortical areas in a group of migraine patients in neurofeedback therapy in comparison with a group of patients using drug therapy.

We also found a focal decrease in the theta band in the left central–parietal leads (C3 and P3) and a local delta decrease in the left temporal and right occipital leads (T3 and O2). Past research has also shown that individuals with chronic pain exhibit increased beta and decreased alpha activity, with additional increased theta/delta [43]. Much evidence suggests the involvement of delta oscillations during sustained pain as a reflex of autonomic processes linked to efforts of homeostatic processes [44]. However, our study is more specifically focused on a phase of the menstrual period that begins with pain and not the pain itself.

Xie et al. [45], suggested that EEG activity in migraine patients is related to a predisposition to painful or high-risk stimuli. In line with Xie et al. [45], we observed power differences between the considered groups in the frontal channels (see Figure S1).

Some authors suggest that migraine is related to a lack of habituation to environmental stimuli due to inadequate information processing [46]. Our study supports the suggestion that the differences in EEG patterns in women with migraine could be involved in pain related to migraine and the regulation of input processing. In other words, chronic pain can change the responsiveness of brain regulatory systems and the emotional processing of somatosensory information; this could be a reason why we found that in somatosensory areas such as C3 and C4, the delta and beta band were larger in the control group than the experimental group. This type of EEG pattern could be related to allostatic load, which also includes the participation of the hypothalamic–pituitary–gonadal axis and the negative feedback regulation of the endogenous opioid system during the release of luteinizing hormone. The cyclic surges of gonadal hormones may directly alter neuronal, glial, and astrocyte function throughout the brain [47].

The irregular EEG patterns found in our study could be used not only for diagnosis criteria, but also for the improvement of pharmacological and psychological therapeutic targets.

Future work should consider EEG data as a part of the physiological changes related to the menstrual cycle in relation to behavioral and emotional traits. As there is increasing evidence that migraine symptoms could be ameliorated by a multimodal approach that includes behavioral interventions like biofeedback, cognitive behavioral therapy, and relaxation, at first, women may consider the use of medication, which may be complemented by an integrated approach that includes exercise, relaxation, and biofeedback. Future studies should investigate the effect of such interventions as reflected in neurophysiological data.

It is also important to address the recognition of premonitory symptoms in migraine, because most patients in the present work were able to correctly predict the onset of migraine headaches. The pain prediction may represent an essential treatment paradigm, where the risk of headache is treated prior to the experience of pain.

6. Conclusions

Finally, we state that more conclusive research is needed to understand the physiology of migraine in greater detail. Similarly, research involving various levels of analysis is needed in order to be more explanatory and less descriptive. It is important to consider that the sample size may have been a limitation in our study. We suggest that the sample size should be increased to have external validation. In addition, carrying out hemisphere correlation studies and coherence analysis will be essential to analyze the diffuse asymmetry of EEG patterns in patients with migraine.

Even though the signs and symptoms of migraine guide the clinical course of diagnosis, the electroencephalographic data can improve or increase the sensitivity of the diagnosis according to the specific phases of the menstrual cycle. Therefore, the results derived from EEG can function as biomarkers of the condition of the patient.

Supplementary Materials: The following supporting information can be downloaded at: https://www.mdpi.com/article/10.3390/app13137443/s1, Figure S1: Distribution of the mean absolute power for delta, theta, alpha, and beta bands for each location of scalp electrodes in the 10–20 system.

Author Contributions: Conceptualization, H.J.P.-G. and V.R.-M.; Methodology, I.M.-B., C.T., D.R., G.G.-A. and V.A.L.-C.; Software, I.M.-B., C.T. and D.R.; Validation, H.J.P.-G. and I.M.-B.; Formal analysis, H.J.P.-G., I.M.-B., C.T. and D.R.; Investigation, H.J.P.-G., V.R.-M., I.M.-B., O.M.-D., C.T., D.R., G.G.-A. and V.A.L.-C.; Data curation, O.M.-D., I.M.-B., C.T., D.R., G.G.-A. and V.A.L.-C.; Writing—original draft, H.J.P.-G., V.R.-M. and O.M.-D.; Writing—review & editing, I.M.-B., C.T., D.R. and V.A.L.-C.; Visualization, I.M.-B.; Supervision, I.M.-B.; Project administration, H.J.P.-G.; Funding acquisition, H.J.P.-G. All authors have read and agreed to the published version of the manuscript.

Funding: This work was supported by Vicerrectoría de Investigación y Estudios de Postgrado de la Benemérita Universidad Autónoma de Puebla: VIEP-BUAP PEGH-EDH-20 (HJ-PG), Mexico.

Institutional Review Board Statement: Protection of human subjects. The authors declare that the procedures followed were in accordance with the regulations of the relevant clinical research ethics committee and with those of the Code of Ethics of the World Medical Association (Declaration of Helsinki). Confidentiality of data. The authors declare that they have followed the protocols of their work center for the publication of patient data. Right to privacy and informed consent. The authors have obtained the written informed consent of the patients or subjects mentioned in the article. The corresponding author has this document.

Informed Consent Statement: The studies involving human participants were reviewed and approved by Local Ethics Committee of Maestría en Diagnóstico y Rehabilitación Neuropsicológica from the Faculty of Psychology of Benemérita Universidad Autónoma de Puebla (DCECEN 254546 OMD). The patients/participants provided their written informed consent to participate in this study.

Data Availability Statement: The original contributions presented in the study are included in the article, further inquiries can be directed to the corresponding author.

Acknowledgments: Financial support was provided by the Mexican funding agency Consejo Nacional de Ciencia y Tecnología (CONACyT 388199/254546 to OMD).

Conflicts of Interest: The authors declare no conflict of interest.

References

1. Kim, J.H.; Suh, S.I.; Seol, H.Y.; Oh, K.; Seo, W.K.; Yu, S.W.; Park, K.W.; Koh, S.B. Regional grey matter changes in patients with migraine: A voxel-based morphometry study. *Cephalalgia* **2008**, *28*, 598–604. [CrossRef]
2. Salomone, S.; Caraci, F.; Capasso, A. Migraine: An Overview. *Open Neurol. J.* **2009**, *3*, 64–71. [CrossRef]
3. Lipton, R.B.; Bigal, M.E.; Scher, A.I.; Stewart, W.F. The global burden of migraine. *J. Headache Pain* **2003**, *4* (Suppl. S1), s3. [CrossRef]
4. Mannix, L.K. Menstrual-related pain conditions: Dysmenorrhea and migraine. *J. Women's Health* **2008**, *17*, 879–891. [CrossRef] [PubMed]
5. Spierings, E.L. Pathogenesis of the migraine attack. *Clin. J. Pain* **2003**, *19*, 255–262. [CrossRef] [PubMed]
6. Bigal, M.E.; Lipton, R.B.; Stewart, W.F. The epidemiology and impact of migraine. *Curr. Neurol. Neurosci. Rep.* **2004**, *4*, 98–104. [CrossRef]
7. Bigal, M.E.; Liberman, J.N.; Lipton, R.B. Obesity and migraine: A population study. *Neurology* **2006**, *66*, 545–550. [CrossRef] [PubMed]
8. Kelman, L. Migraine changes with age: IMPACT on migraine classification. *Headache J. Head Face Pain* **2006**, *46*, 1161–1171. [CrossRef]
9. Di Clemente, L.; Coppola, G.; Magis, D.; Fumal, A.; De Pasqua, V.; Di Piero, V. Interictal habituation deficit of the nociceptive blink reflex: An endophenotypic marker for presymptomatic migraine? *Brain* **2007**, *130*, 765–770. [CrossRef]

10. Capuano, A.; De Corato, A.; Lisi, L.; Tringali, G.; Navarra, P.; Russo, C.D. Proinflammatory-activated trigeminal satellite cells promote neuronal sensitization: Relevance for migraine pathology. *Molecular Pain.* **2009**, *5*, 1744–8069. [CrossRef] [PubMed]
11. Granella, F.; Sances, G.; Allais, G.; Nappi, R.E.; Tirelli, A.; Benedetto, C.; Brundu, B.; Facchinetti, F.; Nappi, G. Characteristics of menstrual and nonmenstrual attacks in women with menstrually related migraine referred to headache centres. *Cephalalgia* **2004**, *24*, 707–771. [CrossRef]
12. MacGregor, E.A.; Frith, A.; Ellis, J.; Aspinall, L.; Hackshaw, A. Incidence of migraine relative to menstrual cycle phases of rising and falling estrogen. *Neurology* **2006**, *67*, 2154–2158. [CrossRef]
13. Pinkerman, B.; Holroyd, K. Menstrual and nonmenstrual migraines differ in women with menstrually-related migraine. *Cephalalgia* **2010**, *30*, 1187–1194. [CrossRef] [PubMed]
14. Silberstein, S.; Patel, S. Menstrual migraine: An updated review on hormonal causes, prophylaxis and treatment. *Expert Opin. Pharmacother.* **2014**, *15*, 2063–2070. [CrossRef] [PubMed]
15. Pavlović, J.M.; Allshouse, A.A.; Santoro, N.F.; Crawford, S.L.; Thurston, R.C.; Neal-Perry, G.S.; Lipton, R.B.; Derby, C.A. Sex hormones in women with and without migraine: Evidence of migraine-specific hormone profiles. *Neurology* **2016**, *87*, 49–56. [CrossRef]
16. Nappi, R.E.; Tiranini, L.; Sacco, S.; De Matteis, E.; De Icco, R.; Tassorelli, C. Role of Estrogens in Menstrual Migraine. *Cells* **2022**, *11*, 1355. [CrossRef] [PubMed]
17. Vetvik, K.G.; MacGregor, E.A. Menstrual migraine: A distinct disorder needing greater recognition. *Lancet Neurol.* **2021**, *20*, 304–315. [CrossRef]
18. Rosenberg, J.; Alter, M.; Byrne, T. Practice parameter: The electroencephalogram in the evaluation of headache. Report of the Quality Standards Subcommittee of the American Academy of Neurology. *Neurology* **1995**, *45*, 1411–1413.
19. Golla, M.D.; Winter, A.L. Analysis of cerebral response to flickering in patients complaining of an episodic headache. *Electroencephalogr. Clin. Neurophysiol.* **1959**, *11*, 539–549. [CrossRef]
20. Walser, H.; Isler, H. Frontal Intermittent Rhythmic Delta Activity, Impairment of Consciousness and Migraine. *Headache J. Head Face Pain* **1981**, *22*, 74–80. [CrossRef]
21. Schoenen, J.; Jamart, B.; Delwaide, P.J. Cartographie electroencephalographique dans les migraines en periodes critique et intercritique. *Electroencephalogr. Neurophysiol. Clinic.* **1987**, *17*, 289–299. [CrossRef] [PubMed]
22. Nyrke, T.; Kangasniemi, P.; Lang, H. Alpha Rhythm in Classical Migraine (Migraine with Aura): Abnormalities in the Headache-Free Interval. *Cephalalgia* **1990**, *10*, 177–182. [CrossRef] [PubMed]
23. Bjork, M.H.; Stovner, L.J.; Nilsen, B.M.; Stjern, M.; Hagen, K.; Sand, T. The occipital alpha rhythm related to the "migraine cycle" and headache burden: A blinded, controlled longitudinal study. *Clin. Neurophysiol.* **2009**, *120*, 464–471. [CrossRef]
24. O'Hare, L.; Menchinelli, F.; Durrant, S.J. Resting-state alpha-band oscillations in migraine. *Perception* **2018**, *47*, 379–396. [CrossRef]
25. Silberstein, S.D. The role of sex hormones in headache. *Neurology* **1992**, *42*, 37–42.
26. Becker, D.; Creutzfeldt, O.; Schwibbe, M.; Wuttke, W. Changes in physiological, eeg and psychological parameters in women during the spontaneous menstrual cycle and following oral contraceptives. *Psychoneuroendocrinology* **1982**, *7*, 75–90. [CrossRef]
27. Solis-Ortiz, S.; Ramos, J.; Arce, C.; Guevara, M.A.; Corsi-Cabrera, M. EEG oscillations during menstrual cycle. *Int. J. Neurosci.* **1994**, *76*, 279–292. [CrossRef]
28. Baehr, E.; Rosenfeld, P.; Miller, L.; Baehr, R. Premenstrual dysphoric disorder and changes in frontal alpha asymmetry. *Int. J. Psychophysiol.* **2004**, *52*, 159–167. [CrossRef]
29. Haraguchi, R.; Hoshi, H.; Ichikawa, S.; Hanyu, M.; Nakamura, K.; Fukasawa, K.; Poza, J.; Rodríguez-González, V.; Gómez, C.; Shigihara, Y. The Menstrual Cycle Alters Resting-State Cortical Activity: A Magnetoencephalography Study. *Front. Hum. Neurosci.* **2021**, *15*, 652789. [CrossRef]
30. Pletzer, B.; Harris, T.A.; Scheuringer, A.; Hidalgo-Lopez, E. The cycling brain: Menstrual cycle-related fluctuations in hippocampal and frontostriatal activation and connectivity during cognitive tasks. *Neuropsychopharmacology* **2019**, *44*, 1867–1875. [CrossRef]
31. Hidalgo-Lopez, E.; Pletzer, B. Fronto-striatal changes along the menstrual cycle during working memory: Effect of sex hormones on activation and connectivity patterns. *Psychoneuroendocrinology* **2021**, *125*, 105108. [CrossRef]
32. Geschwind, N.; Behan, P. Left-handedness: Association with immune disease, migraine, and developmental learning disorder. *Proc. Natl. Acad. Sci. USA* **1982**, *79*, 5097–5100. [CrossRef]
33. General Assembly of the World Medical Association. World Medical Association Declaration of Helsinki: Ethical principles for medical research involving human subjects. *J. Am. Coll. Dent.* **2014**, *81*, 14–18.
34. Olesen, J. International classification of headache disorders. *Lancet Neurol.* **2018**, *14*, 396–397. [CrossRef]
35. MacGregor, E.A. Oestrogen and attacks of migraine with and without aura. *Lancet Neurol.* **2004**, *3*, 354–361. [CrossRef]
36. Allais, G.; Gabellari, I.C.; Burzio, C.; Rolando, S.; De Lorenzo, C.; Mana, O.; Benedetto, C. Premenstrual syndrome and migraine. *Neurol. Sci.* **2012**, *33*, 111–115. [CrossRef]
37. Jonkman, E.J.; Lelieveld, M.H.J. EEG computer analysis in patients with migraine. *Electroencephalogr. Clin. Neurophysiol.* **1981**, *52*, 652–655. [CrossRef]
38. Polich, J.; Ehlers, C.L.; Dalessio, D.J. Pattern shift visual evoked responses and EEG in migraine. *Headache J. Head Face Pain* **1986**, *26*, 451–456. [CrossRef]
39. Drake, M.E.; Du Bois, C.; Huber, S.J.; Pakanis, A.; Denio, L.S. EEG spectral analysis and time domain descriptors in headache. *Headache* **1988**, *28*, 201–203. [CrossRef]

40. Hughes, J.R.; Robbins, L.D. Brain mapping in migraine. *Clin. Electroencephalogr.* **1990**, *21*, 14–24. [CrossRef]
41. Parain, D.; Samson-Dollfus, D. Electroencephalograms in basilar artery migraine. *Electroencephalogr. Clin. Neurophysiol.* **1984**, *58*, 392–399. [CrossRef]
42. Walker, J.E. QEEG-guided neurofeedback for recurrent migraine headaches. *Clin. EEG Neurosci.* **2011**, *42*, 59–61. [CrossRef]
43. Jensen, M.P. A neuropsychological model of pain: Research and clinical implications. *J. Pain* **2010**, *11*, 2–12. [CrossRef]
44. Knyazev, G.G. EEG delta oscillations as a correlate of basic homeostatic and motivational processes. *Neurosci. Biobehav. Rev.* **2012**, *36*, 677–695. [CrossRef]
45. Xie, J.; Lu, Y.; Li, J.; Zhang, W. Alpha neural oscillation of females in the luteal phase is sensitive to high risk during sequential risk decisions. *Behav. Brain Res.* **2021**, *413*, 113427. [CrossRef]
46. Coppola, G.; Pierelli, F.; Schoenen, J. Is the cerebral cortex hyperexcitable or hyperresponsive in migraine? *Cephalalgia* **2007**, *27*, 1429–1439. [CrossRef]
47. Borsook, D.; Maleki, N.; Becerra, L.; McEwen, B. Understanding Migraine through the Lens of Maladaptive Stress Responses: A Model Disease of Allostatic Load. *Neuron* **2012**, *73*, 219–234. [CrossRef]

Disclaimer/Publisher's Note: The statements, opinions and data contained in all publications are solely those of the individual author(s) and contributor(s) and not of MDPI and/or the editor(s). MDPI and/or the editor(s) disclaim responsibility for any injury to people or property resulting from any ideas, methods, instructions or products referred to in the content.

Article

Using Mental Shadowing Tasks to Improve the Sound-Evoked Potential of EEG in the Design of an Auditory Brain–Computer Interface

Koun-Tem Sun, Kai-Lung Hsieh * and Shih-Yun Lee

Department of Information and Learning Technology, National University of Tainan, 33, Sec. 2, Shu-Lin St., Tainan 70005, Taiwan
* Correspondence: d10355002@gm2.nutn.edu.tw; Tel.: +886-6-213-3111 (ext. 773)

Abstract: This study proposed an auditory stimulation protocol based on Shadowing Tasks to improve the sound-evoked potential in an EEG and the efficiency of an auditory brain–computer interface system. We use stories as auditory stimulation to enhance users' motivation and presented the sound stimuli via headphones to enable the user to concentrate better on the keywords in the stories. The protocol presents target stimuli with an oddball P300 paradigm. To decline mental workload, we shift the usual Shadowing Tasks paradigm: Instead of loudly repeating the auditory target stimuli, we ask subjects to echo the target stimuli mentally as it occurs. Twenty-four healthy participants, not one of whom underwent a BCI use or training phase before the experimental procedure, ran twenty trials each. We analyzed the effect of the auditory stimulation based on the Shadowing Tasks theory with the performance of the auditory BCI system. We also evaluated the judgment effectiveness of the three ERPs components (N2P3, P300, and N200) from five chosen electrodes. The best average accuracy of post-analysis was 78.96%. Using component N2P3 to distinguish between target and non-target can improve the efficiency of the auditory BCI system and give it good practicality. We intend to persist in this study and involve the protocol in an aBCI-based home care system (HCS) for target patients to provide daily assistance.

Keywords: EEG; ERPs; BCI; Shadowing Tasks

1. Introduction

Many people with severe motor paralysis, such as spinal cord injury, locked-in syndrome (LIS), and amyotrophic lateral sclerosis (ALS), who have lost communication skills, cannot express their thoughts freely. Yet, most functions of their brain and senses are without dysfunction [1–5]. Over the past few decades, many supportive tools have been developed (including brain–computer interface (BCI) systems) in dramatic proliferation [6–13]. In addition, an auditory BCI (aBCI) is a helpful tool for people with severe motor paralysis at the end stage or who cannot stare at the screen [14–16]. However, the speed and accuracy of contemporary auditory BCIs are slower and lower than those of visual modality BCIs [17].

The user of a stimulus-driven BCI system has to choose to focus on one stimulus out of the numerous stimuli presented at the same time, which evokes a specific event-related potential (ERP) pattern [15,18,19], including components P300 (P3), and N200 (N2), as shown in Figure 1. The oddball paradigm is usually used to elicit the components of ERPs. BCI application extracts the feature values of the data from the monitored electroencephalograph (EEG) simultaneously. Then, the features are classified and generated to a resulting command immediately. The application of EEG is in a broad scope for clinical [20] and non-clinical applications, such as transport, entertainment [21], and education [22]. For example, using EEG equipment in the applications of ML-based disease diagnosis or mental workload prediction is familiar.

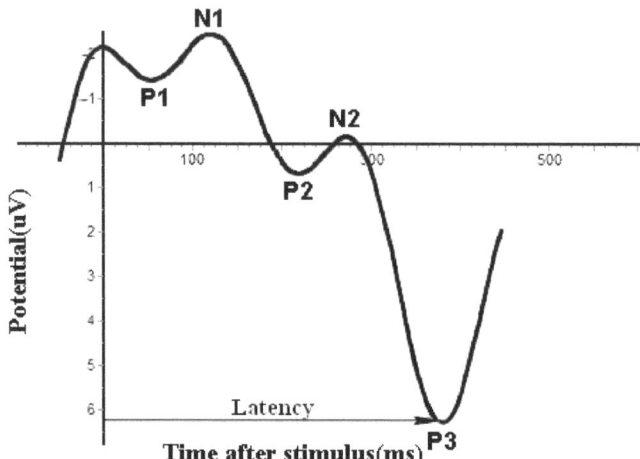

Figure 1. ERP components after the onset of an audio stimulus, including the P300 (labeled P3) and N200 (labeled N2). Generally, the Y-axis scales are often upside down in ERP research.

Sutton proposed that external stimuli would evoke a human brainwave fluctuation, called event-related potentials (ERP), in 1965 [23]. Specific physical events or psychological events trigger these time-dependent potentials of brainwave fluctuations [23,24]. A BCI system based on ERP obtains potentials on the surface of the cortex [13,15,25]. The ERP-based BCI system learns the basics of a user's brain system via the ERPs obtained from the user's brain rhythm [24,26]. The potentials of the user's brainwaves are amplified and recorded by the EEG device [19,27,28]. The ERP-based BCI system accepts these signals from EEG and filters these EEG data. Then, the BCI system uses signal accumulation and averaging methods to extract the specific features of ERP components, then classify and interpret them [16,24,29]. Finally, these signals are converted into instructions and output to the devices to help the user perform [9,12,13,30,31].

A P300 (P3) peak usually appears around 300–400 ms after the stimulus presentation. That is, P3 is a higher deflection peak of an ERP [16,19,28,29,32]. Additionally, an N200 (N2) trough often comes about 200 ms later than the onset of a target stimulus. So, N2 is a lower deflection trough of an ERP [9,16,31,33]. If the user focuses on detecting the targets, the P300 and N200 waves facilely come about [18,34,35]. Usually, the P3 potentials of the non-target stimuli are lower than that of the target stimulus. The situation is just the opposite for N2 [31,36]. In addition, the ERP component latencies deepen the difficulty of discriminating between the target and non-target stimuli [29].

A stimulus-driven aBCI system plays the sound stimuli through either headphones or speakers. The users of a stimulus-driven aBCI must focus only on the sound they want to hear (target) while ignoring the others (non-targets) [17]. For example, if the user pays attention to the sound played in the left earphone and ignores the sound in the right earphone, then the sound played from the left channel of the headphones is the target sound stimulus at that moment. The difference in the responses that the EEG gains allows the system to group the ERPs into target and non-target. The aBCI system then captures and classifies these signals and interprets the discriminative features of the ERP components.

However, it is not a simple task for the user of an aBCI to pay attention to one of the two or more different voices played simultaneously [31,37]. The noise ratio (SNR) of a non-invasive BCI is lower than in the case of invasive technology [38]. Thus, ERP responses of an auditory BCI system were less class-discriminative between attended and unattended stimuli than those of a visual BCI (vBCI) [39,40].

Therefore, numerous studies have used diverse stimulation methods to increase the classification accuracy for better-quality interfacing applications. Most aBCI systems use the Auditory

Steady-State Response (ASSR) [38], N200 [9,26], or P300 [16,31,39] modalities to interpret sound stimuli. Hybrid systems that combine two aBCI modalities or at least one aBCI system with another scheme seek to improve the system's performance [41–44]. Several studies based on ASSR have explored the impact of natural and synthetic sound sources on aBCI [45] to reduce users' mental workload. Using more than two loudspeakers to present the spatial directionality allows the users to accept more than two options simultaneously [43,46,47]. In addition to these stimulations, the work of Marassi et al. contrasts two ways of using the aBCI: passively counting the presented target sound stimuli or simply mentally repeating them when they occur [31]. Further, aBCI performance does not depend solely on the aBCI system. Several studies indicate that the users' mood, attention, and motivation could influence aBCI performance and P300 electrophysiology. Such factors may contribute to inter-individual differences [31,42,48].

However, conventional aBCI has not been practical because of its lack of high accuracy and reliability. The users have to suffer a substantial workload due to these auditory approaches employing more complex interfaces for system efficiency [39,42]. Furthermore, the structure of the human hearing system and the user's attention may be critical factors [49].

A user who wants to do something via an aBCI system must be able to listen attentively to the sound stimuli of the desired option. However, when sound enters one human ear, the tone will be transmitted immediately to the other ear through the human hearing system. With the time difference in binaural hearing, people can identify the position where the sound came from and be alerted to the direction of danger [49]. Yet, this critical feature of the hearing functions makes it difficult for the user to concentrate on listening to the target sound entering one of the ears.

In addition, the degree of user concentration (selective and continuous attention) also affects the accuracy of a BCI system [50,51]. In da Silva-Sauer et al. [51], the authors showed that when the user's attention declines, the accuracy of a BCI falls. When a person needs to pay attention to a particular sound source, they activate the control of selective attention. Thus, many studies of a stimulus-driven BCI system ask subjects to maintain a high degree of concentration during the experiment [52]. Lakey et al. [28] found that using a short mindfulness meditation induction (MMI) could maintain the user's attention and improve the performance of P300-based BCI systems.

Thus, two factors affect the performance of an aBCI system: whether the target sound stimulus can attract the user's attention and whether the user can easily distinguish the target sound stimulus from non-target stimuli. Therefore, we propose a strategy to maintain the user's concentration during the sound stimuli playing to improve the accuracy of our aBCI. We introduce a novel auditory paradigm to solve the problem caused by the human hearing system and the user's attention: using mental Shadowing Tasks to improve EEG's sound-evoked potential of the target stimuli to enhance the aBCI system's efficiency.

At the beginning of our study on an aBCI system, the sound stimuli consisted of periodic click sounds, such as beep, dang, and bleep. However, the effectiveness of such a sound stimulus model was mediocre. The accuracy was equal to or below the chance level. According to the work of Baykara et al., motivation influences P300 amplitude and the performance of a BCI system [48]. If the sound stimuli are monotonous and repetitive, such as beeps, they cannot trigger the users' motivation [45]. Because the discriminative features of the ERP components in our former aBCI system research were severely disorganized, we had to use other sound sources.

Story sound is friendlier for the users than simple periodic click tones since it has a friendlier alternating (musical) temporal structure. So, we hypothesize that the application of the novel stimuli to the auditory BCI will result in a more comfortable interfacing experience. Therefore, we created a prototype of the aBCI system using audio story stimulation and the Shadowing Tasks mechanism [37,53] to carry out the subsequent experiment.

Shadowing Tasks is an experimental technique performed via headphones. Participants are required to repeat the target stimuli aloud immediately after hearing a sentence, word, or phrase. Usually, non-target stimuli appear in the background simultaneously [54].

Cherry's Shadowing Tasks present two distinct auditory messages to the participant's right and left ears and asks the participant to pay attention to the target sound heard in one of the two ears and repeat the sound [37,54].

Shadowing Tasks require the user to have the ability to recognize the target sound from two simultaneously heard messages. The ability to separate the target sound from the noisy background sounds is affected by many variables, such as the speaker's gender, the direction the sound comes from, the pitch, and the speed of speech. Thus, in the Shadowing Tasks, the subjects must engage in selective attention to enable them to focus on the target sound stimuli.

This study adopted Cherry's approach to delivering sound stimuli: stories with the target stimulus. Different story sounds are played to the user's left and right ears in synchronization through headphones. The participant was required to pay attention to the target sound from the left or right headphones [41]. To reduce the mental workload, we asked the participants to mentally repeat the target stimuli, not repeat them aloud [31].

In this study, we incorporated mental Shadowing Tasks to maintain the user's concentration during the presentation of sound stimuli. We aim to confirm whether the mental Shadowing Tasks can improve the EEG's sound-evoked potential of the target stimuli and enhance the aBCI system's efficiency. In addition, we also evaluated the judgment effectiveness of the three ERPs components (N2P3, P300, and N200) across five chosen electrodes.

2. Materials and Methods

2.1. Participants

The participants were 24 healthy people aged 20–22, 7 females. All participants were volunteers and had no head injuries, history of neurological defects, mental illness, or drug treatments. All participants had normal hearing. No participants had used BCI or received training ahead of the experimental procedure. Before participating in the experiment, all subjects signed the Informed Consent Form approved by the Human Research Ethics Committee at National Cheng Kung University. The experimental procedure ended immediately if a participant withdrew during the test, and we dropped the data.

2.2. The aBCI System

2.2.1. The Prototype of the aBCI System Module

Figure 2 shows the experimental setup. There are signal acquisition, signal processing, and application in the prototype.

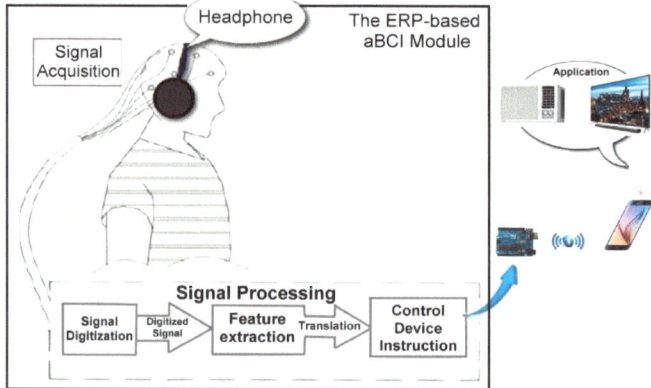

Figure 2. The prototype of the aBCI system. The upper left presents the concept of signal acquisition, the lower the module of signal processing, and the upper right the module of signal application for subsequent development.

The experiment used short stories compiled by the author as sound stimuli. There is a keyword in each short story that appears seven times, and the user must pay close attention and mentally repeat the keyword. The ERP-based aBCI module receives the signal data of the participant's brainwaves via non-invasive EEG equipment that includes 32 channels [15,25]. Thus, the aBCI module can discern the participant's choice and then export the command signal to the application.

2.2.2. The Stimulation Trials Using Audio Story

Based on the Shadowing Tasks, the study needed sound stimuli in the form of a story. We designed twelve audio stories to be the sound stimuli to help participants test the aBCI module using Shadowing Tasks. The twelve audio stories, six recorded by male pronunciation and six by female, were all recorded in mono sound channel. These stories consisted of approximately 117–138 Chinese characters, with playtimes of 45–60 s. Figure 3 shows two of the twelve audio stories (original text written in Chinese).

Lisa's story
Likes to grow flowers and grass. **Lisa** walked into the holiday flower market and bought an air pineapple. **Lisa** asked the boss how to take care of the air pineapple. **Lisa** brought the air pineapple home with joy. **Lisa** hung the air pineapple on the balcony and watered it every day. **Lisa** expects that it can bloom beautiful flowers, but **Lisa** has been raising it for more than three months, and she still does not see the buds! **Lisa** still believes that one day the air pineapple will bloom.

Jack's story
Jack is a boy who loves small car toys very much. **Jack** has collected a variety of model car toys from an early age. **Jack**'s room is full of various model cars. **Jack** loves the matchbox car very much. **Jack** also participated in the model car friendship group. People who know **Jack** say that he is an ideal person because **Jack** wanted to run a logistics company in the future to provide superfast and convenient services.

Figure 3. Two of the twelve audio stories (original text written in Chinese). Each story file includes a keyword (words rendered in boldface) that appears seven times in the story.

Each audio story includes one keyword (words rendered in boldface in Figure 3). The keyword appears seven times in the story, following the principles of the oddball paradigm [35,43], and the inter-stimulus interval (ISI) between the two keywords was 5–8 s. Thus, the user can easily focus on the target stimuli (the keyword appears seven times) in the story he wants to hear by mentally repeating the keyword, not all the words. The unequal ISI avoids anticipatory psychology from the user. The onset times of the keywords in the different stories are all different, to prevent mutual interference.

2.2.3. ERP Trial Features

In the study, the auditory BCI prototype first searched out sound-evoked potentials of the P3 peak and the N2 trough out of the ERPs gained from EEG after every trial. Thus, the system can obtain an N2P3 potential: the P300 potential minus the N200 potential. We used the values of N2P3, P300, and N200 to interpret the discriminative features and identify the best system accuracy in the post-analysis.

Our aBCI system provides two options for the user in every trial: one audio story via the left earphone (L) and the other via the right earphone (R). If the participant focuses on the audio story in the right earphone, then the figure of the ERP components in this trial resembles Figure 4. In Figure 4, the green circles mark the locations of the P3 peaks in the ERP feature, while the red marks indicate the N2 troughs. The P300 potential minus the N200 potential in the same curve is the N2P3 potential. The results using any one component of ERPs, N2P3, P300, and N200, to distinguish which audio story the participant listened to all indicate that the participant focused on the option from the right channel in this trial. Thus, the red curve (R_2) in Figure 4 is what the participant chose in this trial.

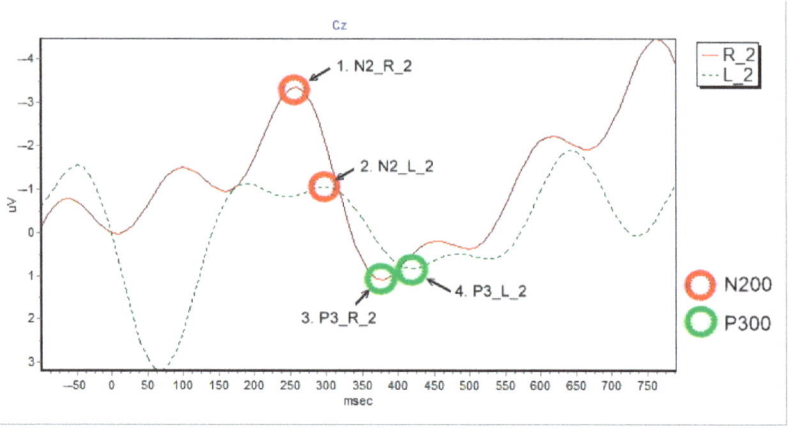

Figure 4. Average waveforms at Cz of the second stage of the 4th test run from participant N04: The red circles mark the locations of N200 (marks 1 and 2), while the green circles mark the locations of P300 (marks 3 and 4).

2.3. Experimental Program

2.3.1. Experimental Equipment

The EEG equipment produced by Braintronics B.V. Company for obtaining the user's brainwave data contains CONTROL-1132, a control unit, and ISO-1032CE, a 32-channel amplifier. In addition, the prototype used a PCI-1713 card to shift the data from analog to digital. The signals of brainwave acquisition used MATLAB's ERPLAB. Additionally, we used Borland C++ Builder to develop the aBCI module.

2.3.2. Data Collection

We made the impedance remain below 10 kΩ in the EEG equipment and set the sampling rate at 500 Hz. According to Peschke et al. [55], the connection between hearing and language processing is found in the Broca area (Cz and Fz) in the human brain by fMRI during the non-word Shadowing Tasks. The work of De Vos et al. [56] gained distinct P300 data from electrode Pz. Further, T3 and T4 of the 10/20 location system lie in the chief auditory cortex [49]. Thus, in this study, the EEG device obtains the user's brainwaves via electrodes T3, T4, Pz, Cz, and Fz on their scalp [16], as shown in Figure 5. Electrode FP2 is grounded, and the reference potential gains from electrodes A1 and A2. Every electrode is Ag/AgCl wet electrode, and the electrode locations refer to the International 10–20 Location System [57–59].

The ISO-1032CE in EEG equipment amplifies the brainwave signal and records the EEG potentials. The control unit, CONTROL-1132, uses a 0.3–15 Hz band-pass filter to filter the signal. Then, the converter card, PCI-1713, shifts the data from analog to digital, and finally, the aBCI system receives all the EEG signals to find the stimulus focused by the user.

For noise processing, there are two parts. The first is to filter the blink noise. So, there is electrode Fp2 around the eye. If the EMG signal is detected, the system will discard the signal. The second is the AC signal of the power supply. EEG hardware equipment has filtering and voltage stabilization functions to filter. For heartbeat noise, we remove it with relative potential between the sampling electrodes and the reference electrodes (electrodes A1 and A2 around the ears). Therefore, the system can remove most of the noise. Finally, based on the principle of event-related potentials: the potentials obtained by multiple stimulations and then averaged, the system can deal with the remaining small part of the noise and gain a stable and reliable electroencephalogram.

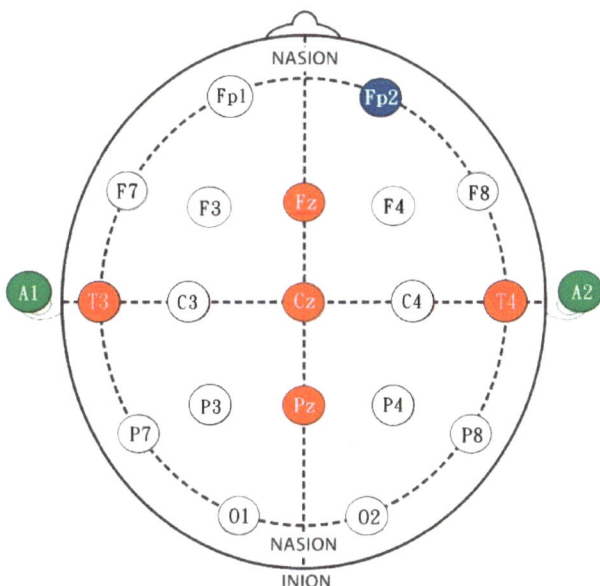

Figure 5. The positions of the electrodes based on the International 10–20 Location System. The green circles, A1 and A2, represent the reference electrodes. The ground electrode is the blue circle, FP2.

2.3.3. Data Processing

1. Stimuli presentation: The system synchronously plays two different audio stories via the left and the right headphones as the stimuli of the aBCI experiment.
2. ERPs acquisition: One keyword appears seven times in each audio story file. The system ignores first time the keyword appears and then obtains the subject's brainwaves the remaining six onset times of the keyword. Therefore, six ERP segments were retrieved one by one inside −100 to 800 ms based on each onset time of six keywords. Then, the aBCI system uses signal accumulation and averaging methods to treat the six ERP segments for every option to gain the ERP features.
3. ERP features interpretation: After the processing of ERPs acquisition, our aBCI system thus finds out P3 and N2 potential and calculates the N2P3 potential. Then, the system would determine which audio story was focused on by the user during the trial after it estimated the component potential for each option with each other.

2.4. Experimental Procedure

We use Figure 2 to explain the experimental prototype of the aBCI. The participant sits comfortably ahead of the aBCI system before the experimental procedure. The first preparation step before the test was to explain the test scheme, audio stories, and how to mentally repeat the keyword in the target story to the participant. Then, the experimenter attached electrodes to the participant's scalp, helped the user to put on the headphones, and checked the headphone volume. After completing all the preparations, the experimental procedure of 20 trials (10 test runs, there are 2 trials in each test) began immediately, up to 30mins. That is, each test run includes two trial stages. Each participant must perform the test run ten times. They must make one choice on each trial (two selections on a test run). Figure 6 shows the flowchart of the test procedure.

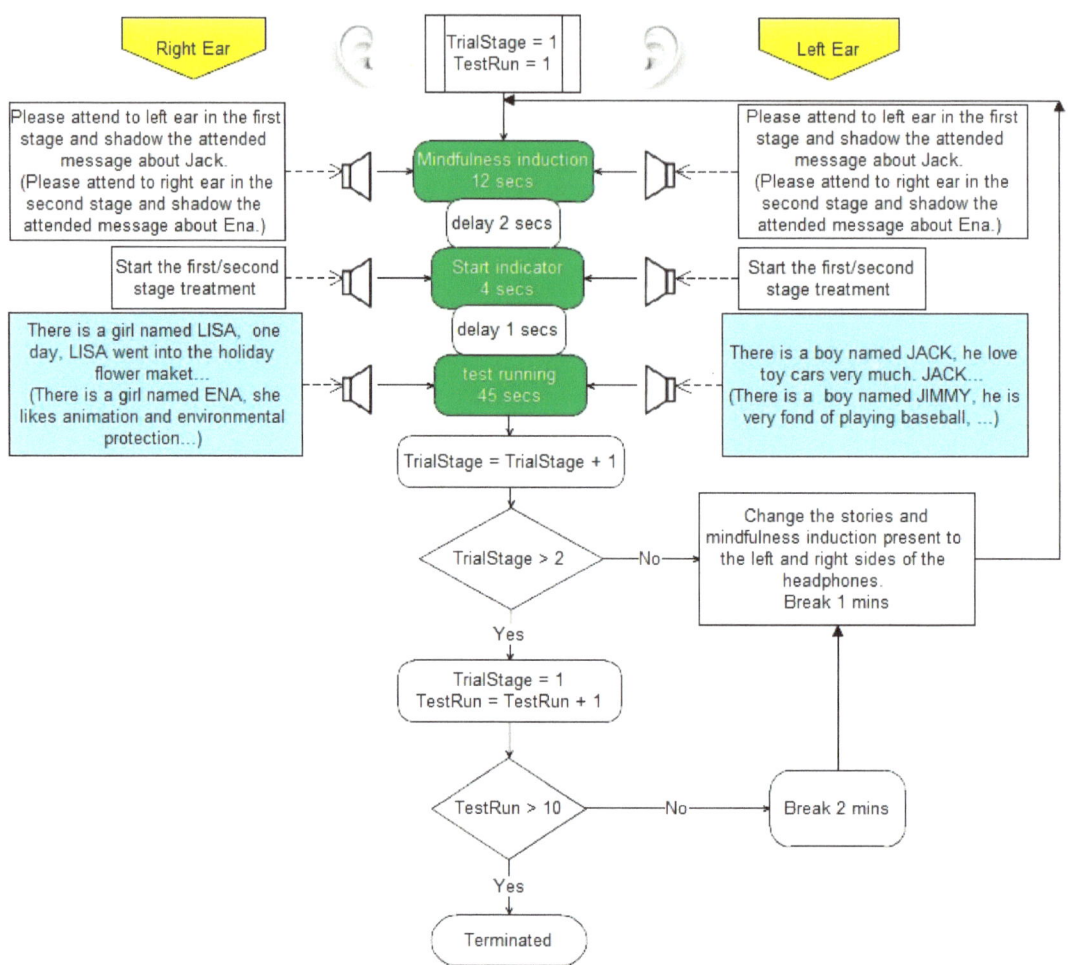

Figure 6. The flowchart of the test procedure. The audio stories in blue squares were played in Chinese by the system during the operation.

Before each trial, the experimenter specified one audio story as the target stimulus. There is an additional instruction: a mindfulness induction, played before each trial stage to help the participant focus on the target sound stimuli. During each trial, the participant must pay attention to the target story and mentally repeat the keyword in the story. The system then obtained the ERPs of both options from EEG equipment. Next, our aBCI system determined the sound-evoked potentials of every ERP component for both options. The choice having the highest potential should be the one the participant was paying attention to during the operation. Then, the system outputted the results of each trial with R or L. If the user focuses on the audio story played through the right earphone, the code is R; otherwise, the code is L. The system will judge it as correct when the output result comes together with the target specified by the experimenter. For example, the experimenter asks the participant to listen to the audio story of the right channel (R), and if the output is R, it is correct. Therefore, after each test run (two trial stages), the user can choose which one he prefers from four options (LL, LR, RL, RR).

2.5. System Evaluation

2.5.1. Information Transfer Rate

We evaluated our proposed system by computing the classification accuracy and the information transfer rate (ITR). ITR is quite valuable for estimating an aBCI system. We refer to the work of Wolpaw et al. to do the bit-rate/min calculation [31,60], as follows:

$$\text{ITR (bit} - \text{rate/min)} = M\{\log_2 N + P\log_2 P + (1-P)\log_2 [\frac{(1-P)}{(N-1)}]\} \quad (1)$$

where M indicates the number of choices made in a minute, N is the number of options, and P is the classification accuracy.

2.5.2. Neural Network

To determine the best identification for each participant, we used artificial intelligent technology neural network (NN) [61], a multi-layer neural network (Figure 7), to learn and analyze the output data of the five electrodes. The output of the BCI system is the input data of the neural network. Each electrode has two pieces of data: the sound-evoked potential obtained from stimulation on the left and right ears via the headphones. Therefore, the system will generate ten data from the five electrodes after every trial. Additionally, the expected output of the multi-layer neural network is to L (=0) or R (=1).

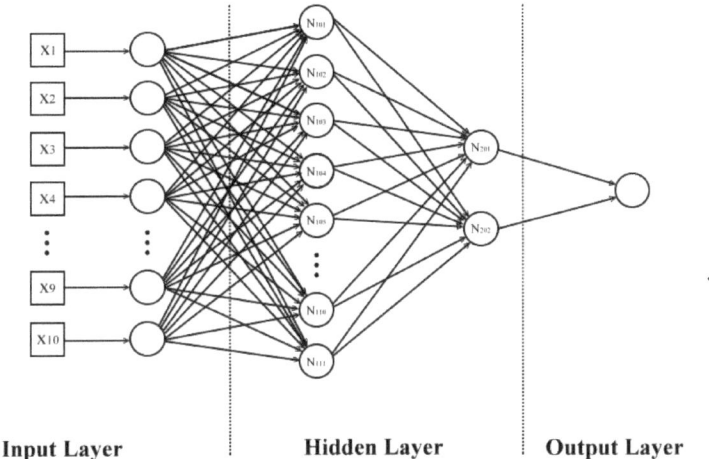

Figure 7. The structure of neural network to learn the data from the five electrodes.

There are 11 neurons (N101 to N111) in Hidden Layer 1 and two (N201 and N202) in Hidden Layer 2. We used the Keras API with parameters set as activation = 'sigmoid', model.compile: loss = 'categorical_crossentropy', optimizer = 'adam', metrics = ['accuracy'].

We use a 5-fold cross-validation method to train and validate the performance of the aBCI system. The NN model divides the data from 20 trials of each participant into five data sets. In each iteration of the cross-validation, four data sets out of five are used as training data, while the remaining data set is the testing data. Additionally, the NN model uses the gradient descent method to update the weights, the cross entropy function (Equation (3)) is used as the loss function (Equation (4)), and the learning rate is set as 0.001. After 10,000 iterations of training, the neural network achieved a high classification accuracy for training data. Then, the testing data is used to evaluate the accuracy of the model. Finally, the averaged accuracy among the five testing data sets obtained.

The activation function of each neuron is defined as sigmoid function:

$$f_w(x) = \frac{1}{1 + e^{(-w^T x + b)}} \tag{2}$$

where w is the weight vector, x is the input vector, and b is the bias.

The cross entropy function is:

$$H(p, q) = -\sum_x p(x) \log q(x) \tag{3}$$

where $p(x)$ is the target distribution and $q(x)$ is the predicted matching distribution.

So, the loss function is:

$$L(w) = -\frac{1}{n} \sum_{i=1}^{n} [y_i \log f_w(x_i) + (1 - y_i) \log(1 - f_w(x_i))] \tag{4}$$

where n is the number of training data, and when y_i (desired output) $\approx f_w(x_i)$ (NN output), $L(w)$ has a minimum value.

3. Results

Twenty-four healthy people were involved in the study and completed all experimental processes. Component N2P3 was used to analyze the EEG output online. The experimental results were analyzed as follows.

3.1. Discriminating the Sound-Evoked Potential in EEG

The system outputs the potential data (such as Table 1), the figures of the ERPs' features (such as Figure 4), and the results using N2P3 to distinguish the EEG data online. Figure 4 illustrates how to discriminate the characteristics of EEG data with a ERPs figure. Table 1 explains the scheme for determining the value of components N2P3, P300, and N200 via sound-evoked potentials. In Figure 4, the red curve presents the ERP evoked for the sound from the right ear, and the green-dot curve shows the ERP induced for the sound from the left ear.

Table 1. ERP values (μV) of options obtained from the Cz, 2nd stage, the 4st test run of N04.

Options	N2P3	P300	N200	Target Specified	Result On Line	Result Off Line
R	4.4627 *	1.0947 *	−3.3680 *	✓	✓	N2P3, P300, N200
L	1.0026	−0.1217	−1.1242			

*: the potential accepted between two options.

Figure 4 and Table 1 show that the P300 value (1.0947 μV at 376 msec) of the red curve (R option) is higher potential than that of the green-dot curve (L option), and the N200 value (−3.3680 μV at 258 msec) of the red curve (R option) is lower potential than that of the green-dot curve (L option). In Table 1, the R option (red curve) exhibits a higher N2P3 value (4.4627 μV). Therefore, the red curve is the dominant option. We thus know that the subject was focusing on the sound from the right channel of the headphones. Hence, the user's choice is R, which is also the target specified by the experimenter. That is, the user made the right choice during this trial.

3.2. Accuracy Analyses of Experimental Results

We used audio story stimulation and mental Shadowing Tasks to accomplish the experiment. The following accuracy analyses of the data obtained from electrodes show the experimental results using Shadowing Tasks.

3.2.1. Accuracy Analyses for All Output Data

The system used five electrodes (T3, T4, Pz, Cz, and Fz) to gain the potential data of the participants' brainwaves (ERPs). The system also outputs the classification results from the EEG data of five electrodes based on each component of the ERPs (Tables 2–4). Additionally, Figure 8 shows the potential difference between target and non-target.

Table 2. The average accuracies by component N200 of the experimental trial using mental Shadowing Tasks.

Unit: %

Subjects	T3	T4	Fz	Cz	Pz	Average
N01	50.00	50.00	70.00 *	60.00	45.00	55.00
N02	60.00	65.00	75.00	80.00 *	70.00	70.00
N03	55.00	45.00	50.00	60.00 *	50.00	52.00
N04	55.00	65.00	85.00 *	75.00	60.00	68.00
N05	40.00	65.00 *	60.00	60.00	35.00	52.00
N06	55.00	40.00	60.00	55.00	65.00 *	55.00
N07	60.00	65.00 *	60.00	60.00	50.00	59.00
N08	60.00	65.00 *	60.00	50.00	55.00	58.00
N09	55.00 *	50.00	55.00 *	55.00 *	55.00 *	54.00
N10	40.00	45.00	70.00 *	70.00 *	65.00	58.00
N11	50.00 *	40.00	50.00 *	50.00 *	50.00 *	48.00
N12	50.00	35.00	65.00 *	45.00	50.00	49.00
N13	70.00	70.00	70.00	75.00 *	45.00	66.00
N14	45.00	80.00 *	50.00	60.00	65.00	60.00
N15	45.00	70.00 *	65.00	65.00	50.00	59.00
N16	50.00	65.00 *	45.00	40.00	40.00	48.00
N17	45.00	80.00 *	70.00	70.00	70.00	67.00
N18	50.00	65.00	70.00	75.00 *	65.00	65.00
N19	70.00 *	60.00	70.00 *	70.00 *	65.00	67.00
N20	45.00	55.00	75.00 *	60.00	45.00	56.00
N21	55.00	55.00	70.00 *	65.00	60.00	61.00
N22	75.00 *	55.00	60.00	55.00	45.00	58.00
N23	60.00	65.00 *	55.00	55.00	50.00	57.00
N24	75.00 *	50.00	75.00 *	50.00	60.00	62.00
Average	54.79	58.33	63.96 *	60.83	54.58	58.50

*: the best accuracy around the electrodes.

Table 3. The average accuracies by component P300 of the experimental trial using mental Shadowing Tasks.

Unit: %

Subjects	T3	T4	Fz	Cz	Pz	Average
N01	75.00 *	75.00 *	60.00	75.00 *	75.00 *	72.00
N02	40.00	60.00 *	35.00	30.00	35.00	40.00
N03	65.00	65.00	50.00	60.00	70.00 *	62.00
N04	70.00 *	55.00	45.00	50.00	55.00	55.00
N05	70.00 *	45.00	45.00	50.00	55.00	53.00
N06	60.00	40.00	50.00	60.00	65.00 *	55.00
N07	55.00	55.00	60.00	55.00	65.00 *	58.00
N08	50.00	60.00	80.00	65.00	85.00 *	68.00
N09	65.00 *	60.00	50.00	65.00 *	55.00	59.00
N10	60.00	55.00	65.00 *	55.00	65.00 *	60.00

Table 3. Cont.

						Unit: %
Subjects	T3	T4	Fz	Cz	Pz	Average
N11	65.00	55.00	80.00 *	60.00	60.00	64.00
N12	55.00	65.00 *	60.00	60.00	60.00	60.00
N13	55.00 *	55.00 *	55.00 *	55.00 *	40.00	52.00
N14	75.00	50.00	55.00	80.00 *	75.00	67.00
N15	70.00	60.00	75.00	80.00	85.00 *	74.00
N16	70.00	60.00	55.00	65.00	75.00*	65.00
N17	65.00 *	60.00	60.00	60.00	55.00	60.00
N18	65.00	45.00	70.00 *	60.00	70.00 *	62.00
N19	65.00	75.00 *	55.00	60.00	65.00	64.00
N20	75.00	65.00	75.00	85.00 *	85.00 *	77.00
N21	70.00 *	70.00 *	60.00	45.00	55.00	60.00
N22	55.00	70.00 *	55.00	60.00	55.00	59.00
N23	70.00 *	60.00	55.00	70.00 *	55.00	62.00
N24	35.00	75.00 *	55.00	50.00	45.00	52.00
Average	62.50	59.79	58.54	60.63	62.71 *	60.83

*: the best accuracy around the electrodes.

Table 4. The average accuracies by component N2P3 of the experimental trial using mental Shadowing Tasks.

						Unit: %
Subjects	T3	T4	Fz	Cz	Pz	Average
N01	80.00	65.00	75.00	85.00 *	65.00	74.00
N02	80.00 *	75.00	60.00	65.00	65.00	69.00
N03	75.00 *	55.00	70.00	55.00	70.00	65.00
N04	75.00	75.00	80.00	90.00 *	85.00	81.00
N05	75.00 *	60.00	65.00	70.00	50.00	64.00
N06	55.00	50.00	65.00	75.00	80.00 *	65.00
N07	65.00	60.00	60.00	65.00	75.00 *	65.00
N08	65.00	70.00	75.00	80.00 *	75.00	73.00
N09	70.00 *	60.00	65.00	70.00 *	70.00 *	67.00
N10	70.00	85.00 *	75.00	80.00	80.00	78.00
N11	65.00	45.00	70.00 *	60.00	65.00	61.00
N12	65.00	60.00	80.00 *	65.00	60.00	66.00
N13	60.00	60.00	60.00	45.00	70.00 *	59.00
N14	65.00	90.00 *	70.00	75.00	90.00 *	78.00
N15	55.00	70.00	75.00	65.00	85.00 *	70.00
N16	80.00 *	60.00	55.00	60.00	80.00 *	67.00
N17	65.00	65.00	75.00	85.00 *	75.00	73.00
N18	55.00	65.00	75.00	85.00	90.00 *	74.00
N19	65.00	90.00 *	70.00	80.00	80.00	77.00
N20	70.00	85.00 *	80.00	80.00	70.00	77.00
N21	70.00	75.00	65.00	80.00 *	55.00	69.00
N22	85.00 *	65.00	75.00	65.00	60.00	70.00
N23	80.00 *	70.00	55.00	65.00	70.00	68.00
N24	35.00	45.00	55.00	60.00 *	50.00	49.00
Average	67.71	66.67	68.75	71.04	71.46 *	69.13

*: the best accuracy around the electrodes.

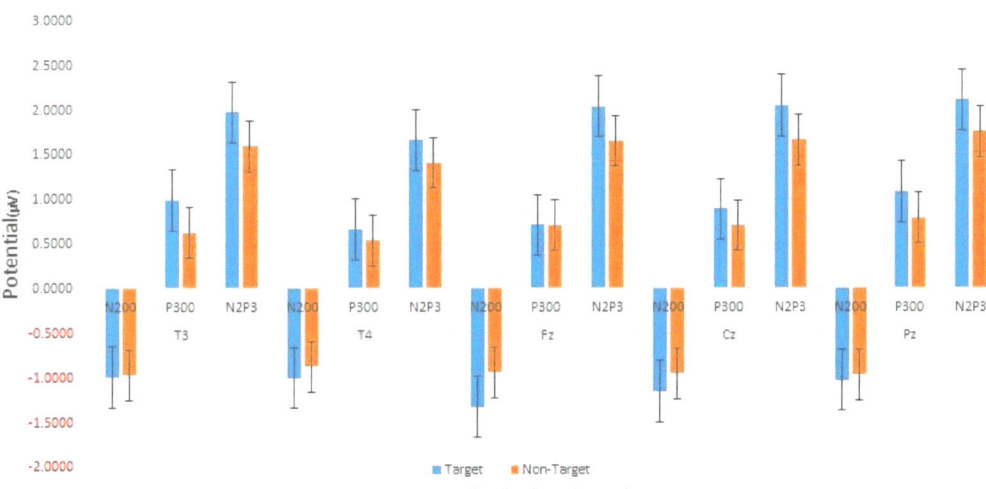

Figure 8. The potential differences between the target and non-target using components N200, P300, and N2P3 of ERPs to classify data from the five electrodes.

Component N2P3 was used to analyze the output of the EEG online. The experimental results in Tables 2–4 demonstrate the correctness of the system's experimental setup: the N2P3 features enabled the best discrimination. Table 5 displays the paired samples t-test results between the target and non-target options regarding accuracy and potential. These results indicate that component N2P3 can discriminate the features of ERPs well. Thus, component N2P3 is the optimal ERP for interpretation.

Table 5. Paired samples t-test results of all trials between the target and non-target options regarding accuracy and potential.

			$\alpha = 0.01$, N = 480			
			Accuracy (%)		Potential (μV)	
Components	Electrode	Case	T-Value	p-Value	T-Value	p-Value
N200	T3		2.335	0.028	−0.1772	0.859
	T4	target	3.366	0.002 *	−1.237	0.217
	Fz	vs.	6.915	0.000 ***	−3.971	0.000 ***
	Cz	non-target	5.159	0.000 ***	−2.249	0.025
	Pz		2.298	0.031 *	−0.576	0.565
P300	T3		5.873	0.000 ***	2.521	0.012
	T4	target	5.113	0.000 ***	1.021	0.308
	Fz	vs.	3.743	0.001 *	0.028	0.977
	Cz	non-target	4.335	0.000 ***	2.011	0.045 *
	Pz		4.663	0.000 ***	2.792	0.005 *
N2P3	T3		7.935	0.000 ***	4.288	0.000 ***
	T4	target	6.496	0.000 ***	3.046	0.002 *
	Fz	vs.	11.327	0.000 ***	4.224	0.000 ***
	Cz	non-target	9.182	0.000 ***	4.522	0.000 ***
	Pz		9.245	0.000 ***	4.160	0.000 ***

* $p < 0.05$; *** $p < 0.001$.

In addition, from the N2P3 results for online output (Table 4), the best-performing electrode Pz and the worst-performing electrode T3 were subjected to paired samples t-test. The mean between Pz and T3 did not reach a significant level (Table 6). So, we inferred that all five points are suitable for sampling electrode points.

Table 6. Paired samples t-test results of average accuracies for every trial between the best and worst electrodes.

	$\alpha = 0.01$, N = 480		
Component	Case	T-Value	p-Value
N2P3	T4 vs. Pz	−1.809	0.084

3.2.2. Accuracy Analyses via Neural Network

The experimental results in Tables 2–4 show that the average accuracy (correct rate) is not consistent across the five electrodes of each participant, and there are high individual differences among the users. Therefore, we use a NN technology for classification and identification to identify the best prediction function for each participant and perform the validation with a 5-fold cross-validation method (Table 7). The accuracy rendered in boldface is the best among the three components of the ERPs for each participant, respectively.

Table 7. Average accuracies based on the NN analysis.

			Unit: %
Subjects	N200	P300	N2P3
N01	70.00	95.00 *	90.00
N02	85.00 *	55.00	85.00 *
N03	60.00	75.00	80.00 *
N04	70.00	60.00	85.00 *
N05	65.00	65.00	70.00 *
N06	65.00	60.00	85.00 *
N07	70.00 *	70.00 *	70.00 *
N08	60.00	80.00 *	80.00 *
N09	60.00	65.00	80.00 *
N10	80.00	70.00	85.00 *
N11	50.00	80.00 *	65.00
N12	55.00	70.00	75.00 *
N13	75.00 *	65.00	75.00 *
N14	70.00	75.00	80.00 *
N15	70.00	90.00 *	75.00
N16	45.00	70.00	75.00 *
N17	80.00	55.00	85.00 *
N18	75.00	80.00	90.00 *
N19	70.00	65.00	85.00 *
N20	65.00	80.00	85.00 *
N21	65.00	85.00 *	85.00 *
N22	60.00	65.00	80.00 *
N23	70.00 *	65.00	70.00 *
N24	65.00 *	50.00	60.00
Average	66.67	70.42	78.96 *

*: the best accuracy around the components.

The average accuracy across all the participants via the NN analysis using the data from component N2P3 was 78.96%, which is better than the accuracies of the other two components.

3.2.3. Analysis of the Average Accuracies of the ERP Components

Table 7 shows the average accuracies of the three ERP components using NN analysis. The average accuracy of using the N2P3 component to discriminate the data from the same electrodes is better than that obtained using the P300 or N200 components. Each electrode gained the same results, as shown in Table 8. As expected, after the computer calculates the results, the average accuracy using the NN analysis is better than that of each electrode as well.

Table 8. Comparison of the average accuracy of three ERP components (N2P3, P300, and N200).

Dependent Variable: Average Accuracies						Unit: %
Components	T3	T4	Fz	Cz	Pz	NN Technology
N200	54.79	58.33	63.96	60.83	54.58	66.67
P300	62.50	59.79	58.54	60.63	62.71	70.42
N2P3	67.71	66.67	68.75	71.04	71.46	78.96

We also verify whether the difference among average accuracies of three components calculated by the NN technology (Table 7) reaches significance, as shown in Table 9.

Table 9. Multiple comparison of the average accuracies calculated by the NN technology.

Dependent Variable: Average Accuracies from the Analysis of NN Technology			
Electrode(I)	Electrode(J)	Mean Discrepancy(I-J)	p-Value
N2P3	N200	12.29167 ***	0.000 ***
	P300	8.54167 *	0.011 *
P300	N200	3.75000	0.400

* $p < 0.05$; *** $p < 0.001$.

3.2.4. Effect of Gender Voice Differences on Accuracy

In the experimental process, the system simultaneously sent a pair of audio stories through the left and right sides of the headphones for participants to listen to during each trial. There are twelve audio stories, six recorded in a male voice and six in a female voice. The auditory stimuli may or may not be the same-gender voice combination. A different-gender voice combination (DG) sends one male voice and one female voice simultaneously. The same-gender voice combination (SG) presents two male (or female) voices simultaneously. Next, we compared whether the auditory stimuli of different-gender voices or same-gender voices affected the accuracy, as shown in Table 10.

Table 10. The average accuracies using different-gender (DG) and same-gender (SG) voices.

						Unit: %
Subjects	N200		P300		N2P3	
	DG	SG	DG	SG	DG	SG
N01	46.00	64.00	82.00	62.00	88.00	60.00
N02	68.00	72.00	48.00	32.00	72.00	66.00
N03	44.00	60.00	60.00	64.00	60.00	70.00
N04	62.00	74.00	60.00	50.00	82.00	80.00
N05	46.00	58.00	52.00	54.00	60.00	68.00
N06	62.00	48.00	44.00	66.00	64.00	66.00
N07	64.00	54.00	56.00	60.00	60.00	70.00
N08	60.00	56.00	68.00	68.00	70.00	76.00

Table 10. Cont.

	N200		P300		N2P3	Unit: %
Subjects	DG	SG	DG	SG	DG	SG
N09	62.00	46.00	64.00	54.00	72.00	62.00
N10	56.00	60.00	66.00	54.00	86.00	70.00
N11	66.00	30.00	56.00	72.00	64.00	58.00
N12	34.00	64.00	60.00	60.00	52.00	80.00
N13	66.00	66.00	62.00	42.00	64.00	54.00
N14	52.00	68.00	84.00	50.00	78.00	78.00
N15	60.00	58.00	78.00	70.00	78.00	62.00
N16	44.00	52.00	64.00	66.00	60.00	74.00
N17	62.00	72.00	68.00	52.00	72.00	74.00
N18	58.00	72.00	66.00	58.00	70.00	78.00
N19	70.00	64.00	62.00	66.00	80.00	74.00
N20	48.00	64.00	82.00	72.00	78.00	76.00
N21	58.00	64.00	54.00	66.00	64.00	74.00
N22	48.00	68.00	66.00	52.00	62.00	78.00
N23	66.00	48.00	68.00	56.00	76.00	60.00
N24	62.00	62.00	60.00	44.00	56.00	42.00
Average	56.83	60.17	63.75	57.92	69.50	68.75
t-test	$p = 0.2827$		$p = 0.0353$ *		$p = 0.7764$	

* $p < 0.05$.

After the paired samples t-test, there are no significant difference between the accuracies of different-gender voice combinations and same-gender voice combinations using components N200 and N2P3 except for P300 ($p = 0.0353$ * < 0.05). Additionally, the paired samples *t*-test results of the correct chosen R and the correct chosen L for all trials show no significant difference between the accuracies (Table 11).

Table 11. Paired samples t-test results of average accuracies for every trial between correct selected R and correct selected L.

	$\alpha = 0.01$, N = 480		
	Case	T-Value	p-Value
N200	correct selected R vs. correct selected L	1.066	0.292
P300	correct selected R vs. correct selected L	−0.639	0.525
N2P3	correct selected R vs. correct selected L	−0.289	0.774

This result implies that accuracy is not affected by the gender voice combination played in each headphone if the system uses component N2P3 to distinguish EEG output data.

3.2.5. Effect of the Different Gender of Subjects on Accuracy

Twenty-four healthy people, seven females, were involved in the study and completed all experimental processes. Next, we compared whether the different gender of subjects affected the accuracy, as shown in Table 12.

After the independent samples t-test, there are no significant difference between the accuracies of different gender of subjects using components P300 and N2P3 to distinguish except for using component N200 (electrodes Fz, Cz, and Pz). This result implies that accuracy is not affected by the gender of users if the system uses component N2P3 to distinguish EEG output data.

Table 12. Independent samples t-test results of all trials between girls and boys.

Components	Electrode	Case	T-Value	p-Value
		$\alpha = 0.01$, N = 340 for Boys and 140 for Girls		
N200	T3	boys vs. girls	−1.193	0.246
	T4	boys vs. girls	−0.327	0.746
	Fz	boys vs. girls	−4.405	0.000 ***
	Cz	boys vs. girls	−2.348	0.028 *
	Pz	boys vs. girls	−2.116	0.045 *
	NN	boys vs. girls	−1.398	0.176
P300	T3	boys vs. girls	0.564	0.590
	T4	boys vs. girls	−1.284	0.212
	Fz	boys vs. girls	0.586	0.564
	Cz	boys vs. girls	1.328	0.226
	Pz	boys vs. girls	0.973	0.341
	NN	boys vs. girls	0.709	0.486
N2P3	T3	boys vs. girls	0.983	0.336
	T4	boys vs. girls	−1.600	0.124
	Fz	boys vs. girls	−0.203	0.841
	Cz	boys vs. girls	−1.788	0.088
	Pz	boys vs. girls	0.201	0.842
	NN	boys vs. girls	−1.303	0.206

* $p < 0.05$; *** $p < 0.001$.

3.3. Bit-Rate Analysis

Table 13 shows the bit rate of the trials from these five electrodes and the NN technology. The average bit rate of the study is lower than that of other studies [13].

Table 13. Comparison of the bit-rate of three ERP components (N2P3, P300, and N200).

Dependent Variable: Average Bit-Rate						
Components	T3	T4	Fz	Cz	Pz	NN Technology
N200	0.0114	0.0345	0.0977	0.0585	0.0104	0.1401
P300	0.0781	0.0477	0.0363	0.0563	0.0808	0.2123
N2P3	0.1585	0.1401	0.1782	0.2260	0.2353	0.4418

4. Discussion

Almost all researchers in this field have tried to promote the efficiency of their aBCI systems through various methods [17,41,45,47,48]. So, improving the efficiency of our aBCI is a primary task in the study as well. The accuracy of an aBCI system is deeply affected by three primary factors. First, is the stimulus appropriate? Second, are the positions of electrodes to obtain the brainwave data appropriate? Finally, should the system select which ERP components (N2P3, P300, and N200) to interpret the data of the user's brainwaves online to achieve the best system efficiency?

Related research has sprung up like mushrooms after rain to solve the above problems. We reviewed several studies similar to this one to discuss their strengths and weaknesses, as shown in Table 14. The further discussion showed in the text description after the table.

Because of human hearing function, two factors affect the performance of the BCI system: whether the target sound stimulus attracts the user's attention and whether the user can easily distinguish the target sound stimulus from the non-target stimuli. The work of Domingos et al. also showed that performing an attention task in an intermittent noisy or silent room had different results. It could be that humans are unaware of all the noise surrounding them every second, but if we deprive them of it, it can be worse [21].

Table 14. A comparison of the advantages and drawbacks of the proposed method with other studies.

References	Stimulation Modality	Electrodes	Subjects	Advantages	Drawbacks
[46]	P300 Spatial real, virtual sounds	Cpz, Poz, P3, P4, P5, P6, Cz, Pz in 10/10	9 HS	Both stimuli types generate different event-related potential response patterns allowing for their separate classification.	1. Too few people participated in the experiment. 2. This analysis was more complicated, based on 8 electrodes.
[48]	P300 Spatial vs. non-spatial	F3, Fz, F4, T7, C3, Cz, C4, T8, Cp3, Cp4, P3, Pz, P4, PO7, PO8, Oz	16 HS	Training improves performance in an auditory BCI paradigm. Motivation influences performance and P300 amplitude.	1. This analysis was more complicated based on 16 electrodes. 2. Average accuracy < 80%
[17]	P300 Spatial auditory	32 channels in the extended 10–20 system	9HS	ErrP-based error correction can be used to make a substantial improvement in the performance of aBCIs.	1. Too few people participated in the experiment. 2. This analysis was more complicated, based on 32 electrodes.
[41]	ASSR+P300 Earphone auditory	Fz, Cz, Pz, P3, P4, Oz, T3 and T4	10 HS	The average accuracy of the hybrid system is better than that of P300 or ASSR alone.	1. Too few people participated in the experiment. 2. This analysis was more complicated, based on a hybrid system.
[45]	ASSR Earphone auditory	Cz, Oz, T7, and T8	6 HS	The average classification accuracies online were excellent, more than 80%.	1. Too few people participated in the experiment. 2. This analysis was more complicated, based on the ASSR method.
[31]	P300 Headphone auditory	Fz, Cz, Pz, Oz, P3, P4, PO7, PO8	10 HS	Mental repetition can be a simpler alternative to the mental count to reduce the mental workload.	1. Too few people participated in the experiment. 2. This analysis was offline.
[16]	Speakers	19 channels	12HS	Multi-loudspeaker patterns through vowel and numeral sound stimulation provided an accuracy greater than 85% of the average accuracy.	1. Too few people participated in the experiment. 2. This analysis was more complicated, based on 19 electrodes.
The proposed method	P300 Headphone auditory	T3, T4, Fz, Cz, Pz	24HS	The method of mental shadowing tasks helps the user focus on the option he wants with ease to reduce the mental workload.	Average accuracy = 78.69%, and it will be better if the accuracy rate can be higher.

We decided to use audio stories as sound stimulation to improve user motivation and use headphones to deliver the audio stories to enable the user to concentrate on and repeat the keywords in the stories. In our previous work [14], we invited seven participants familiar with a BCI system to test our aBCI system. We found the discriminative features in ERPs from the aBCI system are traceable, as shown in Figure 4. The features of P300 and N200 using audio story stimulation and the Shadowing Tasks mechanism were more distinct than those using other methods. These results encouraged us to perform a subsequent study of the aBCI system.

In addition, to a decline in mental workload, we shift the usual Shadowing Tasks paradigm [14]: instead of loudly repeating the auditory target stimuli, we ask subjects to echo the target stimuli mentally as it occurs [31]. We call this approach, different from those of previous studies, the mental Shadowing Tasks mechanism.

This study used twelve audio stories, six recorded by male and six by female voices. The number of audio stories is greater than that of our previous work [14]. According to the post-analysis, we found that the gender of the story voice did not affect the participants' attention or the average accuracies using N2P3 to distinguish the EEG data, as shown in Table 10.

Most aBCI studies use more than eight electrodes to collect data [17,31,38,41,47,48,62]. To reduce user discomfort, we used only three electrodes, Pz, Cz, and Fz, to sample the user's brainwaves via EEG equipment in our previous work [14]. In this study, referencing [16,17,49] and passing through multiple experiments, we selected five electrodes to collect the data: T3, T4, Pz, Cz, and Fz. From Tables 5 and 6, we inferred that all five points are suitable for sampling electrode points.

In the study, we invited 24 participants who had never used a BCI system to test our aBCI system. Tables 2–4 show the average accuracies of EEG data from each electrode for each participant. While the average accuracies are better than chance, the average accuracies are not consistent across the five electrodes of each participant, and there are high individual differences among the participants as well. Therefore, we used a neural network for classification and identification to identify the best prediction function for each user to improve the user's accuracy when using the system. Table 7 shows the average accuracies via the NN analysis. NN analysis does raise all accuracies. The average accuracies for most participants using the data from component N2P3 were the highest among the three components via the NN analysis. The average accuracy across all participants was 78.96% ((921 + 974)/2400, shown in Table 15).

Table 15. The frequency distribution of sound stimuli in the right and left ear using the NN technology to classify the data gained from component N2P3.

N2P3		Specified Condition		Total
		R	L	
Classification result	R	921	226	1147
	L	279	974	1253
Total		1200	1200	

The output of the designed experiment comes from one of two options, R or L. It is different from the example of a medical test for diagnosing a condition that is positive or not. So, we set the right/left ear stimulus as the target and the left/right ear as the non-target, set different thresholds for the values of ERPs, and then obtained a sequence of confusion matrices and the ROC curve (Figure 9). Based on the ROC curve, we find the best cut-off score (accuracy) = 78.32%, slightly lower than that obtained by ANN (=78.96%).

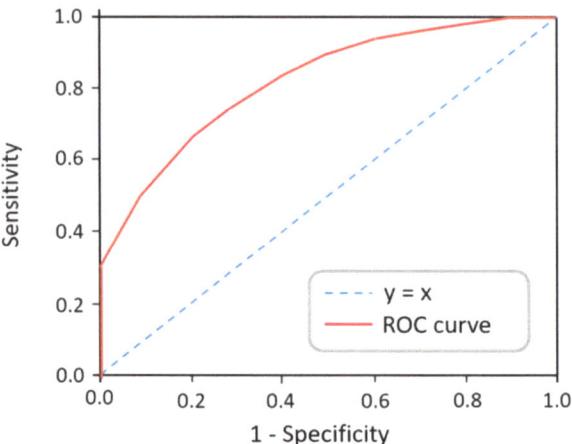

Figure 9. The ROC curve by setting the right ear stimulus as the target and the left ear stimulus as the non-target.

Further, the best result of confusion matrices is the precision = 80.02%, sensitivity = 76.75%, specificity = 79.89%, and recall = 76.62% [63].

In addition to auditory stimulation and sampling electrodes, another factor affecting the performance of a BCI is the interpretation of the ERPs. This study analyzed the data of each electrode using three ERP components (N2P3, P300, and N200). Which ERP component should the aBCI system use to interpret the data of the user's brainwaves online? Table 8 shows the average accuracies across all participants for each electrode (including the average accuracies obtained by the NN analysis). This analysis showed that the average accuracy using component N2P3 to discriminate is the best. Table 8 also indicates that no matter which electrode is used, as long as the system uses component N2P3 to discriminate the data, we obtain better results. This result is consistent with the results of our vBCI experimental [13]. According to Table 9, the average accuracies of using the three components to discriminate the data from the same electrodes is significantly different. The accuracy using component N2P3 remains the best.

Searching for suitable ERP features and algorithms to raise the information transfer rate (ITR) attracts the most activity in BCI research [64]. Based on Höhne et al., the stimuli should exist long enough to evoke the sound-evoked potential. However, if the system uses much longer stimulation, the inter-command intervals will be extended, resulting in lower information–transfer–rates (ITR) [38]. Unacceptably low ITR in this study, as shown in Table 13, is one of the items that the subsequent systems need to improve. Increasing ITR will thus be a target of future research. Perhaps the number of keywords should be reduced from 7 to 5 or 4, or the stories shortened.

Finally, in this study, we noticed that the occurrence of the ERPs components varies from person to person, as shown in Figures 4 and 8. Ref. [65] implemented a thorough test for the audio-visual, visual, and auditory spatial speller paradigms. One of the results is the latencies of auditory-based P300 peaks were longer than those of visual-based P300 peaks, from 250ms to 600ms. Those latencies both occur in target and non-target stimulation. The sampling time of the P300 and N200 components in our aBCI system is different from that of the ERPs diagram of our vBCI system [13], similar to the works of Chang et al. [65] and Marassi et al. [31]. The sampling time of brainwaves for some users is delayed and varies from person to person (Figure 10). That is also an issue for our follow-up research.

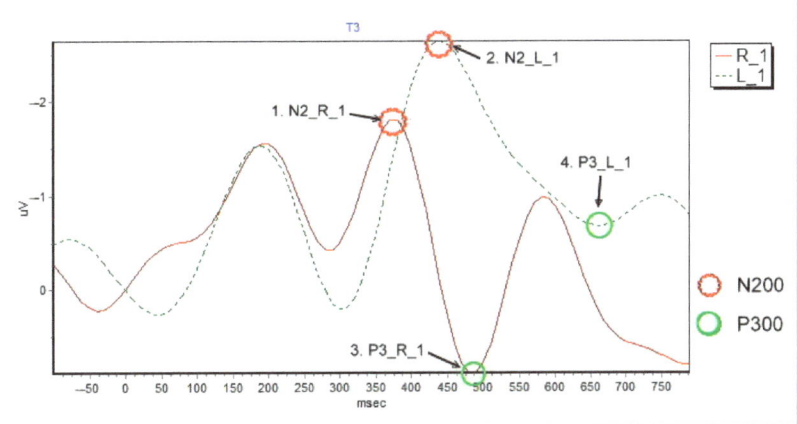

Figure 10. The delay situation of the components P300 and N200 in ERPs: The latencies of auditory P300 (marks 3 and 4) and N200 (marks 1 and 2) responses are longer than the theoretical latencies. The average waveforms at T3 is the first stage output of the 5th test run from participant N01.

5. Conclusions

The shadowing tasks of cognitive science can effectively improve users' concentration. This study applied this principle to help the user focus more on the target stimuli during using the auditory BCI, which improved the recognition accuracy of this system. So, the shadowing tasks approach is a primary innovation of this study.

The shadowing tasks proposed by Cherry elicited increased bilateral activation predominantly in the superior temporal sulci [55]. Additionally, mental repetition can be a simpler alternative to the mental count to reduce the cognitive workload [31]. Therefore, we proposed using mental shadowing tasks to increase the sound-evoked potential of EEG. Further, motivation influences performance and P300 amplitude [48], so we use audio stories to cause and enhance the motive of users.

Patients, such as those at the terminal stage of ALS, cannot use the visual-based BCI system due to the functional loss of muscle activities such as eye movement [15]. Our aBCI system wants to address this situation. Hence, this study focuses on the aBCI paradigm based on the Shadowing Tasks. We hope to develop an auditory BCI home care system.

The study adopts an event-related potential paradigm that combines motion-onset and oddball presentation. This ERP pattern uses components P300 and N200 [9] and the component N2P3 that we identified [13,14].

We compare the average accuracy of each electrode to confirm the performance of the data from each electrode. We also contrasted the interpretation capabilities of components N2P3, P3, and N2. Our results show that the accuracy improves. The efficiency is better than the efficiency obtained using the sound stimuli consisting of periodic click sounds. The average accuracy of each subject exceeded the theoretical chance levels (Table 7). When we interpret the data gained from component N2P3 via the NN technology, the average accuracy reaches 78.96%. Farther, the average accuracies for five users out of eight exceeded 80%. The preliminary results for audio story stimulation with mental Shadowing Tasks are a step forward compared with current state-of-the-art aBCI applications. This result encourages us to conduct future research into aBCI systems with Shadowing Task paradigms for possible inclusion in practical online applications.

Finally, the lower ITR needs to be improved. Therefore, research on more efficient stimuli types for BCI based on the mental Shadowing Tasks is necessary. Future studies should include the stories of the mental Shadowing Tasks, the sampling time of the component P300 and N200 optimization for handicapped or bedridden subjects, and the algorithms to increase information transfer rates. We intend to persist in this study and involve the protocol in an aBCI-based home care system (HCS) for target patients to provide daily assistance without gaze control with their environment.

Author Contributions: Conceptualization, K.-T.S. and K.-L.H.; methodology, K.-T.S. and K.-L.H.; software, K.-L.H. and S.-Y.L.; validation, K.-T.S. and S.-Y.L.; formal analysis, K.-T.S. and K.-L.H.; investigation, K.-T.S., K.-L.H. and S.-Y.L.; resources, K.-L.H. and S.-Y.L.; data curation, K.-L.H. and S.-Y.L.; writing—original draft preparation, K.-L.H.; writing—review and editing, K.-T.S.; supervision, K.-T.S.; project administration, K.-T.S.; funding acquisition, K.-T.S. All authors have read and agreed to the published version of the manuscript.

Funding: This research received no external funding.

Institutional Review Board Statement: The study protocol was approved by the Human Research Ethics Committee at National Cheng Kung University (protocol code: 107-349-2 and date of approval: 16 April 2019).

Informed Consent Statement: All subjects involved in the study had signed the informed consent.

Data Availability Statement: Post-analysis data and raw data are available; please email: d10355002@gm2.nutn.edu.tw or klhsieh@gmail.com.

Conflicts of Interest: The authors declare no conflict of interest.

References

1. Van Es, M.A.; Hardiman, O.; Chio, A.; Al-Chalabi, A.; Pasterkamp, R.J.; Veldink, J.H.; van den Berg, L.H. Amyotrophic Lateral Sclerosis. *Lancet* **2017**, *390*, 2084–2098. [CrossRef]
2. Kiernan, M.C.; Vucic, S.; Cheah, B.C.; Turner, M.R.; Eisen, A.; Hardiman, O.; Burrell, J.R.; Zoing, M.C. Amyotrophic Lateral Sclerosis. *Lancet* **2011**, *377*, 942–955. [CrossRef] [PubMed]
3. Turner, M.R.; Agosta, F.; Bede, P.; Govind, V.; Lulé, D.; Verstraete, E. Neuroimaging in Amyotrophic Lateral Sclerosis. *Biomark. Med.* **2012**, *6*, 319–337. [CrossRef] [PubMed]
4. Kiernan, M.C.; Vucic, S.; Talbot, K.; McDermott, C.J.; Hardiman, O.; Shefner, J.M.; Al-Chalabi, A.; Huynh, W.; Cudkowicz, M.; Talman, P.; et al. Improving Clinical Trial Outcomes in Amyotrophic Lateral Sclerosis. *Nat. Rev. Neurol.* **2021**, *17*, 104–118. [CrossRef] [PubMed]
5. Vahsen, B.F.; Gray, E.; Thompson, A.G.; Ansorge, O.; Anthony, D.C.; Cowley, S.A.; Talbot, K.; Turner, M.R. Non-Neuronal Cells in Amyotrophic Lateral Sclerosis—From Pathogenesis to Biomarkers. *Nat. Rev. Neurol.* **2021**, *17*, 333–348. [CrossRef]
6. Huang, T.W. Design of Chinese Spelling System Based on ERPs. Master's Thesis, National University of Tainan, Tainan, Taiwan, 2011.
7. Sun, K.T.; Huang, T.W.; Chen, M.C. Design of Chinese Spelling System Based on ERP. In Proceedings of the 11th IEEE International Conference on Bioinformatics and Bioengineering, BIBE 2011, Taichung, Taiwan, 24–26 October 2011; pp. 310–313.
8. Liu, Y.-H.; Wang, S.-H.; Hu, M.-R. A Self-Paced P300 Healthcare Brain-Computer Interface System with SSVEP-Based Switching Control and Kernel FDA + SVM-Based Detector. *Appl. Sci.* **2016**, *6*, 142. [CrossRef]
9. Hong, B.; Guo, F.; Liu, T.; Gao, X.; Gao, S. N200-Speller Using Motion-Onset Visual Response. *Clin. Neurophysiol.* **2009**, *120*, 1658–1666. [CrossRef] [PubMed]
10. Yin, E.; Zhou, Z.; Jiang, J.; Chen, F.; Liu, Y.; Hu, D. A Speedy Hybrid BCI Spelling Approach Combining P300 and SSVEP. *IEEE Trans. Biomed. Eng.* **2014**, *61*, 473–483. [CrossRef]
11. Wolpaw, J.R.; Birbaumer, N.; McFarland, D.J.; Pfurtscheller, G.; Vaughan, T.M. Brain-Computer Interfaces for Communication and Control. *Clin. Neurophysiol.* **2002**, *113*, 767–791. [CrossRef]
12. Martínez-Cagigal, V.; Santamaría-Vázquez, E.; Gomez-Pilar, J.; Hornero, R. Towards an Accessible Use of Smartphone-Based Social Networks through Brain-Computer Interfaces. *Expert Syst. Appl.* **2019**, *120*, 155–166. [CrossRef]
13. Sun, K.T.; Hsieh, K.L.; Syu, S.R. Towards an Accessible Use of a Brain-Computer Interfaces-Based Home Care System through a Smartphone. *Comput. Intell. Neurosci.* **2020**, *2020*, 16–18. [CrossRef] [PubMed]
14. Hsieh, K.L.; Sun, K.T. Auditory Brain Computer Interface Design. In Proceedings of the 2017 International Conference on Applied System Innovation (ICASI), Sapporo, Japan, 13–17 May 2017; pp. 11–14.
15. Matsumoto, Y.; Makino, S.; Mori, K.; Rutkowski, T.M. Classifying P300 Responses to Vowel Stimuli for Auditory Brain-Computer Interface. In Proceedings of the 2013 Asia-Pacific Signal and Information Processing Association Annual Summit and Conference, APSIPA 2013, Kaohsiung, Taiwan, 29 October–1 November 2013; pp. 1–5.
16. Borirakarawin, M.; Punsawad, Y. Event-Related Potential-Based Brain-Computer Interface Using the Thai Vowels' and Numerals' Auditory Stimulus Pattern. *Sensors* **2022**, *22*, 5864. [CrossRef]
17. Zeyl, T.; Yin, E.; Keightley, M.; Chau, T. Improving Bit Rate in an Auditory BCI: Exploiting Error-Related Potentials. *Brain-Comput. Interfaces* **2016**, *3*, 75–87. [CrossRef]

18. Aydin, E.A.; Bay, O.F.; Guler, I. P300-Based Asynchronous Brain Computer Interface for Environmental Control System. *IEEE J. Biomed. Health Inform.* **2018**, *22*, 653–663. [CrossRef] [PubMed]
19. Abiri, R.; Borhani, S.; Sellers, E.W.; Jiang, Y.; Zhao, X. A Comprehensive Review of EEG-Based Brain-Computer Interface Paradigms. *J. Neural Eng.* **2019**, *16*, 011001. [CrossRef] [PubMed]
20. Islam, M.S.; Hussain, I.; Rahman, M.M.; Park, S.J.; Hossain, M.A. Explainable Artificial Intelligence Model for Stroke Prediction Using EEG Signal. *Sensors* **2022**, *22*, 9859. [CrossRef]
21. Domingos, C.; da Silva Caldeira, H.; Miranda, M.; Melicio, F.; Rosa, A.C.; Pereira, J.G. The Influence of Noise in the Neurofeedback Training Sessions in Student Athletes. *Int. J. Environ. Res. Public Health* **2021**, *18*, 13223. [CrossRef]
22. Cheng, P.W.; Tian, Y.J.; Kuo, T.H.; Sun, K.T. The Relationship between Brain Reaction and English Reading Tests for Non-Native English Speakers. *Brain Res.* **2016**, *1642*, 384–388. [CrossRef]
23. Sutton, S.; Braren, M.; Zubin, J.; John, E.R. Evoked-Potential Correlates of Stimulus Uncertainty. *Science* **1965**, *150*, 1187–1188. [CrossRef]
24. Kappenman, E.S.; Farrens, J.L.; Zhang, W.; Stewart, A.X.; Luck, S.J. ERP CORE: An Open Resource for Human Event-Related Potential Research. *Neuroimage* **2021**, *225*, 117465. [CrossRef]
25. Jamil, N.; Belkacem, A.N.; Ouhbi, S.; Lakas, A. Noninvasive Electroencephalography Equipment for Assistive, Adaptive, and Rehabilitative Brain–Computer Interfaces: A Systematic Literature Review. *Sensors* **2021**, *21*, 4754. [CrossRef] [PubMed]
26. Gamble, M.L.; Luck, S.J. N2ac: An ERP Component Associated with the Focusing of Attention within an Auditory Scene. *Psychophysiology* **2011**, *48*, 1057–1068. [CrossRef] [PubMed]
27. Regan, D. *Human Brain Electrophysiology: Evoked Potentials and Evoked Magnetic Fields in Science and Medicine*; Elsevier: New York, NY, USA, 1989.
28. Lakey, C.E.; Berry, D.R.; Sellers, E.W. Manipulating Attention via Mindfulness Induction Improves P300-Based Brain-Computer Interface Performance. *J. Neural Eng.* **2011**, *8*, 025019. [CrossRef] [PubMed]
29. Picton, T.W. The P300 Wave of the Human Event-Related Potential. *J. Clin. Neurophysiol.* **1992**, *9*, 456–479. [CrossRef] [PubMed]
30. Panicker, R.C.; Puthusserypady, S.; Sun, Y. Adaptation in P300 Braincomputer Interfaces: A Two-Classifier Cotraining Approach. *IEEE Trans. Biomed. Eng.* **2010**, *57*, 2927–2935. [CrossRef]
31. Marassi, A.; Budai, R.; Chittaro, L. A P300 Auditory Brain-Computer Interface Based on Mental Repetition. *Biomed. Phys. Eng. Express* **2018**, *4*, 035040. [CrossRef]
32. Hoffmann, U.; Vesin, J.M.; Ebrahimi, T.; Diserens, K. An Efficient P300-Based Brain-Computer Interface for Disabled Subjects. *J. Neurosci. Methods* **2008**, *167*, 115–125. [CrossRef]
33. Patel, S.H.; Azzam, P.N. Characterization of N200 and P300: Selected Studies of the Event-Related Potential. *Int. J. Med. Sci.* **2005**, *2*, 147–154. [CrossRef]
34. Donchin, E.; Spencer, K.M.; Wijesinghe, R. The Mental Prosthesis: Assessing the Speed of a P300-Based Brain- Computer Interface. *IEEE Trans. Rehabil. Eng.* **2000**, *8*, 174–179. [CrossRef]
35. Halgren, E.; Marinkovic, K.; Chauvel, P. Generators of the Late Cognitive Potentials in Auditory and Visual Oddball Tasks. *Electroencephalogr. Clin. Neurophysiol.* **1998**, *106*, 156–164. [CrossRef]
36. Zhang, R.; Wang, Q.; Li, K.; He, S.; Qin, S.; Feng, Z.; Chen, Y.; Song, P.; Yang, T.; Zhang, Y.; et al. A BCI-Based Environmental Control System for Patients with Severe Spinal Cord Injuries. *IEEE Trans. Biomed. Eng.* **2017**, *64*, 1959–1971. [CrossRef] [PubMed]
37. Cherry, E.C. Some Experiments on the Recognition of Speech, with One and with Two Ears. *J. Acoust. Soc. Am.* **1953**, *25*, 975–979. [CrossRef]
38. Matsumoto, Y.; Nishikawa, N.; Yamada, T.; Makino, S.; Rutkowski, T.M. Auditory Steady-State Response Stimuli Based BCI Application-the Optimization of the Stimuli Types and Lengths. In Proceedings of the Signal and Information Processing Association Annual Summit and Conference (APSIPA), Kaohsiung, Taiwan, 29 October–1 November 2013; pp. 285–308.
39. Höhne, J.; Tangermann, M. Towards User-Friendly Spelling with an Auditory Brain-Computer Interface: The CharStreamer Paradigm. *PLoS ONE* **2014**, *9*, e102630. [CrossRef]
40. Sosulski, J.; Hübner, D.; Klein, A.; Tangermann, M. Online Optimization of Stimulation Speed in an Auditory Brain-Computer Interface under Time Constraints. *arXiv* **2021**, arXiv:2109.06011.
41. Kaongoen, N.; Jo, S. A Novel Hybrid Auditory BCI Paradigm Combining ASSR and P300. *J. Neurosci. Methods* **2017**, *279*, 44–51. [CrossRef] [PubMed]
42. Lu, Z.; Li, Q.; Gao, N.; Yang, J.; Bai, O. Happy Emotion Cognition of Bimodal Audiovisual Stimuli Optimizes the Performance of the P300 Speller. *Brain Behav.* **2019**, *9*, e01479. [CrossRef]
43. Oralhan, Z. A New Paradigm for Region-Based P300 Speller in Brain Computer Interface. *IEEE Access* **2019**, *7*, 106618–106627. [CrossRef]
44. Lu, Z.; Li, Q.; Gao, N.; Yang, J.; Bai, O. A Novel Audiovisual P300-Speller Paradigm Based on Cross-Modal Spatial and Semantic Congruence. *Front. Neurosci.* **2019**, *13*, 1040. [CrossRef]
45. Heo, J.; Baek, H.J.; Hong, S.; Chang, M.H.; Lee, J.S.; Park, K.S. Music and Natural Sounds in an Auditory Steady-State Response Based Brain–Computer Interface to Increase User Acceptance. *Comput. Biol. Med.* **2017**, *84*, 45–52. [CrossRef]
46. Nishikawa, N.; Makino, S.; Rutkowski, T.M. Spatial Auditory BCI Paradigm Based on Real and Virtual Sound Image Generation. In Proceedings of the 2013 Asia-Pacific Signal and Information Processing Association Annual Summit and Conference, APSIPA 2013, Kaohsiung, Taiwan, 29 October–1 November 2013.

47. Chang, M.; Mori, K.; Makino, S. Spatial Auditory Two-Step Input Japanese Syllabary Brain-Computer Interface Speller. *Procedia Technol.* **2014**, *18*, 25–31. [CrossRef]
48. Baykara, E.; Ruf, C.A.; Fioravanti, C.; Käthner, I.; Simon, N.; Kleih, S.C.; Kübler, A.; Halder, S. Effects of Training and Motivation on Auditory P300 Brain-Computer Interface Performance. *Clin. Neurophysiol.* **2016**, *127*, 379–387. [CrossRef] [PubMed]
49. Moller, A.R. *Hearing: Anatomy, Physiology, and Disorders of the Auditory System*, 2nd ed.; Academic Press: San Diego, CA, USA, 2006; ISBN 978-0-12-372519-6.
50. Lobato, B.Y.M.; Ramirez, M.R.; Rojas, E.M.; Moreno, H.B.R.; Soto, M.D.C.S.; Nuñez, S.O.V. Controlling a Computer Using BCI, by Blinking or Concentration. In Proceedings of the 2018 International Conference on Algorithms, Computing and Artificial Intelligence; Association for Computing Machinery, New York, NY, USA, 21–23 December 2018.
51. Da Silva-Sauer, L.; Valero-Aguayo, L.; dela Torre-Luque, A.; Ron-Angevin, R.; Varona-Moya, S. Concentration on Performance with P300-Based BCI Systems: A Matter of Interface Features. *Appl. Ergon.* **2016**, *52*, 325–332. [CrossRef] [PubMed]
52. Da Silva Souto, C.; Lüddemann, H.; Lipski, S.; Dietz, M.; Kollmeier, B. Influence of Attention on Speech-Rhythm Evoked Potentials: First Steps towards an Auditory Brain-Computer Interface Driven by Speech. *Biomed. Phys. Eng. Express* **2016**, *2*, 325–332. [CrossRef]
53. McDermott, J.H. The Cocktail Party Problem. *Curr. Biol.* **2009**, *19*, R1024–R1027. [CrossRef] [PubMed]
54. Revlin, R. *Human Cognition: Theory and Practice*; Worth Publishers: New York, NY, USA, 2007; ISBN 9780716756675.
55. Peschke, C.; Ziegler, W.; Kappes, J.; Baumgaertner, A. Auditory-Motor Integration during Fast Repetition: The Neuronal Correlates of Shadowing. *Neuroimage* **2009**, *47*, 392–402. [CrossRef] [PubMed]
56. DeVos, M.; Gandras, K.; Debener, S. Towards a Truly Mobile Auditory Brain-Computer Interface: Exploring the P300 to Take Away. *Int. J. Psychophysiol.* **2014**, *91*, 46–53. [CrossRef] [PubMed]
57. Jurcak, V.; Tsuzuki, D.; Dan, I. 10/20, 10/10, and 10/5 Systems Revisited: Their Validity as Relative Head-Surface-Based Positioning Systems. *Neuroimage* **2007**, *34*, 1600–1611. [CrossRef]
58. Wagner, A.; Ille, S.; Liesenhoff, C.; Aftahy, K.; Meyer, B.; Krieg, S.M. Improved Potential Quality of Intraoperative Transcranial Motor-Evoked Potentials by Navigated Electrode Placement Compared to the Conventional Ten-Twenty System. *Neurosurg. Rev.* **2022**, *45*, 585–593. [CrossRef]
59. Ng, C.R.; Fiedler, P.; Kuhlmann, L.; Liley, D.; Vasconcelos, B.; Fonseca, C.; Tamburro, G.; Comani, S.; Lui, T.K.-Y.; Tse, C.-Y.; et al. Multi-Center Evaluation of Gel-Based and Dry Multipin EEG Caps. *Sensors* **2022**, *22*, 8079. [CrossRef]
60. Wolpaw, J.R.; Birbaumer, N.; Heetderks, W.J.; McFarland, D.J.; Peckham, P.H.; Schalk, G.; Donchin, E.; Quatrano, L.A.; Robinson, C.J.; Vaughan, T.M. Brain-Computer Interface Technology: A Review of the First International Meeting. *IEEE Trans. Rehabil. Eng.* **2000**, *8*, 164–173. [CrossRef]
61. Abiodun, O.I.; Jantan, A.; Omolara, A.E.; Dada, K.V.; Mohamed, N.A.; Arshad, H. State-of-the-Art in Artificial Neural Network Applications: A Survey. *Heliyon* **2018**, *4*, e00938. [CrossRef] [PubMed]
62. An, X.; Höhne, J.; Ming, D.; Blankertz, B. Exploring Combinations of Auditory and Visual Stimuli for Gaze-Independent Brain-Computer Interfaces. *PLoS ONE* **2014**, *9*, e111070. [CrossRef] [PubMed]
63. Ahsan, M.M.; Luna, S.A.; Siddique, Z. Machine-Learning-Based Disease Diagnosis: A Comprehensive Review. *Healthcare* **2022**, *10*, 541. [CrossRef]
64. Dornhege, G.; Blankertz, B.; Curio, G.; Müller, K.-R. Increase Information Transfer Rates in BCI by CSP Extension to Multi-Class. In Proceedings of the NIPS 2003, Vancouver, BC, Canada, 8–13 December 2003.
65. Chang, M.; Nishikawa, N.; Struzik, Z.R.; Mori, K.; Makino, S.; Mandic, D.; Rutkowski, T.M. Comparison of P300 Responses in Auditory, Visual and Audiovisual Spatial Speller BCI Paradigms. *arXiv* **2013**, arXiv:1301.6360. [CrossRef]

Disclaimer/Publisher's Note: The statements, opinions and data contained in all publications are solely those of the individual author(s) and contributor(s) and not of MDPI and/or the editor(s). MDPI and/or the editor(s) disclaim responsibility for any injury to people or property resulting from any ideas, methods, instructions or products referred to in the content.

Communication

Predicting the Onset of Freezing of Gait Using EEG Dynamics

Alka Rachel John [1,*], Zehong Cao [2], Hsiang-Ting Chen [3], Kaylena Ehgoetz Martens [4], Matthew Georgiades [5], Moran Gilat [6], Hung T. Nguyen [7], Simon J. G. Lewis [5] and Chin-Teng Lin [1]

1. Australian Artificial Intelligence Institute, Faculty of Engineering and Information Technology, University of Technology Sydney, Ultimo, Sydney 2007, Australia
2. STEM, Mawson Lakes Campus, University of South Australia, Adelaide 5001, Australia
3. School of Computer Science, University of Adelaide, Adelaide 5005, Australia
4. Department of Kinesiology, University of Waterloo, Waterloo, ON N2L 3G1, Canada
5. Parkinson's Disease Research Clinic, Brain and Mind Centre, University of Sydney, Sydney 2006, Australia
6. Department of Rehabilitation Sciences, KU Leuven, 3000 Leuven, Belgium
7. Faculty of Science, Engineering and Technology, Swinburne University of Technology, Hawthorn 3122, Australia
* Correspondence: alkarachel.john@student.uts.edu.au

Abstract: Freezing of gait (FOG) severely incapacitates the mobility of patients with advanced Parkinson's disease (PD). An accurate prediction of the onset of FOG could improve the quality of life for PD patients. However, it is imperative to distinguish the possibility of the onset of FOG from that of voluntary stopping. Our previous work demonstrated the neurological differences between the transition to FOG and voluntary stopping using electroencephalogram (EEG) signals. We employed a timed up-and-go (TUG) task to elicit FOG in PD patients. Some of these TUG tasks had an additional voluntary stopping component, where participants stopped walking based on verbal instruction to "stop". The performance of the convolutional neural network (CNN) in identifying the transition to FOG from normal walking and the transition to voluntary stopping was explored. To the best of our knowledge, this work is the first study to propose a deep learning method to distinguish the transition to FOG from the transition to voluntary stop in PD patients. The models, trained on the EEG data from 17 PD patients who manifested FOG episodes, considering a short two-second transition window for FOG occurrence or voluntary stopping, achieved close to 75% classification accuracy in distinguishing transition to FOG from the transition to voluntary stopping or normal walking. Our results represent an important step toward advanced EEG-based cueing systems for smart FOG intervention, excluding the potential confounding of voluntary stopping.

Keywords: freezing of gait; Parkinson's disease; voluntary stopping; convolutional neural network; EEGNet; Shallow ConvNet; Deep ConvNet

1. Introduction

Freezing of gait (FOG) is a gait impairment resulting from neurodegeneration in advanced Parkinson's disease (PD) patients. Nieuwboer and Giladi [1] defined FOG as the "inability to deal with concurrent cognitive, limbic and motor inputs, causing an interruption of locomotion". This episodic gait difficulty causes the patients to suddenly experience the feeling that their feet are "stuck to the ground" [2] while walking or initiating gait. This increases the risk of falling, negatively affecting the patient's quality of life.

FOG can be triggered by simple activities such as gait initiation, walking through a doorway, encountering obstacles in the pathway, or even performing a dual-task while walking [3,4]. Therefore, accurate and timely detection of FOG can significantly enhance the quality of life for PD patients. An automatic prediction of FOG can provide neurologists with relevant indicators about the condition and its evolution [5]. Furthermore, freezing episodes can be mitigated or prevented with external intervention, such as visual or auditory cues, activated by predicting the onset of FOG.

FOG detection is still a widely researched topic, with attempts made by several combinations of devices and algorithms. The first automatic detection of FOG was proposed by Moore et al. [6] using frequency-based features from accelerometer signals. This work was extended to improve the detection accuracy and developed as a FOG monitoring system with smartphones and wearable accelerometers [7]. Accelerometers [8,9], gyroscopes [8] and inertial data [10,11] have been employed for automatic FOG detection [9].

Several researchers attempted the early detection of FOG as it benefits intervention strategies. Handojoseno et al. [12] detected the onset of FOG based on EEG wavelet energy and entropy features. The onset of FOG was also detected by a sensor placed on the lower limb of the patient [13]. Electrocardiogram and skin conductance were used to predict the onset of FOG [14].

An efficient FOG detection system should not only be able to predict the onset of FOG, but it should also be able to distinguish involuntary stopping from the transition to voluntary stopping. The potential of brain dynamics in discerning the onset of freezing in PD patients has already been established [12]. However, earlier researchers relied on a 5 s window to discern the transition to freezing from normal walking [15,16].

Recently, we discovered that EEG signatures for transition to FOG are distinct from the intention to stop [17]. We observed an increase in the delta, theta, and beta power at the central region during the transition to freezing compared to normal walking. The transition to voluntary stopping was observed to show increased EEG power at the frontal, central, parietal, and occipital regions compared to the transition to FOG. Accurate detection of the transition to FOG from potential confounding transitions to voluntary stopping and normal walking is still challenging because of the complexity of designing handcrafted features.

In order to determine the differences in brain activity during the transition to freezing and the transition to stopping, we employed the classical timed up-and-go (TUG) task [18] in our experiment. The TUG protocol involves the participants starting from a sitting position to standing before walking towards an identified point and then turning back there to return to the starting position. This sequence of steps followed in the TUG protocol elicits freezing episodes, particularly when performed in that order [19]. In order to incorporate the transition to stopping in our experiment, we included TUG trials with voluntary stop conditions. In the TUG trials with voluntary stop, the participants were verbally instructed to "stop". We contrasted the brain dynamics of the patients while walking normally, transitioning to freezing, transitioning to voluntary stopping, and during freezing episodes and voluntarily stopping. Discerning the brain dynamics during the transition to freezing from normal walking or the transition to voluntary stopping could pave the way towards improved therapeutics that accurately predict the possibility of freezing while excluding potentially confounding voluntary stopping instances.

Our aim is to perform automatic feature learning and distinguish between normal gait, transition to FOG, and transition to voluntary stopping. Deep learning (DL) methods are feature learning methods that are not constrained by the engineering ability of handcrafted features or the complexity of the data representation. ConvNets are a type of feed-forward deep neural network, which typically combines convolutional layers with traditional dense layers to reduce the number of weights composing the model. The proposed system eliminates the need to extract features and feature selection manually. We evaluated three classical convolutional neural network (CNN) models: EEGNet [20], Shallow ConvNet [21], and Deep ConvNet [21] to detect FOG.

2. Materials and Methods

2.1. Subjects

Seventeen patients from the Parkinson's Disease Research Clinic at the Brain and Mind Centre, University of Sydney, participated in this study. The University of Sydney Ethics Committee provided ethics approval for this experiment (HREC approval number: 2014/255). All participants for the study were chosen based on the score for the third item on the self-reported FOG questionnaire and assessment of a clinical specialist. The mean age of the participants was 64 ± 7.25 years, and none had any depression or dementia, as assessed by neurologist Simon J. G. Lewis using the DSM-IV criteria. Furthermore, the participants had a Mini-Mental State Examination (MMSE) score ≥ 24 and fulfilled the UK Parkinson's Disease Society Brain Bank (UKPDSBB) criteria [15,17]. Furthermore, these participants had a varying severity and frequency of freezing and, when in their practically defined off period, having withdrawn PD medications overnight, had an MDS Unified Parkinson's Disease Rating Scale III stage of 40.10 ± 12.21 and a Hoehn and Yahr stage of 2.34 ± 0.73.

2.2. Experimental Design

The patients were in their off state, having had no medications for at least 12 h when they participated in this study. They performed the TUG task, starting with the participants seated. The participants were instructed to stand up and walk towards a target location in a large corridor. The target location was marked on the floor using a box with dimensions of 0.6 m × 0.6 m, positioned 6 m away from the starting position to allow for multiple FOG episodes. The participants were instructed to turn within the marked box. In the TUG tasks, the participants were asked to perform either a 180° or a 540° turn. Turning within a box elicits freezing episodes in PD patients. The researcher initially demonstrated the task and the direction to turn within the box, and the participants followed the researcher's example. The experiment was video recorded and reviewed by two clinical researchers to identify freezing episodes.

As described in [17], we considered two variants of the TUG task: the classical TUG task and the TUG task with a voluntary stopping element. As shown in Figure 1A, the classical TUG task was employed as it elicits freezing in PD patients [19]. We considered two seconds immediately preceding the freezing episode as the transition to freezing. The period of two seconds before this transition period was regarded as normal walking.

In the TUG tasks with the voluntary stopping element (Figure 1B), the researcher guided the participant in the voluntary stopping by providing verbal instructions such as "stop" and "walk". In these TUG trials, the target box was located 10 m away from the starting position. The box was located 10 m from the starting position in these TUG tasks so as to prevent participants from anticipating exactly when they might receive the instruction to stop walking. Furthermore, verbal instructions to stop walking were generally provided to the participants while they were walking back to the chair after turning inside the box. The participants were required to stop as soon as they heard the researcher say "stop", and they resumed walking when the researcher said "walk", usually in 5–10 s. We defined two seconds when the researcher said "stop" and the participants were preparing to stop walking as the transition to voluntary stop. We also considered two seconds before the "stop" instruction as normal walking. The participants were randomly asked to perform a standard TUG task or TUG task with voluntary stopping to avoid any habituation effects.

Even though we strived to have an equal number of normal walking, transition to FOG, and voluntary stopping, this was not accomplished. The well-being of the patients was given the top priority, and we stopped the experiment for any PD patients who expressed difficulty in continuing with the experiment. Hence, we could not collect an equal number of trials for normal walking, transitions to FOG, or voluntary stopping. Table 1 shows the number of normal walking trials, the number of transitions to FOG, and the transitions to voluntary stopping for each participant.

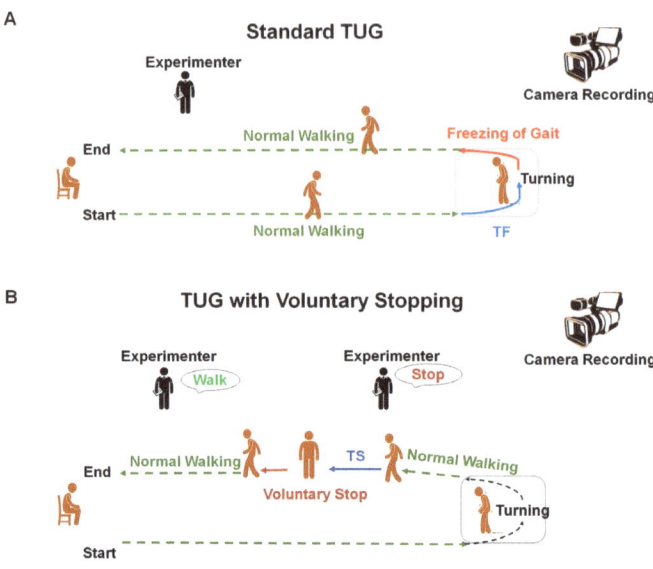

Figure 1. Experimental paradigm of (**A**) standard TUG task. (**B**) TUG task with verbal instructions to "stop" and "walk" to facilitate voluntary stopping. TF denotes the transition to FOG, and TS denotes the transition to voluntary stopping.

Table 1. Participant-based count of normal walking, transition to FOG, and transition to voluntary stopping.

Subject No.	No. of Normal Walking Epochs	No. of Transition to FOG Epochs	No. of Transition Voluntary Stopping Epochs
1	11	8	3
2	12	8	4
3	1	1	0
4	33	33	0
5	5	2	3
6	8	5	3
7	7	1	6
8	15	11	4
9	8	8	0
10	23	23	0
11	5	0	5
12	30	24	6
13	3	0	3
14	15	15	0
15	33	26	7
16	7	1	6
17	17	12	5

2.3. Equipment

The EEG data were collected from the participants using a 32-channel BioSemi Active-Two system (Biosemi Systems, Amsterdam, The Netherlands). The placement of the electrodes was per the International 10–20 system. The patient's skin was prepared by washing with 70% isopropyl alcohol, and data were recorded at a 500 Hz sampling rate. The clinical researchers used ELAN tagging software [22] to tag the precise time of each freezing episode, and the events' information was later imported to EEG manually.

2.4. EEG Processing

The EEGLAB toolbox [23] was used for processing EEG data, as shown in Figure 2. The raw EEG was band-pass filtered between 1 and 30-Hz to eliminate low- and high-frequency noises. The line noise was removed with the pop_cleanline function in EEGLAB. Further, channels with at least three seconds of flatlines were corrected with clean_flatlines functions, and all channels were cleaned with clean_channels. There were 3 ± 0.5 channels removed on average, and these channels were interpolated. Afterwards, normal walking, transition to FOG, and transition to voluntary stopping trials were extracted to provide input to the deep learning models.

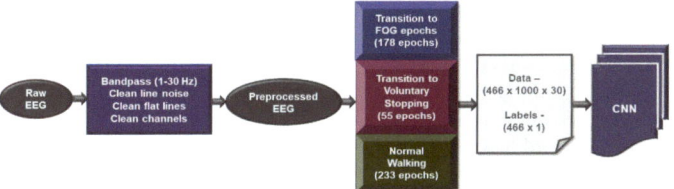

Figure 2. EEG preprocessing and feature extraction for DL models.

In this study, a total of 178 trials of transition to FOG episodes, 55 transitions to voluntary stopping, and 233 trials of normal walking were extracted from the continuous EEG data of 17 subjects. These transitions to FOG were extracted from both standard TUG and TUG tasks for voluntary stopping based on the occurrence of FOG episodes, and each trial was 2 seconds in length. These data were reformatted to a matrix with the shape number of trials × time points × number of channels format ($466 \times 1000 \times 30$) before providing it to the DL models.

We also performed two-class classifications with the transitions to FOG and the transition to voluntary stopping. This matrix was with the shape number of trials × time points × number of channels format ($233 \times 1000 \times 30$) before providing to the DL models.

We performed a grid search to select the optimal hyperparameters for the three CNN models: EEGNet, Shallow ConvNet, and Deep ConvNet. The data were shuffled and randomly divided into three separate sets, the training (60%), validation (20%), and testing sets (20%). The performances of these models were obtained by 5-fold cross-validation. We also performed leave-one-subject-out cross-validation to evaluate the performance of these models.

3. Results and Discussion

Classification Performance

For two-class classification, all models achieved acceptable performances with high sensitivity and specificity, as shown in Table 2. Leave-one-subject-out (LOSO) classification results for the transition to FOG vs. transition to voluntary stopping are shown in Table 3. The undesirable coh-kappa values might be due to the unbalanced classes. Further, all models achieve acceptable performance for three-class classification with high sensitivity and specificity, as shown in Table 4. LOSO classification results for the three-class problem are shown in Table 5.

Our models were trained with a relatively small dataset with an unbalanced number of trials in the three classes, which might have adversely affected the performance of these data-hungry models. Deep ConvNet performed better than EEGNet or Shallow ConvNet because of its greater depth, while EEGNet and Shallow ConvNet might have been more susceptible to noise from the raw EEG data, as they are compact and shallower, degrading the features and resulting in poorer performance.

In earlier works, the period of four seconds prior to freezing was considered as the transition to freezing [15]; however, we considered a shorter window for the transition to freezing [17]. This shorter transition period might have more clinical practicality as brain dynamics dynamically change within a short period. Therefore, by employing a sliding window of two seconds, our results demonstrate that we might be able to identify the transition to freezing from normal walking or the transition to voluntary or intentional stopping. Further, we considered the raw EEG data with minimal data processing to allow for real-time prediction of the onset of FOG. However, employing sophisticated advancements in EEG analysis methods might further improve classification accuracies.

We employed CNN models to detect the transition to FOG or the onset of FOG from the potential confounding intention to stop and normal walking conditions. The distinct brain dynamics during the transitions to FOG and voluntary stopping episodes were exploited in a system design. These classification models are valuable in developing compensatory systems that preserve and advance alternate neural pathways to assist gait in PD patients. Therefore, with an accurate and reliable prediction of freezing, cueing strategies to redirect attention or prompt movement can help alleviate gait impairment in PD patients [24].

Further, the advances in wearable technology have made it possible to deliver a comfortable cueing system for PD patients [25,26]. However, despite the development of several FOG prediction models [14–16], accurately and reliably detecting the onset of freezing remains an open challenge. It is also crucial to avoid the confounding transitions to voluntarily stopping or normal walking to ensure a robust prediction of freezing onset.

Our findings demonstrated the potential of EEG data in distinguishing FOG onset from normal gait or initiation of voluntary stopping. Our results will pave the way toward therapeutic prediction and mitigation of freezing in PD patients. Further, these results aid and promote investigations of intentional stopping during gait, as a reliable prediction of intention could be valuable for motor rehabilitation.

Table 2. Five-fold classification performance for transition to FOG vs. transition to voluntary stopping.

Model	Accuracy	F1-Score	Coh-Kappa	Sensitivity	Specificity
EEGNet	88.09 ± 4.25%	80.09 ± 4.62%	68.30 ± 2.50%	94.42 ± 4.65%	96.21 ± 3.52%
Shallow ConvNet	89.9 ± 2.31%	89.21 ± 3.94%	70.11 ± 3.91%	96.49 ± 2.97%	94.36 ± 3.60%
Deep ConvNet	92.28 ± 2.70%	93.02 ± 2.03%	72.94 ± 2.27%	96.89 ± 2.04%	96.91 ± 2.09%

Table 3. LOSO classification performance for transition to FOG vs. transition to voluntary stopping.

Model	Accuracy	F1-Score	Coh-Kappa	Sensitivity	Specificity
EEGNet	87.28 ± 5.89%	87.61 ± 5.53%	69.19 ± 4.37%	84.89 ± 5.72%	84.16 ± 4.71%
Shallow ConvNet	87.92 ± 4.3%	82.16 ± 3.02%	71.14 ± 4.84%	86.23 ± 3.71%	85.55 ± 4.62%
Deep ConvNet	87.83 ± 5.35%	84.81 ± 5.86%	70.6 ± 5%	86.37 ± 3.31%	84.72 ± 2.49%

Table 4. Five-fold classification performance for transition to FOG vs. transition to voluntary stopping vs. normal walking.

Model	Accuracy	F1-Score	Coh-Kappa	Sensitivity	Specificity
EEGNet	71.92 ± 5.64%	69.49 ± 5.38%	52.57 ± 4.63%	87.8 ± 5.90%	84.02 ± 4.06%
Shallow ConvNet	73.68 ± 3.87%	73.53 ± 3.76%	57.14 ± 4.53%	89.28 ± 4.59%	86.2 ± 3.37%
Deep ConvNet	75.43 ± 1.48%	72.52 ± 1.44%	58.11 ± 1.64%	92.85 ± 1.70%	75.86 ± 1.75%

Table 5. LOSO Classification performance for transition to FOG vs. transition to voluntary stopping vs. normal walking.

Model	Accuracy	F1-Score	Coh-Kappa	Sensitivity	Specificity
EEGNet	70.85 ± 3.25%	70.79 ± 3.86%	52.54 ± 5.89%	83.83 ± 5.65%	82.80 ± 4.13%
Shallow ConvNet	73.45 ± 3.69%	72.84 ± 3.61%	54.43 ± 4.92%	88.91 ± 5.08%	86.34 ± 5.62%
Deep ConvNet	74.65 ± 4.19%	71.54 ± 4.7%	57.52 ± 3.42%	91.18 ± 5.04%	74.46 ± 4.79%

4. Conclusions

In this study, we investigated the application of CNN to an end-to-end classification of transitions to FOG, voluntary stopping, and normal walking. The model automatically learns the discriminative features for classifying normal walking, transitions to FOG, and voluntary stop. Furthermore, the convolutional neural network approach removed the need for feature extraction and selection. This research is the first of its kind, and the reported classification model could pave the way to detecting the onset of FOG precisely and effectively. As the transitions to FOG can be accurately distinguished from the transition to voluntary stopping with just a two-second window, this could enable appropriate interventions (e.g., cueing) to help the patient avoid freezing. Further, a larger dataset can improve the performance of the models, and future studies should investigate the real-time FOG detection performance. This work will expedite the development of future therapeutic interventions that can reliably predict freezing episodes in PD patients. Future interventions for FOG must diligently eliminate all false positives from the confounding voluntary stopping.

Author Contributions: Conceptualization, H.-T.C., K.E.M., M.G. (Matthew Georgiades), M.G. (Moran Gilat), H.T.N., S.J.G.L. and C.-T.L.; Methodology, A.R.J., K.E.M., M.G. (Matthew Georgiades), M.G. (Moran Gila), S.J.G.L. and C.-T.L.; Formal analysis, A.R.J. and Z.C.; Investigation, K.E.M., M.G. (Matthew Georgiades), H.T.N., S.J.G.L. and C.-T.L.; Data curation, K.E.M., M.G. (Matthew Georgiades), M.G. (Moran Gilat) and C.-T.L.; Writing—original draft, A.R.J.; Writing—review & editing, A.R.J., Z.C., H.-T.C., K.E.M., H.T.N., S.J.G.L. and C.-T.L.; Supervision, C.-T.L.; Project administration, K.E.M., H.T.N., S.J.G.L. and C.-T.L.; Funding acquisition, H.T.N., S.J.G.L. and C.-T.L. All authors have read and agreed to the published version of the manuscript.

Funding: This research was partly funded by the Australian Research Council (ARC) under discovery grants DP180100670, DP180100656 and DP210101093. The research was also partially sponsored by the Australia Defence Innovation Hub under Contract No. P18-650825, US Office of Naval Research Global under Cooperative Agreement Number ONRG - NICOP - N62909-19-1- 058, and AFOSR – DST Australian Autonomy Initiative agreement ID10134, and AFOSR Grant No. FA2386-22-1-0042. We also thank the NSW Defence Innovation Network and the NSW State Government of Australia for financial support in part of this research through grant DINPP2019 S1-03/09.

Institutional Review Board Statement: The University of Sydney Ethics Committee provided ethics approval for this study (HREC approval number: 2014/255).

Informed Consent Statement: Informed consent was obtained from all subjects involved in the study.

Data Availability Statement: The data will be made available upon reasonable request to the corresponding author.

Conflicts of Interest: The authors declare no conflict of interest.

References

1. Nieuwboer, A.; Giladi, N. Characterizing freezing of gait in Parkinson's disease: Models of an episodic phenomenon. *Mov. Disord.* **2013**, *28*, 1509–1519. [CrossRef] [PubMed]
2. Giladi, N.; Kao, R.; Fahn, S. Freezing phenomenon in patients with parkinsonian syndromes. *Mov. Disord. Off. J. Mov. Disord. Soc.* **1997**, *12*, 302–305. [CrossRef] [PubMed]

3. Bloem, B.R.; Hausdorff, J.M.; Visser, J.E.; Giladi, N. Falls and freezing of gait in Parkinson's disease: A review of two interconnected, episodic phenomena. *Mov. Disord. Off. J. Mov. Disord. Soc.* **2004**, *19*, 871–884. [CrossRef] [PubMed]
4. Moore, O.; Peretz, C.; Giladi, N. Freezing of gait affects quality of life of peoples with Parkinson's disease beyond its relationships with mobility and gait. *Mov. Disord. Off. J. Mov. Disord. Soc.* **2007**, *22*, 2192–2195. [CrossRef]
5. Del Din, S.; Godfrey, A.; Mazzà, C.; Lord, S.; Rochester, L. Free-living monitoring of Parkinson's disease: Lessons from the field. *Mov. Disord.* **2016**, *31*, 1293–1313. [CrossRef]
6. Moore, S.T.; MacDougall, H.G.; Ondo, W.G. Ambulatory monitoring of freezing of gait in Parkinson's disease. *J. Neurosci. Methods* **2008**, *167*, 340–348. [CrossRef]
7. Mazilu, S.; Hardegger, M.; Zhu, Z.; Roggen, D.; Tröster, G.; Plotnik, M.; Hausdorff, J.M. Online detection of freezing of gait with smartphones and machine learning techniques. In Proceedings of the 2012 6th International Conference on Pervasive Computing Technologies for Healthcare (PervasiveHealth) and Workshops, San Diego, CA, USA, 21–24 May 2012; pp. 123–130.
8. Zhao, Y.; Tonn, K.; Niazmand, K.; Fietzek, U.M.; D'Angelo, L.T.; Ceballos-Baumann, A.; Lueth, T.C. Online FOG identification in Parkinson's disease with a time-frequency combined algorithm. In Proceedings of the 2012 IEEE-EMBS International Conference on Biomedical and Health Informatics, Hong Kong, China, 5–7 January 2012; pp. 192–195.
9. Tripoliti, E.E.; Tzallas, A.T.; Tsipouras, M.G.; Rigas, G.; Bougia, P.; Leontiou, M.; Konitsiotis, S.; Chondrogiorgi, M.; Tsouli, S.; Fotiadis, D.I. Automatic detection of freezing of gait events in patients with Parkinson's disease. *Comput. Methods Programs Biomed.* **2013**, *110*, 12–26. [CrossRef]
10. Rodríguez-Martín, D.; Samà, A.; Pérez-López, C.; Català, A.; Arostegui, J.M.M.; Cabestany, J.; Bayés, A.; Alcaine, S.; Mestre, B.; Prats, A.; et al. Home detection of freezing of gait using support vector machines through a single waist-worn triaxial accelerometer. *PLoS ONE* **2017**, *12*, e0171764. [CrossRef]
11. Samà, A.; Rodríguez-Martín, D.; Pérez-López, C.; Català, A.; Alcaine, S.; Mestre, B.; Prats, A.; Crespo, M.C.; Bayés, A. Determining the optimal features in freezing of gait detection through a single waist accelerometer in home environments. *Pattern Recognit. Lett.* **2018**, *105*, 135–143. [CrossRef]
12. Handojoseno, A.A.; Shine, J.M.; Nguyen, T.N.; Tran, Y.; Lewis, S.J.; Nguyen, H.T. The detection of Freezing of Gait in Parkinson's disease patients using EEG signals based on Wavelet decomposition. In Proceedings of the 2012 Annual International Conference of the IEEE Engineering in Medicine and Biology Society, San Diego, CA, USA, 28 August–1 September 2012; pp. 69–72.
13. Coste, C.A.; Sijobert, B.; Pissard-Gibollet, R.; Pasquier, M.; Espiau, B.; Geny, C. Detection of freezing of gait in Parkinson disease: Preliminary results. *Sensors* **2014**, *14*, 6819–6827. [CrossRef]
14. Mazilu, S.; Calatroni, A.; Gazit, E.; Mirelman, A.; Hausdorff, J.M.; Tröster, G. Prediction of freezing of gait in Parkinson's from physiological wearables: An exploratory study. *IEEE J. Biomed. Health Inform.* **2015**, *19*, 1843–1854. [CrossRef] [PubMed]
15. Handojoseno, A.M.A.; Shine, J.M.; Nguyen, T.N.; Tran, Y.; Lewis, S.J.G.; Nguyen, H.T. Analysis and Prediction of the Freezing of Gait Using EEG Brain Dynamics. *IEEE Trans. Neural Syst. Rehabil. Eng.* **2015**, *23*, 887–896. [CrossRef] [PubMed]
16. Handojoseno, A.M.A.; Shine, J.M.; Nguyen, T.N.; Tran, Y.; Lewis, S.J.G.; Nguyen, H.T. Using EEG spatial correlation, cross frequency energy, and wavelet coefficients for the prediction of Freezing of Gait in Parkinson's Disease patients. In Proceedings of the 2013 35th Annual International Conference of the IEEE Engineering in Medicine and Biology Society (EMBC), Osaka, Japan, 3–7 July 2013; pp. 4263–4266.
17. Cao, Z.; John, A.R.; Chen, H.T.; Martens, K.E.; Georgiades, M.; Gilat, M.; Nguyen, H.T.; Lewis, S.J.; Lin, C.T. Identification of EEG dynamics during freezing of gait and voluntary stopping in patients with Parkinson's disease. *IEEE Trans. Neural Syst. Rehabil. Eng.* **2021**, *29*, 1774–1783. [CrossRef] [PubMed]
18. Morris, S.; Morris, M.E.; Iansek, R. Reliability of measurements obtained with the Timed "Up and Go" Test in people with Parkinson disease. *Phys. Ther.* **2001**, *81*, 810–818. [CrossRef]
19. Zampieri, C.; Salarian, A.; Carlson-Kuhta, P.; Aminian, K.; Nutt, J.G.; Horak, F.B. The instrumented timed up and go test: Potential outcome measure for disease modifying therapies in Parkinson's disease. *J. Neurol. Neurosurg. Psychiatry* **2010**, *81*, 171–176. [CrossRef]
20. Lawhern, V.J.; Solon, A.J.; Waytowich, N.R.; Gordon, S.M.; Hung, C.P.; Lance, B.J. EEGNet: A compact convolutional neural network for EEG-based brain–computer interfaces. *J. Neural Eng.* **2017**, *15*, 056013. [CrossRef]
21. Schirrmeister, R.T.; Springenberg, J.T.; Fiederer, L.D.J.; Glasstetter, M.; Eggensperger, K.; Tangermann, M.; Hutter, F.; Burgard, W.; Ball, T. Deep learning with convolutional neural networks for EEG decoding and visualization. *Hum. Brain Mapp.* **2017**, *38*, 5391–5420. [CrossRef]
22. Brugman, H.; Russel, A.; Nijmegen, X. Annotating Multi-mediaMulti-modal Resources with ELAN. In Proceedings of the 4th International Conference on Language Resources and Language Evaluation (LREC), Lisbon, Portugal, 26–28 May 2004; pp. 2065–2068.
23. Delorme, A.; Makeig, S. EEGLAB: An open source toolbox for analysis of single-trial EEG dynamics including independent component analysis. *J. Neurosci. Methods* **2004**, *134*, 9–21. [CrossRef]
24. Morris, M.E.; Iansek, R.; Kirkwood, B. A randomized controlled trial of movement strategies compared with exercise for people with Parkinson's disease. *Mov. Disord.* **2009**, *24*, 64–71. [CrossRef]
25. Espay, A.J.; Bonato, P.; Nahab, F.B.; Maetzler, W.; Dean, J.M.; Klucken, J.; Eskofier, B.M.; Merola, A.; Horak, F.; Lang, A.E.; et al. Technology in Parkinson's disease: Challenges and opportunities. *Mov. Disord.* **2016**, *31*, 1272–1282. [CrossRef]

26. Ginis, P.; Nieuwboer, A.; Dorfman, M.; Ferrari, A.; Gazit, E.; Canning, C.G.; Rocchi, L.; Chiari, L.; Hausdorff, J.M.; Mirelman, A. Feasibility and effects of home-based smartphone-delivered automated feedback training for gait in people with Parkinson's disease: A pilot randomized controlled trial. *Park. Relat. Disord.* **2016**, *22*, 28–34. [CrossRef] [PubMed]

Disclaimer/Publisher's Note: The statements, opinions and data contained in all publications are solely those of the individual author(s) and contributor(s) and not of MDPI and/or the editor(s). MDPI and/or the editor(s) disclaim responsibility for any injury to people or property resulting from any ideas, methods, instructions or products referred to in the content.

Article

Effect of Carotid Stenosis Severity on Patterns of Brain Activity in Patients after Cardiac Surgery

Irina Tarasova *, Olga Trubnikova, Darya Kupriyanova, Irina Kukhareva, Irina Syrova, Anastasia Sosnina, Olga Maleva and Olga Barbarash

Department of Clinical Cardiology, Research Institute for Complex Issues of Cardiovascular Diseases, 6, Sosnoviy Blvd., 650002 Kemerovo, Russia
* Correspondence: iriz78@mail.ru

Abstract: Background: The negative effects of high-grade carotid stenosis on the brain are widely known. However, there are still insufficient data on the brain state in patients with small carotid stenosis and after isolated or combined coronary and carotid surgery. This EEG-based study aimed to analyze the effect of carotid stenosis severity on associated brain activity changes and the neurophysiological test results in patients undergoing coronary artery bypass grafting (CABG) with or without carotid endarterectomy (CEA). Methods: One hundred and forty cardiac surgery patients underwent a clinical and neuropsychological examination and a multichannel EEG before surgery and 7–10 days after surgery. Results: The patients with CA stenoses of less than 50% demonstrated higher values of theta2- and alpha-rhythm power compared to the patients without CA stenoses both before and after CABG. In addition, the patients who underwent right-sided CABG+CEA had generalized EEG "slowdown" compared with isolated CABG and left-sided CABG+CEA patients. Conclusions: The on-pump cardiac surgery accompanied by specific re-arrangements of frequency–spatial patterns of electrical brain activity are dependent on the degree of carotid stenoses. The information obtained can be used to optimize the process of preoperative and postoperative management, as well as the search for neuroprotection and safe surgical strategies for this category of patients.

Keywords: carotid stenosis; brain electrical activity; EEG; postoperative cognitive dysfunction; coronary artery bypass grafting; carotid endarterectomy

1. Introduction

According to the World Health Organization, cardiovascular diseases (CVD), mainly associated with atherosclerosis, are the leading causes of death worldwide, including in Russia [1]. The Siberian region shows less favorable CVD epidemiology. Various climatic and ecological conditions of the region contribute to the high prevalence of this pathology [2]. Atherosclerosis often affects multiple vascular basins simultaneously. Significant atherosclerotic lesions of several vascular basins determine the severity of the disease, making it difficult to choose the optimal treatment strategy and calling into question the positivity of the prognosis, in particular, coronary artery disease.

The global population of elderly people has been increasing every year, and the ageing of the population has posed new and complex challenges for health professionals not only to increase life expectancy but also to maintain its quality. A high standard of quality of life cannot be reached without preserving a person's intellectual functions. It is known that with age, cognitive functions diminish, and cognitive impairment (CI) develops in the form of memory loss, attention and executive impairment, etc. [3–5].

Cognitive disorders associated with cerebral and coronary atherosclerosis (vascular CI) are widespread among older persons and are more severe than age-related cognitive changes [6–8]. Previous studies have revealed significant interactions between cognitive disorders developing in the elderly and senile age, atherosclerotic changes in cerebral vessels, and accompanying disorders of cerebral blood flow [9,10]. There is evidence that age-related structural and functional changes in arteries, arterioles and capillaries lead to dysregulation of cerebral blood flow and ischemia, leading to disruption of the blood–brain barrier. Additionally, metabolic disorders are developed with reduced delivery of energy substrates to neurons and excretion of by-products of the protein breakdown, increasing neuroinflammation and paracrine regulation dysfunction [11,12]. It is suggested that the atherosclerotic remodeling of the brain vessels can lead to an accelerated progression of brain dysfunction [11]. In this case, carotid artery (CA) stenosis is one of the factors affecting self-regulation of brain perfusion [13]. It has been found that patients with vascular CI often show a decrease in blood flow velocity in the cerebral cortex, especially in the frontal and parietal regions [14,15]. These brain regions are known to be the watersheds of the blood supply, at the boundaries between the vascular pools [16–18]. These zones are more disadvantaged than any other brain region in the case of systolic and/or diastolic dysfunction of the left ventricle, valvular pathology and atrial fibrillation accompanying cardiovascular pathology, as well as during cardiac surgery [11,19].

There is a wide variety of epidemiological and clinical data on vascular CI, but only a few studies have examined changes in the neurophysiological parameters of cardiac surgery patients [20,21]. At the same time, early manifestations of vascular and postoperative CI are subclinical and are detected only using an extended neurophysiological examination. In this regard, careful attention should be paid to the identification of objective and sensitive criteria for early diagnosis of CI in cardiac surgery patients. It is generally accepted that the electroencephalogram (EEG) rhythms reflect the activity of the neural network to be placed under recording electrode [22,23]. As a consequence, the changes in EEG rhythms may be early indicators of structural and functional abnormalities in neural networks associated with vascular and postoperative CI.

Previous studies have shown that the frequency–spatial pattern of brain electrical activity in patients with vascular CI has specific features [8,24,25]. The association between poststroke alpha slowing and CI, which may be mediated by attentional dysfunction, was revealed [24]. Al-Qazzaz et al. [25] studied the discriminatory characteristics of patients with vascular CI and healthy individuals using non-linear EEG analysis methods. It was found that the degree of EEG irregularity and complexity was significantly lower in patients with vascular CI compared to control subjects. We previously showed that a theta activity increase in the frontal and occipital sites, as well as high theta/alpha ratios, may be considered as the earliest EEG markers of vascular cognitive disorders [8]. Moretti et al. proposed several promising EEG markers that could be important in the differential diagnosis of vascular and neurodegenerative CI. The alpha3/alpha2 and theta/gamma indices showed prognostic significance for the progression of the neurodegenerative type of CI [26–28]. The changes in the electrical activity of neurons in the post-stroke period proved to be promising in the search for prognostic markers of clinical recovery in patients with ischemic brain damage. A study by Zappasodi et al. found that a bilateral increase in low-frequency activity and a decrease in hemispheric asymmetry in the acute phase of a unilateral stroke in the middle cerebral artery basin predicts a worse functional outcome in the future [29].

However, there is still insufficient information on the modification of the brain electrical activity in cardiac surgery patients. Cardiac surgery has been shown to be associated with local or diffuse brain damage [21,30–32]. It is assumed that chronic cerebral ischemia in patients with cardiovascular diseases, as well as episodes of acute ischemia that occur during on-pump cardiac surgery, can contribute to specific changes in the brain's electrical activity. Our previous studies have shown that EEG patterns associated with coronary artery bypass grafting (CABG) have specific features, depending on the presence of preoperative CI or cognitive decline in the early postoperative period [21,33]. We found that the presence of early POCD was accompanied by negative postoperative dynamics of EEG parameters with the increase in low-frequency activity. Skhirtladze-Dworschak et al. found that the occurrence of nonconvulsive status epilepticus after open cardiac surgery is associated with mitigating secondary brain injury [34].

Thus, recent studies have shown that the patterns of brain activity are associated with perioperative brain damage in cardiac surgery patients. However, the role of the severity of carotid stenosis in the development of the postoperative changes in brain activity and cognitive functions is uncertain. It has previously been shown that hemodynamically significant stenoses of CA (70–99%) can be a risk factor for brain damage during cardiac surgery [31,35]. However, little is known about the effects of small stenoses of CA (<50%) on the state of the brain in cardiac surgery patients. There have been several research studies into the negative effects of asymptomatic stenosis of CA on the state of the brain after cardiac surgery [33,35]. This has resulted in the perception that CA stenoses of less than 50% are hemodynamically insignificant. Therefore, this has led to insufficient attention being paid to preoperative management and intraoperative brain protection in patients with CA stenoses of less than 50%.

There are some data in the literature about the serious neurological complications (stroke, postoperative delirium, etc.) that occur in the group of patients with hemodynamically significant stenoses [36,37]. Research studies about the brain activity changes associated with postoperative cognitive decline in patients with stenoses of the coronary and carotid arteries are rare, especially after simultaneous cardiac surgery. It is important to note that the intraoperative episodes of brain ischemia during combined coronary and carotid revascularization does not necessarily lead to brain damage such as stroke. Meanwhile, less pronounced, diffuse ischemic brain damage may have a significantly higher frequency. Further, this may lead to a decline in cognitive functions and complicate the postoperative management of patients undergoing combined cardiac surgery.

In this paper, we will analyze the effect of carotid stenosis severity on associated EEG changes and the results of neurophysiological examination, including the frequency and structure of CI, in patients undergoing cardiac surgery (isolated CABG and combined CABG and carotid endarterectomy (CEA)) in the early postoperative period.

2. Materials and Methods

2.1. Subjects

This study was a prospective, observational cohort investigation. From a cohort of patients who underwent on-pump coronary surgery in the clinic of the Research Institute for Complex Issues of Cardiovascular Diseases, a sample of 140 subjects was selected. All of the patients met the study criteria and signed an informed consent form. The isolated CABG group consisted of 86 patients, 29 of whom had unilateral CA stenoses of less than 50%. The CABG+CEA group were divided into two groups: the group of left-sided CEA+CABG (n = 30) and the group of right-sided CEA+CABG (n = 24) (see Figure 1).

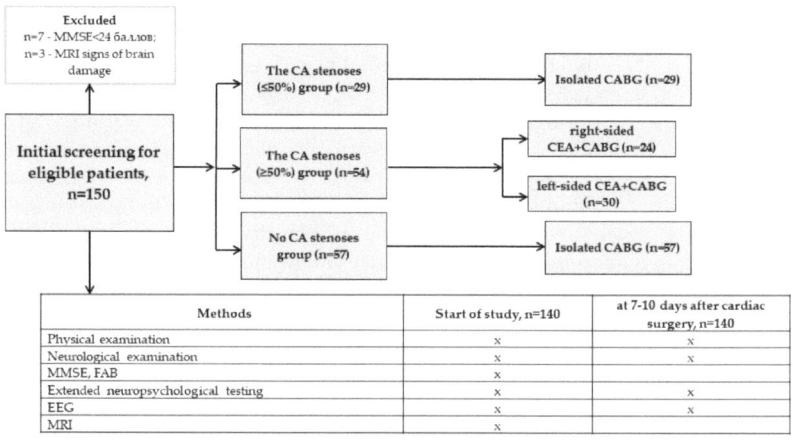

Figure 1. Overview of the study design.

The inclusion criteria were as follows: Aged between 45 and 74 years and elective isolated on-pump CABG or combined CABG and CEA. Only right-handed subjects were included in the study to avoid any influence on cognitive status and EEG data regarding the factor of laterality.

The exclusion criteria were the presence of pathological changes in the central nervous system, as indicated by the results of multi-layered spiral computed tomography; depressive symptoms, as identified by the Beck Depression Inventory (BDI-II) (sum scores ≥ 8); dementia, as indicated by the Mini-Mental State Examination (MMSE) (sum scores ≤ 24) and Frontal Assessment Battery (FAB) (sum scores ≤ 11); life-threatening arrhythmias; functional class IV heart failure, according to the New York Heart Association (FC NYHA IV) guidelines; chronic obstructive pulmonary disease; malignant pathology; diseases of the central nervous system; brain injury. Patients receiving anxiolytic therapy were also excluded from the study.

All patients underwent standardized physical, neurological, and instrumental examinations. The examiners were blind to the cognitive status of the patients. The severity of the coronary lesions was assessed using the findings of coronary angiography (Innova 3100; GE Medical Systems, Carrollton, TX, USA). Carotid artery ultrasound and echocardiography with estimation of the left ventricular ejection fraction (LVEF) were performed with the Vivid 7 ultrasound machine (GE Medical Systems).

The patients received baseline and symptomatic therapy before and after surgery, consistent with the general principles of treatment for the patients with CAD, chronic heart failure, and hypertension (National Recommendations, 2020) (see Table 1).

Table 1. The clinical and anamnestic characteristics of the patients before cardiac surgery (n = 140).

Variable	Value
Age, years, Me (Q25; Q75)	59 (56; 64)
Mini-mental state, scores, Me (Q25; Q75)	27 (26; 28)
Frontal assessment battery, scores, Me (Q25; Q75)	16 (15; 17)
BDI-II, scores, Me (Q25; Q75)	3 (2; 4)
Educational attainment, years, n (%)	
8–10	101 (72)
≥ 15	39 (28)
Functional class of angina, n (%)	
I-II	94 (67)
III	46 (33)
Functional class NYHA, n (%)	
I-II	109 (78)
III	31 (22)
History of myocardial infarction, n (%)	104 (74)
Fraction of left ventricle ejection, %, Me (Q25; Q75)	58 (54; 62)
Type 2 of diabetes mellitus, n (%)	48 (34)
Carotid arteries stenoses, n (%)	
One-sided $\leq 50\%$	29 (21)
One-sided 70–99%	7 (5)
Two-sided $\geq 50\%$	47 (34)
History of stroke, n (%)	15 (11)
Cardiopulmonary bypass time, min, Me (Q25; Q75)	90 (83; 97)
Aorta cross-clamping time, min, Me (Q25; Q75)	68 (56; 50)
Medication, n (%)	
ACEi	124 (89%)
Statin	94 (67%)
Beta-blockers	137 (98%)
Antiplatelet drugs	135 (96%)
CCB	59 (42%)
Nitrates	23 (16%)

ACEi, angiotensin-converting enzyme inhibitor; CCB, calcium channel blockers; NYHA, heart failure by the New York Heart Association.

All surgical interventions in patients of the isolated CABG and CABG+CEA groups with the use of cardiopulmonary bypass, normothermia and 25–30% hemodilution were carried out. In almost all cases, a blood pharmaco-cold cardioplegia was used. The standard anesthesia and infusion scheme was performed for all types of procedures. All stages of the surgery were accompanied by invasive hemodynamic control and real-time monitoring of cerebral cortex oxygenation (rSO2) (INVOS 3100; Somanetics, Troy, MI, USA). For simultaneous intervention (CABG+CEA), the initial stage of surgery was endarterectomy with arterial plasty and a xenopericardial patch.

2.2. Neurophysiological Assessment

The patients were assessed at baseline (1–3 days before surgery) and 7–10 days after surgery.

The cognitive screening and the extended neuropsychological test battery to evaluate three functional cognitive domains (psychomotor and executive function, attention and short-term memory) were conducted. Parallel test versions were used in repeated measurements in order to minimize learning effects. The neuropsychological test battery has been previously described [33,38]. Postoperative cognitive decline after CABG was determined by a 20% decrease in the cognitive score compared to baseline in 20% of the tests [31].

EEGs were recorded via a 62-channel Quik-cap (NeuroScan, El Paso, TX, USA). The scalp locations of the electrodes were based on the modified 10/10 System, and a nose bridge electrode was used as a reference. Bipolar eye movement electrodes were applied to the canthus and cheek bone to monitor eye movement artifacts. The EEGs were recorded using an NEUVO-64 system (NeuroScan, El Paso, TX, USA) in the eyes-closed and eyes-open conditions, in a dimly lit, soundproof, electrically shielded room, and recording lengths were about 10 min. The amplifier bandwidths were 1.0 to 50.0 Hz, and EEGs were digitized at 1000 Hz. The data were analyzed off-line using the Neuroscan 4.5 software program (Compumedics, TX, USA). We performed visual inspections for eye movements, electromyographic interferences, and other artifacts. Artifact-free EEG fragments were divided into 2 s epochs and underwent Fourier transformations. For each subject, the EEG power values were averaged within the theta1 (4–6 Hz), theta2 (6–8 Hz), alpha1 (8–10 Hz), and alpha2 (10–13 Hz) ranges [39]. The EEG power values of each channel for every subject in each band were obtained. The next step was the clustering of data recorded in 56 leads into 5 electrode zones symmetrically in the left and right hemispheres: frontal, central, parietal, occipital and temporal. The midline sites (Fpz, Fz, etc.) were excluded. The clustering of nearby electrodes was conducted to increase statistical significance.

2.3. Statistical Analysis

All data were analyzed using STATISTICA 10.0 (StatSoft, Tulsa, OK, USA). The normality of the distribution of clinical and demographic parameters was tested using the Kolmogorov–Smirnov test. Most of the clinical parameters as well as cognitive indicators were not normally distributed and were analyzed using the Wilcoxon and Mann–Whitney tests. EEG data were normalized using the logarithm transformation and further analysis of the EEG data was carried out using a repeated-measures ANOVA. Levene's test was used to assess the equality of variances for EEG variables. The Greenhouse–Geisser correction of statistical significance was used in ANOVA. Post hoc pairwise comparisons for groups of patients were performed using Newman–Keuls multiple comparison tests.

3. Results

3.1. The Effect of Small Stenoses CA ($\leq$50%) on the Postoperative Neurophysiological Changes in On-Pump CABG Patients

3.1.1. Neurophysiological Data

This analysis included 86 patients who had undergone isolated CABG. According to the results of the preoperative examination, they were divided into two groups: those with CA stenoses of less than 50% (n = 29) and those without stenoses (n = 57).

The postoperative period was standard in all the patients, without adverse cardiovascular events (intraoperative and postoperative heart attacks, strokes, life-threatening arrhythmias, bleeding, etc.).

POCD occurred in 22 (76.0%) patients with CA stenoses and in 32 (61%) patients without stenoses after isolated CABG (OR = 1.99, 95% CI = 0.77–5.18, Z = 1,42, p = 0,15). Thus, the incidence of POCD had a tendency of an increasing number of cases in the CA stenoses group.

The POCD structure consisted of a decrease in the psychomotor and executive function, as well as short-term memory in both groups. At the same time, the patients with CA stenoses made more errors in the tests of executive functions ($p \leq 0.05$), and patients without stenoses had more missed signals in the same tests. In the domain of short-term memory, between-group differences were obtained in the 10-nonsense-syllable memorizing test (p = 0.04).

3.1.2. EEG Data

For the next stage of the analysis, a repeated-measures ANOVA with a between-subjects factor of GROUP (two levels: with CA stenoses of less than 50%/without stenoses), and within-subjects factors of EXAMINATION TIME (two levels: before/after surgery), AREA (five levels: frontal, central, parietal, occipital and temporal), and LATERALITY (two levels: left/right hemisphere) was conducted. The significant factors and interactions associated with the GROUP factor are found in the theta2, alpha1 and alpha2 EEG ranges.

The statistically significant interactions of the factors GROUP × EXAMINATION TIME ($F_{1,84}$ = 4.95, p = 0.03) and GROUP × EXAMINATION TIME × AREA × LATERALITY ($F_{4,336}$ = 3.54, p = 0.02) were found in the theta2 range of EEG resting state with eyes closed. The patients with CA stenoses had higher values of the theta2-rhythm power at 7–10 days after CABG in comparison to the patients without stenoses (Figure 2). In addition, the CA stenoses group had higher values of rhythm power in the left hemisphere in the frontal and centroparietal cortical regions and in the right hemisphere in all sites, except for the occipital regions.

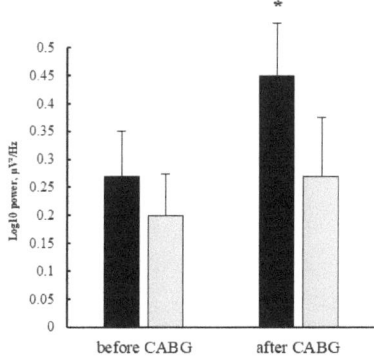

Figure 2. Differences in the theta2-rhythm power of EEG resting state with eyes closed in patients who underwent on-pump CABG, depending on the presence of CA stenoses less than 50%: dark columns—the patients with CA stenoses, light columns—the patients without CA stenoses, error bars denote SE, *—p < 0.05 Newman–Keuls multiple comparison test.

The significance of the GROUP factor was obtained in the theta2, alpha1 and alpha2 frequency ranges of EEG resting state with eyes open ($F_{1,84}$ = 4.68, p = 0.034; $F_{1,84}$ = 3.88, p = 0.05 and $F_{1,84}$ = 4.96, p = 0.029, respectively). The patients with CA stenoses had higher power values of these rhythms compared to patients without stenoses before and after cardiac surgery.

Additionally, the analysis of EEG resting state with open eyes revealed a statistically significant interaction of the factors GROUP × EXAMINATION TIME × AREA × LATERALITY ($F_{4,336}$ = 2.77, p = 0.04) in the alpha2 frequency range. Before surgery, the power of rhythm was higher in the right frontal (p = 0.04) and central (p = 0.025) areas in patients with CA stenoses compared to patients without stenoses. There were no between-group differences in the left hemisphere. After CABG, the patients with CA stenoses had higher power values in the frontal (p = 0.03 and p = 0.036, respectively), central (p = 0.007 and p = 0.01, respectively) and parietal (p = 0.02 and p = 0.019, respectively) regions of the left and right hemispheres (Figure 3).

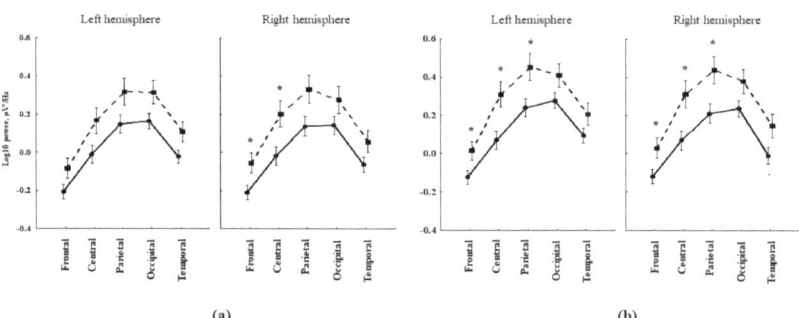

Figure 3. Lateral differences in the alpha2-rhythm power changes of EEG resting state with eyes open in patients after on-pump CABG, depending on the presence of CA stenoses less than 50%: (**a**) before cardiac surgery; (**b**) after surgery; solid lines—the patients without stenoses, dashed lines—the patients with CA stenoses, error bars denote SE, *—p < 0.05 Newman–Keuls multiple comparison test.

Thus, the presence of CA stenoses of less than 50% in patients who underwent on-pump CABG was associated with more pronounced signs of EEG of brain dysfunction. Both before and after CABG, the patients with CA stenoses demonstrated higher values of theta2- and alpha-rhythm power compared to the patients without CA stenoses.

3.2. The Postoperative Neurophysiological Status Changes in the Patients after Combined On-Pump CABG and CEA

3.2.1. Neurophysiological Data

This analysis included 111 patients who have undergone combined coronary and carotid artery revascularization or isolated CABG. According to the results of the preoperative examination, they were divided into three groups: the group of left-sided CEA+CABG (n = 30), the group of right-sided CEA+CABG (n = 24), and the group of isolated CABG (n = 57). The patients with combined coronary and carotid surgery had significant CA stenoses as assessed by digital angiography (NASCET criteria).

No adverse cardiovascular events (myocardial infarction, stroke, death, and repeated unplanned revascularization) were observed in the patients in the early postoperative period for simultaneous CABG+CEA or isolated CABG. In this cohort, POCD occurred in 34 (63.0%) patients with CABG+CEA, and in 32 (61%) patients with isolated CABG (OR = 1.33, 95% CI = 0.62–2.84, p = 0.59). Significant between-group differences were detected for the psychomotor and executive function indicators. At 7–10 days after surgery, the psychomotor speed in two neurodynamic tests was higher in the CABG group than in the group with CABG+CEA (p = 0.0002 and p = 0.005, respectively). In addition, the CABG patients had better indicators of executive control in the same tests at 7–10 days after surgery compared to the patients with CABG+CEA (p = 0.0004 and p = 0.02, respectively).

3.2.2. EEG Data

A repeated-measures ANOVA with a between-subjects factor of GROUP (three levels: CABG+left-sided CEA/CABG+right-sided CEA/isolated CABG) and within-subjects factors of EXAMINATION TIME (two levels: before/after surgery), AREA (five levels: frontal, central, parietal, occipital and temporal), and LATERALITY (two levels: left/right hemisphere) was conducted. The significant factors and interactions associated with the GROUP factor are found in EEG resting state with eyes closed in the theta1 frequency range.

There was a significant factor in EXAMINATION TIME—$F_{1,108} = 46.6, p \leq 0.0001$. It was found that the theta1 power increased after surgery at 7–10 days of the postoperative period as compared with the preoperative level both in the CABG patients and in the two CABG+CEA groups. This effect was more pronounced in CABG+right-sided CEA patients ($p = 0.0001$); they differed also from the isolated CABG group at 7–10 days after surgery ($p = 0.026$) (Figure 4).

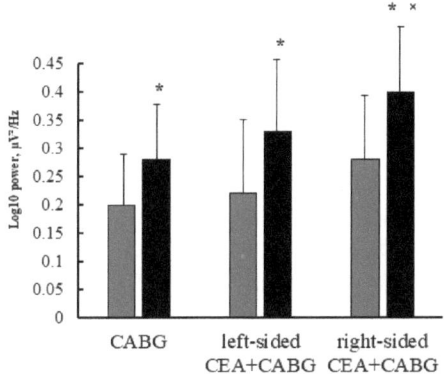

Figure 4. The postoperative theta1 rhythm power changes in the patients undergoing isolated CABG and simultaneous intervention (CABG+CEA): grey columns—the CABG patients, dark columns—the CABG+CEA patients; error bars denote SE, *—$p < 0.05$ Newman–Keuls multiple comparison test for the postoperative indicators in comparison to preoperative levels, ×—$p < 0.05$ Newman–Keuls multiple comparison test for the postoperative indicators in CABG+CEA group as compared to CABG group.

The interaction of factors GROUP × LATERALITY ($F_{2,108} = 3.22, p = 0.04$) was also significant. The left-sided CEA+CABG patients demonstrated the fewest lateral differences of theta1 power. The isolated CABG and CABG+right-sided CEA patients had higher theta1 power values in the left hemisphere as compared to the right one. This effect was more pronounced in CABG+right-sided CEA patients ($p = 0.0004$ and $p = 0.00008$, respectively) (Figure 4).

Another significant interaction of factors GROUP × EXAMINATION TIME × AREA × LATERALITY ($F_{8,432} = 2.15, p = 0.048$) was revealed. The theta1 power differences between the patients who underwent isolated CABG and right-sided CEA+CABG were found. Before surgery, the right-sided CEA+CABG patients had higher theta1 power values than isolated CABG patients only in the frontal cortical regions in both hemispheres ($p = 0.001$ and $p = 0.047$, respectively). After surgery, the between-group differences were more pronounced in the left hemisphere. The right-sided CEA+CABG patients had higher theta1 power values than isolated CABG patients in all cortical regions, except occipital. In the right hemisphere, the between-group differences were only in the frontal, central and temporal regions, as seen in Figure 5.

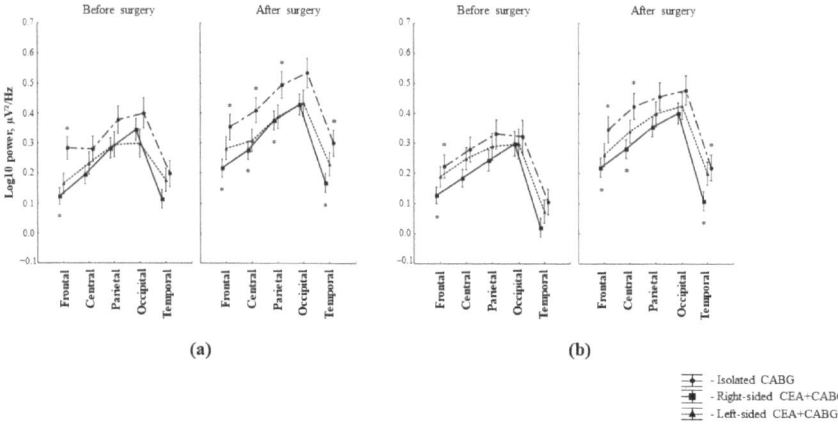

Figure 5. The topography of the postoperative theta1 rhythm power changes in the patients undergoing isolated CABG and simultaneous intervention (CABG+CEA): (**a**) left hemisphere; (**b**) right hemisphere; error bars denote SE, *—$p < 0.05$ Newman–Keuls multiple comparison test for between-group differences.

4. Discussion

As found in our study, the frequency of POCD was higher in patients with less than 50% CA stenoses in comparison to the patients without them (76% vs. 61%). However, the frequency was comparable in combined coronary and carotid surgery and isolated CABG (63% vs. 61%). The POCD structure both in the patients with CA stenosis of less than 50% and in patients with hemodynamically significant CA stenoses (70–99%) consisted of executive function decline, which was determined as the non-successful performance of neurodynamic tasks in these patients. Previously, it has been shown that for the correct assessment of the signal sequence in neurodynamic tests, a high level of indicative activity is required. This causes increased brain energy consumption [40]. It can be assumed that highly organized cognitive activity is disrupted by the deterioration of cerebral blood flow in patients with CA stenoses.

We also demonstrated that the patients with CA stenosis of less than 50% had more pronounced signs of brain dysfunction as compared with patients without stenoses. These changes were diffuse and expressed as higher power values of resting state EEG in the frequency band from 6 to 13 Hz. Earlier, it has been shown that an increase in the slow rhythm power is associated with a decrease in the level of cortical activation and may be a reflection of chronic cerebral ischemia [41,42]. It should be noted that these pathological EEG signs were observed in patients with CA stenosis of less than 50% already in the preoperative period and persisting after surgery. One of the possible causes of neurological complications in patients with hemodynamically insignificant CA stenoses may be the instability of small atherosclerotic plaques with the development of vasoconstrictor and procoagulant effects [43]. There is an assumption that the atherosclerosis in patients with multiple vascular lesions may proceed more aggressively [44]. We may propose that such patients probably develop a more pronounced systemic inflammatory response associated with cardiopulmonary bypass. Earlier experiments showed that the combined effect of ischemia and hypoxia induces an increase in the production of pro-inflammatory cytokines (TNF-α, IL-1β and IL-6) in the brain, which contributes to damage and increased permeability of the blood–brain barrier, and as a consequence, the development of brain edema [45,46]. In addition, cerebral blood flow autoregulation may be disrupted more often in patients with CA stenoses, leading to the decrease in the brain's resistance to acute ischemia and hypoperfusion associated with cardiopulmonary bypass [36,47]. The state of the circle of Willis and the density of leptomeningeal collaterals also contribute to brain hemodynamic parameters [48,49]. On the other hand, the interaction between macro- and microcirculation requires attention in regard to postoperative neurophysiological changes in the patients after cardiac surgery. Earlier, it has been found that carotid atherosclerosis, white matter hyperintensities and lacunar infarction are associated with and commonly contribute to the deterioration of neurological function [50,51].

Additionally, one conclusion we reached was that patients with CA stenoses of less than 50% are vulnerable to the effects of the factors that accompany cardiac surgery using cardiopulmonary bypass compared with patients without CA lesions. The presence of even hemodynamically insignificant stenoses in cardiac surgery patients makes it possible to include them in the group at increased risk of brain damage in the perioperative period. This category of patients should be considered as requiring more careful preoperative management, the use of methods of perioperative protection of the brain, the choice of safe strategies for myocardial revascularization and the involvement of methods of cognitive rehabilitation.

A next finding of our study was that the patients who underwent right-sided CABG+CEA are characterized by the most pronounced theta power changes and generalized "slow-down" of the EEG compared with patients who underwent isolated CABG and left-sided CABG+CEA.

It has been recently reported that severe carotid stenosis can disturb the hemodynamic balance, illustrated by blood flow laterality [52]. As shown by the results in our work, a contralateral stenosis of the CA was observed in 86% of cases in patients who underwent CABG+right-sided CEA. Our study showed that the right hemisphere was more vulnerable intraoperatively. In the study by M. Hedberg and K.G. Engström [53], it was shown that a stroke occurs more often in the right than in the left hemisphere in the early postoperative period of cardiac surgery.

Therefore, the results of the study lead us to conclude that on-pump cardiac surgery is a traumatic brain event, regardless of the type of intervention. Bilateral CA lesion increases the severity of cortical dysfunction in the postoperative period, which requires the use of complex brain protection methods. At the same time, it is worth noting that combined CABG and CEA surgery in comparison with isolated CABG does not lead to more significant brain damage. This fact is an additional argument that makes the strategy of one-stage revascularization of the brain and heart justified.

A set of characteristics of the resting EEG, including a postoperative theta power increase and generalized "slowdown", was obtained in our study. This is a universal brain response to damage, indicating an imbalance between cortical and subcortical structures and a decrease in the functional activity of the cerebral cortex [8,22,23,41,42]. The topography of postoperative EEG activity disturbances included the frontal, temporal and parieto-occipital regions. It is assumed that patients with cardiovascular diseases are most susceptible to ischemic changes in the frontal regions of the brain, which plays a key role in the executive function, action planning and working memory [3–5,40]. At the same time, neurodegenerative brain damage, first of all, is detected in the hippocampus and adjacent areas of the brain (cingulate and temporo-parietal cortex) [23,27,28]. Recent studies of cognitive disorders in a cohort of cardiovascular disease patients have shown that it is difficult to differentiate neurodegenerative and ischemic patterns of brain damage; to a greater extent, researchers are inclined to a mixed etiology of cognitive deficits associated with both the progression of atherosclerotic changes in brain vessels and age-related neurodegenerative changes [7,8,54].

5. Conclusions

The high frequency of cognitive decline in the postoperative period in patients who underwent cardiac surgery with the use of cardiopulmonary bypass and the ambiguity of the mechanisms underlying the development of brain damage encourage further study of this phenomenon in a cohort of patients with cardiovascular diseases. Our results show that an integrated approach using modern methods of neuropsychological testing and computerized EEG allows for timely diagnosis of postoperative cognitive disorders and can be useful in determining the effectiveness and safety of cardiac surgery. We showed that cardiac surgical interventions with cardiopulmonary bypass are associated with a high risk of episodes of brain ischemia. This may be accompanied by specific rearrangements of frequency–spatial patterns of electrical brain activity, dependent on the degree of damage to coronary and carotid arteries. The information obtained can be used to optimize the process of preoperative management and the search for anesthesiologic brain protection and safe surgical techniques and strategies for myocardial revascularization, as well as postoperative rehabilitation of this category of patients.

Author Contributions: Conceptualization, I.T., O.T. and O.B.; methodology, I.T. and O.T.; validation, I.T. and O.T.; formal analysis, I.T. and D.K.; investigation, D.K., I.K., I.S. and A.S.; data curation, D.K., I.K., I.S., A.S. and O.M.; writing—original draft preparation, I.T.; writing—review and editing, O.T. and O.B.; project administration, O.T.; funding acquisition, O.B. All authors have read and agreed to the published version of the manuscript.

Funding: The authors declare that this study received funding from the Federal State Ministry of Science and Education of Russian Federation (The fundamental theme No. 122012000364-5 dated 20 January 2022). The funder was not involved in the study design, collection, analysis, interpretation of data, the writing of this article or the decision to submit it for publication.

Institutional Review Board Statement: The study was conducted in accordance with the Declaration of Helsinki, and approved by the Institutional Ethics Committee of the Research Institute for Complex Issues of Cardiovascular Diseases (01/2011-2520). The study was registered on the ClinicalTrials.gov (NCT05172362).

Informed Consent Statement: Informed consent was obtained from all subjects involved in the study.

Data Availability Statement: Not applicable.

Conflicts of Interest: The authors declare no conflict of interest. The funder had no role in the design of the study; in the collection, analyses, or interpretation of data; in the writing of the manuscript; or in the decision to publish the results.

References

1. Benjamin, E.J.; Muntner, P.; Alonso, A.; Bittencourt, M.S.; Callaway, C.W.; Carson, A.P.; Chamberlain, A.M.; Chang, A.R.; Cheng, S.; Das, S.R.; et al. Heart disease and stroke statistics—2019 update: A report from the American heart association. *Circulation* **2019**, *139*, e56–e528. [CrossRef]
2. Efimova, E.V.; Konobeevskaya, I.N.; Maksimenko, G.V.; Karpov, R.S. smoking and cardiovascular mortality in tomsk inhabitants as a typical city of western siberia. *Cardiovasc. Ther. Prev.* **2017**, *16*, 115–121. [CrossRef]
3. Ngandu, T.; Lehtisalo, J.; Solomon, A.; Levälahti, E.; Ahtiluoto, S.; Antikainen, R.; Bäckman, L.; Hänninen, T.; Jula, A.; Laatikainen, T.; et al. A 2 year multidomain intervention of diet, exercise, cognitive training, and vascular risk monitoring versus control to prevent cognitive decline in at-risk elderly people (FINGER): A randomised controlled trial. *Lancet* **2015**, *385*, 2255–2263. [CrossRef] [PubMed]
4. Sanchez-Lopez, J.; Silva-Pereyra, J.; Fernández, T.; Alatorre-Cruz, G.C.; Castro-Chavira, S.A.; González-López, M.; Sánchez-Moguel, S.M. High levels of incidental physical activity are positively associated with cognition and EEG activity in aging. *PLoS ONE* **2018**, *13*, e0191561. [CrossRef]
5. Gallagher, M.; Okonkwo, O.C.; Resnick, S.M.; Jagust, W.J.; Benzinger, T.L.; Rapp, P.R. What are the threats to successful brain and cognitive aging? *Neurobiol. Aging* **2019**, *83*, 130–134. [CrossRef] [PubMed]
6. Alosco, M.L.; Spitznagel, M.B.; Sweet, L.H.; Josephson, R.; Hughes, J.; Gunstad, J. Atrial Fibrillation Exacerbates Cognitive Dysfunction and Cerebral Perfusion in Heart Failure. *Pacing Clin. Electrophysiol.* **2014**, *38*, 178–186. [CrossRef] [PubMed]
7. Gorelick, P.B.; Counts, S.E.; Nyenhuis, D. Vascular cognitive impairment and dementia. *Biochim. et Biophys. Acta (BBA)-Mol. Basis Dis.* **2015**, *1862*, 860–868. [CrossRef] [PubMed]
8. Tarasova, I.V.; Trubnikova, O.A.; Barbarash, O.L. EEG and Clinical Factors Associated with Mild Cognitive Impairment in Coronary Artery Disease Patients. *Dement. Geriatr. Cogn. Disord.* **2018**, *46*, 275–284. [CrossRef]
9. Hsu, C.L.; Best, J.R.; Davis, J.C.; Nagamatsu, L.S.; Wang, S.; Boyd, L.A.; Hsiung, G.R.; Voss, M.W.; Eng, J.J.; Liu-Ambrose, T. Aerobic exercise promotes executive functions and impacts functional neural activity among older adults with vascular cognitive impairment. *Br. J. Sports Med.* **2017**, *52*, 184–191. [CrossRef]
10. Rosenberg, G.A. Binswanger's disease: Biomarkers in the inflammatory form of vascular cognitive impairment and dementia. *J. Neurochem.* **2018**, *144*, 634–643. [CrossRef]

11. de la Torre, J. The Vascular Hypothesis of Alzheimer's Disease: A Key to Preclinical Prediction of Dementia Using Neuroimaging. *J. Alzheimer's Dis.* **2018**, *63*, 35–52. [CrossRef] [PubMed]
12. Fulop, G.A.; Tarantini, S.; Yabluchanskiy, A.; Molnar, A.; Prodan, C.I.; Kiss, T.; Csipo, T.; Lipecz, A.; Balasubramanian, P.; Farkas, E.; et al. Role of age-related alterations of the cerebral venous circulation in the pathogenesis of vascular cognitive impairment. *Am. J. Physiol. Circ. Physiol.* **2019**, *316*, H1124–H1140. [CrossRef]
13. Anazodo, U.C.; Shoemaker, K.; Suskin, N.; Essali, T.; Wang, D.; Lawrence, K.S.S. Impaired Cerebrovascular Function in Coronary Artery Disease Patients and Recovery Following Cardiac Rehabilitation. *Front. Aging Neurosci.* **2016**, *7*, 224. [CrossRef] [PubMed]
14. Sun, Y.; Cao, W.; Ding, W.; Wang, Y.; Han, X.; Zhou, Y.; Xu, Q.; Zhang, Y.; Xu, J. Cerebral Blood Flow Alterations as Assessed by 3D ASL in Cognitive Impairment in Patients with Subcortical Vascular Cognitive Impairment: A Marker for Disease Severity. *Front. Aging Neurosci.* **2016**, *8*, 211. [CrossRef] [PubMed]
15. Bangen, K.J.; Werhane, M.L.; Weigand, A.J.; Edmonds, E.C.; Delano-Wood, L.; Thomas, K.; Nation, D.A.; Evangelista, N.D.; Clark, A.L.; Liu, T.T.; et al. Reduced Regional Cerebral Blood Flow Relates to Poorer Cognition in Older Adults With Type 2 Diabetes. *Front. Aging Neurosci.* **2018**, *10*, 270. [CrossRef]
16. Suzuki, H.; Matsumoto, Y.; Ota, H.; Sugimura, K.; Takahashi, J.; Ito, K.; Miyata, S.; Furukawa, K.; Arai, H.; Fukumoto, Y.; et al. Hippocampal Blood Flow Abnormality Associated With Depressive Symptoms and Cognitive Impairment in Patients With Chronic Heart Failure. *Circ. J.* **2016**, *80*, 1773–1780. [CrossRef]
17. Li, Y.; Li, M.; Zhang, X.; Yang, S.; Fan, H.; Qin, W.; Yang, L.; Yuan, J.; Hu, W. Clinical features and the degree of cerebrovascular stenosis in different types and subtypes of cerebral watershed infarction. *BMC Neurol.* **2017**, *17*, 166. [CrossRef]
18. Wåhlin, A.; Nyberg, L. At the Heart of Cognitive Functioning in Aging. *Trends Cogn. Sci.* **2019**, *23*, 717–720. [CrossRef]
19. Bunch, T.J.; Galenko, O.; Graves, K.G.; Jacobs, V.; May, H.T. Atrial Fibrillation and Dementia: Exploring the Association, Defining Risks and Improving Outcomes. *Arrhythmia Electrophysiol. Rev.* **2019**, *8*, 8–12. [CrossRef]
20. Tschernatsch, M.; Juenemann, M.; Alhaidar, F.; El Shazly, J.; Butz, M.; Meyer, M.; Gerriets, T.; Schönburg, M.; Schramm, P. Epileptic seizure discharges in patients after open chamber cardiac surgery—A prospective prevalence pilot study using continuous electroencephalography. *Intensiv. Care Med.* **2020**, *46*, 1418–1424. [CrossRef]
21. Tarasova, I.; Razumnikova, O.; Trubnikova, O.; Mezentsev, Y.; Kupriyanova, D.; Barbarash, O. Neurophysiological correlates of postoperative cognitive disorders. *Zhurnal Nevrol. i psikhiatrii im. S.S. Korsakova* **2021**, *121*, 18–23. [CrossRef] [PubMed]
22. Başar, E.; Gölbaşı, B.T.; Tülay, E.; Aydın, S.; Başar-Eroğlu, C. Best method for analysis of brain oscillations in healthy subjects and neuropsychiatric diseases. *Int. J. Psychophysiol.* **2016**, *103*, 22–42. [CrossRef]
23. Babiloni, C.; Barry, R.J.; Başar, E.; Blinowska, K.J.; Cichocki, A.; Drinkenburg, W.H.; Klimesch, W.; Knight, R.T.; da Silva, F.L.; Nunez, P.; et al. International Federation of Clinical Neurophysiology (IFCN)—EEG research workgroup: Recommendations on frequency and topographic analysis of resting state EEG rhythms. Part 1: Applications in clinical research studies. *Clin. Neurophysiol.* **2020**, *131*, 285–307. [CrossRef] [PubMed]
24. Schleiger, E.; Wong, A.; Read, S.; Rowland, T.; Finnigan, S. Poststroke QEEG informs early prognostication of cognitive impairment. *Psychophysiology* **2017**, *54*, 301–309. [CrossRef]
25. Al-Qazzaz, N.K.; Ali, S.H.B.M.; Ahmad, S.A.; Islam, M.S.; Escudero, J. Discrimination of stroke-related mild cognitive impairment and vascular dementia using EEG signal analysis. *Med. Biol. Eng. Comput.* **2017**, *56*, 137–157. [CrossRef] [PubMed]
26. Moretti, D.V. Electroencephalography-driven approach to prodromal Alzheimer's disease diagnosis: From biomarker integration to network-level comprehension. *Clin. Interv. Aging* **2016**, *11*, 897–912. [CrossRef] [PubMed]
27. Moretti, D.V. Increase of EEG Alpha3/Alpha2 Power Ratio Detects Inferior Parietal Lobule Atrophy in Mild Cognitive Impairment. *Curr. Alzheimer Res.* **2018**, *15*, 443–451. [CrossRef]
28. Moretti, D.V. Theta and alpha EEG frequency interplay in subjects with mild cognitive impairment: Evidence from EEG, MRI, and SPECT brain modifications. *Front. Aging Neurosci.* **2015**, *7*, 31. [CrossRef]
29. Zappasodi, F.; Pasqualetti, P.; Rossini, P.M.; Tecchio, F. Acute Phase Neuronal Activity for the Prognosis of Stroke Recovery. *Neural Plast.* **2019**, *2019*, 1971875. [CrossRef]
30. Pérez-Belmonte, L.M.; Román-Terán, C.M.S.; Jiménez-Navarro, M.; Barbancho, M.A.; García-Alberca, J.M.; Lara, J.P. Assessment of Long-Term Cognitive Impairment After Off-Pump Coronary-Artery Bypass Grafting and Related Risk Factors. *J. Am. Med. Dir. Assoc.* **2015**, *16*, 263.e9–263.e11. [CrossRef]
31. Bhamidipati, D.; Goldhammer, J.E.; Sperling, M.R.; Torjman, M.C.; McCarey, M.M.; Whellan, D.J. Cognitive Outcomes After Coronary Artery Bypass Grafting. *J. Cardiothorac. Vasc. Anesthesia* **2017**, *31*, 707–718. [CrossRef] [PubMed]
32. Hood, R.; Budd, A.; Sorond, F.A.; Hogue, C.W. Peri-operative neurological complications. *Anaesthesia* **2018**, *73*, 67–75. [CrossRef] [PubMed]
33. Trubnikova, O.A.; Tarasova, I.V.; Mamontova, A.S.; Syrova, I.D.; Maleva, O.V.; Barbarash, O.L. A role of carotid stenoses in the structure of early postoperative cognitive dysfunction in patients underwent coronary artery bypass grafting. *Zhurnal Nevrol. i psikhiatrii im. S.S. Korsakova* **2014**, *114*, 36–42. [PubMed]
34. Skhirtladze-Dworschak, K.; Felli, A.; Aull-Watschinger, S.; Jung, R.; Mouhieddine, M.; Zuckermann, A.; Tschernko, E.; Dworschak, M.; Pataraia, E. The Impact of Nonconvulsive Status Epilepticus after Cardiac Surgery on Outcome. *J. Clin. Med.* **2022**, *11*, 5668. [CrossRef]
35. Kara, I.; Erkin, A.; Saclı, H.; Demirtas, M.; Percin, B.; Diler, M.S.; Kırali, K. The Effects of Near-Infrared Spectroscopy on the Neurocognitive Functions in the Patients Undergoing Coronary Artery Bypass Grafting with Asymptomatic Carotid Artery Disease: A Randomized Prospective Study. *Ann. Thorac. Cardiovasc. Surg.* **2015**, *21*, 544–550. [CrossRef]

36. Hori, D.; Ono, M.; Adachi, H.; Hogue, C.W. Effect of carotid revascularization on cerebral autoregulation in combined cardiac surgery. *Eur. J. Cardio-Thoracic Surg.* **2015**, *49*, 281–287. [CrossRef]
37. Irqsusi, M.; Vannucchi, A.; Beckers, J.; Kasseckert, S.; Waldhans, S.; Moosdorf, R.G.H.; Vogt, S. Early Results of Surgical Simultaneous Therapy for Significant Carotid Artery Stenosis and Heart Disease. *Thorac. Cardiovasc. Surg.* **2017**, *66*, 261–265. [CrossRef]
38. Trubnikova, O.A.; Tarasova, I.V.; Moskin, E.G.; Kupriyanova, D.S.; Argunova, Y.A.; Pomeshkina, S.A.; Gruzdeva, O.V.; Barbarash, O.L. Beneficial Effects of a Short Course of Physical Prehabilitation on Neurophysiological Functioning and Neurovascular Biomarkers in Patients Undergoing Coronary Artery Bypass Grafting. *Front. Aging Neurosci.* **2021**, *13*, 699259. [CrossRef]
39. Volf, N.; Tarasova, I. The influence of reward on the performance of verbal creative tasks: Behavioral and EEG effects. *Hum. Physiol.* **2013**, *39*, 302–308. [CrossRef]
40. Tam, N.D. Improvement of Processing Speed in Executive Function Immediately following an Increase in Cardiovascular Activity. *Cardiovasc. Psychiatry Neurol.* **2013**, *2013*, 212767. [CrossRef]
41. Snyder, S.M.; Hall, J.R.; Cornwell, S.L.; Falk, J.D. Addition of EEG improves accuracy of a logistic model that uses neuropsychological and cardiovascular factors to identify dementia and MCI. *Psychiatry Res.* **2011**, *186*, 97–102. [CrossRef] [PubMed]
42. Shibata, T.; Musha, T.; Kosugi, Y.; Kubo, M.; Horie, Y.; Kuwayama, N.; Kuroda, S.; Hayashi, K.; Kobayashi, Y.; Tanaka, M.; et al. Altered Neuronal Activity Topography Markers in the Elderly with Increased Atherosclerosis. *Front. Aging Neurosci.* **2017**, *9*, 216. [CrossRef] [PubMed]
43. Qiao, R.; Huang, X.; Qin, Y.; Li, Y.; Davis, T.P.; Hagemeyer, C.E.; Gao, M. Recent advances in molecular imaging of atherosclerotic plaques and thrombosis. *Nanoscale* **2020**, *12*, 8040–8064. [CrossRef] [PubMed]
44. Manolis, A.S.; Manolis, T.A. Patients with Polyvascular Disease: A Very High-Risk Group. *Curr. Vasc. Pharmacol.* **2022**, *20*, 475–490. [CrossRef] [PubMed]
45. Song, T.-T.; Bi, Y.-H.; Gao, Y.-Q.; Huang, R.; Hao, K.; Xu, G.; Tang, J.-W.; Ma, Z.-Q.; Kong, F.-P.; Coote, J.H.; et al. Systemic pro-inflammatory response facilitates the development of cerebral edema during short hypoxia. *J. Neuroinflammation* **2016**, *13*, 63. [CrossRef]
46. Lu, Y.; Chang, P.; Ding, W.; Bian, J.; Wang, D.; Wang, X.; Luo, Q.; Wu, X.; Zhu, L. Pharmacological inhibition of mitochondrial division attenuates simulated high-altitude exposure-induced cerebral edema in mice: Involvement of inhibition of the NF-κB signaling pathway in glial cells. *Eur. J. Pharmacol.* **2022**, *929*, 175137. [CrossRef] [PubMed]
47. Chen, J.-W.; Lin, C.-H.; Hsu, R.-B. Mechanisms of early and delayed stroke after systematic off-pump coronary artery bypass. *J. Formos. Med. Assoc.* **2015**, *114*, 988–994. [CrossRef]
48. Lan, L.; Leng, X.; Ip, V.; Soo, Y.; Abrigo, J.; Liu, H.; Fan, F.; Ma, S.H.; Ma, K.; Ip, B.Y.; et al. Sustaining cerebral perfusion in intracranial atherosclerotic stenosis: The roles of antegrade residual flow and leptomeningeal collateral flow. *J. Cereb. Blood Flow Metab.* **2018**, *40*, 126–134. [CrossRef]
49. Liu, H.; Wang, D.; Leng, X.; Zheng, D.; Chen, F.; Wong, L.K.S.; Shi, L.; Leung, T.W.H. State-of-the-Art Computational Models of Circle of Willis With Physiological Applications: A Review. *IEEE Access* **2020**, *8*, 156261–156273. [CrossRef]
50. Fang, H.; Leng, X.; Pu, Y.; Zou, X.; Pan, Y.; Song, B.; Soo, Y.O.Y.; Leung, T.W.H.; Wang, C.; Zhao, X.; et al. Hemodynamic Significance of Middle Cerebral Artery Stenosis Associated With the Severity of Ipsilateral White Matter Changes. *Front. Neurol.* **2020**, *11*, 214. [CrossRef]
51. Li, L.; He, S.; Liu, H.; Pan, M.; Dai, F. Potential risk factors of persistent postural-perceptual dizziness: A pilot study. *J. Neurol.* **2021**, *269*, 3075–3085. [CrossRef] [PubMed]
52. Zarrinkoob, L.; Wåhlin, A.; Ambarki, K.; Eklund, A.; Malm, J. Quantification and mapping of cerebral hemodynamics before and after carotid endarterectomy, using four-dimensional flow magnetic resonance imaging. *J. Vasc. Surg.* **2021**, *74*, 910–920.e1. [CrossRef] [PubMed]
53. Hedberg, M.; Engström, K.G. Stroke after cardiac surgery—Hemispheric distribution and survival. *Scand. Cardiovasc. J.* **2012**, *47*, 136–144. [CrossRef]
54. van der Velpen, I.F.; Yancy, C.W.; Sorond, F.A.; Sabayan, B. Impaired Cardiac Function and Cognitive Brain Aging. *Can. J. Cardiol.* **2017**, *33*, 1587–1596. [CrossRef] [PubMed]

Disclaimer/Publisher's Note: The statements, opinions and data contained in all publications are solely those of the individual author(s) and contributor(s) and not of MDPI and/or the editor(s). MDPI and/or the editor(s) disclaim responsibility for any injury to people or property resulting from any ideas, methods, instructions or products referred to in the content.

Article

Delirium after Spinal Surgery: A Pilot Study of Electroencephalography Signals from a Wearable Device

Soo-Bin Lee [1], Ji-Won Kwon [2], Sahyun Sung [3], Seong-Hwan Moon [2] and Byung Ho Lee [2,*]

1 Department of Orthopedic Surgery, Catholic Kwandong University International St. Mary's Hospital, 25 Simgok-ro 100beon-gil, Seo-gu, Incheon 22711, Korea
2 Department of Orthopedic Surgery, Yonsei University College of Medicine, 50 Yonsei-ro, Seodaemun-gu, Seoul 03722, Korea
3 Department of Orthopedic Surgery, Ewha Womans University Seoul Hospital, 260 Gonghang-daero, Gangseo-gu, Seoul 07804, Korea
* Correspondence: bhlee96@yuhs.ac; Tel.: +82-2-2228-2180

Abstract: Postoperative delirium after spinal surgery in elderly patients has been a recent concern. However, there has not been a study of delirium after spinal surgery based on electroencephalography (EEG) signals from a compact wearable device. We aimed to analyze differences in EEG signals from a wearable device in patients with and without delirium after spinal surgery. Thirty-seven patients who underwent cervical or lumbar decompression and instrumented fusion for degenerative spinal disease were included. EEG waves were collected from a compact wearable device, and percentage changes from baseline to within 1 week and 3 months after surgery were compared between patients with and without delirium. In patients with delirium, the anxiety- and stress-related EEG waves—including the H-beta (19.3%; $p = 0.003$) and gamma (18.8%; $p = 0.006$) waves—and the tension index (7.8%; $p = 0.011$) increased, and the relaxation-related theta waves (-23.2%; $p = 0.016$) decreased within 1 week after surgery compared to the non-delirium group. These results will contribute to understanding of the EEG patterns of postoperative delirium and can be applied for the early detection and prompt treatment of postoperative delirium after spinal surgery.

Keywords: postoperative delirium; spinal surgery; electroencephalography; wearable device

1. Introduction

Delirium is a clinical syndrome with core symptoms of inattention and acute cognitive dysfunction that often fluctuate [1]. The incidence of postoperative delirium after spinal surgery is reported to be 4.5–24.3% [2–5] and has been found to be associated with prolonged hospital stays and increased costs of care, postoperative functional deterioration, and increased mortality [6,7]. Up to USD 82.4 billion is spent annually on medical costs associated with delirium in the United States alone [8]. Well-known risk factors include older age, duration of surgery, and blood loss during surgery [9]. As global life expectancy is continuously increasing [10], and considering the characteristics of spinal surgery [11–13]—including a long surgical duration and increased risk of blood loss—the individual, societal, and financial burdens of postoperative delirium after spinal surgery are expected to increase [14].

Early diagnosis and treatment of delirium can reduce length of stay, in-hospital morbidity, and healthcare costs [6]. A diagnosis of delirium is primarily made by trained psychiatrists based on the Diagnostic and Statistical Manual of Mental Disorders, 5th Edition (DSM-5). For other medical personnel to detect and evaluate delirium easily, an assessment tool called the Confusion Assessment Method (CAM) has been developed [15], and for the case of intensive care unit (ICU) patients, the Confusion Assessment Method for the ICU (CAM-ICU) has been developed and validated. Although several biomarkers have been studied for an objective diagnostic test for delirium, further research is required to apply them in clinical practice [16,17].

Electroencephalography (EEG) has attracted attention as an objective diagnostic tool for delirium [18]. EEG waves are known to correspond to certain mental functions and states. Delta waves are associated with deep sleep stages [19]. Theta waves reflect relaxation, drowsiness, and meditation. Alpha waves reflect relaxation and reduced anxiety [20]. Sensorimotor rhythm (SMR) waves reflect active, busy, or anxious thinking. M-beta waves indicate anxiety and performance [21]. H-beta waves are associated with significant stress, anxiety, and arousal [22]. Gamma waves reflect stress and conscious perception [23,24]. However, the inconvenience of testing methods using EEG recording devices remains a limitation. To the best of our knowledge, there has not been a study of delirium after spinal surgery based on EEG signals from a compact wearable device. In this study, we aimed to analyze EEG signals from such a device in patients with and without delirium both before and after spinal surgery.

2. Materials and Methods

Ethical approval for this study was obtained from the Institutional Review Board (IRB) of the corresponding author's hospital (Yonsei University IRB and Ethics Committee: 4-2018-0709). All methods were performed in accordance with the Declaration of Helsinki and Yonsei University's institutional guidelines. Informed consent was obtained from all subjects and/or their legal guardians.

From April 2019 to April 2020, 37 patients at least 60 years of age who underwent cervical or lumbar spinal surgery were prospectively enrolled. The enrolled patients were limited to elective surgery for degenerative cervical or lumbar spine disease, excluding trauma or tumor cases. The main diagnoses included cervical spondylotic radiculomyelopathy (6 patients), adjacent segment disease (4 patients), degenerative spondylolisthesis (7 patients), lumbar spinal stenosis (16 patients), and degenerative scoliosis (4 patients). All included patients underwent spinal decompression and instrumented spinal fusion, and all procedures were performed by a single surgeon (B.H.L.).

The patients were divided into two groups: with and without delirium (the delirium and non-delirium groups, respectively). The patients who presented with delirious behavior and were diagnosed with delirium by a psychiatrist within 1 week after surgery were classified as the delirium group. The detection of delirium symptoms was carried out by ward nurses, residents, and caregivers jointly observing the patient 24 h per day. The Confusion Assessment Method (CAM), which is the most common assessment tool, was used, and the following four factors were assessed: (1) acute onset, (2) inattention, (3) disorganized thinking, and (4) altered levels of consciousness. The evaluation of delirium was performed every 1–2 h while checking the patient's condition. After the detection of delirium symptoms, the patient's delirium status was accurately diagnosed and treated through emergency consultation with a psychiatrist or psychiatric residents in the hospital. All patients in the delirium group had fluctuating symptoms during the hospitalization period and recovered their normal mental status by the time of discharge. For all patients in the two groups, immediate postoperative pain was controlled with intravenous patient-controlled analgesia (PCA) containing 1 mg of fentanyl for 2–3 days. Oral analgesics, such as non-steroidal anti-inflammatory drugs (NSAIDs), pregabalin, and gabapentin, were administered at the recommended dose or less, regardless of the groups. The doses of pregabalin and gabapentin were increased to 75 mg BID and 100 mg TID, respectively, as needed. The average length of stay ranged from 7 to 10 days depending on the general condition and recovery of delirium.

2.1. Evaluation of EEG Signals from a Wearable Device

A wearable device for acquiring EEG signals (model: Amp GS5001; SOSO H&C [25], Kyungpook University, Daegu, Korea) was used for this study. This device was used in three previous published studies on surgeons' and nurses' mental stress [26–28]. The device consists of a headband, a main board, two dry electrodes as EEG sensors, and a reference electrode. Two electrodes were positioned at the prefrontal 1 and 2 (Fp1 and Fp2) sites according to the International 10–20 EEG system [29]. The reference electrode was placed on the right earlobe (Figure 1).

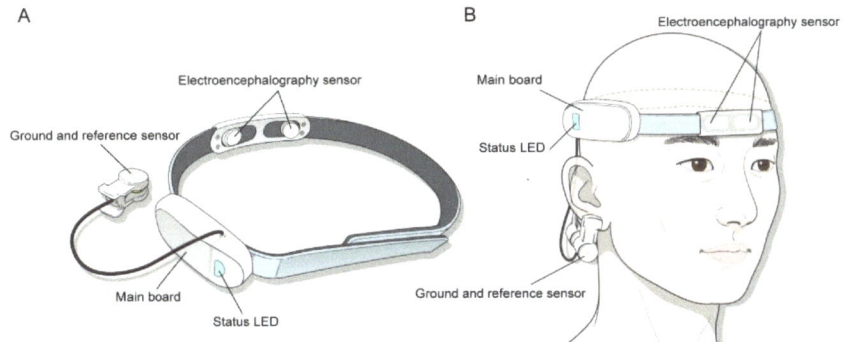

Figure 1. (**A**) Profile of the wearable device. (**B**) The device shown on the patient's head.

The cortical activity of the frontal lobe was measured with the patient in a resting state with eyes open and closed for 3 min each. To reject artifacts caused by eyeball movement, a neurologist specializing in EEG performed a visual inspection. EEG data were reanalyzed using MATLAB R2012b (MathWorks, Inc., Natick, MA, USA) software, and a bandpass filter of 1–50 Hz was applied to the fast Fourier transform. Data analysis using the software was performed by SOSO H&C after anonymization, and the company was blinded to the study. The frequency power was calculated as the square of the amplitudes for delta (0.5–3.5 Hz), theta (4–7 Hz), alpha (8–12 Hz), SMR (12–15 Hz), mid-range beta (M-beta, 15–20 Hz), high-range beta (H-beta, 20–30 Hz), and gamma waves (30–50 Hz) [30]. Relative frequency power was calculated as the ratio of the corresponding frequency powers to the whole frequency (Figure 2). In addition, we calculated the tension index using the following formula from SOSO H&C based on references for tension [27,31]:

$$\text{Tension} = \{[\text{Log}(\text{H-Beta}/\text{Alpha}) + 1.0843]/2.058993\} \times 99 + 1 \tag{1}$$

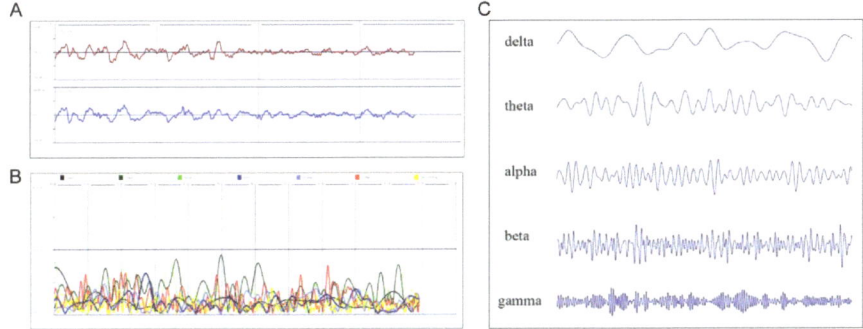

Figure 2. (**A**) Raw and (**B**) reanalyzed EEG signals, according to the frequency bands displayed in the MATLAB software. (**C**) EEG waveforms, according to the frequency bands. EEG: electroencephalography.

All patients were assessed three times: preoperatively (baseline), within 1 week postoperatively, and 3 months postoperatively. All data measurements were performed in the same manner. The patients waited 1 min while wearing the device in a resting state, and the EEG data were collected with the eyes open and closed for 3 min each. The patients were seated, in a distraction-free environment, and a research nurse who was familiar with the measurement methods measured the EEG data using the device. For the baseline and 3 months after surgery, the measurements were performed in an empty outpatient clinic room, while the measurement within 1 week after surgery was performed in a ward room. Baseline EEG data were obtained one day before surgery for admitted patients. Data were then gathered 1–7 days after surgery when the patients were diagnosed with delirium by psychiatrists. The average presentation time of delirium symptoms was 2 days after surgery. Through consultation with psychiatrists and prompt communication with ward nurses, research nurses, and residents, EEG signals in patients with delirium were measured immediately after the presentation of delirium symptoms with the consent of the patients' guardians. If the measurement was impossible because the patient was excessively hyperactive and unable to cooperate, the measurement was performed after waiting until the agitation subsided. After discharge, data were gathered at the three-month outpatient follow-up visit. Percentage changes in data from baseline to within 1 week after surgery and from baseline to 3 months after surgery were compared between the delirium and non-delirium groups.

2.2. Statistical Analyses

The study population characteristics are presented as means ± standard deviations for continuous variables and frequencies (percentages) for categorical variables. For comparisons between the delirium and non-delirium groups, the independent two sample *t*-test was used for continuous variables, while Fisher's exact test was used for categorical variables. We applied a linear mixed model (LMM) to assess the interaction according to the time and group of EEG signals measured over three times. Statistical analyses were performed using SAS (version 9.4; SAS Institute, Cary, NC, USA) and R version 4.1.3 (http://www.r-project.org (accessed on 23 September 2022). The statistical significance of the interaction p-value was set to $p < 0.15$, and other than that the significance threshold was $p < 0.05$.

3. Results

All enrolled patients were sent to the general ward after surgery; no serious postoperative complications except for delirium were observed. Postoperative delirium developed in 6 of 37 patients (16.2%): 4 females (67%) and 2 males (33%). The mean age of these patients was 72.7 ± 3.0 years. Affected patients presented symptoms of hyperactive delirium, including rambling, restlessness, hallucinations, and aggressiveness. The demographics and surgical characteristics of the delirium and non-delirium groups are shown in Table 1. The Mental component Summary (MCS) of the 36-Item Short-Form Health Survey (SF-36) was marginally lower in the delirium group than in the non-delirium group ($p = 0.061$). There were no significant differences in age, gender, American Society of Anesthesiologists class, history of dementia, physical component summary of the SF-36, surgical site, surgical duration, volume of intraoperative blood loss, or total units of transfused packed red blood cells.

According to the results of the LMM, the interactions between group and time were significant in the theta waves ($p = 0.083$), H-beta waves ($p = 0.014$), gamma waves ($p = 0.023$), and tension index ($p = 0.105$), while the other variables were not significant. The results of performing a group post hoc analysis of these variables are shown in Table 2. In patients with delirium, H-beta and gamma waves and the tension index increased, while theta waves decreased, within 1 week after surgery. Patients in the delirium group showed significantly more changes in H-beta (19.3%; $p = 0.003$), gamma (18.8%; $p = 0.006$), and theta waves (-23.2%; $p = 0.016$), as well as the tension index (7.8%; $p = 0.011$), within 1 week after surgery from baseline compared to patients in the non-delirium group. In the graph of H-beta, gamma, and theta waves and tension index, patterns of peaks and troughs 1 week after surgery followed by recovery to baseline levels by 3 months after surgery were noted in the delirium group, whereas a somewhat flat pattern was noted in the non-delirium group (Figure 3).

Table 1. Demographics and surgical characteristics of the enrolled patients.

	Delirium (*n* = 6)	Non-Delirium (*n* = 31)	*p*-Value
Age (years)	72.7 ± 3.0 (71–78)	70.1 ± 6.4 (60–84)	0.353
Gender			
Female	4 (67)	17 (55)	0.680
Male	2 (33)	14 (45)	
ASA class			
I	0	1 (3)	
II	3 (50)	13 (42)	>0.999
III	3 (50)	17 (55)	
History of dementia	1 (17)	1 (3)	0.302
SF-36			
PCS	27 ± 12	33 ± 14	0.313
MCS	40 ± 19	53 ± 16	0.061
Surgical site			
Cervical	1 (17)	6 (19)	>0.999
Thoracolumbar	5 (83)	25 (81)	
Surgical duration (min)	198 ± 52	201 ± 56	0.934
Intraoperative blood loss (mL)	317 ± 177	465 ± 258	0.178
Total units of transfused packed RBCs	0.3 ± 0.7	1.0 ± 1.6	0.308

Values are reported as the mean ± standard deviation (range) or number (%). ASA: American Society of Anesthesiologists class; SF-36: 36-Item Short-Form Health Survey; PCS: Physical Component Summary; MCS: Mental Component Summary; RBCs: red blood cells.

Table 2. Percentage changes in EEG signals from baseline to within 1 week and to 3 months after surgery in the delirium and non-delirium groups.

	Percentage Change from Baseline to within 1 Week after Surgery (%)			Percentage Change from Baseline to 3 Months after Surgery (%)		
	Delirium	Non-Delirium	*p*-Value	Delirium	Non-Delirium	*p*-Value
Delta waves	−26.2 ± 33.0	31.5 ± 71.7	0.082	−20.2 ± 52.6	30.7 ± 78.0	0.169
Theta waves	−23.2 ± 18.0	5.9 ± 25.0	0.016	−14.0 ± 30.5	5.5 ± 26.1	0.144
Alpha waves	−6.2 ± 10.9	0.4 ± 9.4	0.126	1.6 ± 13.1	2.1 ± 15.9	0.932
SMR waves	15.5 ± 19.6	0.1 ± 10.8	0.155	19.5 ± 24.1	−0.9 ± 14.9	0.134
M-beta waves	19.1 ± 21.4	−1.3 ± 12.3	0.097	20.4 ± 24.7	−0.1 ± 15.7	0.017
H-beta waves	19.3 ± 14.6	−1.6 ± 13.5	0.003	14.7 ± 22.8	−0.6 ± 15.5	0.064
Gamma waves	18.8 ± 20.0	−2.4 ± 14.3	0.006	12.8 ± 24.1	−1.7 ± 16.4	0.099
Tension index	7.8 ± 7.2	−0.8 ± 6.9	0.011	3.3 ± 10.8	−0.9 ± 8.8	0.334

Values are reported as the mean ± standard deviation. SMR: sensorimotor rhythm.

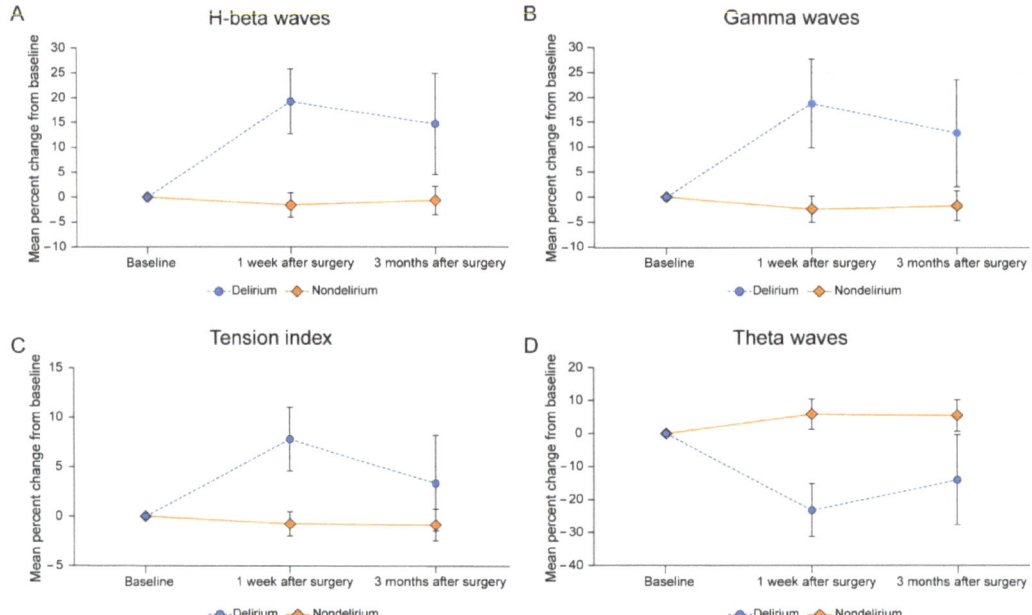

Figure 3. Mean percentage changes in (**A**) H-beta waves, (**B**) gamma waves, (**C**) tension index, and (**D**) theta waves within 1 week and 3 months after surgery compared to baseline in patients with and without delirium. The error bars shown are standard errors of the mean.

4. Discussion

Postoperative delirium after spinal surgery in elderly patients has been a recent concern, and several attempts to achieve effective and objective diagnoses of delirium have been reported. However, to the best of our knowledge, there has not been a study of delirium after spinal surgery based on EEG signals from a compact wearable device. Using EEG signals from such a device, we found greater changes in H-beta, gamma, and theta waves and tension index within 1 week after surgery from baseline in the delirium group compared to the non-delirium group. Based on these results, EEG analysis using a wearable device appears to be an effective diagnostic tool and may be applicable during the early diagnosis of delirium after spinal surgery.

While postoperative delirium has been an increasing concern for spinal surgeons recently, it remains an elusive concept and difficult to diagnose [32]. A diagnosis of delirium is based on the patient's clinical history, behavioral observations, and cognitive assessments [1]. It is not easy for spinal surgeons and nurses to recognize delirium, and even psychiatrists may overlook or misdiagnose this disorder. Although there are clinical assessment tools for delirium, novel diagnostic methods—such as biomarkers and wearable accelerometer devices—are being actively studied [16,33–35].

The association between EEG changes and delirium has been studied, but there are few published reports. EEG has been mainly used to differentiate delirium from non-convulsive status epilepticus or other psychiatric conditions [36,37]. Bispectral index monitoring and adjustments of the depth of anesthesia appear to correlate with reduced postoperative delirium [38,39]. Since Koponen et al. first reported reduced alpha waves and increased theta and delta waves in patients with delirium [40], other studies have followed. In a study of 28 patients with delirium who underwent cardiothoracic surgery, an increase in delta waves from the frontal and parietal lobes was the only difference between patients with and without delirium [18]. Urdanibia-Centelles et al. reported high mean global field power—mainly driven by delta wave activity—in septic patients with delirium and proposed continuous EEG monitoring for its diagnosis [41]. However, Oh et al. pointed out that the routine use of EEG monitoring for delirium screening is time-consuming and inefficient [17].

In the present study, we found differences in specific EEG waves between patients with and without delirium after spinal surgery. H-beta and gamma waves, which are known to be related to stress [24], increased in patients with delirium. Moreover, the tension index—which consists of alpha and H-beta waves—increased, whereas theta waves decreased, in patients with delirium. These results are somewhat different from those of published studies. Most studies of delirium were performed in ICUs [18,41,42]; however, our patients did not require ventilators or ICU care. We believe that patients in the ICU are more likely to show subdued and hypoactive features than those in the general ward, possibly due to the severity of their illness [43]. Therefore, slow waves such as delta and theta waves are more likely to be predominant in ICU patients. On the other hand, restless and agitated states, which are commonly seen in hyperactive delirium, correlate well with features of H-beta and gamma waves [44,45]. Confused, anxious, and irritable behaviors are commonly seen in patients with delirium in the general ward and are a concern for most spinal surgeons and nurses. We believe that our results accurately reflect these aspects of patients with delirium.

In the analysis of demographic and surgical risk factors, we found marginally significant differences only in the MCS of the SF-36 between the two groups. Considering the preliminary nature and small sample size of this study, we believe that this is a potentially important finding. The MCS reflects emotional problems, vitality, mental health, and social functioning. MCS scores can vary from 0 to 100, with higher MCS scores indicating better mental health [46]. Patients with delirium in our study had lower MCS scores, which could be related to the mental vulnerability identified in the changing EEG wave patterns.

This study has a number of major strengths. By measuring EEG signals immediately after the onset of delirium, not only the medical staff but also the patients' guardians were able to accurately recognize the patients' condition. Second, this study highlights a novel EEG wave pattern in patients with delirium after spinal surgery. The major features of anxiety and stress in patients with delirium were identified, which will likely lead to a deeper understanding of the mental health of patients undergoing spinal surgery. Third, to the best of our knowledge, this was the first study to investigate delirium after spinal surgery using a compact wearable device. Numerous applications of this device can be used in the future, such as monitoring the mental status of patients who undergo spinal surgery, including those in the hospital or even after discharge to home, which would never be possible with traditional EEG monitoring. In addition, further analyses of mental health according to the type of spinal disease and surgery needed could be performed more easily.

Our study has several limitations. Due to the preliminary nature of this study and the low incidence rate of postoperative delirium, the number of enrolled patients with and without delirium was relatively small. Further large-scale studies are needed to confirm our results. Furthermore, we did not evaluate the effects of baseline cognitive function, medication use, and severity of illness on the outcome measures. Because EEG changes can be affected by many other factors, further analyses of additional factors should be conducted. Finally, the correlation between EEG and the type and duration of delirium was not analyzed in this study and needs to be investigated in the future.

5. Conclusions

In the delirium group, H-beta and gamma waves and the tension index increased, while theta waves decreased, compared to the non-delirium group 1 week after surgery. Our data show that EEG signals from a wearable device have the potential to be used to screen patients for delirium, which may lead to earlier diagnoses and treatments for delirium after spinal surgery. Future large-scale studies are needed to investigate the effects of factors related to delirium, such as baseline cognitive function, medication use, and severity of illness.

Author Contributions: Conceptualization, S.-B.L. and B.H.L.; methodology, J.-W.K.; software, S.S.; validation, S.-H.M.; formal analysis, S.-B.L.; investigation, B.H.L.; resources, J.-W.K.; data curation, S.S.; writing—original draft preparation, S.-B.L.; writing—review and editing, S.-B.L.; visualization, S.-B.L.; supervision, B.H.L.; project administration, B.H.L.; funding acquisition, B.H.L. All authors have read and agreed to the published version of the manuscript.

Funding: This research received no external funding.

Institutional Review Board Statement: The study was conducted in accordance with the Declaration of Helsinki and approved by the Institutional Review Board (or Ethics Committee) of Yonsei University (protocol code 4-2018-07-09, 6 September 2018).

Informed Consent Statement: Informed consent was obtained from all subjects involved in the study.

Data Availability Statement: The data presented in this study are available upon request from the corresponding author.

Acknowledgments: We would like to thank SOSO H&C for analyzing the data, Dong Su Jang for providing illustrations, and Hye Sun Lee and Juyeon Yang, for statistically analyzing the data.

Conflicts of Interest: The authors declare no conflict of interest.

References

1. Fong, T.G.; Tulebaev, S.R.; Inouye, S.K. Delirium in elderly adults: Diagnosis, prevention and treatment. *Nat. Rev. Neurol.* **2009**, *5*, 210–220. [CrossRef] [PubMed]
2. Song, K.J.; Ko, J.H.; Kwon, T.Y.; Choi, B.W. Etiology and related factors of postoperative delirium in orthopedic surgery. *Clin. Orthop. Surg.* **2019**, *11*, 297–301. [CrossRef] [PubMed]
3. Seo, J.S.; Park, S.W.; Lee, Y.S.; Chung, C.; Kim, Y.B. Risk factors for delirium after spine surgery in elderly patients. *J. Korean Neurosurg. Soc.* **2014**, *56*, 28–33. [CrossRef] [PubMed]
4. Lee, J.K.; Park, Y.S. Delirium after spinal surgery in korean population. *Spine* **2010**, *35*, 1729–1732. [CrossRef] [PubMed]
5. Kawaguchi, Y.; Kanamori, M.; Ishihara, H.; Abe, Y.; Nobukiyo, M.; Sigeta, T.; Hori, T.; Kimura, T. Postoperative delirium in spine surgery. *Spine J.* **2006**, *6*, 164–169. [CrossRef] [PubMed]
6. Nazemi, A.K.; Gowd, A.K.; Carmouche, J.J.; Kates, S.L.; Albert, T.J.; Behrend, C.J. Prevention and management of postoperative delirium in elderly patients following elective spinal surgery. *Clin. Spine Surg.* **2017**, *30*, 112–119. [CrossRef] [PubMed]
7. Gleason, L.J.; Schmitt, E.M.; Kosar, C.M.; Tabloski, P.; Saczynski, J.S.; Robinson, T.; Cooper, Z.; Rogers, S.O., Jr.; Jones, R.N.; Marcantonio, E.R.; et al. Effect of delirium and other major complications on outcomes after elective surgery in older adults. *JAMA Surg.* **2015**, *150*, 1134–1140. [CrossRef]
8. Kinchin, I.; Mitchell, E.; Agar, M.; Trepel, D. The economic cost of delirium: A systematic review and quality assessment. *Alzheimers Dement.* **2021**, *17*, 1026–1041. [CrossRef]
9. Kang, T.; Park, S.Y.; Lee, J.H.; Lee, S.H.; Park, J.H.; Kim, S.K.; Suh, S.W. Incidence & risk factors of postoperative delirium after spinal surgery in older patients. *Sci. Rep.* **2020**, *10*, 9232.
10. GBD 2019 Demographics Collaborators. Global age-sex-specific fertility, mortality, healthy life expectancy (hale), and population estimates in 204 countries and territories, 1950–2019: A comprehensive demographic analysis for the global burden of disease study 2019. *Lancet* **2020**, *396*, 1160–1203. [CrossRef]
11. Kulkarni, A.G.; Patel, J.Y.; Asati, S.; Mewara, N. "Spine surgery checklist": A step towards perfection through protocols. *Asian Spine J.* **2022**, *16*, 38–46. [CrossRef] [PubMed]
12. Inoue, T.; Mizutamari, M.; Hatake, K. Surgical invasiveness of single-segment posterior lumbar interbody fusion: Comparing perioperative blood loss in posterior lumbar interbody fusion with traditional pedicle screws, cortical bone trajectory screws, and percutaneous pedicle screws. *Asian Spine J.* **2021**, *15*, 856–864. [CrossRef]
13. Hasegawa, T.; Ushirozako, H.; Yamato, Y.; Yoshida, G.; Yasuda, T.; Banno, T.; Arima, H.; Oe, S.; Yamada, T.; Ide, K.; et al. Impact of spinal correction surgeries with osteotomy and pelvic fixation in patients with kyphosis due to osteoporotic vertebral fractures. *Asian Spine J.* **2021**, *15*, 523–532. [CrossRef] [PubMed]
14. Morino, T.; Hino, M.; Yamaoka, S.; Misaki, H.; Ogata, T.; Imai, H.; Miura, H. Risk factors for delirium after spine surgery: An age-matched analysis. *Asian Spine J.* **2018**, *12*, 703–709. [CrossRef] [PubMed]
15. Inouye, S.K.; van Dyck, C.H.; Alessi, C.A.; Balkin, S.; Siegal, A.P.; Horwitz, R.I. Clarifying confusion: The confusion assessment method. A new method for detection of delirium. *Ann. Intern. Med.* **1990**, *113*, 941–948. [CrossRef]
16. Robinson, T.N.; Raeburn, C.D.; Angles, E.M.; Moss, M. Low tryptophan levels are associated with postoperative delirium in the elderly. *Am. J. Surg.* **2008**, *196*, 670–674. [CrossRef]
17. Oh, S.T.; Park, J.Y. Postoperative delirium. *Korean J. Anesthesiol.* **2019**, *72*, 4–12. [CrossRef]
18. Van der Kooi, A.W.; Zaal, I.J.; Klijn, F.A.; Koek, H.L.; Meijer, R.C.; Leijten, F.S.; Slooter, A.J. Delirium detection using eeg: What and how to measure. *Chest* **2015**, *147*, 94–101. [CrossRef]

19. Tatum, W.O.t.; Husain, A.M.; Benbadis, S.R.; Kaplan, P.W. Normal adult eeg and patterns of uncertain significance. *J. Clin. Neurophysiol.* **2006**, *23*, 194–207. [CrossRef]
20. Cahn, B.R.; Polich, J. Meditation states and traits: Eeg, erp, and neuroimaging studies. *Psychol. Bull.* **2006**, *132*, 180–211. [CrossRef]
21. Kirmizi-Alsan, E.; Bayraktaroglu, Z.; Gurvit, H.; Keskin, Y.H.; Emre, M.; Demiralp, T. Comparative analysis of event-related potentials during go/nogo and cpt: Decomposition of electrophysiological markers of response inhibition and sustained attention. *Brain Res.* **2006**, *1104*, 114–128. [CrossRef] [PubMed]
22. Pfurtscheller, G.; Lopes da Silva, F.H. Event-related eeg/meg synchronization and desynchronization: Basic principles. *Clin. Neurophysiol.* **1999**, *110*, 1842–1857. [CrossRef]
23. Herrmann, C.S.; Demiralp, T. Human eeg gamma oscillations in neuropsychiatric disorders. *Clin. Neurophysiol.* **2005**, *116*, 2719–2733. [CrossRef]
24. Minguillon, J.; Lopez-Gordo, M.A.; Pelayo, F. Stress assessment by prefrontal relative gamma. *Front. Comput. Neurosci.* **2016**, *10*, 101. [CrossRef] [PubMed]
25. SOSO H&C. Available online: http://www.soso-g.co.kr/ (accessed on 23 September 2022).
26. Kwon, J.W.; Lee, S.B.; Sung, S.; Park, Y.; Ha, J.W.; Kim, G.; Suk, K.S.; Kim, H.S.; Lee, H.M.; Moon, S.H.; et al. Which factors affect the stress of intraoperative orthopedic surgeons by using electroencephalography signals and heart rate variability? *Sensors* **2021**, *21*, 4016. [CrossRef] [PubMed]
27. Kwon, J.W.; Sung, S.; Lee, S.B.; Lee, H.M.; Moon, S.H.; Lee, B.H. Intraoperative real-time stress in degenerative lumbar spine surgery: Simultaneous analysis of electroencephalography signals and heart rate variability: A pilot study. *Spine J.* **2020**, *20*, 1203–1210. [CrossRef] [PubMed]
28. Sung, S.; Kwon, J.-W.; Kim, J.-E.; Lee, Y.-J.; Lee, S.-B.; Lee, S.-K.; Moon, S.-H.; Lee, B.H. Real-time stress analysis affecting nurse during elective spinal surgery using a wearable device. *Brain Sci.* **2022**, *12*, 909. [CrossRef]
29. Klem, G.H.; Luders, H.O.; Jasper, H.H.; Elger, C. The ten-twenty electrode system of the international federation. The international federation of clinical neurophysiology. *Electroencephalogr. Clin. Neurophysiol. Suppl.* **1999**, *52*, 3–6.
30. Abhang, P.A.; Gawali, B.W.; Mehrotra, S.C. Technical aspects of brain rhythms and speech parameters. In *Introduction to Eeg and Speech-Based Emotion Recognition*; Academic Press: Cambridge, MA, USA, 2016; pp. 51–79.
31. Xu, R.; Zhang, C.; He, F.; Zhao, X.; Qi, H.; Zhou, P.; Zhang, L.; Ming, D. How physical activities affect mental fatigue based on eeg energy, connectivity, and complexity. *Front. Neurol.* **2018**, *9*, 915. [CrossRef]
32. Deksnyte, A.; Aranauskas, R.; Budrys, V.; Kasiulevicius, V.; Sapoka, V. Delirium: Its historical evolution and current interpretation. *Eur. J. Intern. Med.* **2012**, *23*, 483–486. [CrossRef]
33. Davoudi, A.; Manini, T.M.; Bihorac, A.; Rashidi, P. Role of wearable accelerometer devices in delirium studies: A systematic review. *Crit. Care Explor.* **2019**, *1*, e0027. [CrossRef] [PubMed]
34. Cerejeira, J.; Batista, P.; Nogueira, V.; Vaz-Serra, A.; Mukaetova-Ladinska, E.B. The stress response to surgery and postoperative delirium: Evidence of hypothalamic-pituitary-adrenal axis hyperresponsiveness and decreased suppression of the gh/igf-1 axis. *J. Geriatr. Psychiatry Neurol.* **2013**, *26*, 185–194. [CrossRef] [PubMed]
35. Osse, R.J.; Tulen, J.H.; Hengeveld, M.W.; Bogers, A.J. Screening methods for delirium: Early diagnosis by means of objective quantification of motor activity patterns using wrist-actigraphy. *Interact. Cardiovasc. Thorac. Surg.* **2009**, *8*, 344–348. [CrossRef] [PubMed]
36. Oh, E.S.; Fong, T.G.; Hshieh, T.T.; Inouye, S.K. Delirium in older persons: Advances in diagnosis and treatment. *JAMA* **2017**, *318*, 1161–1174. [CrossRef] [PubMed]
37. Tu, T.M.; Loh, N.K.; Tan, N.C. Clinical risk factors for non-convulsive status epilepticus during emergent electroencephalogram. *Seizure* **2013**, *22*, 794–797. [CrossRef] [PubMed]
38. Whitlock, E.L.; Torres, B.A.; Lin, N.; Helsten, D.L.; Nadelson, M.R.; Mashour, G.A.; Avidan, M.S. Postoperative delirium in a substudy of cardiothoracic surgical patients in the bag-recall clinical trial. *Anesth. Analg.* **2014**, *118*, 809–817. [CrossRef]
39. Chan, M.T.; Cheng, B.C.; Lee, T.M.; Gin, T.; Group, C.T. Bis-guided anesthesia decreases postoperative delirium and cognitive decline. *J. Neurosurg. Anesthesiol.* **2013**, *25*, 33–42. [CrossRef]
40. Koponen, H.; Partanen, J.; Paakkonen, A.; Mattila, E.; Riekkinen, P.J. Eeg spectral analysis in delirium. *J. Neurol. Neurosurg. Psychiatry* **1989**, *52*, 980–985. [CrossRef]
41. Urdanibia-Centelles, O.; Nielsen, R.M.; Rostrup, E.; Vedel-Larsen, E.; Thomsen, K.; Nikolic, M.; Johnsen, B.; Moller, K.; Lauritzen, M.; Benedek, K. Automatic continuous eeg signal analysis for diagnosis of delirium in patients with sepsis. *Clin. Neurophysiol.* **2021**, *132*, 2075–2082. [CrossRef]
42. Hunter, A.; Crouch, B.; Webster, N.; Platt, B. Delirium screening in the intensive care unit using emerging qeeg techniques: A pilot study. *AIMS Neurosci.* **2020**, *7*, 1–16. [CrossRef]
43. la Cour, K.N.; Andersen-Ranberg, N.C.; Weihe, S.; Poulsen, L.M.; Mortensen, C.B.; Kjer, C.K.W.; Collet, M.O.; Estrup, S.; Mathiesen, O. Distribution of delirium motor subtypes in the intensive care unit: A systematic scoping review. *Crit. Care* **2022**, *26*, 53. [CrossRef] [PubMed]
44. Hayashi, T.; Okamoto, E.; Nishimura, H.; Mizuno-Matsumoto, Y.; Ishii, R.; Ukai, S. Beta activities in eeg associated with emotional stress. *Int. J. Intell. Comput. Med. Sci. Image Process.* **2009**, *3*, 57–68. [CrossRef]

45. Díaz, M.H.; Cid, F.M.; Otárola, J.; Rojas, R.; Alarcón, O.; Cañete, L. Eeg beta band frequency domain evaluation for assessing stress and anxiety in resting, eyes closed, basal conditions. *Procedia Comput. Sci.* **2019**, *162*, 974–981. [CrossRef]
46. Hays, R.D.; Morales, L.S. The rand-36 measure of health-related quality of life. *Ann. Med.* **2001**, *33*, 350–357. [CrossRef] [PubMed]

Article

Detrending Moving Average, Power Spectral Density and Coherence: Three EEG-Based Methods to Assess Emotion Irradiation during Facial Perception

Mariia Chernykh [1,*], Bohdan Vodianyk [2], Ivan Seleznov [2,3], Dmytro Harmatiuk [2], Ihor Zyma [1], Anton Popov [2,4] and Ken Kiyono [3,*]

1. Department of Human and Animal Physiology, Educational and Scientific Center "Institute of Biology and Medicine", Taras Shevchenko National University of Kyiv, 01033 Kyiv, Ukraine
2. Department of Electronic Engineering, Igor Sikorsky Kyiv Polytechnic Institute, 03056 Kyiv, Ukraine
3. Division of Bioengineering, Graduate School of Engineering Science, Osaka University, Osaka 560-0043, Japan
4. Faculty of Applied Sciences, Ukrainian Catholic University, 79000 Lviv, Ukraine
* Correspondence: mariia.chernykh@knu.ua (M.C.); kiyono@bpe.es.osaka-u.ac.jp (K.K.)

Abstract: Understanding brain reactions to facial expressions can help in explaining emotion-processing and memory mechanisms. The purpose of this research is to examine the dynamics of electrical brain activity caused by visual emotional stimuli. The focus is on detecting changes in cognitive mechanisms produced by negative, positive, and neutral expressions on human faces. Three methods were used to study brain reactions: power spectral density, detrending moving average (DMA), and coherence analysis. Using electroencephalogram (EEG) recordings from 48 subjects while presenting facial image stimuli from the International Affective Picture System, the topographic representation of the evoked responses was acquired and evaluated to disclose the specific EEG-based activity patterns in the cortex. The theta and beta systems are two key cognitive systems of the brain that are activated differently on the basis of gender. The obtained results also demonstrate that the DMA method can provide information about the cortical networks' functioning stability, so it can be coupled with more prevalent methods of EEG analysis.

Keywords: EEG; emotions; facial perception; power spectral density; coherence; detrended moving average; DMA

1. Introduction

The recognition and evaluation of human facial expressions are among the most evolutionary essential processes. Apart from that, this particular type of visual stimuli is characterized by a broad spectrum of attributes, rendering them a complex system of features for human perceptual processing. Recent studies emphasized that our ability to detect and process human faces arises from the interplay between inherited predisposition and acquired experience-based skills [1], both modulating our perception. Interestingly enough, three-month-old children do not show differential neural responses to fearful and neutral faces, regardless of gaze direction, which reflects the gradual development of discrimination abilities in the field of facial perception [2]. Consequently, the function of the human brain during facial processing is currently seen as a dynamic Bayesian-based system that percepts and processes sensory inputs and integrates them with previously formed templates.

Brain studies in humans have also shown that facial processing is modulated by the affective salience of faces, especially those with expressions of fear; however, other social cues are involved as well. Even though traditional models suggest facial expression and identity are processed in distinctive areas, the current findings highlight that emotion processing can have a strong influence on facial recognition and memory mechanisms [3].

When it comes to EEG data evaluation, several techniques of analysis are traditionally used to reveal distinctive aspects of neural oscillation. Among them, the power spectral density (PSD) algorithm highlights an increase or decrease in the power of the particular oscillation range that accompanies evoked response to a broad range of stimulus types, reflecting cortical activation processes [4,5]. Withal coherence allows for bringing up dynamic interactions between functional brain areas on the basis of distant signal synchronization processes while executing a specific task [6].

However, putting together the complete picture of induced neurodynamics requires not only oscillatory power and emergence synchronicity evaluation, but also the assessment of the temporal stability of the effective neural associations. That is why the study of the potential applications of nonlinear EEG analysis methods taking into account the nonstationarity of EEG time series is relevant and can also provide clues for investigating subtle mechanisms on the basis of the long-range fractal scaling properties of the signal.

Various scaling methods exist to quantify the fractal scaling properties of biomedical time series; the most commonly used method so far is detrended fluctuation analysis (DFA) [7]. Nevertheless, recent developments in scaling analysis techniques suggest that the DFA methodology possesses some crucial drawbacks that lead to the inaccurate estimation of the scaling properties of a signal [8]. A novel method of scaling analysis, detrended moving average (DMA) [9] analysis, was introduced, and its theoretical foundation was developed [10–12]. DMA has huge potential for identifying the long- and short-term correlations of the time series, and does not have all the drawbacks of DFA.

The DFA and DMA methods were both used in various applications in regard to EEG analysis, such as emotional specification from musical stimuli [13], the quantifying depth of anaesthesia [14,15], analyzing human sleep [16], and assessing human emotions [17]. However, in most of the cases, the nonlinear properties of EEG signals are described in a standalone manner or used as one of the features in training the machine-learning algorithms. So far, there has been no research jointly describing the relation between conventional measures in neurophysiology with the fractal scaling features of the EEG time series.

In this work, we aimed to combine the widely used methods of EEG data evaluation such as PSD and coherence with a detrending moving average algorithm fitted to highlight long-range spatiotemporal correlations within EEG data in order to display different neurophysiological aspects of visual perception. Thus, comparing the results provided by widely used methods of EEG data analysis with the results provided by the DMA algorithm, a mathematical apparatus that is new in the field of electrophysiological research was carried out. The data used for the analysis consisted of EEG-based short-term variations of neural oscillations corresponding to the perception of neutral facial expressions, which were primed by the faces of either positive or negative emotional valence.

2. Materials and Methods

2.1. Subjects and Data Collection

The study included data collected from 48 volunteers who were the students of Taras Shevchenko National University of Kyiv (29 females) aged 18–24 (Mean age = 21, SD = 1.76). The experimental design was approved by the Institutional Ethics Committee of the Educational and Scientific Center Institute of Biology and Medicine, Taras Shevchenko National University of Kyiv. Each of the participants received and filled out a written informed consent in accordance with the Declaration of Helsinki (Helsinki, Finland, June 1964), the Convention for the protection of Human Rights and Dignity of the Human Being with regard to the Application of Biology and Medicine: Convention on Human Rights and Biomedicine (Oviedo, 4 April 1997), and the Declaration of the Principles on Tolerance (28th session of the General Conference of UNESCO, Paris, 16 November 1995). The list of major exclusion criteria consisted of addictive behaviors, mentions of mental or neurological cases in the patient's clinical history, the use of psychiatric medications, untreated vision impairments, and color blindness.

Both the image demonstration and EEG recordings were performed using the Neurocom electroencephalographic complex in conformity with a purposely designed template. The data acquisition was performed using 19 active recording channels applied to the scalp according to the international "10–20%" system.

2.2. Experiment Design

The stimulation material was organized into four 3 min long image sequences (2 emotional and 2 neutral), and each separate stimulus was demonstrated for 5 s. With the view to perform emotional priming, the image series were displayed as follows: each emotional array preceded a neutral one; the series with positive facial expressions was shown first to avoid emotional trace overlap. The experimental paradigm implied the following:

1. EEG recording while the subjects rest with closed and open eyes;
2. EEG recording during emotional stimuli perception (positive and negative facial expressions);
3. EEG recording during neutral stimuli perception, modulated by positive images (n1) and negative images (n2).

Stimuli were taken from the International Affective Pictures System (IAPS) [18] according to the average values of their emotional valence. Thus, the parameters of the stimuli were as follows:

1. neutral faces (M = 4.22, SD = 1.64 to M = 5.84, SD = 1.62);
2. positive faces (M = 6.94, SD = 1.42 to M = 8.03, SD = 1.13);
3. negative faces (M = 1.82, SD = 1.64 to M = 3.91, SD = 1.62).

The EEG data were subdivided into a range of sub-bands, allowing for evaluating the induced responses generated by the human cortex. Therefore, the decisive list of the EEG spectral ranges included the following constituents:

1. 3.5–5.8 Hz (θ_1);
2. 5.9–7.4 Hz (θ_2);
3. 7.5–9.4 Hz (α_1);
4. 9.5–10.7 Hz (α_2);
5. 10.8–13.5 Hz (α_3);
6. 13.6–25 Hz (β_1);
7. 25.1–40 Hz (β_2).

All the analytical techniques mentioned in this study were individually applied to each EEG sub-band.

2.3. Estimation of Power Spectral Density

Power spectral density (PSD) analysis is a widely used frequency-based method of EEG power distribution study [19]. In this work, PSD was calculated from EEG data during the processing of various visual stimuli. First, fast Fourier transform was computed and then squared to obtain the PSD estimate. The mean normalized PSD for every EEG channel in the frequency ranges mentioned above was determined by adding power values in a specific frequency range selected for further analysis. Lastly, the value of total powers in every range was divided by the total spectral power of the corresponding EEG channel.

2.4. Detrended Moving Average Analysis

The detrending moving average (DMA) method is used to estimate the scaling exponent of long-range correlated series [10], including EEG during cognitive workload [20]. The DMA algorithm is widely applied to evaluate the long- and short-term correlations of random time series, both one- and high-dimensional, in time and spatial domains [21].

DMA analysis consists of the following steps: First, the time series of each EEG channel are filtered using a 4-th order Butterworth band-pass filter with a pass band corresponding to the required frequency range described in the experiment design. The values of the cutoff frequencies were chosen directly from specified EEG bands: θ_1 [3.5, 5.8], θ_2 [5.9, 7.4], α_1 [7.5, 9.4], α_2 [9.5, 10.7], α_3 [10.8, 13.5], β_1 [13.6, 25], β_2 [25.1, 40] Hz. As the second step, the Hilbert transform was applied to each filtered signal to obtain the analytic signal, and then the envelope was estimated as the absolute value of the analytic signal [22].

Afterwards, we integrated the transformed signal as follows:

$$y[i] = \sum_{j=1}^{i} x[j], \tag{1}$$

In the next step, we calculated the fluctuation function $F(s)$ of the integrated series:

$$F(s) = \sqrt{\frac{1}{N} \sum_{i=1}^{N} \left(y[i] - \widetilde{y}_{SG}^{(m,s)}[i]\right)^2}, \tag{2}$$

where $\widetilde{y}_{SG}^{(m,s)}$ represents the m-th order Savitzky–Golay smoothing filter [23] for $\{y[i]\}$, and s is window length of the Savitzky–Golay filter. In our approach, we used the fourth Savitzky–Golay filter that corresponded to the centered DMA [24].

A linear relationship on a double-logarithmic plot of $F(s)$ against scales s indicates the power-law scaling range. In this scaling range, fluctuations can be characterized by a scaling exponent α. Scaling exponent α was derived from the slope of the linear part of the relation between $log(F(s))$ against scales $log(s)$, where the linear part was empirically chosen to fit the aim of the research for all the subjects with different types of stimulus. As a result, the values of scaling exponents were obtained for all channels in the defined frequency bands. Depending on its value, the meaning of the scaling exponent is interpreted as follows:

1. $\alpha < 0.5$—long-range anticorrelated signal;
2. $\alpha = 0.5$—uncorrelated signal;
3. $\alpha > 0.5$—long-range correlated signal;

2.5. Coherence Analysis

Coherence analysis is generally recognized as a standard metric for EEG data analysis. The coherence of EEG activity in different brain regions is used to measure the synchronicity of oscillations in two distinctive areas of the neocortex (namely, functional connectivity) for different functional states and processes [25], and emotions [26]. In the present study, coherence is used to indicate statistically significant similarity of neural source activity in a particular frequency range in two distinct cortical regions.

In this work, the coherence was calculated according to the following pipeline. The pairs of different EEG channels were chosen for every frequency sub-band, and the coherence was calculated [27] using the entire recordings. The validity of the coherence coefficient between pairs of electrodes was addressed by conducting surrogate data analysis [28]. The phase randomization technique was applied to obtain the surrogate of the EEG signal. For one of the EEG signals, the Fourier Transform was acquired, and then the random number from range $-\pi \ldots +\pi$ was added to the phase of each harmonic component. Afterwards, the inverse Fourier transform was performed, and the coherence was calculated between one initial EEG signal and the surrogate EEG. Each pair of electrodes underwent 100 iterations of the procedure. The t-test ($p < 0.05$) was performed to define coherence coefficients, which differed significantly from the surrogate data. If the coefficient was significantly different, it was chosen for further analysis; otherwise, it was neglected. As a result, we obtained the valid EEG coherence values between the electrode pairs for every frequency range and pair of experiments.

2.6. Statistical Tests

The Wilcoxon and Mann–Whitney tests were used to perform statistical tests of significance for the PSD, DMA, and coherence parameters of EEG. These tests were selected to identify the channels, and frequency ranges were the median values of the analyzed parameters were different for the two experimental sets of EEG data (n1, n2). Through the Wilcoxon signed-rank test, two generalized n1/n2 trials without gender splitting were tested. The Mann–Whitney rank test was used to test female/male groups for the same set of trials.

Likewise, statistical tests were undertaken for DMA scaling exponents. If the median value of the scaling exponent for the n1 EEG signal was statistically distinct from that of the n2 one in a given channel, the significance was obtained for that channel. As all of the frequency bands chosen for this study were generated from the same EEG recording, it may be concluded that all of the frequency bands were evaluated together. Thus, the p-value correction was performed to eliminate the multiple-comparisons problem. To address this issue, p-values were corrected using the Holm–Sidak correction [29].

The schematic head figures with EEG channels were used to visualize the obtained results. On these figures, ">0" was assigned to the associated EEG channel if the difference between medians of the PSD and scaling exponents was larger than 0, "<0" was assigned if the difference was less than 0, and "0" was assigned if there were no significant differences between medians. All values associated with a particular color on the heat map facilitated result perception.

For coherence analysis, the same type of statistical analysis was applied. The visualization of the coherence statistical test was performed in the following manner: if the difference between medians was larger than zero, the red line connected two electrodes labels of the EEG channel; if the difference was less than zero, a blue line was assigned; and if the differences were not significant, no visual changes were applied to the head figure. The Wilcoxon test was used in situations where experiments were compared without gender division, but the Mann–Whitney test was used when gender division was present.

3. Results

To visualize the results, topographical head heat maps were used. The color bar represents the main features of signals obtained by normalization between the minimal and maximal values.

Therefore, the topographical plots represent the statistically significant differences in PSD values that are induced by distinctive types of visual stimulation. These data were also reflected within the head maps.

First, the analysis of power spectral density values reflected the neurodynamics related to the visual stimulus perception and processing when subjects were presented with emotional stimuli (Figure 1).

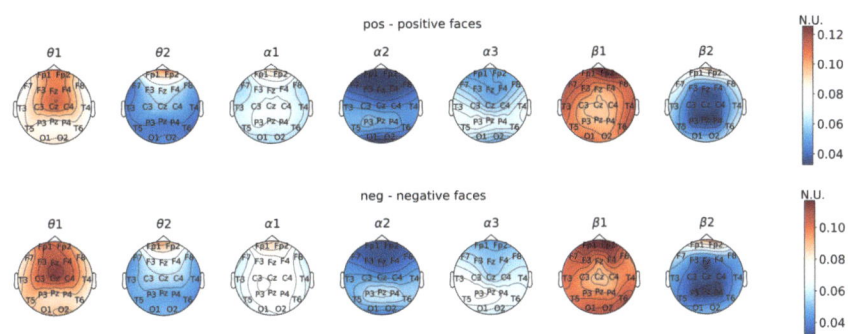

Figure 1. Topographic distribution of PSD values during the perception of positive and negative faces.

A comparison of the EEG data recorded during the demonstration of the first and second series of neutral images revealed the development of a well-defined focus of activation in the left anterior parietal cortical area within the θ_2 sub-band when the subjects were processing neutral faces preceded by negative images (Figure 2).

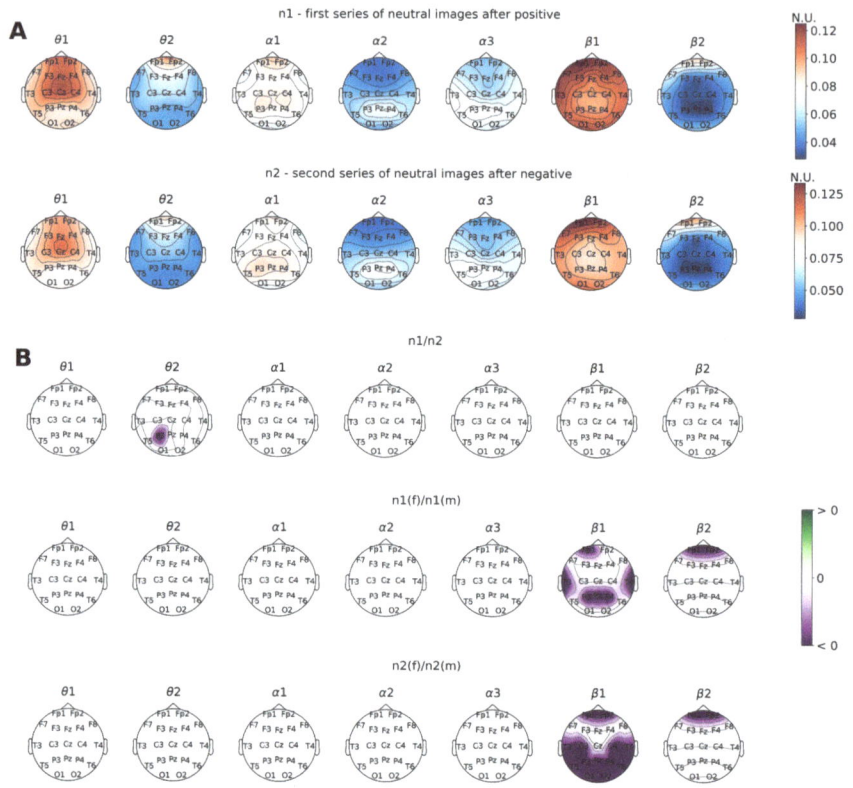

Figure 2. Topographic distribution (**A**) and regions of statistically significant differences (**B**) in PSD values during perception of neutral faces, which were preceded by positive (n1) and negative (n2) faces.

At the same time, the comparative analysis between male and female subject groups highlighted some of the gender-specific differences in information processing that were primarily visible within the β EEG band. Thus, enhanced activation processes were located within the prefrontal and frontal cortical areas in the β_2 sub-band. Apart from that, men also tended to have generally increased activation distributed among the majority of cortical regions in the β_1 sub-band.

The analysis of coherence alterations in the paradigm of our study showed the expected generalized multiple connections within low and medium values (0.3–0.7) (Figure 3). The maximal number of connections was observed in the θ_1, θ_2, α_1, and β_2 EEG sub-bands in the case of positive stimulus perception. Exposure to visual stimuli with negative emotional coloring was accompanied by a less pronounced generalization of distant EEG synchronization (a smaller number of functional connections in the neocortex), with the maximum represented in the θ_1, θ_2, and α_1 EEG bands. (Figure 3). Lastly, no dominant foci of connections were revealed in either positive or negative emotional stimuli.

Regarding the high values of coherence coefficients ($0.7 < K < 1$), it can be concluded that, during exposure to both emotional series, there was a central-right-sided association observed in the frontal, central, and parietal–posterior temporal zones within all EEG bands (Figure 3) In addition, there was a distant bilateral connection formed in the temporal and occipital regions within the θ_1, θ_2, and α_1 sub-bands.

Moreover, coherence topography appeared to be similar in the two emotional series. However, positive facial perception induced an additional left-hemispheric posterior–frontal–temporal association (F7–T3–T5) in the θ_1, θ_2, and α_1 sub-bands alongside a symmetrical network of connections in the right hemisphere (F8–T4–T6) in the θ-band. The perception of negative stimuli was accompanied by the chain of distant connections described above, complemented by frontotemporal relations in the right hemisphere within the α_1 sub-band, and posterior temporal connections in the α_2 and α_3 bands of oscillations. Lastly, the left hemisphere's rear frontal and temporal cortical areas were interconnected within α_2 EEG sub-bands (Figure 3).

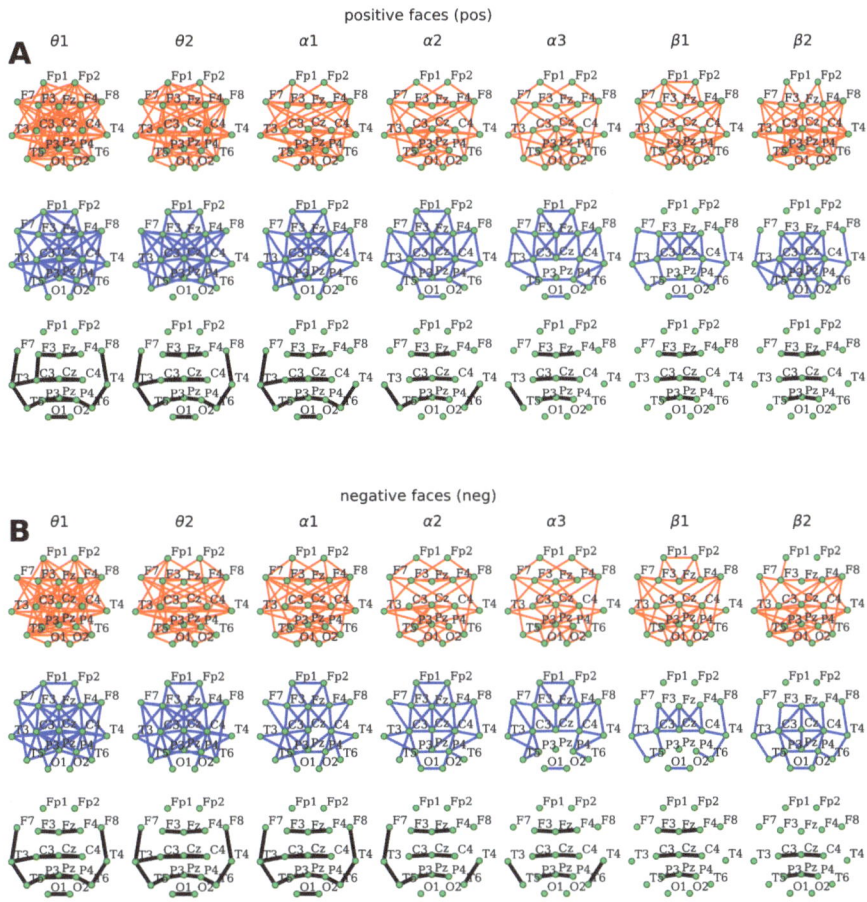

Figure 3. Topographical maps of coherent connections of different strength (red—$0.3 < K < 0.5$, blue—$0.5 < K < 0.7$, black—$0.7 < K < 1.0$) within EEG sub-bands during emotional facial perception: reaction to positive faces (**A**), reaction to negative faces (**B**).

Likewise, the topography of distant coherent connections distribution ($0 < K < 1$) during both neutral emotion-primed series perception appeared to be generally similar to the one observed in the case of emotional stimulus processing. Moreover, the composition of connections was almost identical for the groups of low-, middle-, and high-coherence levels (Figure 4).

Nonetheless, several differences were observed for the group with high coherence levels ($0.7 < K < 1$). The perception of neutral faces modulated by negative stimuli, compared to the series of positive-primed ones, was characterized by the absence of the left hemispheric connection within frontocentral (F3–C3) cortical areas in the θ_1 sub-band and the right hemispheric temporocentral (C4–T4) relation in θ_2 (Figure 4). In addition, there was an interaction between the left posterior-temporal and parietal regions (T5–P3) observed in the α_3 EEG sub-band during the perception of neutral stimuli preceded by negative ones (Figure 4). Otherwise, no significant differences were revealed in the topography of distant EEG-synchronization distribution.

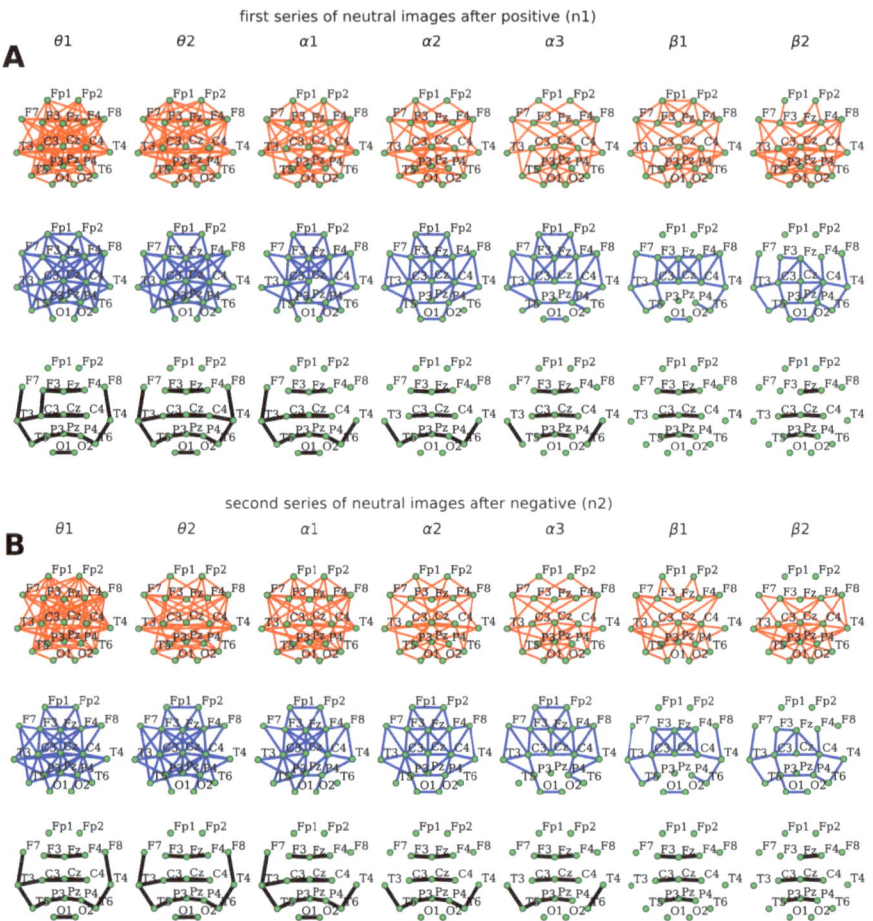

Figure 4. Topographical maps of coherent connections of different strength (red—$0.3 < K < 0.5$, blue—$0.5 < K < 0.7$, black—$0.7 < K < 1.0$) within EEG sub-bands during emotional facial perception: reaction to the first set of neutral faces after positive faces (**A**); reaction to the second set of neutral faces after negative faces (**B**).

Furthermore, the analysis of the topographical distribution of distant connections in our study from the gender perspective showed a general decrease in connection strength ($K < 0.7$) throughout the consequent stages of the experiment. Moreover, in almost all cases, changes in coherence levels were mainly located in the orbitofrontal cortex and precentral areas cortical areas ($F3; Fz; C3; Cz; C4$) within the right hemisphere in almost all EEG ranges (Figure 5). At the same time, a differential effect of the presented facial valence on the topography and the number of distant connections were observed. For instance, the second series of neutral faces processing was characterized by fewer coherent connections than the first neutral series was. On the other hand, the demonstration of emotional faces was associated with increased functional connections, especially when the subjects were presented with positive stimuli. In addition, enhanced coherence levels were observed in the female group within the β_2 sub-band during all types of experimental stimulation. For instance, positive stimuli elicited the emergence of symmetrical connections between the left central (C3) and posterior frontal areas (F7; F8) ($K < 0.5$), while the presentation of negative faces resulted in left hemispheric temporal association (T3–T5) ($K > 0.7$). In the case of neutral stimulus processing, connections within the left frontal area $Fp1$–Fz (in n1 series) and $Fp1$–$F3$ (in n2 series) were observed ($K < 0.7$). Lastly, the values of the coherence coefficient were generally higher ($0.5 < K < 0.7$) for bands of low-frequency EEG oscillations ($\theta_1, \theta_2; \alpha_1$), and lower ($K < 0.5$) for high-frequency oscillations ($\alpha_3; \beta_1, \beta_2$) (Figure 5).

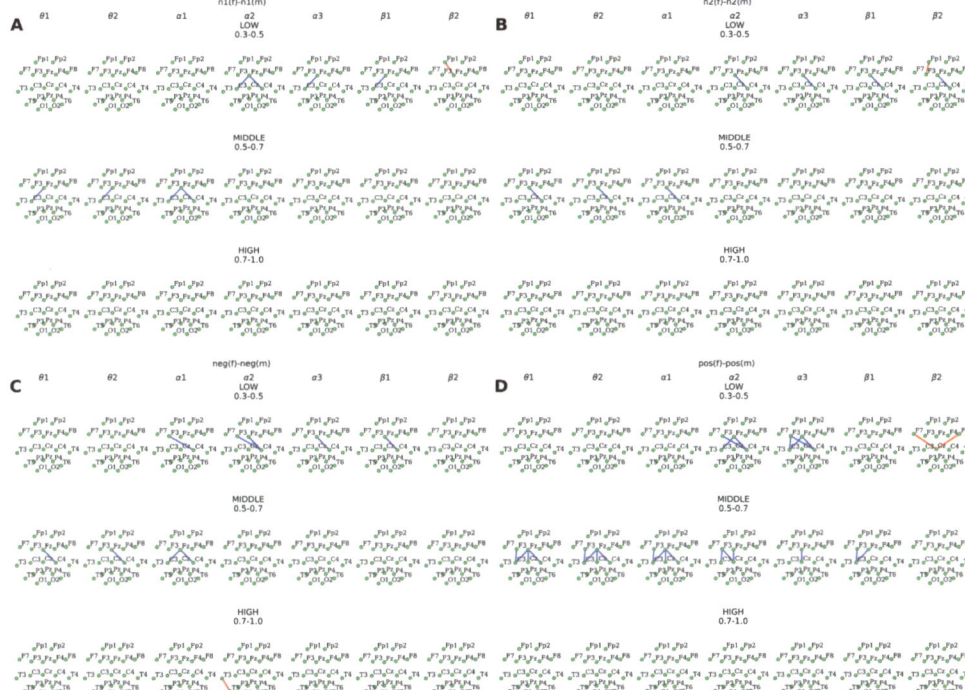

Figure 5. Topographical distribution of statistically significant coherent connection differences. (**A**) n1 (f)/ n1(m). (**B**) n2(f)/n2(m). (**C**) neg (f)/ neg (m). (**D**) pos (f)/ pos (m). Red line—the difference between medians is larger than zero; blue line—the difference is less than zero; no line—there is no significant differences.

At the average DMA analysis level, no statistically significant differences were obtained for the trials with emotional stimuli (Figure 6).

Interestingly, the DMA algorithm results marked the tendencies mentioned above on the intergroup level from a slightly different perspective. In this case, the male group showed an occipitoparietal activation region within the β_1 EEG sub-band while processing neutral images preceded by both positive and negative emotional stimuli. This fact once again stands for the general activation of the cognitive beta network. At the same time, the female group was characterized by an extensive network of connections among the temporal, central, and frontal regions in the α_3 sub-band when neutral faces were presented after the negative ones.

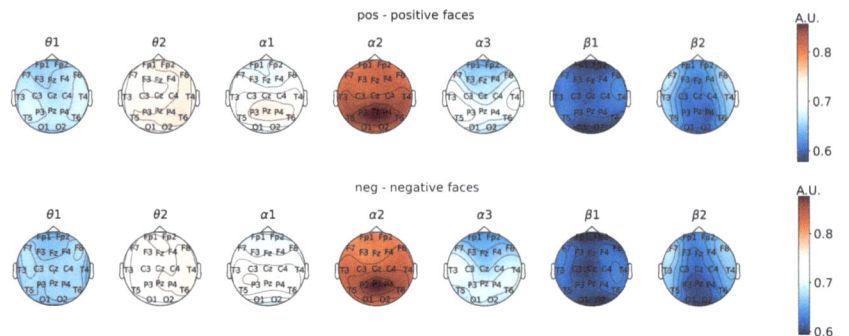

Figure 6. Topographical distribution of DMA scaling exponent values.

At the general level, DMA analysis displayed areas of statically significant connections in the left temporal and frontal cortex when subjects were exposed to neutral facial expressions after the positive series (Figure 7).

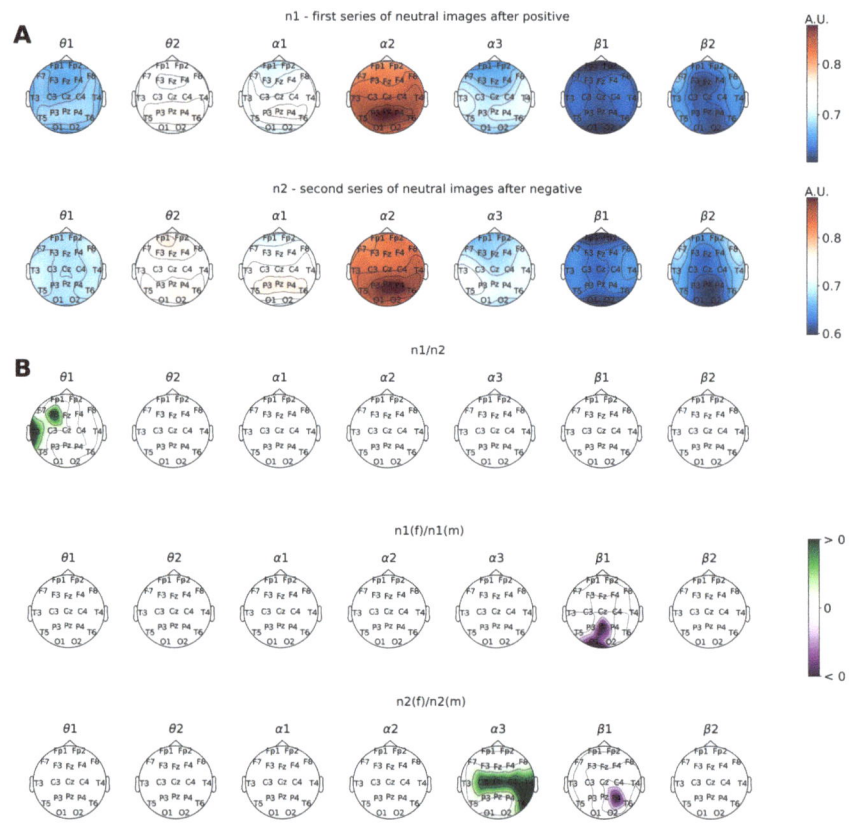

Figure 7. Topographical distribution (**A**) and regions of statistically significant differences (**B**) in DMA scaling exponent values.

4. Discussion

Regarding the PSD results, the involvement of left parietal cortical regions might represent the effect of negative emotional priming on association and verbalization-related neurodynamics, while a positive background enhances memorization [30]. The prefrontofrontal locus of activation in β_1 might highlight the increase in emotional regulation, internal attention, and verbalization in the male group [31]. In comparison, the generalized enhanced activation within the β_1 sub-band in the male group can be seen as evidence of expanded spatial-attention-related processes in the visual domain.

In general, obtained results correspond to current accumulated data. They indicate a greater reactivity of the high-frequency part of the EEG spectral parameters, especially in response to the negative modality of demonstrated faces. Apart from that, emphasis was placed on the high significance of the theta EEG oscillations in the anterior cortical areas alongside high-frequency bands in the temporal, parietal and occipital regions (presumably including the fusiform gyrus), which play a crucial role in the mechanisms of facial perception and processing [32–34].

Unlike the PSD algorithm, the use of coherence analysis, being indicative of the oscillatory processes synchronicity in different cortical areas, allowed for identifying even subtler information exchange in the brain under the conditions of our examination. Thus, it was possible to show the functional correlation of the topographically distinct cortical regions during the perception of emotional faces. However, in contrast to the available data in the literature [33], we observed coherence alterations in almost all EEG bands, generally over all cortical regions, especially in the middle range of coherence coefficient values, which seems entirely consistent, taking into account modern concepts regarding the complex nature of mechanisms underlying the perception of human faces, especially with a different emotional modality. The introduction of a separate analytical technique for the functional connectivity assessment at high coherence levels ($K > 0.7$) allowed for identifying an area with a changing rate of excitability depending on emotional modality. We observed the formation of a correlation hub in the prefrontal region (including the temporal and parietal areas, which are seen as the projections of the amygdala and cingular cortex). This finding highlights the role of this cerebral region in terms of sensory signal processing, cognitive behavioral strategy setting, decision making, and, more importantly, the actualization of attention processes. It turned out that increased reactivity to negative facial expressions also radiates its effect on the perception of neutral faces in the case of emotional priming.

The introduction of the gender parameter into the analysis also allowed for showing the locus of distant coherence changes at the group level. It was revealed that gender differences appear to be manifested in all experimental sessions. In this instance, the main hub of interest included coherent coupling between the prefrontal and anterior parietal regions in the left hemisphere, and these functional associations were less prominent in the female group. In addition, the general decrease in coherent connections during the perception of positive images is much more topographically expressed, indicating a more activating component of negative perception. The observation that image perception is accompanied by changes in functional association mainly in the left hemisphere may be due to the fact cortical areas of this hemisphere are involved in the categorical (semantic) analysis of information, which is a necessary stage of perception.

Notably, the only EEG range where coherence coefficient levels were significantly high in the female group was the β_2 sub-band. Within this oscillatory range, we detected a functional association between the anterior temporal and posterior frontal regions of both hemispheres, and the vertex, which might indicate the activation of working memory mechanisms during perceptual task execution [35].

The tendencies brought up by the DMA-based analysis corresponded to the PSD results. The activation region within the temporofrontal area in the left hemisphere might reflect the remaining positive experience trail alongside the intensification of inner attention mechanisms [36] and working memory activation traces against the background of positive emotional experience. The occipitoparietal area of coherent coupling in the male group might support the general activation of the cognitive beta network (neocortical mechanisms). At the same time, the pronounced pattern of coherent associations in the female group within the α_3 sub-band might represent not only the specific activation of working memory mechanisms, but also the lack of the downstream control and suppressive function of attention upon emotions and mental imagery and semantic activations [37,38]. Lastly, generally enhanced levels of the DMA coefficient were observed in the central parietal cortical area and supposedly the fusiform gyrus, which are known for their role in facial perception and recognition [32,33].

Lastly, in the framework of this work, we proceeded from notions of the physiological and functional heterogeneity of the brain's rhythmic activity. In the context of this paradigm, all EEG bands can be divided into sub-bands, which allows for studying the mechanisms of cognitive functioning more subtly. For instance, strong evidence exists that the θ EEG band has a heterogeneous physical and functional structure. However, at present, most authors analyze the range as a whole (4–7.5 Hz), focusing separately on changes in its individual sub-bands depending on the current functional state of the brain [39,40].

At the same time, there are several studies in which the θ band of oscillations is primordially divided into two sub-bands associated with the activation of the emotion- and memory-related mechanisms, respectively [41–43].

In the context of this work, we expected to highlight the prognostic significance of dividing the θ range into two sub-bands under the conditions of using various types of EEG analysis. Our results show the legitimacy of separating the theta range into at least two subranges (Figures 2–7) that indicate different cognitive response mechanisms.

The same opinion ambiguity is seen when it comes to α band analysis. We built our study on the research evidence, which suggests functional variety in the α EEG band. Therefore, α_1—(8–9 Hz) is considered to play an inherent role as a correlate of memory processes; α_2—(10–11 Hz) is connected to the attention phenomenon; and α_3—(12–13 Hz) is associated with the processes of inner speech and logical thinking [44–47].

The study of brain dynamics, taking into account frequency sub-bands, can provide nontrivial information about neural networks functioning during mental and cognitive activation, especially when using modern nonlinear analytical algorithms.

5. Conclusions

The current study emphasized different aspects of facial perception and processing neurodynamics using methods of particular traits of EEG signals: power spectral distribution, coherence, and long-range DMA-derived connections. The main gender-specific difference consisted in the activation of the two major systems of cognitive function, namely, emotional limbic-associated (theta system) and cognitive cortical-associated (beta system). We can conclude that the DMA algorithm is suitable for oscillatory stability evaluation, which is particularly important considering the topographical overlap established by all three analytical algorithms. Thus, it can be combined with more common methods of EEG data analysis, such as PSD and coherence, to form a more complete picture of cortical functioning.

Author Contributions: Conceptualization, I.Z. and M.C.; methodology, I.Z., M.C., K.K. and A.P.; software, I.S., B.V. and D.H.; validation, I.Z. and K.K.; formal analysis, M.C., I.Z., A.P. and K.K.; data curation, M.C., B.V. and I.S.; writing—original draft preparation, M.C., I.Z., B.V. and I.S.; writing—review and editing, A.P. and K.K.; visualization, M.C., B.V., I.S. and D.H.; supervision, I.Z., K.K. and A.P.; project administration, A.P.; funding acquisition, K.K. All authors have read and agreed to the published version of the manuscript.

Funding: This work was partially supported by JSPS KAKENHI Grant No. 20K20659.

Institutional Review Board Statement: The study was conducted in accordance with the Declaration of Helsinki, and approved by the Institutional Ethics Committee of the Educational and Scientific Center "Institute of Biology and Medicine", Taras Shevchenko National University of Kyiv (#14-510, 21 May 2022).

Informed Consent Statement: Informed consent was obtained from all subjects involved in the study.

Data Availability Statement: The data presented in this study are available on request from the corresponding author.

Conflicts of Interest: The authors declare no conflict of interest.

References

1. Wang, P.; Gauthier, I.; Cottrell, G. Are Face and Object Recognition Independent? A Neurocomputational Modeling Exploration. *J. Cogn. Neurosci.* **2016**, *28*, 558–574. [CrossRef]
2. Di Lorenzo, R.; Munsters, N.M.; Ward, E.K.; de Jonge, M.; Kemner, C.; van den Boomen, C. Is It Fear? Similar Brain Responses to Fearful and Neutral Faces in Infants with a Heightened Likelihood for Autism Spectrum Disorder. *J. Autism Dev. Disord.* **2021**, *51*, 961–972. [CrossRef]
3. Acunzo, D.; MacKenzie, G.; van Rossum, M.C.W. Spatial attention affects the early processing of neutral versus fearful faces when they are task-irrelevant: A classifier study of the EEG C1 component. *Cogn. Affect. Behav. Neurosci.* **2019**, *19*, 123–137. [CrossRef]
4. Soleymani, M.; Pantic, M.; Pun, T. Multimodal emotion recognition in response to videos. *IEEE Trans. Affect. Comput.* **2012**, *3*, 211–223. [CrossRef]
5. Kortelainen, J.; Vayrynen, E.; Seppanen, T. High-frequency electroencephalographic activity in left temporal area is associated with pleasant emotion induced by video clips. *Comput. Intell. Neurosci.* **2015**, *2015*, 762769. [CrossRef]
6. González-Garrido, A.A.; Gómez-Velázquez, F.R.; Salido-Ruiz, R.A.; Espinoza-Valdez, A.; Vélez-Pérez, H.; Romo-Vazquez, R.; Gallardo-Moreno, G.B.; Ruiz-Stovel, V.D.; Martínez-Ramos, A.; Berumen, G. The analysis of EEG coherence reflects middle childhood differences in mathematical achievement. *Brain Cogn.* **2018**, *124*, 57–63. [CrossRef]
7. Peng, C.-K.; Havlin, S.; Stanley, H.; Goldberger, A.L. Quantification of scaling exponents and crossover phenomena in nonstationary heartbeat time series. *Chaos Interdiscip. J. Nonlinear Sci.* **1995**, *5*, 82–87. [CrossRef]
8. Kiyono, K.; Tsujimoto, Y. Nonlinear filtering properties of detrended fluctuation analysis. *Phys. Stat. Mech. Its Appl.* **2016**, *462*, 807–815. [CrossRef]
9. Alessio, E.; Carbone, A.; Castelli, G.; Frappietro, V. Second-order moving average and scaling of stochastic time series. *Phys. Condens. Matter* **2002**, *27*, 197–200. [CrossRef]
10. Carbone, A.; Kiyono, K. Detrending moving average algorithm: Frequency response and scaling performances. *Phys. Rev. E* **2016**, *93*, 063309. [CrossRef]
11. Kiyono, K.; Tsujimoto, Y. Time and frequency domain characteristics of detrending-operation-based scaling analysis: Exact DFA and DMA frequency responses. *Phys. Rev. E* **2016**, *94*, 012111. [CrossRef] [PubMed]
12. Holl, M.; Kiyono, K.; Kantz, H. Theoretical foundation of detrending methods for fluctuation analysis such as detrended fluctuation analysis and detrending moving average. *Phys. Rev. E* **2019**, *99*, 033305. [CrossRef] [PubMed]
13. Sengupta, S.; Biswas, S.; Nag, S.; Sanyal, S.; Banerjee, A.; Sengupta, R.; Ghosh, D. Emotion specification from musical stimuli: An EEG study with AFA and DFA. In Proceedings of the 2017 4th International Conference on Signal Processing and Integrated Networks (SPIN), Noida, India, 2–3 February 2017; pp. 596–600. [CrossRef]
14. Jospin, M.; Caminal, P.; Jensen, E.W.; Litvan, H.; Vallverdú, M.; Struys, M.M.; Vereecke, H.E.M.; Kaplan, D.T. Detrended Fluctuation Analysis of EEG as a Measure of Depth of Anesthesia. *IEEE Trans. Biomed. Eng.* **2007**, *54*, 840–846. [CrossRef]
15. Nguyen-Ky, T.; Wen, P.; Li, Y. An Improved Detrended Moving-Average Method for Monitoring the Depth of Anesthesia. *IEEE Trans. Biomed. Eng.* **2010**, *57*, 2369–2378. [CrossRef] [PubMed]
16. Lee, J.S.; Yang, B.H.; Lee, J.H.; Choi, J.H.; Choi, I.G.; Kim, S.B. Detrended fluctuation analysis of resting EEG in depressed outpatients and healthy controls. *Clin. Neurophysiol.* **2007**, *118*, 2489–2496. [CrossRef]
17. Choong, W.Y.; Khairunizam, W.; Omar, M.I.; Murugappan, M.; Abdullah, A.H.; Ali, H.; Bong, S.Z. EEG-Based Emotion Assessment using Detrended Flunctuation Analysis (DFA). *J. Telecommun. Electron. Comput. Eng. (JTEC)* **2018**, *10*, 105–109.
18. Lang, P.J.; Bradley, M.M.; Cuthbert, B.N. International affective picture system (IAPS): Technical manual and affective ratings. *NIMH Cent. Study Emot. Atten.* **1997**, *1*, 3.
19. Duan, W.; Chen, X.; Wang, Y.J.; Zhao, W.; Yuan, H.; Lei, X. Reproducibility of power spectrum, functional connectivity and network construction in resting-state EEG. *J. Neurosci. Methods* **2021**, *348*, 108985. [CrossRef]
20. Seleznov, I.; Zyma, I.; Kiyono, K.; Tukaev, S.; Popov, A.; Chernykh, M.; Shpenkov, O. Detrended Fluctuation, Coherence, and Spectral Power Analysis of Activation Rearrangement in EEG Dynamics During Cognitive Workload. *Front. Hum. Neurosci.* **2019**, *13*, 270. [CrossRef]
21. Gu, G.-F.; Zhou, W.-X. Detrending moving average algorithm for multi-fractals. *Phys. Rev. E* **2010**, *82*, 011136. [CrossRef]
22. Feldman, M. Hilbert transform in vibration analysis. *Mech. Syst. Signal Process.* **2011**, *25*, 735–802. 2010.07.018. [CrossRef]
23. Savitzky, A.; Golay, M.J. Smoothing and differentiation of data by simplified least squares procedures. *Anal. Chem.* **1964**, *36*, 1627–1639. [CrossRef]
24. Tsujimoto, Y.; Miki, Y.; Shimatani, S.; Kiyono, K. Fast algorithm for scaling analysis with higher-order detrending moving average method. *Phys. Rev. E* **2016**, *93*, 053304. [CrossRef] [PubMed]

25. Basharpoor, S.; Heidari, F.; Molavi, P. EEG coherence in theta, alpha, and beta bands in frontal regions and executive functions. *Appl. Neuropsychol. Adult* **2021**, *28*, 310–317. [CrossRef] [PubMed]
26. Wu, X.; Zheng, W.-L.; Lu, B.-L. Identifying functional brain connectivity patterns for EEG-based emotion recognition. In Proceedings of the 2019 9th International IEEE/EMBS Conference on Neural Engineering (NER), San Francisco, CA, USA, 20–23 March 2019; pp. 235–238.
27. GitLab. Available online: https://gitlab.com/BVod/diplom-eeg-v1.0/-/blob/coherence/coh.py (accessed on 2 February 2022).
28. Faes, L.; Pinna, G.D.; Porta, A.; Maestri, R.; Nollo, G. Surrogate data analysis for assessing the significance of the coherence function. *IEEE Trans. Biomed. Eng.* **2004**, *51*, 1156–1166. [CrossRef] [PubMed]
29. Šidák, Z. Rectangular Confidence Regions for the Means of Multivariate Normal Distributions. *J. Am. Stat. Assoc.* **1967**, *62*, 626–633. [CrossRef]
30. Schneider, J.M.; Abel, A.D.; Ogiela, D.A.; Middleton, A.E.; Maguire, M.J. Developmental differences in beta and theta power during sentence processing. *Dev. Cogn. Neurosci.* **2016**, *19*, 19–30. [CrossRef]
31. Taran, S.; Bajaj, V. Emotion recognition from single-channel EEG signals using a two-stage correlation and instantaneous frequency-based filtering method. *Comput. Methods Programs Biomed.* **2019**, *173*, 157–165. [CrossRef]
32. Shu, N.; Liu, Y.; Duan, Y.; Li, K. Hemispheric Asymmetry of Human Brain Anatomical Network Revealed by Diffusion Tensor Tractography. *BioMed Res. Int.* **2015**, *2015*, 908917. [CrossRef]
33. Güntekin, B.; Başar, E. A review of brain oscillations in perception of faces and emotional pictures. *Neuropsychologia* **2014**, *58*, 33–51. [CrossRef]
34. Aktürk, T.; de Graaf, T.A.; Abra, Y.; Şahoğlu-Göktaş, S.; Özkan, D.; Kula, A.; Güntekin, B. Event-related EEG oscillatory responses elicited by dynamic facial expression. *Biomed. Eng. Online* **2021**, *20*, 1–17. [CrossRef] [PubMed]
35. Sehatpour, P.; Molholm, S.; Javitt, D.C.; Foxe J.J. Spatiotemporal dynamics of human object recognition processing: An integrated high-density electrical mapping and functional imaging study of "closure" processes. *NeuroImage* **2006**, *29*, 605–618. [CrossRef]
36. Pomper, U.; Ansorge, U. Theta-Rhythmic Oscillation of Working Memory Performance. *Psychol. Sci.* **2021**, *32*, 1801–1818. [CrossRef] [PubMed]
37. Senoussi, M.; Verbeke, P.; Desender, K.; De Loof, E.; Talsma, D.; Verguts, T. Theta oscillations shift towards optimal frequency for cognitive control. *Nat. Hum. Behav.* **2022**, *6*, 1000–1013. [CrossRef]
38. Yin, Z.; Wang, Y.; Dong, M.; Wang, Y.; Ren, S.; Liang, J. Short-range and long-range neuronal oscillatory coupling in multiple frequency bands during face perception. *Int. J. Psychophysiol.* **2020**, *152*, 26–35. [CrossRef] [PubMed]
39. Korotkova, T.; Ponomarenko, A.; Monaghan, C.K.; Poulter, S.P.; Cacucci, F.; Wills, T.; Hasselmo, M.E.; Colin Lever, C. Reconciling the different faces of hippocampal theta: The role of theta oscillations in cognitive, emotional and innate behaviors. *Neurosci. Biobehav. Rev.* **2018**, *85*, 65–80. [CrossRef]
40. Lu, L.; Ren, Y.; Yu, T.; Liu, A.; Wang, S.; Tan, L.; Zeng, J.; Feng, Q.; Lin, R.; Liu, Y.; et al. Control of locomotor speed, arousal, and hippocampal theta rhythms by the nucleus incertus. *Nat. Commun.* **2020**, *11*, 262. [CrossRef]
41. Krause, C.M.; Viemerö, V.; Rosenqvist, A.; Sillanmäki, L.; Åström, T. Relative electroencephalographic desynchronization and synchronization in humans to emotional film content: An analysis of the 4–6, 6–8, 8–10 and 10–12 Hz frequency bands. *Neurosci. Lett.* **2000**, *286*, 9–12. [CrossRef]
42. Teplan, M.; Krakovská, A.; Stolc, S. EEG responses to long-term audio–visual stimulation. *Int. J. Psychophysiol.* **2006**, *59*, 81–90. [CrossRef]
43. Razumnikova, O.M. Gender differences in hemispheric organization during divergent thinking: An EEG investigation in human subjects. *Neurosci. Lett.* **2004**, *362*, 193–195. [CrossRef] [PubMed]
44. Vogt, F.; Klimesch, W.; Doppelmayr, M. High-frequency components in the alpha band and memory performance. *J. Clin. Neurophysiol.* **1998**, *15*, 167–172. [CrossRef] [PubMed]
45. Feshchenko, V.A.; Reinsel, R.A.; Veselis, R.A. Multiplicity of the alpha rhythm in normal humans. *J. Clin. Neurophysiol.* **2001**, *18*, 331–344. [CrossRef] [PubMed]
46. Sadaghiani, S.; Scheeringa, R.; Lehongre, K.; Morillon, B.; Giraud, A.-L.; Kleinschmidt, A. Intrinsic Connectivity Networks, Alpha Oscillations, and Tonic Alertness: A Simultaneous Electroencephalography/Functional Magnetic Resonance Imaging Study. *J. Neurosci.* **2010**, *30*, 10243–10250. . [CrossRef] [PubMed]
47. Aftanas, L.I.; Golocheikine, S.A. Human anterior and frontal midline theta and lower alpha reflect emotionally positive state and internalized attention: High-resolution EEG investigation of meditation. *Neurosci. Lett.* **2001**, *310*, 57–60. [CrossRef]

Article

Promise for Personalized Diagnosis? Assessing the Precision of Wireless Consumer-Grade Electroencephalography across Mental States

Amedeo D'Angiulli [1,2,*], Guillaume Lockman-Dufour [1] and Derrick Matthew Buchanan [1]

[1] NICER Lab, Carleton University, Ottawa, ON K1S5B6, Canada; guillaumelockmandufo@cmail.carleton.ca (G.L.-D.); matthewbuchanan@cmail.carleton.ca (D.M.B.)
[2] Department of Neuroscience, Carleton University, Ottawa, ON K1S5B6, Canada
* Correspondence: amedeo.dangiulli@carleton.ca

Citation: D'Angiulli, A.; Lockman-Dufour, G.; Buchanan, D.M. Promise for Personalized Diagnosis? Assessing the Precision of Wireless Consumer-Grade Electroencephalography across Mental States. *Appl. Sci.* **2022**, *12*, 6430. https://doi.org/10.3390/app12136430

Academic Editors: Fabio La Foresta and Serena Dattola

Received: 30 March 2022
Accepted: 17 June 2022
Published: 24 June 2022

Publisher's Note: MDPI stays neutral with regard to jurisdictional claims in published maps and institutional affiliations.

Copyright: © 2022 by the authors. Licensee MDPI, Basel, Switzerland. This article is an open access article distributed under the terms and conditions of the Creative Commons Attribution (CC BY) license (https://creativecommons.org/licenses/by/4.0/).

Abstract: In the last decade there has been significant growth in the interest and application of using EEG (electroencephalography) outside of laboratory as well as in medical and clinical settings, for more ecological and mobile applications. However, for now such applications have mainly included military, educational, cognitive enhancement, and consumer-based games. Given the monetary and ecological advantages, consumer-grade EEG devices such as the Emotiv EPOC have emerged, however consumer-grade devices make certain compromises of data quality in order to become affordable and easy to use. The goal of this study was to investigate the reliability and accuracy of EPOC as compared to a research-grade device, Brainvision. To this end, we collected data from participants using both devices during three distinct cognitive tasks designed to elicit changes in arousal, valence, and cognitive load: namely, Affective Norms for English Words, International Affective Picture System, and the *n*-Back task. Our design and analytical strategies followed an ideographic person-level approach (electrode-wise analysis of vincentized repeated measures). We aimed to assess how well the Emotiv could differentiate between mental states using an Event-Related Band Power approach and EEG features such as amplitude and power, as compared to Brainvision. The Emotiv device was able to differentiate mental states during these tasks to some degree, however it was generally poorer than Brainvision, with smaller effect sizes. The Emotiv may be used with reasonable reliability and accuracy in ecological settings and in some clinical contexts (for example, for training professionals), however Brainvision or other, equivalent research-grade devices are still recommended for laboratory or medical based applications.

Keywords: Emotiv EPOC; brainvision; brain-computer interface BCI; reliability; accuracy; arousal; valence; mental load; consumer-grade EEG; research-grade EEG

1. Introduction

1.1. Emotiv EPOC Wireless EEG Device in Context: Current Research and Applications

Since its inception, the field of electroencephalography (EEG) and its potential uses have developed considerably. In laboratory research settings, EEG has been used extensively for cognitive neuroscience research, and in medical settings, it is considered the most robust method of diagnosing seizures and epilepsy [1,2]. In the last decade, however, there has been significant growth in the interest and application of using EEG outside of the laboratory and medical settings [3]. EEG has become the most widely used neuroimaging technique for brain-computer interfaces (BCI). Some of these extended uses of EEG include military operations such as controlling weapons or drones [4–8], educational classroom applications such as monitoring student's attention/other mental states or helping them engage with material [9–13], cognitive enhancement such as increasing cognitive load or focus [12,14,15], and consumer based games such as computer games or physical toys controlled via brain waves [2,15–19].

One challenge in developing outside lab EEG applications, however, is that traditional research- and medical-grade EEG devices are costly (~US$50,000+), typically stationary, fully wired, and time consuming to properly set up, in that electrode application may typically take 30–60 min depending on the type and number of wet (gel-based) sensors [3]. For example, one research-grade EEG device is the Brainvision EEG headset. Its most typical configuration includes 32 channels and must be connected by wires to an amplifier, a computer, and a monitor, for data input and analysis. Its electrodes require the use of electro-conductive gel prior to the experiment, which allows it to minimize impedance and be more sensitive to the electrical activity produced by the brain. While these factors are a strength of Brainvision in regard to the accuracy and reliability of data collection, they can also be interpreted as limitations in other capacities. For instance, all of these factors make using traditional EEG challenging in more ecological settings (i.e., educational and training health settings). In response to these challenges, over the last decade or so, wearable, portable, cost-efficient EEG headsets have been developed [3]. To date, the most recent and exhaustive review of consumer grade EEG devices impressively covering applications in cognition, brain computer interfaces, education research, and game development has been published by Sawangjai et al. [20]. Their review dealt primarily with the validity associated with the wireless EEG devices, that is, with the issues related to whether what was measured by wireless consumer-grade devices reflected the same constructs (for example the same EEG or/and ERP components or signatures) or the same entities/features intended to be identified and measured as the ones traditionally obtained with research-grade EEG devices. However, Sawangjai et al. did not explicitly differentiate issues of validity from those of reliability, therefore, their assessment did not adequately address related issues of personalization, these are the central objective and contribution of our work.

A commercial consumer-gradedevice that emerged in response to the limiting factors associated with stationary research-grade systems was the Emotiv EPOC. The EPOC headset is much more affordable; as of fall 2021, the newer 14-channel version of the Emotiv EPOC (now called the Emotiv EPOC X) can be purchased for less than US $1000. These headsets boast much quicker and easier installation and removal times of 3–5 min, in addition to being wireless. This device also has low weight and a battery life of up to 12 hours. Such devices have the advantage of being more user and consumer-friendly making them well-suited for situations where ergonomics and patient or participant movement is involved; as well as situations where scalability is of interest (e.g., having 30 students in a classroom all wearing an Emotiv, which would be highly impractical using a research device like Brainvision). Crucially, what set EPOC apart from the other available commercial devices is that at the time of its release it had one of the largest set of wireless electrodes, making it comparable to the electrode set sizes used in a significant part of literature representing research with traditional laboratory wired EEG equipment.

Although there are some clear advantages to using a device like the Emotiv headsets in consumer/ecological and in some clinical settings, it does not come without its own challenges. Consumer grade EEG devices typically compromise data resolution for shorter setup time, affordability, and portability [21–26]. Since portable headsets have less electrodes and channels than standard research-grade EEGs, scalp coverage and spatial resolution decrease [27]. The use of dry electrodes can also pose a problem as impedance can hinder the detection of electrical activity and lead to lower data values. This can be mitigated by using saline solution, but its effect is minimal compared to electro-conductive gel [24]. Overall, concerns around portable EEGs are centred on the accuracy and reliability of their data as compared to standard, research-grade EEGs.

There is an important growing body of scientific literature investigating the Emotiv under different conditions utilizing various EEG signatures such as the P300 [24] which has demonstrated its poor reliability/accuracy compared to research-grade devices. However, other studies have demonstrated that EPOC can accurately detect P300 and N200 [26,27]. That said, the study from Barham et al. (2017) utilized a modified Emotiv with electrodes that had been upgraded to research-grade quality [26] which may have played a significant role in the success of the Emotiv application. Another study from Badcock et al. (2015) [28] further demonstrated that the morphology of EEG signals such as the P1, N1, P2, N2, and P3 were highly correlated between the Emotiv and research-grade Neuroscan system; however there were some differences in the amplitude of the signal. Needless to say, there remains a disparity in the literature which requires more research to be done to elucidate the truth about the Emotiv. There are so many factors that may influence the success of an Emotiv application: Notably, the study by Duvinage et al. (2012) [24] which reported poor quality of the Emotiv for detecting P300 utilized a visual odd-ball paradigm, whereas the studies that were more successful [26–28] utilized auditory odd-ball paradigms. Therefore, the modality of the task used to elicit the EEG signal may play an important role. However, a more recent study from Fouad (2021) [29] was able to successfully capture and utilize subject's P300 signal with reasonable accuracy for a brain-computer interface speller (a visual paradigm) when combined with machine learning such as support vector machines.

A critical emerging issue related to wireless portable EEG devices such as EPOC is the application to personalized healthcare [30]. Personalization implies that the quality of repeated measurements of desired EEG features have an acceptable to optimal level of **precision** or **reliability** (i.e., *measurement or instrumental precision*) at the level of an individual (i.e., the so-called "N-of-1") (see [31]). Although a dominant mainstream view is that some aspects of the classification and processing of the signal are deemed to be handled by the integration with machine learning on the basis of massive iterative training applied to large within subject repeated measures, the latter types of approaches are no substitutes for the quality of EEG signal acquisition from the individual subject, that is, they do not directly address the possible problems associated with *weak measurement* at the individual level. Therefore, the issues partly require approaches and techniques that address small-N or N-of-1 experimental designs, and neither can be *only* solved with the current mainstream data mining science based on large samples nor with parametrization referenced to a population distribution (see [31]).

The issues in EEG personalization largely overlap with the current debate in the multidisciplinary field of measurement which contrasts the **idiographic** person-oriented approach [32,33] versus the **nomothetic** population distribution referenced approach (for extensive overview and review and discussion within the context of psychological and brain sciences see [34]). Personalized predictions can only be made based on prior data from the individual for whom a prediction is to be made (idiographic data) and not with aggregated data from other individuals (nomothetic data) [35]. This is an implication from the classic ergodicity theorems [36], from which it follows that intra-individual (within subjects) variation (IAV) is equivalent to inter-individual (between subjects) variation (IEV), that is, they are ergodic, only if IAV is homogeneous in time, that is, if there are no fluctuations, trends or other types of time-dependent changes in the IAV time series). If the structure of IAV is heterogeneous, then it can no longer be examined by switching to the IEV perspective—as in most of the approaches relying on population distributions—because the two types of variations are incommensurable. In terms of relevance for neurocognitive processes pertinent to EEG, the non-ergodic implications also apply to functional brain connectivity [37,38] and the underlying dynamics of neural networks (e.g., see [39]).

Thus, employing small-N designs that focus on the individual participant as the *replication unit* and the *control of him/herself* is a sound alternative or/and supplementary tool to assess the precision of personalization EEG devices for the measurement of individual mental states. Indeed, there are alternative approaches that derive and conform to the tradition of behavioral [40] and psychophysical [31,41] experimental designs. One is the numerical method known as Vincentization or Vincentized average [40], in which data from small samples (N = 3–5) are binned so that the derived distribution reflects approximately the same shape as each individual, and a second approach is the item-wise (or by-item) analysis approach [42] in which the dependent variable is aggregated across subjects for a particular item (i.e., stimulus or channel) and consequently the item becomes the unit or case which the statistics is applied to, usually requiring a completely nested within-subjects design. These two methods are complementary and when used together (i.e., item-wise analysis of vincentized data) they offer the opportunity to integrate intra-individual and inter-individual/group variability by empirically constraining or restricting the range of heterogeneity of the combined variance, so that variability can be assumed to be reasonably ergodic.

In summary, there seems to be some disparity in the literature regarding the reliability, reproducibility, and accuracy of the Emotiv EPOC headset. Moreover, the literature is specifically lacking in studies investigating other characteristics of the EEG signal such as power and amplitude. The literature is also lacking in studies investigating the precision of EEG features related to a wider variety of mental states such as arousal, valence, and mental load. Most important, so far there have been only a handful of studies (reviewed above) which compared measurements collected from the *same* subjects, serving as controls of themselves, and done on *both* consumer-grade EPOC Emotiv and research-grade devices.

To start filling some of the gaps in the literature we here report a preliminary empirical study assessing the precision of the Emotiv EPOC consumer device, with a particular eye to implications for EEG personalization. That is, the empirical question we address here is one of **reliability** or **precision**: *can EEG signal measured with EPOC from an individual A, while he/she is in a defined psychological state X, consistently replicate the EEG signal from the same individual A in the same psychological state X when measured with a stationary traditional device such as Brainvision assumed as the "gold standard"*. This is independent from the validity question, that is, whether the EEG signal measured with both types of EEG devices from A can be validly classified as a correlate of the same state X *in the general population*.

1.2. The Present Study

The present study aimed at investigating the accuracy and reliability of consumer-grade portable and wireless EEG as compared to standard, wired research-grade EEG, using the Emotiv EPOC 14-channel headset and the Brainvision 32-channel headset. While, as already mentioned, there are many empirical studies and exhaustive reviews which have focused on addressing the validity of wireless consumer-grade EEG devices including Emotiv EPOC, few have directly addressed the issue of reliability of measurement, independent of the content-validity of the measurement; among the latter small literature, even fewer have compared EEG features in data collected from the same subjects using both types of devices, that is, using a repeated measures (i.e., within-subject sample-matching) design.

Accordingly, to advance current research in the field, the present study contributed a novel approach for the direct assessment of the reliability and accuracy of wireless consumer-grade EEG devices, using as paradigmatic example case Emotiv EPOC. Specifically, we investigated the features of EEG power and amplitude measurements for specific frequency bands of matched electrodes (the 14 channels these headsets share are AF3 (FP1), F7, F3, FC5, T7, P7, O1, O2, P8, T8, FC6, F4, F8 and AF4 (FP2)) between the Emotiv device and the Brainvision device during different mental states in the same participants. In the next sections, we describe the EEG features and the level of psychological states considered.

1.2.1. EEG Features: Event-Related Band Power and Amplitude

Typically, a clear separation is made between the analysis of the EEG in the frequency domain (Spectral analysis; measurements: power, coherence, phase, etc.) and the analysis of the EEG in the time domain, whose most common form is the Event-Related Potentials (ERPs), obtained by averaging activity associated with events of the same nature (reported measurements: amplitude, latency, topography).

However, most recently, hybrid approaches have also been devised in which time and frequency domain EEG features are examined as covariates (for example see [43–46]), particularly, to investigate spectral changes associated with stimulus or task time-course, as opposed to resting brain activity. In line with these approaches, in the present study we focused on two EEG event-locked or task-related features: Event-related band frequency amplitude and power. As described by Klimesch [47], the event-related **amplitude** is the 'magnitude' of an oscillation, reflecting the distance between the maximal positive- and negative-going points (phase) of an oscillatory cycle. The related feature of **power** is the under-the-curve integral of the event-related amplitude response (to a stimulus and/or task) within the entire frequency band during a test period.

1.2.2. Definition of Low vs. High Mental States

Three distinct psychological tasks were used which involve emotional and cognitive processes and have been designed and validated to elicit varying levels of arousal, valence, and mental load: Affective Norms for English Words (ANEW), International Affective Picture System (IAPS), and the n-Back task. These tasks were selected because they demonstrated reasonable validity in distinguishing a relative difference in low as opposed to high psychological states as defined by the subjective self-ratings of the participants and as reflected by the state-dependent changes in EEG correlated with those low vs high levels.

ANEW is a set of 1034 English-language words associated with emotional values [48]. This dataset was made to complement the IAPS database described below [49,50] to assess attentional and emotional processing in the auditory and verbal/linguistic domains. The words are assessed by self-ratings based on arousal, which goes from calm to excited, and valence, which ranges from unpleasant to pleasant. For example, "war" would evoke high arousal and low valence since it is stressful and displeasing, while "rollercoaster" would evoke high arousal and high valence due to its mostly pleasant and exciting nature [49]. Low level as opposed to high level of word arousal and valence has been previously demonstrated to alter cortical activity in ways measurable by EEG [51].

The IAPS is a repository of 956 photos chosen to be standardized visual stimuli in experiments of emotion and attention. Just like the ANEW system, the photos are chosen based on their ability to cause emotions and are primarily classified using valence and arousal [52]. Photos can also be rated based on dominance, but the latter variable was not considered in this study. Similarly to the ANEW, this task can create noticeable differences in EEG in relation to changes from low as compared to high levels of arousal and valence, as previous experiments have been able to detect such differences in brain activity during image processing, depending on pleasantness and unpleasantness [53].

The n-Back task was initially designed to include up to three items [54], and then upgraded to six items [55]. In this task, the participant is exposed to a stream of stimuli and must indicate whether the current stimuli matches the one n trials before [56]. For example, in a 1-back task, participants must indicate whether or not each letter matches the one right before; in a 3-back task, participants must indicate whether or not each letter matches the one 3 letters before; etc. The change in attention from low as opposed to high cognitive or mental load required for this task can be detected by EEG [57].

1.2.3. Hypothesis and Predictions

Our default working hypothesis was that the *wireless consumer-grade device we considered (Emotiv EPOC) should perform as accurately as the selected gold standard stationary device (Brainvision)*. Consequently, we predicted that the low as compared to the high levels in mental psychological states—as defined by psychological self-report in the ANEW (arousal), IAPS (valence) and n-Back (cognitive load)—should show similar within-person and homogeneous-group (i.e., vincentized) patterns of *relative* differences/changes between the corresponding EEG features (amplitude and power) associated with each respective level (i.e., low vs. high mental state). These patterns of changes should occur across all homologous pairs of compared electrodes in both types of devices. Thus, to the extent in which this default prediction would be supported for all, or some of, those psychological states, it would provide an assessment of the degree of reliability of the wireless consumer-grade device.

2. Materials and Methods
2.1. Sample and Data Re-Analysis Approach

The data utilized in this study was derived from a previous study by the same authors [20] which investigated global averages of EEG power across the scalp comparing EEG signal pooled across all electrodes (14 for EPOC vs. 32 for Brainvision). The database used repeated long recording session samples from "professional observers" in the span of 48 hours. Participants were three healthy university graduate students (mean age 24.66 years old). They all gave verbal consent to participate and waived the requirement of signed consent. The study was part of a grant funded pilot project which was retrospectively assessed and approved by Carleton University Institutional Research Ethics Board (section Human Subjects) under the regulation set by the Canadian Tri-Council [58].

All participants were right-handed males with corrected-to-normal or normal vision, and none reported any history of neurological impairment or were currently using psychoactive medications. The average number of hours of sleep on the previous night was 7.5 h. Testing was conducted from 10 a.m. to 4 p.m. with a one-hour break in between on two separate days. The pre-experimental short adult version of the *mood and feelings questionnaire* [5] was administered by an independent research assistant unaware of the hypotheses and goals of the study to screen for mood differences or emotional changes before the experimental sessions between days. No remarkable differences were reported as all participants scored similarly (overall score range: 3–5) in the two days and consistently well below the recommended clinical cut-off (possible maximum = 26; clinical cut-off ≥ 11).

Prior to the experiment, all participants gained familiarity with the actual stimuli and conditions of the tasks during several study sessions, since they were involved in selecting the stimuli and the conditions, they designed the computer programs for the tasks, and they test piloted the delivery and the performance of the task computer programs and the instrumentation. Since all these activities involved extensive exposure and practice over time, it can be assumed the materials of the tasks were overlearned and practice effects, particularly due to the order in which the devices were used, were washed away. This compensated for lack of counterbalancing due to small sample.

The present study utilized a different analytical strategy going beyond global EEG patterns, to focus specifically on the accuracy and reliability of different EEG signals at electrode sites matched between the two devices; these present analyses and results are novel unpublished findings. Thus, the unit of analysis, or "subjects" were the 14 EEG electrodes. This type of within-subject approach is usually traditionally known as by-item or item-wise analysis [42,59] or, in keeping with EEG terminology, an *electrode-wise* analysis.

The item-wise approach corresponds to a random effects model on the items, and a fixed-effects model on subjects. Therefore, the effects from statistical tests can be generalized to new items and task from the same subjects but cannot be used to make a reliable prediction for new subjects that would be generalized to the population [60]. The mean-wise group approach is ordinarily used to obtain the highest possible effect. In contrast, the item-wise approach centers around the pursuit of replicable effects based on weak but stable interindividual correlation coefficients, that is, on relatively homogeneous intraindividual variance. The weak correlation for an item is generally due to excessive interindividual noise. To reduce such noise heterogeneity, in our study the participants' data were first vincentized, that is, they were partitioned according to time-series bins with same data density per interval of time by averaging the participants' quantile functions, as previously done by D'Angiulli et al. (2020) [61]. Successively, bin-by-bin means were estimated for each electrode (i.e., across participants). This procedure insured the definition of group quantiles from which a reliable distribution function could be constructed for the data from each electrode even with N = 3 (see [62]). In essence and more simply, the outcome of the vincentization procedure is obtaining an average distribution that reflects closely the distribution of each individual subject, and as mentioned this optimizes ergodicity in the data. It is of critical importance to point out here that the statistical tests of accuracy were applied to the repeated measures linked with the set of electrodes, hence, the size of the study coincides with the size of the repeated measurements (N = 14) not the actual sample size (number of participants).

In sum, we used an approach through which, by combining electrode-wise and vincentization techniques, we obtained a completely nested within-subject design which tends to homogenize (i.e., limit heterogeneity of) the combined inter- and intraindividual variance. The latter set up therefore permits to draw inferences more confidently at the idiographic or individual person prediction level, which ultimately is the purpose of the main applications of consumer-grade devices such as EPOC.

2.2. Electrophysiological Measures

The portable consumer grade device used in this study was the 14-electrode Emotiv EPOC. The research grade device used in this study was the 32-electrode Brainvision system. EPOC was utilized first followed by Brainvision.

For the Brainvision actiCHamp, EEG activity was first amplified and then sampled at 1 kHz. An online bandpass filter from 1 to 100 Hz was applied. During offline processing, the Brainvision data was downsampled using EEGLAB [63] to 128 Hz to make it compatible with the EPOC sampling rate.

For the EPOC, EEG signals were initially preprocessed as specified by the manufacturer were high-pass filtered with a 0.16 Hz cut-off, pre-amplified and low-pass filtered at an 83 Hz cut-off. The data were then digitized at 2048 Hz. The digitized signal was further filtered using a 5th-order sinc notch filter, and down-sampled to 128 Hz following standard practices used within the community of users (for example, see [64]).

For both systems, the average offline EEG signal was subtracted from each electrode for each time point (offline re-referencing to a common mean reference). The offline averaging was done in a way that allowed separate averages for the **Event-Related Band Potentials (ERBPs)** in each experimental condition (such as the different tasks described below in Sections 2.3 and 2.4) and stimulus type for electrodes with epochs ranging from −200 ms pre-stimulus to 1000ms post-stimulus. The trials where excessive peak-to-peak deflection occurred at non-ocular electrode placements were considered contaminated and excluded from the averages. Following artefact correction and removal, less than 15% of trials were rejected.

To extract the EEG features, we applied the Fast Fourier Transform (FFT) algorithm to transform time recordings into frequency recordings in order to compute the continuous input EEG data for each condition. We tabulated and computed the total spectral content of the electrical activity of each electrode, using a −1024 to 1024 ms time window. We computed FFTs for each participant on a grand average for each condition (high arousal, low arousal, high valence, low valence) on both the ANEW and IAPS tasks, and low and high load for the two versions of the n-Back task. FFT computation was performed through the BESA spectral analysis pipeline (http://wiki.besa.de/index.php?title=BESA_Research_Spectral_Analysis, accessed on 20 May 2022). The FFT was applied to the marked regions of the blocked data windows to examine the frequency content of the signal. FFTs were normalized and the frequency power and amplitude bands from 1–50 Hz were exported for further analysis (FFT was viewed as amplitude and power in frequency bands, and the tabulated results were saved in ASCII data files). The power Band frequencies were defined as follows: Delta (1.0 Hz–4.0 Hz), Theta (4.0 Hz–8.0 Hz), Alpha (8.0 Hz–14.0 Hz), Beta (14.0 Hz–30.0 Hz), Gamma (30.0 Hz–50.0 Hz).

2.3. IAPS and ANEW Procedures

The order in which the three observers carried out the tasks were respectively: (1) ANEW, IAPS, n-Back; (2) n-Back, IAPS, ANEW; (3) IAPS, n-Back, ANEW.

Participants were informed by pre-set computer cues of the expectations for each task, such as whether pictures, words, or letters would be presented and which items required a response. Participants were one meter away from the computer screen where the stimuli were presented. The researcher stood two meters behind the participant and offered verbal instructions. Participants were primed with instructions to experience emotional connection with every presented image or word. Participants viewed every presented image/word for five seconds during a period of 10 min. After each block, participants were allowed a break for as long as required, which generally lasted five minutes (this was consistent between the testing completed with EPOC and the testing completed with Brainvision). During these breaks we checked electrode contact and stability of measurement. The stability of EEG signal from Brainvision was checked and maintained by monitoring the impedance of active electrodes and ensuring that it was kept below 5kOhms; this was also facilitated by the use of electrogel. The stability of EEG signal from EPOC was maintained by removing the device after each task and reapplying saline solution. Head positions were re-measured to maintain consistency between electrode placements each time the device was replaced on the head.

In total, 50 images/words were randomly presented to participants, with a break after the 25th and 50th image/word. After participants confirmed readiness, the experimenter launched the stimulus presentation program. Stimulus presentation began with a central cross fix for 200 ms to help participants focus their gaze and reduce eye movements. After this, image/word presentation occurred for 4500 ms, followed by a 1300 ms delay before the presentation of the cross fix. Images were based on scores of the standardized IAPS/ANEW scale; they were designated as extremely high or low arousal in a block, and extremely high or low valence in another block. Blocks were sets of 50 image/word presentations. Participants were told to stay still while viewing images/words. Images filled a 48.3 cm (19") monitor with a 1024 × 768 pixel resolution, while letters had a font size of 48 points and were shown in the center of the monitor.

2.4. n-Back Procedures

Participants performed a memory task that consisted of remembering a letter based on its position in a sequence of continuously presented letters. A cross-fix would also appear at the center of the screen before each letter's presentation. The variations of this task were the 1-back and the 3-back. In the 1-back, participants were told to press a key if the letter displayed matched the one presented right before. The rationale is identical for the 3-back, where participants must indicate whether or not each letter matches the one shown three letters before. Each letter of size 48 was shown for 500 ms in the center of the monitor. EEG was recorded for all answers, valid and invalid alike. Every participant finished six blocks of the two task variations and performed three of each, alternating between variations and starting with the 1-back. Each block had 26 trials for a total of 78 trials for each variation. Subjects performed three blocks of 13 trials for the 1-back condition, and three blocks of 13 trials for the 3-back condition. A total of 117 ERBP averages for the 1-back condition and 117 ERBP averages for the 3-back condition were obtained. No trials were omitted for the *n*-Back task.

2.5. Data Analysis

The estimated marginal means were charted for the absolute value of frequency band power and its amplitude, which were expressed in microvolts (μV), for each headset's electrodes. The estimated marginal means were plotted across participants though the analysis was performed according to the electrode-wise schema (one-to-one paired matching between homologous electrodes). In order to compare headset accuracy and reliability when using the same channels, the Emotiv EPOC's 14 electrodes were matched to Brainvision's 14 correspondingly-located electrodes. Thus, the 14 electrodes used in Emotiv were AF3, AF4, F3, F4, F7, F8, FC5, FC6, T7, T8, P8, P7, O1 and O2, while the 14 ones used in Brainvision were FP1, FP2, F3, F4, F7, F8, FC5, FC6, T7, T8, P8, P7, O1, O2 and Oz (see Figure 1). (Note that the EPOC system doesn't have electrodes placed along the midline of the head, and it uses the CMS and DRL electrodes as dual reference points, instead of a single ground electrode as in the Brainvision system. Pooled reference electrodes are shown in green in Figure 1)

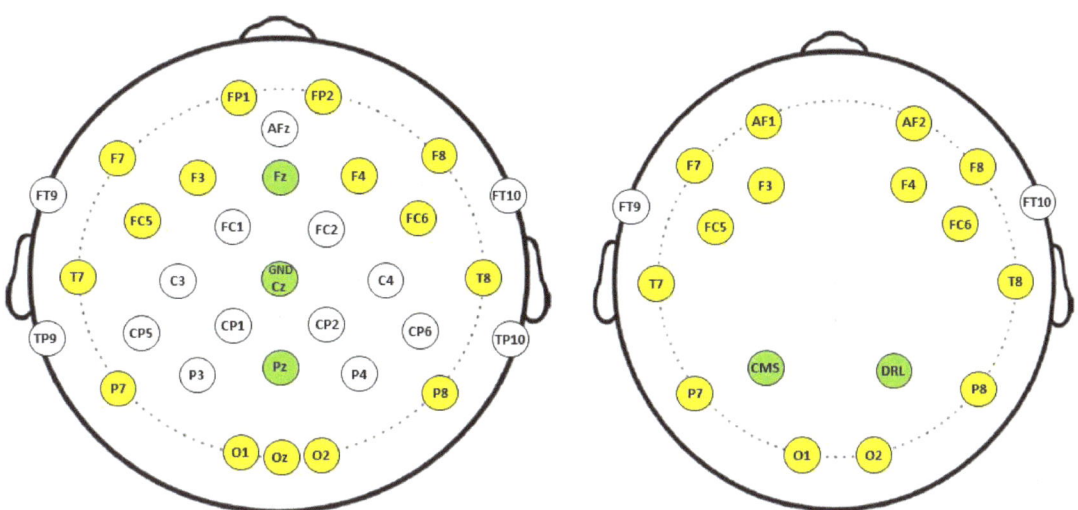

Figure 1. Left Panel: Brainvision (32-electrode) EEG system with ground at Cz. Right Panel: Emotiv EPOC (14-electrode) system with CMS at P3 and DRL at P4.

The estimated marginal means of both amplitude and power were reported in figures to demonstrate visually the detection of arousal, valence and workload, in addition to each EEG's ability to discriminate between high and low psychologically defined states in these three conditions (all these figures are included as Supplementary Materials). For conciseness, we summarize all main findings in the table below (see Table 1A,B).

Table 1. **A**. Grand averaged event-related power amplitudes and nonparametric contrasts for low vs. high mental states in ANEW, IAPS and n-Back Tasks measured by a consumer-grade (EPOC) and a research grade (Brainvision) EEG system. (Statistics were performed on electrode-wise repeated measurements, N = 14). **B**. Grand averaged cumulated power and nonparametric contrasts for low vs. high mental states in ANEW, IAPS and n-Back Tasks measured by a consumer-grade (EPOC) and a research grade (Brainvision) EEG system. (Statistics were performed on electrode-wise repeated measurements, N = 14).

	(A)									
	EPOC					Brainvision				
Band	Mental State		z	p	μ^2	Mental State		z	p	μ^2
	Low	High				Low	High			
				(ANEW, Arousal)						
Delta	2.02	2.16	−1.10	0.271	0.09	10.74	9.34	3.18	0.001	0.72
Theta	1.48	1.57	−1.57	0.116	0.18	8.57	7.044	3.30	<0.001	0.78
Alpha	2.22	2.30	−1.45	0.148	0.15	13.23	11.344	3.30	<0.001	0.78
Beta	2.20	2.11	1.51	0.132	0.16	16.53	18.784	−2.23	0.026	0.36
Gamma	0.89	0.97	−2.04	0.041	0.30	14.94	18.99	−3.23	0.001	0.75
				(ANEW, Valence)						
Delta	1.45	1.77	−2.07	0.039	0.31	6.65	8.19	−3.30	<0.001	0.78
Theta	2.10	1.72	1.30	0.195	0.12	5.62	5.93	−3.30	<0.001	0.78
Alpha	2.13	1.98	0.88	0.379	0.06	10.28	10.51	−1.38	0.167	0.14
Beta	2.50	2.18	2.10	0.036	0.32	12.50	13.55	−3.30	<0.001	0.78
Gamma	0.52	0.70	−3.04	0.002	0.66	11.27	12.21	−3.17	0.002	0.72

Table 1. Cont.

(A)

	EPOC					Brainvision				
Band	Mental State		z	p	μ^2	Mental State		z	p	μ^2
	Low	High				Low	High			
					(IAPS, Arousal)					
Delta	1.96	2.76	−3.30	<0.001	0.78	7.97	7.24	3.17	0.002	0.72
Theta	**1.42**	**1.78**	**−3.11**	**0.002**	**0.69**	**5.80**	**6.00**	**−3.18**	**0.001**	**0.72**
Alpha	2.21	2.48	−2.93	0.003	0.61	8.79	9.03	−1.48	0.140	0.16
Beta	**1.73**	**1.93**	**−3.30**	**<0.001**	**0.78**	**14.57**	**15.39**	**−3.30**	**<0.001**	**0.78**
Gamma	**0.42**	**0.63**	**−3.31**	**<0.001**	**0.78**	**14.53**	**15.22**	**−3.24**	**0.001**	**0.75**
					(IAPS, Valence)					
Delta	1.75	2.80	−3.30	<0.001	0.78	7.66	6.99	3.30	<0.001	0.78
Theta	1.35	2.05	−3.23	0.001	0.75	5.98	5.84	2.00	0.046	0.29
Alpha	**2.04**	**2.78**	**−3.30**	**<0.001**	**0.78**	**10.04**	**10.28**	**−1.92**	**0.055 ***	**0.26**
Beta	1.80	2.30	−3.30	<0.001	0.78	15.07	15.02	0.67	0.506	0.03
Gamma	0.39	0.69	−3.31	<0.001	0.78	14.57	14.15	3.30	<0.001	0.78
					(n-Back 1,3,5)					
Delta	2.70	2.39	2.45	0.014	0.43	2.89	2.72	1.23	0.221	0.11
Theta	1.95	2.05	−1.54	0.124	0.17	3.07	2.95	1.29	0.198	0.12
Alpha	2.75	2.69	0.47	0.638	0.02	3.08	3.07	0.44	0.660	0.01
Beta	2.77	2.75	0.03	0.975	0.00	3.92	4.23	−3.30	<0.001	0.78
Gamma	0.95	0.93	0.59	0.556	0.02	4.25	4.49	−2.86	0.004	0.58
					(n-Back 2,4,6)					
Delta	4.58	2.35	3.30	<0.001	0.78	2.28	2.30	−0.19	0.851	0.00
Theta	2.90	1.98	3.30	<0.001	0.78	2.78	3.11	−2.79	0.005	0.56
Alpha	4.08	2.55	3.30	<0.001	0.78	2.65	2.94	−3.20	0.001	0.73
Beta	3.81	2.72	3.30	<0.001	0.78	4.66	4.64	−2.31	0.021	0.38
Gamma	1.27	0.89	3.30	<0.001	0.78	4.62	4.56	−0.25	0.802	0.00

Note. Estimated marginal means of EEG power amplitude for mental states are reported in μV. *For standard errors please refer to Figures S1–S6 in Supplementary Materials.* Z statistics calculated from Wilcoxon non parametric test for small samples. Rows in bold identify comparisons yielding statistically similar results for both devices. "*" indicates marginal significance.

(B)

	EPOC					Brainvision				
Band	Mental State		z	p	μ^2	Mental State		z	p	μ^2
	Low	High				Low	High			
					(ANEW, Arousal)					
Delta	2.25	1.88	1.85	0.064	0.24	31.11	27.85	2.10	0.035	0.32
Theta	0.58	0.60	−0.46	0.649	0.02	13.38	9.59	3.30	<0.001	0.78
Alpha	1.22	1.04	1.92	0.055 *	0.06	29.82	21.29	3.30	<0.001	0.78
Beta	0.40	0.36	1.69	0.090	0.20	12.63	16.65	−2.17	0.030	0.34
Gamma	0.11	0.11	−0.24	0.812	0.00	9.94	15.09	−3.11	0.002	0.69
					(ANEW, Valence)					
Delta	0.89	1.02	−0.69	0.488	0.03	12.09	20.83	−3.30	<0.001	0.78
Theta	1.17	0.64	1.92	0.055 *	0.26	6.40	6.40	0.21	0.834	0.00
Alpha	0.97	0.60	2.48	0.013	0.44	19.65	18.98	1.41	0.158	0.14
Beta	0.49	032	2.35	0.019	0.39	7.21	8.45	−3.30	<0.001	0.78
Gamma	0.02	0.03	1.00	0.317	0.07	5.25	6.40	−3.30	<0.001	0.78
					(IAPS, Arousal)					
Delta	1.97	2.81	−1.98	0.048	0.28	16.14	13.08	3.30	<0.001	0.78
Theta	**0.54**	**0.70**	**−2.42**	**0.015**	**0.42**	**5.90**	**6.36**	**−2.13**	**0.033**	**0.32**
Alpha	1.27	1.16	0.25	0.807	0.00	11.34	14.77	−2.73	0.006	0.53
Beta	0.21	0.22	−0.92	0.357	0.06	9.57	11.26	−3.30	<0.001	0.78
Gamma	<0.01	<0.01	1.00	0.317	0.07	8.44	9.57	−3.17	0.002	0.72

Table 1. Cont.

	(B)									
	EPOC					Brainvision				
Band	Mental State		z	p	μ^2	Mental State		z	p	μ^2
	Low	High				Low	High			
					(IAPS, Valence)					
Delta	1.60	2.69	−3.02	0.003	0.65	14.10	12.14	2.79	0.005	0.56
Theta	0.47	0.98	−3.05	0.002	0.67	6.29	6.24	0.47	0.638	0.02
Alpha	0.96	1.40	−3.11	0.002	0.69	16.98	17.95	−1.57	<0.001	0.12
Beta	0.24	0.33	−3.22	0.001	0.74	10.59	10.68	−0.18	0.861	0.00
Gamma	0.01	0.01	0.00	1.000	0.00	9.08	8.50	3.30	<0.001	0.78
					(n-Back 1,3,5)					
Delta	3.68	2.70	2.35	0.019	0.39	4.63	3.72	1.41	0.158	0.14
Theta	0.99	1.10	−1.82	0.069	0.24	2.50	2.42	0.47	0.638	0.02
Alpha	1.67	1.65	0.03	0.975	0.00	2.12	1.97	1.16	0.245	0.10
Beta	0.67	0.69	−0.25	0.807	0.00	1.17	1.34	3.18	0.001	0.72
Gamma	0.14	0.15	−1.70	0.090	0.21	1.09	1.24	−2.67	0.008	0.51
					(n-Back 2,4,6)					
Delta	10.48	2.64	3.30	<0.001	0.78	2.53	2.39	0.66	0.510	0.03
Theta	2.45	1.05	3.30	<0.001	0.78	2.37	2.79	−2.61	0.009	0.49
Alpha	4.79	1.50	3.30	<0.001	0.78	1.42	1.69	−2.73	0.006	0.53
Beta	1.16	0.61	3.30	<0.001	0.78	1.71	1.72	−0.16	0.875	0.00
Gamma	0.17	0.11	3.19	0.001	0.73	1.47	1.38	1.95	0.052	0.27

Note. Estimated marginal means of EEG power for mental states are reported in μV. *For standard errors please refer to Figures S1–S6 in Supplementary Materials.* Z statistics calculated from Wilcoxon non parametric test for small samples. Rows in bold identify comparisons yielding statistically similar results for both devices. "*" indicates marginal significance.

Two-tailed paired Wilcoxon Signed Rank Tests were used to evaluate data to determine if there was a significant difference between the measures of each wave type at high and low states. To balance for Type I error likelihood, given the low statistical power (since the item analysis was limited to n = 14, the number of electrode-associated repeated measurements), we adopted a correction of the p-value with a generalization (family-wise average rate of false discovery rate) of the Hochberg-Benjamini procedure [65] with alpha 0.09, which gives a significance threshold of $p = 0.054$. Eta squared (μ^2) was computed as a measure of the effect size of the difference between high and low states. Finally, the value of EPOC electrode measurements was represented proportionally to the value of electrode measurements obtained by Brainvision, as a measure of EEG signal measurement *efficiency*. For example, if for gamma waves EPOC detected a signal amplitude of 1.5 μV while Brainvision detected a signal amplitude of 15 μV, Emotiv would be represented as sensing 10% of what Brainvision did.

Lastly, two confirmatory additional "control" analyses were performed (their results are not reported here but are available upon request). Firstly, we repeated the same analyses done on the vincentized averages for the data relative to each of the three professional observers. The patterns of results were very similar, supporting the rationale that the vincentized group data reflect valid inferences at N-of-1 personalization level. Secondly, we compared the non-parametric statistics reported here to the parametric version of the same analyses based on paired t-tests comparisons; the results were virtually identical using either parametric or nonparametric procedures.

3. Results

Considering both EEG features of amplitude and power, there were only very few instances in which EPOC and Brainvision converged in yielding a significant difference in the same direction, from low to high mental state, in the three tasks (7 out of 30 comparisons for amplitude, on a binomial test $p = 0.0003$; 5 out of 30 comparisons for power, on a binomial test $p = 0.0052$). The convergent comparisons are shown in bold in Table 1A,B.

Sensitivity to detect individual significant changes between low and high mental state level in any frequency band and any direction were comparable in both systems for most tasks and states except ANEW where EPOC performed sensibly worse, picking up changes only for Gamma as opposed to all electrodes as in Brainvision, and yielding about half of the effect size on average. Additionally, the performance of EPOC and Brainvision showed equally few effects for the 1,2,3 version of the n-Back task.

In terms of absolute measurement of EEG signal, EPOC demonstrated a variable performance as compared to Brainvision depending on the task. EPOC's efficiencies for the tasks were computed as the ratio of EPOC's measurements over Brainvision measurements; they are represented graphically in Figure 2. When compared to Brainvision measurements in ANEW (see Figure 2A), EPOC's efficiency in measuring amplitude changes was modest for arousal, generally hovering around 40-20% and below 10% for Gamma, and low for valence 30–20% and below 5% for Gamma. In the IAPS (see Figure 2A), EPOC's efficiency in measuring power was even lower for both arousal and valence across all frequencies ranging between 20 and 10%, with Gamma being below 2% or undetected as compared to Braivision measurements. However, EPOC's efficiency was much higher in the n-Back task (See Figure 2B) whereby it was variedly distributed across bands going from low/modest (i.e., Gamma) to medium (i.e., Theta, Beta), on par with or even exceeding (i.e., Alpha and Beta) Brainvision. Notably, in the 2,4,6 n-Back high load state EPOC outperformed Brainvison in all frequency bands except Gamma.

Finally, when the patterns of changes are considered across the frequency spectrum, as summarized in Table 2, it can be observed that there is minimal overlap between the results obtained with EPOC and those with Brainvision across tasks.

Table 2. Summary of patterns of EEG amplitude and power changes measured with EPOC and Brainvision.

Task	EEG Feature	Mental State	System	Significant Changes from Low to High State
ANEW	Amplitude	Arousal	Epoc	Increase for Gamma
			BV	Decrease for Delta, Theta, and Alpha Increase for Beta and Gamma
		Valence	Epoc	Decrease for Beta Increase for Delta and Gamma
			BV	Increase for all frequency bands
IAPS	Amp	Arousal	Epoc	Increase for all frequency bands
			BV	Decrease for Delta Increase for Theta, Beta and Gamma
		Valence	Epoc	Increase for all frequency bands
			BV	Decrease for Delta, Theta, and Gamma Increase for Alpha *
n-Back	Amp	1-3-5	Epoc	Decrease for Delta
			BV	Increase for Beta and Gamma
		2-4-6	Epoc	Decrease for all frequency bands
			BV	Increase for Theta, Alpha, and Beta
ANEW	Power	Arousal	Epoc	Decrease for Alpha *
			BV	Decrease for Delta, Theta, and Alpha Increase for Beta and Gamma
		Valence	Epoc	Decrease for Theta *, Alpha, and Beta
			BV	Increase for Delta, Beta, and Gamma

Table 2. *Cont.*

Task	EEG Feature	Mental State	System	Significant Changes from Low to High State
IAPS	Power	Arousal	Epoc	Increase for Delta and Theta
			BV	Decrease for Delta Increase for Theta, Alpha, Beta, and Gamma
		Valence	Epoc	Increase for Delta, Theta, Alpha, and Beta
			BV	Decrease for Delta and Gamma Increase for Alpha
n-Back	Power	1-3-5	Epoc	Decrease for Delta
			BV	Decrease for Beta Increase for Gamma
		2-4-6	Epoc	Decrease for all frequency bands
			BV	Decrease for Gamma Increase for Theta and Alpha

Note. "*" indicates marginally significant effects.

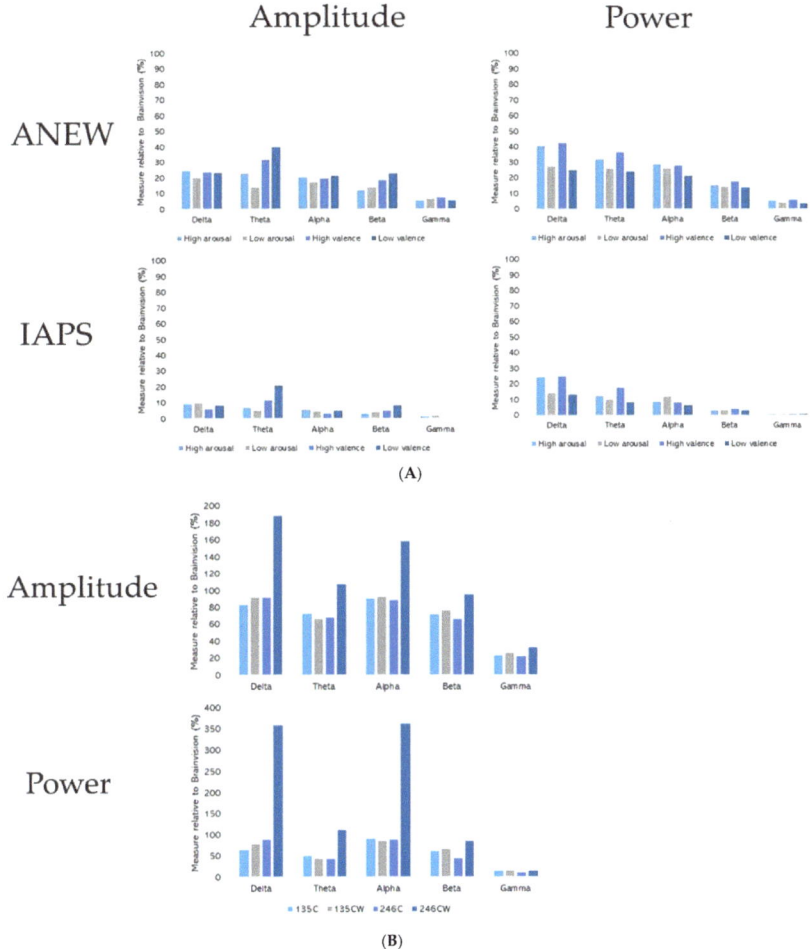

Figure 2. (**A**) Efficiency ratio of EPOC signal strength compared to Brainvision in ANEW and IAPS tasks. (**B**) Efficiency ratio of EPOC signal strength compared to Brainvision in the n-Back tasks.

4. Discussion

The results of our study overall seem to suggest that EPOC is capable of differentiating between high vs low arousal/valence states using the amplitude of different frequency bands, but that the overall signal detected from EPOC was only a fraction of the signal detected by Brainvision. Despite the poorer signal detection, EPOC was still able to differentiate between the mental states via amplitude in most instances, although it had relatively small effect sizes compared to those of Brainvision. For amplitude, it also appears that EPOC most prominently detected differences in low frequency bands such as Delta and Theta. Brainvision on the other hand commonly detected differences of amplitude across all frequency bands.

As above, the same can generally be said about the EPOC for differentiating mental states using EEG power, with a few exceptions. The ability of EPOC to detect Gamma power whatsoever was regularly quite low. Its mean amplitude was 8.7% of the amplitude measured by Brainvision. Even though EPOC was able to identify significant differences between high and low mental states in Gamma waves, the corresponding amplitude values never rose above 1.5 µV, even though Brainvision's measures went up to 13 µV. Since Gamma waves have the lowest amplitude of EEG frequency bands, the difficulty in registering their activity might be caused by the limited sensitivity and limited amount of electrodes on the EPOC headset. Additional factors could also include lower electrode accuracy and higher impedance. EPOC does not use conductive paste but saline solution, which can leave the electrodes more prone to corrosion and dry out more quickly than electrogel or paste.

Unexpectedly, in the instance of the n-Back task 2-4-6 EPOC actually appeared to outperform Brainvision's capability of differentiating high vs low mental load across frequency bands. In this same way, the power detected via EPOC was greater than the power detected by Brainvision. However, Brainvision still performed well in differentiating mental load during the n-Back task with reasonable effect sizes.

Given that the focus of the present study was accuracy and reliability, the tasks we used were selected because they demonstrated validity in distinguishing a relative difference in low and high psychological states as defined by subjective self-ratings of the participants and as reflected by concurrent state-dependent changes in EEG. Our study however has some implications and interesting insights also for research on comparative validity of consumer-grade vs. research-grade EEG devices.

The most EEG recent research focusing on content validity and attempting to relate a particular ERBP feature or signature to a particular mental state has been generally inconclusive, since the findings are contradictory. In particular, the literature on emotional arousal and valence using IAPS yielded as many studies which find a pattern of reduced Alpha power and increased Theta [66] as those which either find null effects, or even show the opposite pattern, i.e., increased Alpha and decreased Delta and/or Theta [67] as well as increased Beta and Gamma [68–70]. One possible plausible conclusion that has been repeatedly proposed is that the specific pattern of results related to EEG power spectral distribution may be therefore dependent on design, stimuli, task, context, and individual differences [71]. The findings also vary according to the electrodes considered, whereby, for example, opposite patterns can be observed in frontal/anterior as compared to parietal/posterior sites. An exhaustive review is outside the scope and space of this paper, but it is reasonable to reach similar conclusions by considering recent meta-analyses on the n-Back task and mental workload in general [69]. The major issue in the context of the present study is overlearning and practice effects in that the professional observers serving as participants in our study underwent a lot of pre-experimental repeated exposure with materials, tasks and both devices. The literature on the effect of the extensive exposure or practice variable on EEG correlates of emotional processing is scant (for example see [72]) but there is evidence for the n-Back task that Alpha and Theta increase with cognitive load [73–75]. Our findings show that both EPOC and Brainvision did not find Alpha desynchronization followed by a concurrent increase of slow frequency oscillations (Delta or Theta), as reported by some abovementioned literature. In our study, Alpha showed occasional increases or null findings while increases in Delta were rather ubiquitous and inconsistent. The most frequent and consistent changes were increases in Theta, Beta and Gamma going from low to high mental states, which is, as we have cursorily mentioned, in keeping with the literature on practiced mental or emotional processing load. This is both interesting as an extension of research on training of cognitive load and working memory capacity as well as, we believe, a relatively new contribution for the area on emotional processing since there are very few event-related EEG studies on repeated exposure to pleasant or unpleasant stimuli.

In our experiment the sample size was very low, and it was impossible to achieve reliable results that generalize to population distributions. However, because we used an ideographic approach (i.e., electrode-wise analysis of vincentized data) which we can defend to be generalized at the level of person prediction, the actual sample size was relatively unimportant for the explicitly stated scope (tests of reliability) addressed in this paper; this last point holds especially because we took several convergent repeated measurements (N = 14), which were the actual basis for our test of accuracy.

Nonetheless, a possible caveat is that, given our very small sample, it was obviously impossible to properly counterbalance learning effects associated with the device used, namely, EPOC was used first and then Brainvision was used on all subjects, potentially, this order might have induced a learning effect favoring Brainvision. Yet, the rate of significant differences between high and low mental states picked up by EPOC and Brainvision did not show any differential order trend, that is, there were no obvious and consistent differences between sensitivity in detecting significant effects between mental states for the first device used or for the second one, and no advantage of one device over the other can be clearly observed or inferred.

Another related possible caveat might be learning effects of the individual, regardless of the device used. However, as we have pointed out all three participants were graduate student professional observers which were abundantly familiar with stimuli, tasks, and devices. Furthermore, all participants carried out the tasks in a different order. Most likely, these two aspects were sufficient to wash out learning effects due to the task sequence during the experiment.

Our findings would suggest that future applications, such as brain-computer interfaces, requiring a more ecological setup should be able to confidently use EPOC for tasks intending to elicit change in valence/arousal/mental load such as ANEW, IAPS, or n-Back. However, in a laboratory or medical setting the researcher or clinician would still benefit from the quality performance found using Brainvision. Future research should consider ways of optimizing the signal quality of the EPOC so as to obtain larger effect sizes when investigating differences of amplitude/power in response to different mental states. For instance, it would be feasible to leverage machine learning models to enhance the accuracy and classification capabilities of the signals obtained via the EPOC [29]. This would be a promising, although more complex, avenue for increasing the efficacy of the Emotiv device. Modifying the hardware of the EPOC may also be feasible and effective, as Barham et al. (2017) demonstrated that upgrading the EPOC with research-grade electrodes was successful, although the EPOC still had significantly more rejected trials than the research-grade device (Neuroscan) used in their study [26].

5. Conclusions

In summary, the consumer-grade wearable EPOC device appears to be able to differentiate between mental states of valence, arousal, and mental load to some degree, but is overall still not as reliable or accurate as the research-grade Brainvision device. However, EPOC appears to be particularly good at differentiating mental load during an n-Back task via frequency band amplitude/power. Overall, this is in line with previous literature [21], but here we have extended these observations to tasks related to valence, arousal, and mental load.

In principle, EEG measured during high vs low arousal, valence, and mental load, should elicit different EEG signatures (e.g., synchronization/desynchronization and corresponding increases/decreases in amplitude/power), as the mental states require differing levels of emotional or cognitive resources [76–82]. This would be expected from a research-grade EEG device like Brainvision. However, due to the limited amount of evidence for this type of outcome using consumer-grade wearable EEG devices like the EPOC [83,84], further research regarding content validity is still required to confirm this expectation by empirical tests performed using a procedure similar to ours (i.e., by comparing homologous repeated measurements using both devices within the same individuals) and not, as currently done in many review papers (e.g., [20]), by simply assuming that finding a EEG feature or signature with the wireless consumer grade device (e.g., P300) in one sample has the same meaning as reports about the same signature or feature in different samples, even if utilizing comparable experimental paradigms.

Finally, our main objective was to demonstrate an initial methodological benchmark for precision, rather than conducting a comprehensive assessment of all existing consumer-grade EEG devices. However, because the two types of EEG devices compared in the present study are among the most "popular" and widely used for research purposes in labs across the world (especially the Emotiv EPOC, see [85]), the present comparison does contribute a heuristic inductive assessment regarding the precision of, presumably, one of the "best" current EEG consumer instrumentation available as compared to an established (i.e., standard) wired stationary counterpart.

Supplementary Materials: The following supporting information can be downloaded at: https://www.mdpi.com/article/10.3390/app12136430/s1.

Author Contributions: Conceptualization, A.D.; methodology, A.D., D.M.B. and G.L.-D.; validation, A.D., D.M.B. and G.L.-D.; formal analysis, G.L.-D., D.M.B. and A.D.; investigation, A.D., D.M.B. and G.L.-D.; resources, A.D.; data curation, G.L.-D., A.D. and D.M.B.; writing—original draft preparation, D.M.B., G.L.-D. and A.D.; writing—review and editing, D.M.B., G.L.-D. and A.D.; supervision, A.D.; project administration, A.D. All authors have read and agreed to the published version of the manuscript.

Funding: This research was funded by Defense R&D Canada, by Thales Research and Technology Canada, and by a research partnership grant from the Department of National Defense of Canada and the Natural Sciences and Engineering Research Council of Canada. Open access publication (APC) was funded by a Carleton University Research Impact Endeavour (CURIE) award to A.D.

Institutional Review Board Statement: Ethical review and approval were waived for this study because this was a secondary analysis of completely anonymized archival data.

Informed Consent Statement: Informed verbal consent was obtained from all subjects involved in the original data collection pre-archived studies.

Data Availability Statement: Data may be made available upon request to the authors provided it is for uses in accordance with international ethical, intellectual property and copyright regulations. Public archival of this data is not possible due to privacy restrictions imposed by the funding agencies.

Acknowledgments: The authors would like to acknowledge Jeremy Grant for his role in assisting with the initial data collection and first analysis of this data which was previously published. We thank three anonymous reviewers (especially Reviewer 1) for the numerous insightful comments, criticisms, and suggestions which have helped to greatly improve the paper.

Conflicts of Interest: The authors declare no conflict of interest. The funders had no role in the design of the study; in the collection, analyses, or interpretation of data; in the writing of the manuscript, or in the decision to publish the results.

References

1. Smith, S.J.M. EEG in the diagnosis, classification, and management of patients with epilepsy. *J. Neurol. Neurosurg. Psychiatry* **2005**, *76*, ii2–ii7. [CrossRef] [PubMed]

2. Vasiljevic, G.A.M.; De Miranda, L.C. Brain–Computer Interface Games Based on Consumer-Grade EEG Devices: A Systematic Literature Review. *Int. J. Hum. Comput. Interact.* **2019**, *36*, 105–142. [CrossRef]
3. TajDini, M.; Sokolov, V.; Kuzminykh, I.; Shiaeles, S.; Ghita, B. Wireless Sensors for Brain Activity—A Survey. *Electronics* **2020**, *9*, 2092. [CrossRef]
4. Barngrover, C.; Althoff, A.; DeGuzman, P.; Kastner, R. A Brain–Computer Interface (BCI) for the Detection of Mine-Like Objects in Sidescan Sonar Imagery. *IEEE J. Ocean. Eng.* **2015**, *41*, 123–138. [CrossRef]
5. Ganga, R.C.; Vijayakumar, P.; Badrinath, P.; Singh, A.R.; Singh, M. Drone control using EEG signal. *J. Adv. Res. Dyn. Control Syst.* **2019**, *11*, 2107–2113.
6. Munyon, C.N. Neuroethics of Non-primary Brain Computer Interface: Focus on Potential Military Applications. *Front. Neurosci.* **2018**, *12*, 696. [CrossRef] [PubMed]
7. Binnendijk, A.; Marler, T.; Bartels, E.M. *Brain-Computer Interfaces: U.S. Military Applications and Implications, An Initial Assessment*; RAND Corporation: Santa Monica, CA, USA, 2020. [CrossRef]
8. Czech, A. Brain-Computer Interface Use to Control Military Weapons and Tools. In *Control, Computer Engineering and Neuroscience*; Springer: Berlin/Heidelberg, Germany, 2021; pp. 196–204. [CrossRef]
9. Hernandez-Cuevas, B.; Egbert, W.; Denham, A.; Mehul, A.; Crawford, C.S. Changing Minds: Exploring Brain-Computer Interface Experiences with High School Students. In Proceedings of the 2020 CHI Conference on Human Factors in Computing Systems, Honolulu, HI, USA, 25–30 April 2020. [CrossRef]
10. Gnedykh, D. Trends and Prospects of Using Brain-Computer Interfaces in Education. *Sib. Psikhologicheskiy Zhurnal* **2021**, 108–129. [CrossRef] [PubMed]
11. Papanastasiou, G.; Drigas, A.; Skianis, C.; Lytras, M. Brain computer interface based applications for training and rehabilitation of students with neurodevelopmental disorders. A literature review. *Heliyon* **2020**, *6*. [CrossRef]
12. Rohani, D.A.; Puthusserypady, S. BCI inside a virtual reality classroom: A potential training tool for attention. *EPJ Nonlinear Biomed. Phys.* **2015**, *3*, 12. [CrossRef]
13. Liu, N.-H.; Chiang, C.-Y.; Chu, H.-C. Recognizing the Degree of Human Attention Using EEG Signals from Mobile Sensors. *Sensors* **2013**, *13*, 10273–10286. [CrossRef]
14. Thomas, K.P.; Vinod, A.P.; Guan, C. Enhancement of attention and cognitive skills using EEG based neurofeedback game. In Proceedings of the 2013 6th International IEEE/EMBS Conference, San Diego, CA, USA, 6–8 November 2013; pp. 21–24. [CrossRef]
15. Vinod, A.P.; Thomas, K.P. *Neurofeedback Games Using EEG-Based Brain–Computer Interface Technology*; The Institution of Engineering and Technology: London, UK, 2018; pp. 301–329. [CrossRef]
16. Sekhavat, Y.A. Battle of minds: A new interaction approach in BCI games through competitive reinforcement. *Multimedia Tools Appl.* **2019**, *79*, 3449–3464. [CrossRef]
17. Paszkiel, S. Using BCI and VR Technology in Neurogaming. In *Signal Processing and Machine Learning for Brain-Machine Interfaces*; Springer: Berlin/Heidelberg, Germany, 2019; pp. 93–99. [CrossRef]
18. Marshall, D.; Coyle, D.; Wilson, S.; Callaghan, M. Games, Gameplay, and BCI: The State of the Art. *IEEE Trans. Comput. Intell. AI Games* **2013**, *5*, 82–99. [CrossRef]
19. Bos, D.P.; Obbink, M.; Nijholt, A.; Hakvoort, G.; Christian, M. Towards multiplayer BCI games. In Proceedings of the Workshop on Multiuser and Social Biosignal Adaptive Games and Playful Applications, BioS-Play, 2010; Available online: http://www.physiologicalcomputing.net/workshops/biosplay2010/BioSPlay_Gurkok%20et%20al%20(Multiplayer%20BCI).pdf (accessed on 20 May 2022).
20. Sawangjai, P.; Hompoonsup, S.; Leelaarporn, P.; Kongwudhikunakorn, S.; Wilaiprasitporn, T. Consumer Grade EEG Measuring Sensors as Research Tools: A Review. *IEEE Sens. J.* **2019**, *20*, 3996–4024. [CrossRef]
21. Buchanan, D.M.; Grant, J.; D'Angiulli, A. Commercial wireless versus standard stationary EEG systems for personalized emotional brain-computer interfaces: A preliminary reliability check. *Neurosci. Res. Notes* **2019**, *2*, 7–15. [CrossRef]
22. Maskeliunas, R.; Damasevicius, R.; Martisius, I.; Vasiljevas, M. Consumer grade EEG devices: Are they usable for control tasks? *PeerJ* **2016**, *4*, e1746. [CrossRef]
23. Nijboer, F.; Van De Laar, B.; Gerritsen, S.; Nijholt, A.; Poel, M. Usability of Three Electroencephalogram Headsets for Brain–Computer Interfaces: A Within Subject Comparison. *Interact. Comput.* **2015**, *27*, 500–511. [CrossRef]
24. Duvinage, M.; Castermans, T.; Petieau, M.; Hoellinger, T.; Cheron, G.; Dutoit, T. Performance of the Emotiv Epoc headset for P300-based applications. *Biomed. Eng. Online* **2013**, *12*, 56. [CrossRef]
25. Duvinage, M.; Castermans, T.; Dutoit, T.; Petieau, M.; Hoellinger, T.; De Saedeleer, C.; Seetharaman, K.; Cheron, G. A P300-based Quantitative Comparison between the Emotiv Epoc Headset and a Medical EEG Device. In *BioMedical Engineering OnLine*; Springer Nature: Berlin, Germany, 2012. [CrossRef]
26. Barham, M.P.; Clark, G.M.; Hayden, M.J.; Enticott, P.; Conduit, R.; Lum, J. Acquiring research-grade ERPs on a shoestring budget: A comparison of a modified Emotiv and commercial SynAmps EEG system. *Psychophysiology* **2017**, *54*, 1393–1404. [CrossRef]
27. Liu, X.; Chao, F.; Jiang, M.; Zhou, C.; Ren, W.; Shi, M. Towards Low-Cost P300-Based BCI Using Emotiv Epoc Headset. In Proceedings of the UK Workshop on Computational Intelligence, Cardiff, UK, 6–8 September 2017; Volume 650, pp. 239–244. [CrossRef]

28. Badcock, N.A.; Preece, K.A.; de Wit, B.; Glenn, K.; Fieder, N.; Thie, J.; McArthur, G. Validation of the Emotiv EPOC EEG system for research quality auditory event-related potentials in children. *PeerJ* **2015**, *3*, e907. [CrossRef]
29. Fouad, I.A. A robust and reliable online P300-based BCI system using Emotiv EPOC + headset. *J. Med Eng. Technol.* **2021**, *45*, 94–114. [CrossRef]
30. Balanou, E.; van Gils, M.; Vanhala, T. State-of-the-Art of Wearable EEG for Personalized Health Applications. *Stud. Health Technol. Inform.* **2013**, *189*, 119–124. [CrossRef] [PubMed]
31. Smith, P.L.; Little, D.R. Small is beautiful: In defense of the small-N design. *Psychon. Bull. Rev.* **2018**, *25*, 2083–2101. [CrossRef] [PubMed]
32. Molenaar, P.C.M. A Manifesto on Psychology as Idiographic Science: Bringing the Person Back Into Scientific Psychology, This Time Forever. *Meas. Interdiscip. Res. Perspect.* **2004**, *2*, 201–218. [CrossRef]
33. Bos, F.M.; Snippe, E.; de Vos, S.; Hartmann, J.A.; Simons, C.J.; van der Krieke, L.; de Jonge, P.B.; Wichers, M. Can we jump from cross-sectional to dynamic interpretationsof networks? Implications for the network perspective in psychiatry. *Psychother. Psychosom.* **2017**, *86*, 175–177. [CrossRef] [PubMed]
34. Molenaar, P.C. On the implications of the classical ergodic theorems: Analysis of developmental processes has to focus on intra-individual variation. *Dev. Psychobiol.* **2007**, *50*, 60–69. [CrossRef]
35. Shah, R.V.; Grennan, G.; Zafar-Khan, M.; Alim, F.; Dey, S.; Ramanathan, D.; Mishra, J. Personalized machine learning of depressed mood using wearables. *Transl. Psychiatry* **2021**, *11*, 1–18. [CrossRef]
36. Birkhoff, G.D. What is the ergodic theorem? *Am. Math. Mon.* **1942**, *49*, 222–226. [CrossRef]
37. Nelson, C.A.; de Haan, M.; Thomas, K.M. *Neuroscience of Cognitivedevelopment: The Role of Experience and the Developing Brain*; Wiley: New York, NY, USA, 2006.
38. Sporns, O. *Networks of the Brain*; MIT Press: Cambridge, MA, USA, 2010.
39. Medaglia, J.D.; Ramanathan, D.M.; Venkatesan, U.M.; Hillary, F.G. The challenge of non-ergodicity in network neuroscience. *Netw. Comput. Neural Syst.* **2011**, *22*, 148–153. [CrossRef]
40. Vincent, S.B. The function of the viborissae in the behavior of the white rat. *Anim. Behav. Monogr.* **1912**, *1*, 84.
41. Atkinson, R.C.; Bower, G.H.; Crothers, E.J. *Introduction to Mathematical Learning Theory*; Wiley: Hoboken, NJ, USA, 1965.
42. Bedny, M.; Aguirre, G.; Thompson-Schill, S.L. Item analysis in functional magnetic resonance imaging. *NeuroImage* **2007**, *35*, 1093–1102. [CrossRef]
43. Makeig, S.; Debener, S.; Onton, J.; Delorme, A. Mining event-related brain dynamics. *Trends Cogn. Sci.* **2004**, *8*, 204–210. [CrossRef] [PubMed]
44. Pfurtscheller, G. and Aranibar, A. Event-related cortical desychronization detected by power measurement of scalp EEG. *Electroencephalograph. Clin. Neurophysiol.* **1977**, *42*, 817–826. [CrossRef]
45. Byczynski, G.; Schibli, K.; Goldfield, G.; Leisman, G.; D'Angiulli, A. EEG Power Band Asymmetries in Children with and without Classical Ensemble Music Training. *Symmetry* **2022**, *14*, 538. [CrossRef]
46. D'Angiulli, A.; Kenney, D.; Pham, D.A.T.; Lefebvre, E.; Bellavance, J.; Buchanan, D.M. Neurofunctional Symmetries and Asymmetries during Voluntary out-of- and within-Body Vivid Imagery Concurrent with Orienting Attention and Visuospatial Detection. *Symmetry* **2021**, *13*, 1549. [CrossRef]
47. Klimesch, W. Alpha-band oscillations, attention, and controlled access to stored information. *Trends Cogn. Sci.* **2012**, *16*, 606–617. [CrossRef]
48. Stevenson, R.A.; Mikels, J.A.; James, T.W. Characterization of the Affective Norms for English Words by discrete emotional categories. *Behav. Res. Methods* **2007**, *39*, 1020–1024. [CrossRef]
49. Bradley, M.; Lang, P.J. *Affective Norms for English Words (ANEW): Instruction Manual and Affective Ratings (Technical Report C-1)*; Gainesv Cent Res Psychophysiology, University of Florida: Gainesville, FL, USA, 1999.
50. Bradley, M.M.; Lang, P.J. *Affective Norms for English Words (ANEW): Instruction Manual and Affective Ratings (Technical Report C-2)*; University of Florida: Gainesville, FL, USA, 2010.
51. Imbir, K.K. Affective Norms for 4900 Polish Words Reload (ANPW_R): Assessments for Valence, Arousal, Dominance, Origin, Significance, Concreteness, Imageability and, Age of Acquisition. *Front. Psychol.* **2016**, *7*, 1081. [CrossRef]
52. Lang, P.J.; Bradley, M.M.; Cuthbert, B.N. *International Affective Picture System (IAPS): Affective Ratings of Pictures and Instruction Manual*; Technical Report A-8; University of Florida: Gainesville, FL, USA, 2008.
53. Hajcak, G.; Dennis, T.A. Brain potentials during affective picture processing in children. *Biol. Psychol.* **2009**, *80*, 333–338. [CrossRef]
54. Kirchner, W.K. Age differences in short-term retention of rapidly changing information. *J. Exp. Psychol.* **1958**, *55*, 352–358. [CrossRef]
55. Mackworth, J.F. Paced memorizing in a continuous task. *J. Exp. Psychol.* **1959**, *58*, 206–211. [CrossRef]
56. Gajewski, P.D.; Hanisch, E.; Falkenstein, M.; Thönes, S.; Wascher, E. What Does the n-Back Task Measure as We Get Older? Relations Between Working-Memory Measures and Other Cognitive Functions Across the Lifespan. *Front. Psychol.* **2018**, *9*, 2208. [CrossRef] [PubMed]
57. Scharinger, C.; Soutschek, A.; Schubert, T.; Gerjets, P. Comparison of the Working Memory Load in N-Back and Working Memory Span Tasks by Means of EEG Frequency Band Power and P300 Amplitude. *Front. Hum. Neurosci.* **2017**, *11*, 6. [CrossRef]

58. Canadian Institutes of Health Research, Natural Sciences and Engineering Research Council of Canada, and Social Sciences and Humanities Research Council, Tri-Council Policy Statement: Ethical Conduct for Research Involving Humans, December 2018. Available online: https://ethics.gc.ca/eng/documents/tcps2-2018-en-interactive-final.pdf (accessed on 1 June 2022).
59. D'Angiulli, A.; Griffiths, G.; Marmolejo-Ramos, F. Neural correlates of visualizations of concrete and abstract words in preschool children: A developmental embodied approach. *Front. Psychol.* **2015**, *6*, 856. [CrossRef] [PubMed]
60. Zhou, X.; Li, M.; Zhou, H.; Li, L.; Cui, J. Item-Wise Interindividual Brain-Behavior Correlation in Task Neuroimaging Analysis. *Front. Neurosci.* **2018**, *12*, 817. [CrossRef]
61. D'Angiulli, A.; Pham, D.A.T.; Leisman, G.; Goldfield, G. Evaluating Preschool Visual Attentional Selective-Set: Preliminary ERP Modeling and Simulation of Target Enhancement Homology. *Brain Sci.* **2020**, *10*, 124. [CrossRef]
62. Genest, C. Vincentization Revisited. *Ann. Stat.* **1992**, *20*, 1137–1142. [CrossRef]
63. Delorme, A.; Makeig, S. EEGLAB: An Open Source Toolbox for Analysis of Single-Trial EEG Dynamics Including Independent Component Analysis. *J. Neurosci. Methods* **2004**, *134*, 9–21. [CrossRef]
64. Badcock, N.A.; Mousikou, P.; Mahajan, Y.; De Lissa, P.; Thie, J.; McArthur, G. Validation of the Emotiv EPOC® EEG gaming system for measuring research quality auditory ERPs. *PeerJ* **2013**, *1*, e38. [CrossRef]
65. Benjamini, Y.; Hochberg, Y. Controlling the False Discovery Rate: A Practical and Powerful Approach to Multiple Testing. *J. R. Stat. Soc. Ser. B* **1995**, *57*, 289–300. [CrossRef]
66. Schubring, D.; Schupp, H.T. Emotion and brain oscillations: High arousal is associated with decreases in alpha-and lower beta-band power. *Cerebral Cortex* **2021**, *31*, 1597–1608. [CrossRef]
67. Güntekin, B.; Femir, B.; Gölbaşı, B.T.; Tülay, E.; Başar, E. Affective pictures processing is reflected by an increased long-distance EEG connectivity. *Cogn. Neurodyn.* **2017**, *11*, 355–367. [CrossRef] [PubMed]
68. Güntekin, B.; Başar, E. Event-related beta oscillations are affected by emotional eliciting stimuli. *Neurosci. Lett.* **2010**, *483*, 173–178. [CrossRef] [PubMed]
69. Strube, A.; Rose, M.; Fazeli, S.; Büchel, C. Alpha-to-beta-and gamma-band activity reflect predictive coding in affective visual processing. *Sci. Rep.* **2021**, *11*, 23492. [CrossRef] [PubMed]
70. Yeo, D.; Choi, J.W.; Kim, K.H. Increased Gamma-band Neural Synchrony by Pleasant and Unpleasant Visual Stimuli. *J. Biomed. Eng. Res.* **2018**, *39*, 94–102.
71. Chikhi, S.; Matton, N.; Blanchet, S. EEG power spectral measures of cognitive workload: A meta-analysis. *Psychophysiology* **2022**, *59*, e14009. [CrossRef]
72. Güntekin, B.; Tülay, E. Event related beta and gamma oscillatory responses during perception of affective pictures. *Brain Res.* **2014**, *1577*, 45–56. [CrossRef]
73. Blacker, K.J.; Negoita, S.; Ewen, J.B.; Courtney, S.M. N-back versus complex span working memory training. *J. Cogn. Enhanc.* **2017**, *1*, 434–454. [CrossRef]
74. Liu, Y.; Ayaz, H.; Onaral, B.; Shewokis, P.A. Neural Adaptation to a Working Memory Task: A Concurrent EEG-fNIRS Study. In Proceedings of the 2015 International Conference on Augmented Cognition, Los Angeles, CA, USA, 7 July–2 August 2015; pp. 268–280. [CrossRef]
75. Gevins, A.; Smith, M.E.; McEvoy, L.; Yu, D. High-resolution EEG mapping of cortical activation related to working memory: Effects of task difficulty, type of processing, and practice. *Cereb. Cortex* **1997**, *7*, 374–385. [CrossRef]
76. Klotzsche, F.; Mariola, A.; Hofmann, S.; Nikulin, V.V.; Villringer, A.; Gaebler, M. Using EEG to Decode Subjective Levels of Emotional Arousal During an Immersive VR Roller Coaster Ride. In Proceedings of the 2018 IEEE Conference on Virtual Reality and 3D User Interfaces (VR), Christchurch, New Zealand, 18–22 March 2018; pp. 605–606. [CrossRef]
77. Sarma, P.; Barma, S. Emotion recognition by distinguishing appropriate EEG segments based on random matrix theory. *Biomed. Signal Process. Control* **2021**, *70*, 102991. [CrossRef]
78. Duma, G.M.; Mento, G.; Semenzato, L.; Tressoldi, P. EEG anticipation of random high and low arousal faces and sounds. *F1000Research* **2019**, *8*, 1508. [CrossRef]
79. Gummadavelli, A.; Kundishora, A.J.; Willie, J.T.; Andrews, J.; Gerrard, J.; Spencer, D.D.; Blumenfeld, H. Neurostimulation to improve level of consciousness in patients with epilepsy. *Neurosurg. Focus* **2015**, *38*, E10. [CrossRef] [PubMed]
80. Aftanas, L.; Golocheikine, S. Human anterior and frontal midline theta and lower alpha reflect emotionally positive state and internalized attention: High-resolution EEG investigation of meditation. *Neurosci. Lett.* **2001**, *310*, 57–60. [CrossRef]
81. Müller-Bardorff, M.; Schulz, C.; Peterburs, J.; Bruchmann, M.; Mothes-Lasch, M.; Miltner, W.; Straube, T. Effects of emotional intensity under perceptual load: An event-related potentials (ERPs) study. *Biol. Psychol.* **2016**, *117*, 141–149. [CrossRef]
82. Boring, M.J.; Ridgeway, K.; Shvartsman, M.; Jonker, T.R. Continuous decoding of cognitive load from electroencephalography reveals task-general and task-specific correlates. *J. Neural Eng.* **2020**, *17*, 056016. [CrossRef] [PubMed]
83. Wang, S.; Gwizdka, J.; Chaovalitwongse, W.A. Using Wireless EEG Signals to Assess Memory Workload in the n-Back Task. *IEEE Trans. Human-Machine Syst.* **2015**, *46*, 424–435. [CrossRef]
84. Kutafina, E.; Heiligers, A.; Popovic, R.; Brenner, A.; Hankammer, B.; Jonas, S.M.; Mathiak, K.; Zweerings, J. Tracking of Mental Workload with a Mobile EEG Sensor. *Sensors* **2021**, *21*, 5205. [CrossRef] [PubMed]
85. Williams, N.S.; McArthur, G.M.; Badcock, N.A. 10 years of EPOC: A scoping review of Emotiv's portable EEG device. *BioRxiv* **2020**. [CrossRef]

Article

Effect of Rehabilitation on Brain Functional Connectivity in a Stroke Patient Affected by Conduction Aphasia

Serena Dattola and Fabio La Foresta *

Department of Civil, Energy, Environmental and Materials Engineering (DICEAM), Mediterranea University of Reggio Calabria, Via Graziella Feo di Vito, 89060 Reggio Calabria, Italy; serena.dattola@unirc.it
* Correspondence: fabio.laforesta@unirc.it

Abstract: Stroke is a medical condition that affects the brain and represents a leading cause of death and disability. Associated with drug therapy, rehabilitative treatment is essential for promoting recovery. In the present work, we report an EEG-based study concerning a left ischemic stroke patient affected by conduction aphasia. Specifically, the objective is to compare the brain functional connectivity before and after an intensive rehabilitative treatment. The analysis was performed by means of local and global efficiency measures related to the execution of three tasks: naming, repetition and reading. As expected, the results showed that the treatment led to a balancing of the values of both parameters between the two hemispheres since the rehabilitation contributed to the creation of new neural patterns to compensate for the disrupted ones. Moreover, we observed that for both name and repetition tasks, shortly after the stroke, the global and local connectivity are lower in the affected lobe (left hemisphere) than in the unaffected one (right hemisphere). Conversely, for the reading task, global and local connectivity are higher in the impaired lobe. This apparently contrasting trend can be due to the effects of stroke, which affect not only the site of structural damage but also brain regions belonging to a functional network. Moreover, changes in network connectivity can be task-dependent. This work can be considered a first step for future EEG-based studies to establish the most suitable connectivity measures for supporting the treatment of stroke and monitoring the recovery process.

Keywords: stroke; conduction aphasia; high-density EEG; brain functional connectivity; rehabilitation

Citation: Dattola, S.; La Foresta, F. Effect of Rehabilitation on Brain Functional Connectivity in a Stroke Patient Affected by Conduction Aphasia. *Appl. Sci.* **2022**, *12*, 5991. https://doi.org/10.3390/app12125991

Academic Editor: Vladislav Toronov

Received: 28 April 2022
Accepted: 10 June 2022
Published: 13 June 2022

Publisher's Note: MDPI stays neutral with regard to jurisdictional claims in published maps and institutional affiliations.

Copyright: © 2022 by the authors. Licensee MDPI, Basel, Switzerland. This article is an open access article distributed under the terms and conditions of the Creative Commons Attribution (CC BY) license (https://creativecommons.org/licenses/by/4.0/).

1. Introduction

Stroke is a leading cause of death and disability. Each year, 14 million people suffer their first stroke worldwide, and 80 million people living in the world have experienced it [1]. Stroke is defined by the World Health Organization as a neurological deficit of cerebrovascular cause that persists beyond 24 h or is interrupted by death within 24 h. The 24-h limit differentiates stroke from transient ischemic attack (TIA), which is a temporary cerebral dysfunction related to stroke symptoms characterized by a swift resolution. Stroke can be classified into two main categories: ischemic, caused by a blockage of the blood flow to the brain, and hemorrhagic, due to the rupture of a blood vessel [2]. Both types lead to the dysfunction of the brain areas affected by the stroke. Signs and symptoms of stroke may include numbness, confusion, difficulty in speaking or understanding speech, and loss of balance or coordination. In most cases, the symptoms affect only one side of the body. The effects of stroke can be very different because they depend on the type, severity and location of the lesions. An early detection associated with proper medical treatment is essential for reducing stroke outcomes. Finally, post-stroke rehabilitation represents a very important process for recovering lost function and relearning the skills of everyday life. Even if complete recovery is unusual, rehabilitation can help the patient to regain independence and reintegrate into community life [3].

Conduction aphasia is an acquired language disorder first hypothesized by Carl Wernicke [4]. Conduction aphasia is characterized by intact comprehension and fluent (but paraphasic) speech production, whereas speech repetition is impaired. Therefore, aphasic people produce paraphasic errors and show word-finding difficulty [5]. Conduction aphasia is considered as a disconnection syndrome since it is due to an interruption of communication between anterior and posterior language areas [6]. Lesions usually involve the left cerebral hemisphere, as reported in [7]. Several studies proved that the mechanisms of neuroplasticity lead to recovery from aphasia [8,9]. Moreover, neurophysiological studies revealed that the intra- and inter-hemispheric rearrangement of the linguistic network occurs following a stroke, which involves linguistic areas [10,11]. Finally, the findings of neural reorganization in patients with conduction aphasia have proven the importance of rehabilitative treatment [12–15].

Graph theory can be conveniently exploited for studying the behavior of brain networks. Brain regions are considered nodes, whereas the edges connecting nodes represent the brain's functional or structural connectivity between the regions [16]. In particular, the brain's functional connectivity concerns the functionally integrated relationship between distant brain regions and is expressed in terms of statistical dependencies in the time domain (correlation) and in the frequency domain (coherence) among neurophysiological measurements. Recently, it has been proven that healthy brain networks have a "small-world" architecture, characterized by clustered local connectivity (functional segregation) and short path lengths (functional integration) between nodes [17,18]. Global and local efficiency of a network are two parameters that quantify the performance of a network in information exchange [19] and are related to path lengths and the clustering coefficient, respectively. The higher global and local efficiency, the more efficient the network. Efficiency measures are also more robust in the case of disconnected graphs.

In the present paper, we propose a case study about a patient who suffered from conduction aphasia due to a left ischemic stroke. The purpose is to compare the brain functional connectivity before and after an intensive rehabilitative treatment. The analysis was performed by means of local and global efficiency measures related to the execution of three tasks. As expected, the results showed that the treatment led to a balancing of the values of both parameters between the two hemispheres.

2. Background

The field of EEG-based studies about functional connectivity measures in stroke patients has not yet been deeply explored. Caliandro et al. studied network reorganization after an acute stroke [20]. From a comparison with healthy subjects, the authors found a bilaterally decreased small-worldness in the delta band and bilaterally increased small-worldness in the alpha2 band, regardless of the side of the ischemic lesion. In the theta band, small-worldness decreases bilaterally only in patients with left hemispheric stroke. The study of seven stroke patients reported in [21] showed that when the lesion is not bilateral, the impaired hemisphere has a higher small-worldness than the healthy hemisphere; when the lesion is bilateral, there is no significant difference between small-worldness of the right and left hemisphere. In [22], the authors studied a group of patients with unilateral stroke and found that the motor imagery of the affected hand showed a significantly lower small-worldness and local efficiency as compared to the unaffected hand. Philips et al. carried out a longitudinal analysis of a group of stroke patients undergoing an intensive rehabilitative treatment. Conversely, the study revealed that a reduction in both global and local efficiency in the 12.5–25 Hz band is associated with motor recovery [23].

3. Materials and Methods

3.1. Case Description

This study is concerned with the case of a 50-year-old right-handed female who suffered from conduction aphasia after a left ischemic stroke involving white matter of the fronto-parietal lobe and left temporo-occipital areas. She had surgery to replace the aortic valve with a mechanical prosthesis one week before the stroke. She arrived at the rehabilitative unit of IRCCS Centro Neurolesi Bonino-Pulejo (Messina) one month after the stroke. Neurological, neuropsychological and logopedic assessments were carried out. The patient was attentive, cooperative, and time- and space-oriented. Neurological examination showed a right facio-brachio-crural hemiparesis. The patient followed a drug therapy of oral anticoagulants and antihypertensive. The rehabilitative treatment combined different types of training. In particular: physiotherapeutic training, including balance and gait exercises, Bobath and task-oriented exercises, robotic rehabilitation; neuropsychological training, which provided psychological support to improve the patient's emotional-behavioral control, strategies to improve patient motivation during rehabilitation, problem-solving strategies; logopedic training, which consisted of sentence repetition therapy, stimulation-facilitation therapy, and group communication treatment. The rehabilitation was carried out every day for a session of 60 min for each type of treatment. The tasks performed during training were different from those used for the assessment before and after the treatment.

EEG data were acquired by means of the 256-channel HydroCel Geodesic Sensor Net, belonging to the Electrical Geodesics (EGI) EEG system (Figure 1). The electrode impedance was kept <50 kΩ, on the basis of EGI guidelines. The reference electrode was Cz, placed in the middle of the scalp. The sampling rate was 250 Hz. The EEG data were band-pass filtered between 1 and 40 Hz by means of EGI's Net Station EEG software and cleaned from artifacts by visual inspection. The signals from sensors placed on the face and the neck were affected by muscle artifacts, so only 173 electrodes from the starting 256 were considered. In addition, the LORETA-KEY software used for our study required only cephalic (no face, no neck) electrodes to be considered. Finally, the EEG recordings were average referenced and segmented into artifact-free non-overlapping epochs of 1 s. HD-EEG were acquired at baseline (T0) and after a rehabilitative treatment of two months (T1). HD-EEG were recorded while the patient was performing specific language tasks displayed on a computer screen in order to set the time and use a standard method without the influence of external stimuli. The task paradigm was created by means of E-prime 3.0, a leading software for designing, collecting and analyzing data for behavioral research. E-Prime provides a complete environment for building experiments with text, images, sound, and videos through an easy-to-use graphical interface. E-prime provides a millisecond accuracy for subject responses and sound onset times. The EEG recordings were carried out during time blocks consisting of three task periods with alternating rest periods. Specifically, the experimental setup included the following tasks: naming, repetition, and reading. During the naming task, the subject had to mention 24 images, each displayed on the computer screen for 3 s. The images were divided into two groups of 12, alternating with 10 s of rest. The total duration of the naming task is 72 s. The repetition task consisted of repeating 16 words of 5 s, played by speakers. The words were divided into two groups of 8, alternating with 10 s of rest. The total duration of the repetition task was 80 s. The reading task consisted of reading 16 words of 3 s, displayed on the computer screen. The words were divided into two groups of 8, alternating with 10 s of rest. The total duration of the naming task was 48 s. Each task at T0 and the corresponding one at T1 have the same duration. The timing of each above-mentioned event was properly set by the operator during the building of the task blocks with E-prime. Finally, E-prime sent markers to the EGI Net Station based on the onset and end time of each group of stimuli.

The whole procedure was conducted conforming to the related guidelines and regulations. As it is a case study, approval by the local Ethics Committee is not required. The patient signed an informed consent form.

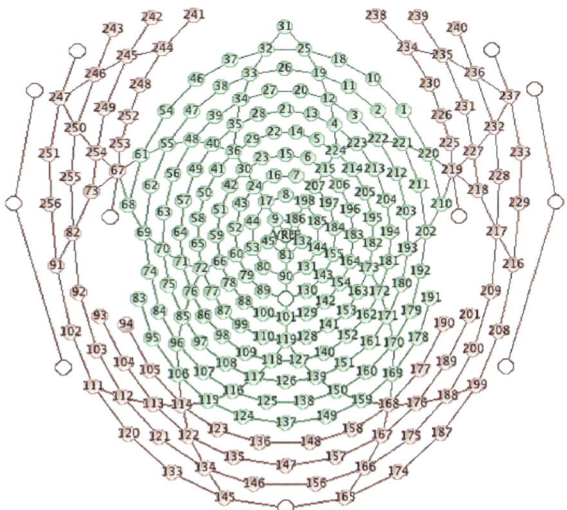

Figure 1. A 2D map of the 256 channel 256-channel HydroCel Geodesic Sensor Net. The considered electrodes are highlighted in green. The electrodes in red were discarded.

3.2. Brain Network Analysis

The functional connectivity of the brain was estimated by the Lagged Linear Connectivity (LLC) parameter, which was computed by the Connectivity Toolbox implemented in the LORETA-KEY software (v20210701). LLC provides a measurement of the statistical dependence among active brain sources for each pair of regions of interest (ROIs) for a specified frequency band [24]. In this case, we computed LLC for all 84 possible ROIs defined by the LORETA-KEY software. The ROIs correspond to distinct Brodmann areas, 42 for each hemisphere (Figure 2). Each ROI consists of a single voxel, the one that is the closest to the center of mass of the ROI. The single centroid voxel is an excellent representative of the ROI. The analysis was performed for the frequency range 1–40 Hz. The EEG signals of naming, repetition and reading tasks were divided into 72, 80 and 48 epochs of 1 s, respectively. LLC was calculated for windows of 2 epochs, so we obtained 36, 40 and 24 connection matrices for the naming, repetition and reading tasks, respectively.

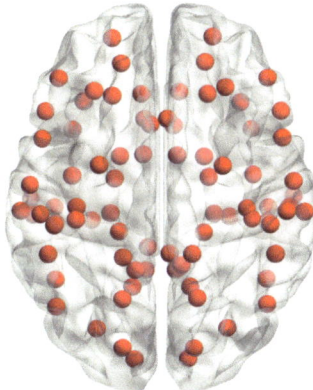

Figure 2. Red points denote the 84 ROIs considered.

Starting from the computation of LLC, the properties of the brain networks were computed by two parameters: global efficiency and local efficiency. The average or global efficiency of a graph G is defined as:

$$E_{glob}(G) = \frac{1}{N(N-1)} \sum_{i \neq j \in G} \frac{1}{d_{ij}} \quad (1)$$

where N is the number of nodes, and d_{ij} is the shortest path length between nodes i and j. The local efficiency is given by the following expression:

$$E_{loc}(G) = \frac{1}{N} \sum_{i \in G} E(G_i) \quad (2)$$

The above-mentioned parameters provide a measure of the efficiency in the information exchanges of a network [19]. For both parameters, higher values mean greater network efficiency.

Data processing was performed in MATLAB environment (R2021b). Global and local efficiency were computed by means of the *Brain Connectivity Toolbox*, a MATLAB toolbox for structural and functional brain connectivity analysis (https://sites.google.com/site/bctnet/, accessed on 18 April 2022).

4. Results

For each task, global and local efficiency were estimated for each hemisphere at time T0 and T1. Figures 3–5 show the boxplots of the global efficiency and local efficiency values. A statistical analysis was performed to assess if the differences between the injured (left) and the non-injured (right) hemisphere at T0 and T1 were significant. For this aim, we chose to perform a nonparametric test since the Shapiro–Wilk test [25] revealed that the global and local efficiency values are in some cases normally distributed and in others not normally distributed. In particular, the Wilcoxon rank-sum test [26] was carried out under the null hypothesis that for each task, the medians of the global and local efficiency values between the left and right hemisphere at time T0 do not differ from the corresponding ones at time T1. The significance level was set at 5% so that the difference between T0 and T1 is statistically significant when the *p*-value is less than 0.05. Table 1 shows the *p*-values derived from the statistical analysis for each task. At T0, for both naming and repetition task, global and local efficiency in the left (impaired) hemisphere are lower than those of the right (unimpaired) hemisphere. For the reading task, conversely, global and local efficiency of the left hemisphere are higher than those of the right hemisphere. All the differences are statistically significant, except for local efficiency related to the naming task. After the rehabilitative treatment, at T1, the global and local efficiency values between the two hemispheres become balanced, and there is no longer a statistically significant difference. This trend reflects the expected behavior, as the rehabilitation contributed to the creation of new neural patterns to compensate for the disrupted ones. The results also suggest that local and global networks of the brain are altered in stroke patients but not always in the same direction. This trend can be due to the effects of stroke, which affect not only the site of structural damage but also distant brain regions that belong to a functional network. The patient also underwent the Aachener Aphasie Test (AAT), a standardized test that provides an assessment of language functioning after brain injury and determines the presence of aphasia [27]. The scores of the AAT at time T0 and T1 show an improvement of the aphasic deficits of the patient (Figure 6). In particular, the batteries of the naming, repetition and written language tests revealed a gain of 17, 9 and 11 points, respectively, after the rehabilitative treatment. Note that the "written language" battery includes the tests related to the reading ability assessment. The AAT results support our findings of the functional connectivity analysis.

Naming Task

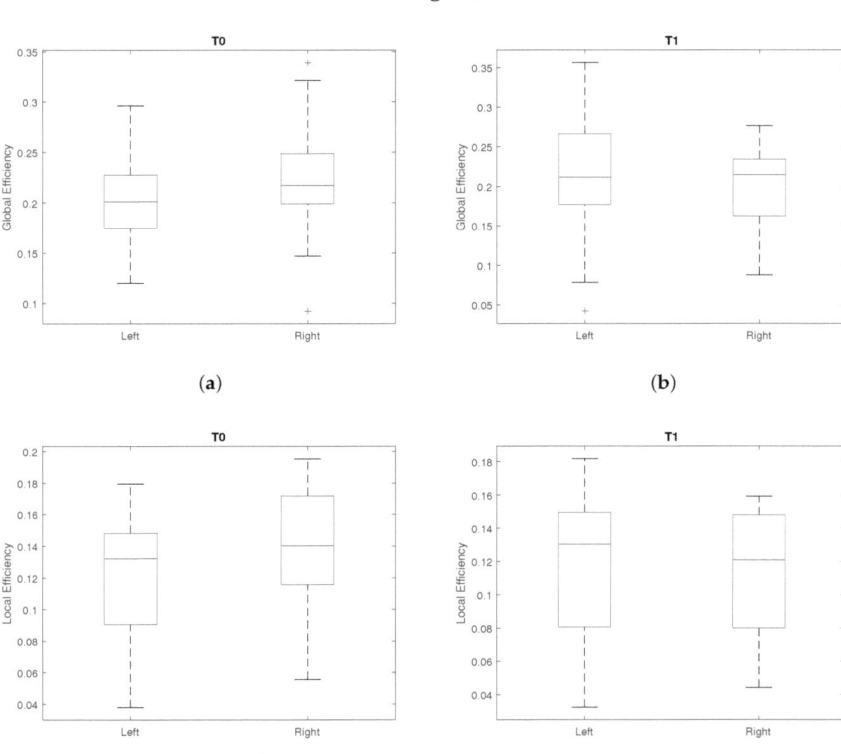

Figure 3. Naming task: boxplot of global efficiency for each hemisphere at time (**a**) T0 and (**b**) T1; boxplot of local efficiency for each hemisphere at time (**c**) T0 and (**d**) T1. On each box, the bottom and the top edges denote the 25th and 75th percentiles, respectively; the line inside the box indicates the median; the "whiskers" extend below and above the box up to the minimum and maximum data values, respectively. The '+' marker symbol outside the whiskers represents the outliers.

Repetition Task

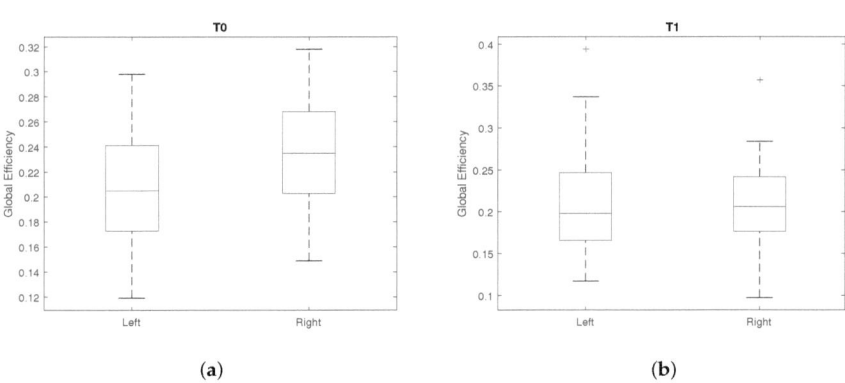

Figure 4. *Cont.*

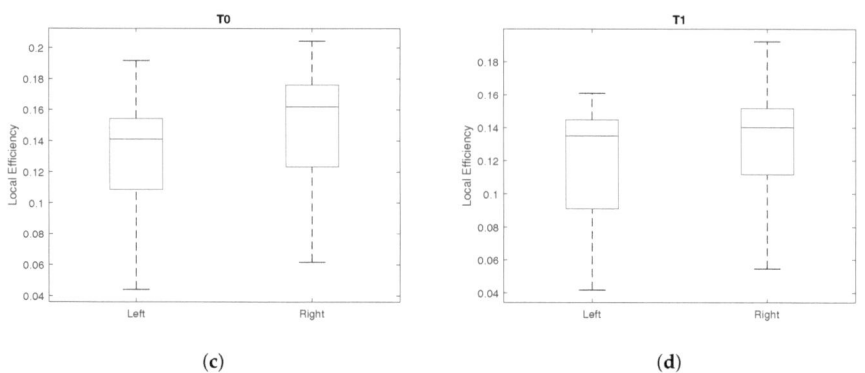

|(c)|(d)|

Figure 4. Repetition task: boxplot of global efficiency for each hemisphere at time (**a**) T0 and (**b**) T1; boxplot of local efficiency for each hemisphere at time (**c**) T0 and (**d**) T1. On each box, the bottom and the top edges denote the 25th and 75th percentiles, respectively; the line inside the box indicates the median; the "whiskers" extend below and above the box up to the minimum and maximum data values, respectively. The '+' marker symbol outside the whiskers represents the outliers.

Reading Task

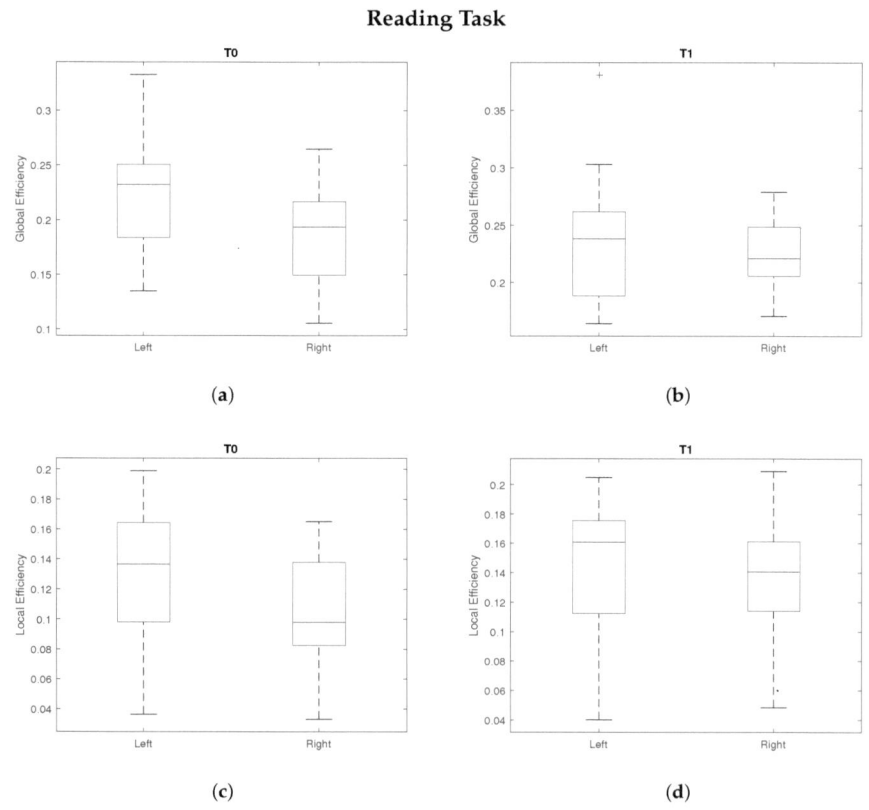

Figure 5. Reading task: boxplot of global efficiency for each hemisphere at time (**a**) T0 and (**b**) T1; boxplot of local efficiency for each hemisphere at time (**c**) T0 and (**d**) T1. On each box, the bottom and the top edges denote the 25th and 75th percentiles, respectively; the line inside the box indicates the median; the "whiskers" extend below and above the box up to the minimum and maximum data values, respectively. The '+' marker symbol outside the whiskers represents the outliers.

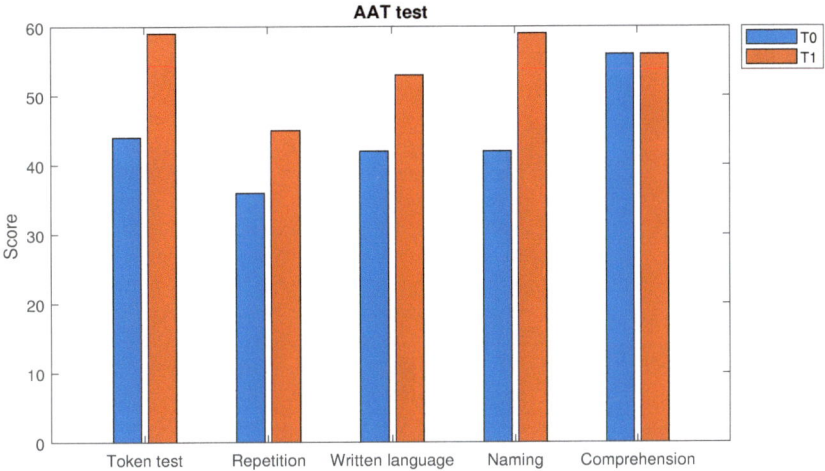

Figure 6. AAT test score at T0 and T1.

Table 1. Results of the Wilcoxon rank-sum test.

Network Parameters	Naming Task p-Value		Repetition Task p-Value		Reading Task p-Value	
	T0	T1	T0	T1	T0	T1
Global Efficiency	0.0456 *	0.6728	0.0136 *	0.5098	0.0193 *	0.3922
Local Efficiency	0.0597	0.5341	0.0023 *	0.1992	0.0033 *	0.1340

* denotes statistically significant differences ($p < 0.05$).

5. Discussion and Conclusions

Stroke is a medical condition that affects the brain and prevents it from functioning properly. Combined with drug therapy, rehabilitative treatment is a precious tool for promoting recovery. The effectiveness of the treatment depends on several factors, such as the therapy type, duration, intensity and beginning treatment early [28]. The treatment can include neuromotor rehabilitation, speech therapy, and cognitive and respiratory rehabilitation. It has been shown that the combination of robotic, psychomotor and cognitive therapy produces positive effects in the rehabilitative process [29]. Moreover, a new effective physiotherapy method for improving movement in post-stroke patients is represented by virtual reality therapy, which allows the subject to interact with an environment within a simulated reality [30]. Therefore, a multidisciplinary approach, based on the cooperation and collaboration between several health professionals, is considered to be a key point for a successful rehabilitative treatment.

In this study, we analyzed the functional connectivity of a left ischemic stroke patient, who suffered from conduction aphasia, before and after a period of intensive rehabilitation. In a human brain, most of the time, the left hemisphere is dominant for language processing [31]. This has been proven by studies that detected a higher activity during language processing in the left hemisphere and a greater probability of linguistic impairment deriving from injuries to the left hemisphere [32–34]. As for aphasia, the role of the right hemisphere is still unclear and debated. Nevertheless, there is some evidence that the right hemisphere plays a facilitatory role in the recovery after the rehabilitative treatment in the subacute stage (up to six months after a stroke) [35].

In our study, among the network parameters that are commonly used to describe the brain networks, we chose the global and local connectivity. We expected that after the treatment, a balance of functional connectivity between the impaired and healthy hemisphere occurred. Actually, this trend was observed during the execution of all the three considered tasks. It is noteworthy to point out that for name and repetition tasks, shortly after the stroke, the global and local connectivity are lower in the affected lobe (left hemisphere) than in the unaffected one (right hemisphere). Conversely, for the reading task, global and local connectivity are higher in the left lobe. This apparently contrasting behavior can be explained as a consequence of the alteration of the brain functional network, which can also concern regions belonging to the unaffected lobe. Moreover, changes in network connectivity can be task-dependent, as reported in previous studies [36,37].

In conclusion, functional connectivity in stroke patients has not yet been sufficiently explored and needs to be further investigated. Our work can be considered a starting point for future in-depth research. However, despite the potential of our research, some limitations in this work need to be addressed. First, this is a case report that considers only one subject, so a longitudinal study involving a cohort of patients would further validate our conclusions. Then, a future study based on more than three tasks would provide a more comprehensive assessment. Our analysis was performed considering the total band of the EEG signal. It would be interesting to perform an analysis for each EEG frequency sub-band (delta, theta, alpha, beta, and gamma). Moreover, other connectivity measures could be tested to find out the most suitable features for the intended purpose. In this way, the potentiality of EEG would be fully exploited for supporting the treatment of stroke and monitoring the recovery process.

Author Contributions: Conceptualization, S.D. and F.L.F.; methodology, S.D. and F.L.F.; formal analysis, S.D.; investigation, S.D.; project administration, F.L.F.; writing—original draft preparation, S.D.; writing—review and editing, F.L.F.; supervision, F.L.F. All authors have read and agreed to the published version of the manuscript.

Funding: This research received no external funding.

Institutional Review Board Statement: Ethical review and approval were waived for this study, due to it is a case study.

Informed Consent Statement: Informed consent was obtained from the subject involved in the study.

Data Availability Statement: Restrictions apply to the availability of these data. Data was obtained from "IRCCS—Centro Neurolesi Bonino-Pulejo" of Messina (Italy) and are available from the authors with the permission of "IRCCS—Centro Neurolesi Bonino-Pulejo" of Messina (Italy).

Acknowledgments: The authors thank the doctors of IRCCS Centro Neurolesi Bonino-Pulejo of Messina (Italy) for their insightful comments and suggestions.

Conflicts of Interest: The authors declare no conflict of interest.

References

1. World Stroke Organization. Annual Report. 2019. Available online: https://www.world-stroke.org (accessed on 11 April 2022).
2. World Health Organization. *Cerebrovascular Disorders: A Clinical and Research Classification*; World Health Organization: Geneva, Switzerland, 1978.
3. Gresham, G.E.; Stason, W.B.; Duncan, P.W. *Post-Stroke Rehabilitation*; Diane Publishing: Darby, PA, USA, 2004; Volume 95.
4. Wernicke, C. *Der Aphasische Symptomencomplex: Eine Psychologische Studie auf Anatomischer Basis*; Cohn & Weigert: Breslau, Poland, 1874.
5. Ardila, A. A review of conduction aphasia. *Curr. Neurol. Neurosci. Rep.* **2010**, *10*, 499–503. [CrossRef]
6. Geschwind, N. Disconnexion syndromes in animals and man. *Brain* **1965**, *88*, 585. [CrossRef]
7. Hickok, G.; Erhard, P.; Kassubek, J.; Helms-Tillery, A.K.; Naeve-Velguth, S.; Strupp, J.P.; Strick, P.L.; Ugurbil, K. A functional magnetic resonance imaging study of the role of left posterior superior temporal gyrus in speech production: Implications for the explanation of conduction aphasia. *Neurosci. Lett.* **2000**, *287*, 156–160. [CrossRef]
8. Angrilli, A.; Elbert, T.; Cusumano, S.; Stegagno, L.; Rockstroh, B. Temporal dynamics of linguistic processes are reorganized in aphasics' cortex: An EEG mapping study. *Neuroimage* **2003**, *20*, 657–666. [CrossRef]
9. Breier, J.I.; Castillo, E.M.; Boake, C.; Billingsley, R.; Maher, L.; Francisco, G.; Papanicolaou, A.C. Spatiotemporal patterns of language-specific brain activity in patients with chronic aphasia after stroke using magnetoencephalography. *Neuroimage* **2004**, *23*, 1308–1316. [CrossRef]

10. Sarasso, S.; Määttä, S.; Ferrarelli, F.; Poryazova, R.; Tononi, G.; Small, S.L. Plastic changes following imitation-based speech and language therapy for aphasia: A high-density sleep EEG study. *Neurorehabilit. Neural Repair* **2014**, *28*, 129–138. [CrossRef]
11. Spironelli, C.; Angrilli, A. Brain plasticity in aphasic patients: Intra-and inter-hemispheric reorganisation of the whole linguistic network probed by N150 and N350 components. *Sci. Rep.* **2015**, *5*, 12541. [CrossRef]
12. Thompson, C.K.; den Ouden, D.B. Neuroimaging and recovery of language in aphasia. *Curr. Neurol. Neurosci. Rep.* **2008**, *8*, 475. [CrossRef]
13. Mattioli, F.; Ambrosi, C.; Mascaro, L.; Scarpazza, C.; Pasquali, P.; Frugoni, M.; Magoni, M.; Biagi, L.; Gasparotti, R. Early aphasia rehabilitation is associated with functional reactivation of the left inferior frontal gyrus: A pilot study. *Stroke* **2014**, *45*, 545–552. [CrossRef]
14. Stefaniak, J.D.; Halai, A.D.; Ralph, M.A.L. The neural and neurocomputational bases of recovery from post-stroke aphasia. *Nat. Rev. Neurol.* **2020**, *16*, 43–55. [CrossRef]
15. Formica, C.; De Salvo, S.; Micchìa, K.; La Foresta, F.; Dattola, S.; Mammone, N.; Corallo, F.; Ciavola, A.; Arcadi, F.A.; Marino, S.; et al. Cortical Reorganization after Rehabilitation in a Patient with Conduction Aphasia Using High-Density EEG. *Appl. Sci.* **2020**, *10*, 5281. [CrossRef]
16. Rubinov, M.; Sporns, O. Complex network measures of brain connectivity: Uses and interpretations. *Neuroimage* **2010**, *52*, 1059–1069. [CrossRef]
17. Achard, S.; Bullmore, E. Efficiency and cost of economical brain functional networks. *PLoS Comput. Biol.* **2007**, *3*, e17. [CrossRef]
18. Bullmore, E.; Sporns, O. Complex brain networks: Graph theoretical analysis of structural and functional systems. *Nat. Rev. Neurosci.* **2009**, *10*, 186–198. [CrossRef]
19. Latora, V.; Marchiori, M. Efficient behavior of small-world networks. *Phys. Rev. Lett.* **2001**, *87*, 198701. [CrossRef] [PubMed]
20. Caliandro, P.; Vecchio, F.; Miraglia, F.; Reale, G.; Della Marca, G.; La Torre, G.; Lacidogna, G.; Iacovelli, C.; Padua, L.; Bramanti, P.; et al. Small-world characteristics of cortical connectivity changes in acute stroke. *Neurorehabilit. Neural Repair* **2017**, *31*, 81–94. [CrossRef]
21. Mammone, N.; De Salvo, S.; Marino, S.; Bonanno, L.; Ieracitano, C.; Dattola, S.; La Foresta, F.; Morabito, F.C. Estimating the Asymmetry of Brain Network Organization in Stroke Patients from High-Density EEG Signals. In *Neural Approaches to Dynamics of Signal Exchanges*; Springer: Singapore, 2020; pp. 475–483.
22. Fallani, F.D.V.; Pichiorri, F.; Morone, G.; Molinari, M.; Babiloni, F.; Cincotti, F.; Mattia, D. Multiscale topological properties of functional brain networks during motor imagery after stroke. *Neuroimage* **2013**, *83*, 438–449. [CrossRef]
23. Philips, G.R.; Daly, J.J.; Príncipe, J.C. Topographical measures of functional connectivity as biomarkers for post-stroke motor recovery. *J. Neuroeng. Rehabil.* **2017**, *14*, 1–16. [CrossRef]
24. Pascual-Marqui, R.D. Instantaneous and lagged measurements of linear and nonlinear dependence between groups of multivariate time series: Frequency decomposition. *arXiv* **2007**, arXiv:0711.1455.
25. Shaphiro, S.; Wilk, M. An analysis of variance test for normality. *Biometrika* **1965**, *52*, 591–611. [CrossRef]
26. Gibbons, J.; Chakraborti, S. *Nonparametric Statistical Inference*; Springer: Cham, Switzerland, 2011.
27. Huber, W.; Poeck, K.; Weniger, D.; Willmes, K. *AAT-Aachener Aphasie Test*; Hogrefe: Göttingen, Germany, 1983.
28. van Meijeren-Pont, W.; Tamminga, S.J.; Fiocco, M.; Avila, A.G.; Volker, G.; Janssen, S.M.; Vlieland, T.P.V.; Oosterveer, D.M.; SCORE Study Group. Patient Activation During the First 6 Months After the Start of Stroke Rehabilitation. *Arch. Phys. Med. Rehabil.* **2022**, *in press*. [CrossRef] [PubMed]
29. Vostrỳ, M.; Zilcher, L. Combined Therapy for Patients after Ischemic Stroke from The Point View of Comprehensive Rehabilitation. *J. Educ. Health Soc.* **2020**, *11*, 119–125. [CrossRef]
30. Montalbán, M.A.; Arrogante, O. Rehabilitation through virtual reality therapy after a stroke: A literature review. *Rev. Cient. Soc. Esp. Enferm. Neurol.* **2020**, *52*, 19–27. [CrossRef]
31. Knecht, S.; Dräger, B.; Deppe, M.; Bobe, L.; Lohmann, H.; Flöel, A.; Ringelstein, E.B.; Henningsen, H. Handedness and hemispheric language dominance in healthy humans. *Brain* **2000**, *123*, 2512–2518. [CrossRef] [PubMed]
32. Mesulam, M.M.; Weintraub, S.; Rogalski, E.J.; Wieneke, C.; Geula, C.; Bigio, E.H. Asymmetry and heterogeneity of Alzheimer's and frontotemporal pathology in primary progressive aphasia. *Brain* **2014**, *137*, 1176–1192. [CrossRef]
33. Fridriksson, J.; Yourganov, G.; Bonilha, L.; Basilakos, A.; Den Ouden, D.B.; Rorden, C. Revealing the dual streams of speech processing. *Proc. Natl. Acad. Sci. USA* **2016**, *113*, 15108–15113. [CrossRef]
34. Mahowald, K.; Fedorenko, E. Reliable individual-level neural markers of high-level language processing: A necessary precursor for relating neural variability to behavioral and genetic variability. *Neuroimage* **2016**, *139*, 74–93. [CrossRef]
35. Cocquyt, E.M.; De Ley, L.; Santens, P.; Van Borsel, J.; De Letter, M. The role of the right hemisphere in the recovery of stroke-related aphasia: A systematic review. *J. Neurolinguist.* **2017**, *44*, 68–90. [CrossRef]
36. Vinehout, K.; Schmit, B.D.; Schindler-Ivens, S. Lower limb task-based functional connectivity is altered in stroke. *Brain Connect.* **2019**, *9*, 365–377. [CrossRef]
37. Grefkes, C.; Nowak, D.A.; Eickhoff, S.B.; Dafotakis, M.; Küst, J.; Karbe, H.; Fink, G.R. Cortical connectivity after subcortical stroke assessed with functional magnetic resonance imaging. *Ann. Neurol.* **2008**, *63*, 236–246. [CrossRef]

Article

Wavelet-Based Multi-Class Seizure Type Classification System

Hezam Albaqami [1,2,*], Ghulam Mubashar Hassan [1] and Amitava Datta [1]

[1] Department of Computer Science and Software Engineering, The University of Western Australia, Perth 6009, Australia; ghulam.hassan@uwa.edu.au (G.M.H.); amitava.datta@uwa.edu.au (A.D.)
[2] Department of Computer Science and Artificial Intelligence, University of Jeddah, Jeddah 201589, Saudi Arabia
* Correspondence: hezam.albaqami@research.uwa.edu.au

Abstract: Epilepsy is one of the most common brain diseases that affects more than 1% of the world's population. It is characterized by recurrent seizures, which come in different types and are treated differently. Electroencephalography (EEG) is commonly used in medical services to diagnose seizures and their types. The accurate identification of seizures helps to provide optimal treatment and accurate information to the patient. However, the manual diagnostic procedures of epileptic seizures are laborious and require professional skills. This paper presents a novel automatic technique that involves the extraction of specific features from epileptic seizures' EEG signals using dual-tree complex wavelet transform (DTCWT) and classifying them into one of the seven types of seizures, including absence, complex-partial, focal non-specific, generalized non-specific, simple-partial, tonic-clonic, and tonic seizures. We evaluated the proposed technique on the TUH EEG Seizure Corpus (TUSZ) ver.1.5.2 dataset and compared the performance with the existing state-of-the-art techniques using the overall F1-score due to class imbalance of seizure types. Our proposed technique achieved the best results of a weighted F1-score of 99.1% and 74.7% for seizure-wise and patient-wise classification, respectively, thereby setting new benchmark results for this dataset.

Keywords: diagnostics; dual-tree complex wavelet transform (DTCWT); electroencephalography (EEG); epilepsy; LightGBM; seizure type classification

Citation: Albaqami, H.; Hassan, G.M.; Datta, A. Wavelet-Based Multi-Class Seizure Type Classification System. *Appl. Sci.* **2022**, *12*, 5702. https://doi.org/10.3390/app12115702

Academic Editors: Serena Dattola and Fabio La Foresta

Received: 4 May 2022
Accepted: 1 June 2022
Published: 3 June 2022

Publisher's Note: MDPI stays neutral with regard to jurisdictional claims in published maps and institutional affiliations.

Copyright: © 2022 by the authors. Licensee MDPI, Basel, Switzerland. This article is an open access article distributed under the terms and conditions of the Creative Commons Attribution (CC BY) license (https:// creativecommons.org/licenses/by/ 4.0/).

1. Introduction

1.1. Background

Epilepsy is the most widespread brain disease among children and adults after stroke [1]. It is defined as "a sudden and recurrent brain malfunction and is a disease that reflects an excessive and hypersynchronous activity of the neurons within the brain" [2]. Over 60 million of the world's population are diagnosed with epilepsy, whose defining feature is recurrent seizures. Such seizure attacks impair the brain's normal functions, leading the patient to be vulnerable and unsafe.

Seizures are medically classified into two main categories—focal seizures or generalized seizures—depending on the extent to which regions of the brain are affected. Focal seizures are seizures that originate and affect a circumscribed region of the brain. Focal seizures are further classified into simple or complex, based on the patient's level of awareness. Generalized seizures, on the other hand, involve most areas of the brain. Based on motor and non-motor symptoms, generalized seizure classifications can be absence, tonic, atonic, clonic, tonic-clonic, or myoclonic seizures [3,4]. Classification of seizure is very essential for accurate diagnosis and treatment.

Identifying the type of seizure, although sometimes difficult, can be done by clinical observation and referencing medical history and demographic information, and is supported by general brain imaging techniques such as EEG, magnetoencephalography (MEG), and fMRI [5,6]. EEG is the most practical and cost-effective tool to diagnose epilepsy currently [7]. Video-EEG monitoring is often required to support the decision for seizure classification [8].

For treatment, seizures can be controlled in most cases (up to 70%) of patients by consuming medication to achieve a steady-state concentration in the blood. Surgical intervention is another option for certain conditions. For up to 20% of epileptic patients, there is no medical treatment that exists to control seizures [2]. The accurate identification of the type of seizures influences medication choice and provides information to patients, families, researchers, and clinicians caring for patients with epilepsy [4,9].

It is a challenging task to classify the type of seizure accurately. Several factors make the classification difficult. Firstly, some types of seizures share the same clinical and EEG symptoms. For instance, it has been shown that even for a highly experienced neurologist, sometimes it is hard to distinguish between focal and generalized seizures [10]. Secondly, in some cases, it is required to perform long-term monitoring (i.e., video-EEG monitoring), which may last for days [7]. Therefore, manual analysis of these long recordings requires a substantial amount of effort and time from neurologists.

In addition, signal interpretation is known to have a low inter-rater agreement, which fully depends on the level of expertise of the expert. Moreover, inter-subject variability significantly adds to the difficulties associated with the diagnosis of an epileptic seizure, leading to a variety of manifestations of the same type of seizures across different patients, and sometimes for the same individual over time. Finally, signal artifacts also hinder the correct interpretation of EEG. With these challenges, in a field that already has a shortage of healthcare experts, computer-aided diagnostic (CAD) methods have great potential to support decision making in the diagnosis of such a critical disease.

1.2. Review of Related Work

A considerable amount of research has been published on automated seizure detection and prediction. However, the automatic classification of seizure types has received little attention due to two main reasons: firstly, the difficulties inherent in the classification problem for seizure types, and secondly, a lack of clinical data [11].

Since the start of this century, considerable research outcomes have focused on the automation of epileptic seizure diagnoses [8,9]. Generally, the procedure of automatic seizure analysis involves two phases: feature extraction and classification [12,13]. Various methods have been proposed for feature extraction over time, including time-domain [14], frequency-domain [13,15], and time-frequency domain [16].

Time-frequency methods became popular due to inclusion of both time and frequency features. Among time-frequency methods, wavelet transform (WT)-based feature extraction is the most promising method to extract robust features from EEG signals [17]. The strategies in wavelet-based feature extraction from EEGs use continuous wavelet transform (CWT) [18], discrete wavelet transform (DWT) [19], wavelet packet decomposition (WPD) [19,20], tunable Q-factor wavelet transform (TQWT) [21,22], and dual-tree wavelet transform (DTCWT) [23].

Regarding the availability of clinical data, it has been observed that in recent years, hospitals and universities have made appreciative efforts to encourage research on the automatic diagnosis of epileptic seizures by generating large volumes of openly available clinical EEG data. One of the most extensive publicly obtainable EEG datasets, the Temple University Hospital EEG Corpus (TUH EEG), is comprised of 14,000 subjects and has more than 25,000 clinical recordings [24]. The Corpus has various subsets, each focusing on different scopes of research interests. The TUH EEG Seizure Corpus (TUSZ) [25], one of the subsets, was created to motivate research on developing high-performance epileptic seizure detection algorithms using advances in machine learning algorithms [25]. This dataset contains manually annotated seizure events based on archival neurologist reports and careful examinations of the signals by students and neurologists from Temple University [25]. The seizure events in the TUSZ are labeled with eight different types of seizures: focal non-specific seizure (FNSZ), generalized non-specific seizure (GNSZ), simple partial seizure (SPSZ), complex partial seizure (CPSZ), absence seizure (ABSZ), tonic seizure (TNSZ), tonic-clonic seizure (TCSZ), and myoclonic seizure (MYSZ). The details of these labels are presented in Table 1. The corpus team continuously updates the corpus, and Table 2 presents the distribution of data for the last two versions of TUSZ.

Table 1. Seizure type descriptions for TUH EEG Seizure Corpus (TUSZ).

Seizure Type	Seizure Description
FNSZ	Focal seizures which cannot be specified with its type.
SPSZ	Partial seizures during consciousness which is specified by clinical signs only.
CPSZ	Partial Seizures during unconsciousness which is specified by clinical signs only.
GNSZ	Generalized seizures which cannot be further specified with its type.
ABSZ	Absence discharges observed on EEG where patient loses consciousness for few seconds (also known as petit mal).
TNSZ	Stiffening of body during seizure (EEG effects disappear).
TCSZ	At first stiffening and then jerking of body (also known as grand mal).
MYSZ	Myoclonous jerks of limbs.

Table 2. Data distribution for different types of seizures in two versions of TUSZ.

Seizure Type	No. of Seizure Events		Duration (s)		No. of Patients	
	ver.1.4.0	ver.1.5.2	ver.1.4.0	ver.1.5.2	ver.1.4.0	ver.1.5.2
FNSZ	992	1836	73,466	121,139	109	150
GNSZ	415	583	34,348	59,717	44	81
CPSZ	342	367	33,088	36,321	34	41
ABSZ	99	99	852	852	13	12
TNSZ	67	62	1271	1204	2	3
TCSZ	50	48	5630	5548	11	12
SPSZ	44	52	1534	2146	2	3
MYSZ	3	3	1312	1312	2	2

To the best of our knowledge, we found only eight published research studies which used TUSZ for the problem of seizure type classification; the summary is presented in Table 3. Regarding the seven (7) types of seizure classification, Roy et al. [9] applied extreme gradient boosting (XGBoost) and KNN to classify the EEG signals into seven classes of seizures. The study reported F1-scores of 85.1% and 90.1% for XGBoost and K-nearest neighbor (KNN), respectively. Similarly, Aristizabal et al. [26] developed a deep learning model known as neural memory networks (NMN) to classify seven types of seizures. The study reported a 94.50% F1-score. In another study related to the seven-class problem, Asif et al. [11] applied a deep learning framework, called SeizureNet with ensemble learning and multiple DenseNets that achieved a 95% F1-score.

Raghu et al. [27] extracted EEG image features using a pretrained Google Inception 3 and classified them using support vector machine (SVM), achieving an accuracy of 88.3% to classify seven types of seizure classes and a normal class. Similarly, in [28], a convolutional neural network (CNN) model, *AlexNet*, is applied to classify EEG images based on the technique of short-time Fourier transform (STFT) to classify seven types of seizure and non-seizure class. The study yielded an accuracy of 84.06%. Liu et al. [8] applied a hybrid bi-linear model consisting of CNN and long short-term memory (LSTM) to classify eight types of seizures. The study reported a 97.4% F1-score.

For the four-class classification of seizures, Wijayanto et al. [29] applied empirical mode decomposition (EMD) to EEGs for feature extraction and quadratic SVM for classification. The study reported an accuracy of 95%. In another study, Ramadhani et al. [30] applied EMD, Mel frequency cepstral coefficients (MFCC), and independent component analysis (ICA) to EEG data for feature extraction and SVM for classification of four classes of seizures and achieved 91.4% accuracy. For three classes of seizure classification, Saric et al. [31]

developed a field programmable gate array (FPGA)-based framework for the classification of generalized and focal epileptic seizures using a feed-forward multi-layer neural network and achieved an accuracy of 95.14%.

In spite of the good performance reported in the aforementioned studies, it is expected that the reported techniques cannot be used in real world situations as the studies either did not report the performance when tested on data from new patients or reported lower performance. Out of the eight studies presented in Table 3, only two studies [9,11] considered the generalization of their proposed techniques. Both studies mentioned a considerable decrease in the performance of their system, where the performance decreased by 45%. This shows that there is still a large gap for advancement for better generalization capability for the classification systems.

It is interesting to observe from Table 3 that the authors of these studies chose a different number of seizure classes, ranging from a three-class problem to an eight-class problem for seizure type classification. The reason behind the choice of the number of classes is not discussed in most of these studies. The authors of [9,11,26–28] excluded the seizure type MYSZ from their experiments because the signals of this type were only recorded from two patients (see Table 2). However, in [8], the authors chose to utilize all seizure types in the dataset regardless of the number of patients. Table 3 presents the investigated seizure types for each study.

It can be observed from Table 1 that in TUSZ, there are six specific types of seizures and two non-specific general types. From a pathological point of view, these types are not completely disjoint but form a hierarchical sub-grouping [4,26]. It has been stated in [7] that when there is inadequate evidence to label the type of seizure confidently, the corpus team tends to label an event as either focal non-specific or generalized non-specific based on the seizure's focality and locality [25]. Both of these types are not medically distinct from one another, whereas SPNS and CPSZ are more specific types of FNSZ, and ABSZ, TNSZ, TNSZ, and MYSZ are more specific types of GNSZ [4,26]. Thus, considering the label FNSZ as a unique type of seizure against the specific focal types CPSZ and SPSZ might cause the classifier not to perform well, and similarly for the classification of GNSZ.

Therefore, in this study, we are considering two different classification problems. In the first problem, each label is considered in the dataset as a unique seizure type, and results are compared with the existing state-of-the-art results. On the other hand, the second problem is the introduction of a new challenge, which is more important pathologically, that deals with the specific seizure type classification to investigate the effect of the non-specific labels in TUSZ (five-class classification).

In order to solve the above mentioned problems, we propose a novel technique that focuses on wavelet-based machine learning methods for automatic seizure type classification in multi-channel EEG recordings. We only utilized EEG data and decomposed the EEG signals into different levels of components using DTCWT to extract specific features from these decomposed components. We used shift-invariant DTCWT for feature extraction from a biomedical signal and its classification, which is done for the first time in the literature for seizure type classification. Moreover, we tested our proposed technique on the largest available seizure EEG database, containing various types of epileptic seizures. In order to ensure the effectiveness and generalization of our technique, we thoroughly tested our proposed technique across subjects in addition to normal testing. The experimental results show that our proposed novel technique performs well for both problems of seizure-type classification.

The rest of this paper is organized as follows: Table 2 discusses information about the data utilized in this research and the details of our proposed technique. Table 3 presents the evaluation methodology and the analyses of the obtained results. A thorough discussion is provided in Table 4. Table 5 concludes the article with a future research plan.

Table 3. Summary of existing state-of-the-art techniques for seizure classification.

Method	No. of Seizure Classes	Classes Considered	Features	Performance (%)
Transfer learning Inceptionv3 [27]	8 *	GNSZ, FNSZ, SPSZ, CPSZ, ABSZ, TNSZ, TCSZ, NORM	SFFT	88.3 Accuracy
AlexNet [28]	8 *	GNSZ, FNSZ, SPSZ, CPSZ, ABSZ, TNSZ, TCSZ, NORM	SFFT	84.06 Accuracy
CNN+LSTM+MLP [8]	8	GNSZ, FNSZ, SPSZ, CPSZ, ABSZ, TNSZ, TCSZ, MYSZ	SFFT	97.40 F1-score
SeizureNet Ensemble CNNs [11]	7	GNSZ, FNSZ, SPSZ, CPSZ, ABSZ, TNSZ, TCSZ	FFT	95 F1-score
Plastic NMN [26]	7	GNSZ, FNSZ, SPSZ, CPSZ, ABSZ, TNSZ, TCSZ	FFT	94.5 F1-score
K-NN [9]	7	GNSZ, FNSZ, SPSZ, CPSZ, ABSZ, TNSZ, TCSZ	FFT	90.1 F1
XGBoost [9]	7	GNSZ, FNSZ, SPSZ, CPSZ, ABSZ, TNSZ, TCSZ	FFT	85.1 F1-score
SVM [30]	4 *	GNSZ, FNSZ, TCSZ, NORM	MFCC+HD+ICA	91.4 Accuracy
FPGA-based ANN [31]	3 *	GNSZ, FNSZ, NORM	CWT	95.14 Accuracy
SVM [29]	4	GNSZ, FNSZ, SPSZ, TNSZ	EMD	95 Accuracy

* Including non-seizure EEG class. +Normal EEGs.

Table 4. EEG channel names included in our study.

#	Channels	#	Channels
1	FP1-F7	2	F7-T3
3	T3-T5	4	T5-O1
5	FP2-F8	6	F8-T4
7	T4-T6	8	T6-O2
9	T3-C3	10	C3-CZ
11	CZ-C4	12	C4-T4
13	FP1-F3	14	F3-C3
15	C3-P3	16	P3-O1
17	FP2-F4	18	F4-C4
19	C4-P4	20	P4-O2

Table 5. Hyperparameters for LightGBM classifier.

Hyperparameter	Value
boosting type	gbdt
Learning_rate	0.2
n_estimators	1500
colsample_bytree	0.13151
importance_type	split
num_leaves	31
subsample	0.8

2. Materials and Methods

2.1. Data

Our study is based on TUSZ ver. 1.5.2 dataset [25], which is the largest publicly available dataset released in 2020. This dataset includes 3050 seizure events, consisting of various seizure morphologies and recorded from over 300 different patients. The TUSZ was collected from archival hospital data at Temple University Hospital (TUH), where clinical EEG data was retrieved and stored in .EDF format [25]. The signals were recorded based on the international 10-20 EEG system. Table 2 presents the details of the distribution of TUSZ. The EEG signals in TUSZ are annotated based on electrographic, electro-clinical, and clinical manifestations. More details about the dataset can be found in [25], and the dataset is available online at the corpus website (https://isip.piconepress.com/projects/tuh_eeg/, accessed on 1 May 2021). The seizure type MYSZ is excluded from our study due to its scarcity in the dataset, as it is recorded from only two patients in the recently released version (see Table 2). As mentioned earlier, this decision is in accordance with previous research studies in the same field [9,11,26–28].

2.2. Proposed Technique

Our proposed technique involves multiple steps which include preprocessing of the data, extracting the important features, and then classifying them. The architecture of our proposed technique is presented in Figure 1, and all the steps are explained below.

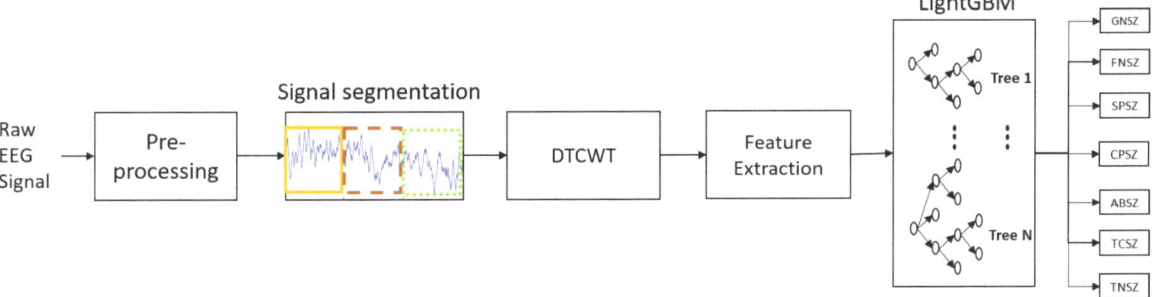

Figure 1. Overall architecture of the proposed technique.

2.2.1. Preprocessing

All the data in TUSZ do not have the same montage and sampling rate, as those recordings were collected from real hospital data (see Section 2.1). As a result, we performed some initial procedures to generalize the input data prior to feature extraction. Firstly, the EEG segments which are exclusively responsible for seizures were extracted from the dataset. This was achieved using the annotation file provided with the dataset, including the start and the stop time of each seizure event. After extracting the seizure events, we used the transverse central parietal (TCP) montage to accentuate spike activity [9]. Montage is a differential view of the data, which consists of differencing the signals collected from two electrodes (e.g., Fp1-F7, F7-T3) [32]. In fact, neurologists are very particular about the type of montage used when interpreting an EEG because it helps in noise reduction [32,33]. Different experiments on montage selection have been done in [32,34], and TCP was found to be the most efficient montage that helps different machine learning algorithms to detect seizures. Secondly, we re-sampled all recordings at 250 Hz, as the EEG recordings in the TUSZ have various sampling rates ranging from 250 Hz to 512 Hz [8]. Finally, we cropped each extracted signal into equally non-overlapped segments such that each segment is of the length of two seconds. This choice was influenced by [9], where the authors investigated different window lengths, and they reported that the two second window length of the signal is the most optimal choice to achieve the best classification results. In summary, we took the following preprocessing steps in sequence to generalize the input data for processing:

1. Used the transverse central parietal (TCP) montage to accentuate spike activity. Table 4 presents the EEG channels considered in our study.
2. Re-sampled all recordings at 250 Hz.
3. Cropped the signal into equally non-overlapped segments such that each segment is of 2 seconds, resulting in 500 data points.

After the initial preprocessing steps, the input data were generalized and ready to be processed for transformation.

2.2.2. Feature Extraction

Wavelet transform (WT) methods have been employed successfully to solve various non-stationary signal problems [17,35,36], including EEG [20]. WT is a spectral estimation method that provides another representation of the signal at different scale components. Wavelet Transform (DWT) is one of the WT's most popular techniques that decomposes a given signal $x[k]$ into a mutually orthogonal set of wavelets through convolution with filter banks. For j levels of decomposition, a signal $x[k]$ is passed through two bandpass filters: high $Hp[.]$ and low $Lp[.]$ starting from level $j = 1$. The output of each level is two down sampled components, approximation $Aj[i]$ and detail $Dj[i]$, which are represented as

$$D_j[i] = \sum_k x[k] \cdot Hp[2 \cdot i - k] \quad (1)$$

$$A_j[i] = \sum_k x[k] \cdot Lp[2 \cdot i - k] \quad (2)$$

The approximation component $Aj[i]$ can be further decomposed into another level of A_{j+1} and D_{j+1} as shown in Figure 2 until the maximum or required level of j is reached.

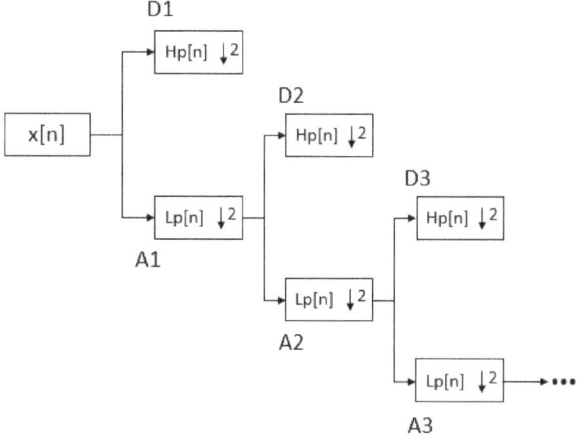

Figure 2. The structure of three-scale level discrete wavelet transform (DWT).

DWT has many successful applications; however, it has some drawbacks, such as insufficient information in high-frequency components, shift-variance, low directionality, and absence of phase shift. Over time, different enhancements have been introduced to cover the shortcomings of DWT. Dual-tree complex wavelet transform (DTCWT) is an extension of DWT which was proposed by Kingsbury [37] and developed later by Selesnick et al. [38]. It uses extra double low-pass filters and an additional two high-pass filters to produce four components at each level which include real and imaginary parts. DTCWT can be imagined as two parallel DWTs, as shown in Figure 3. This transformation is approximately shift-invariant and directionally selective in two and higher dimensions, which are very important in applications such as pattern recognition and signal analysis. Therefore, DTCWT has less shift variance and more directionality as compared to DWT.

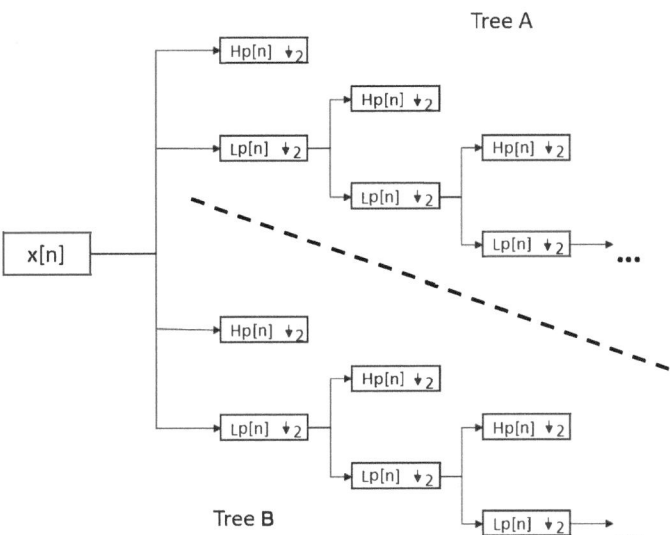

Figure 3. The structure of three-scale level of DTCWT, giving real and imaginary parts of complex coefficients from tree A and tree B. In this study, the EEG signals were decomposed into four levels of decompositions.

In our proposed technique, we decomposed the EEG signals into four levels using DTCWT using the Python library DTCWT [39] (https://github.com/rjw57/dtcwt/tree/0.12.0, accessed on 10 May 2021). The level of decomposition was set manually based on trial and error after experimenting with different levels of decomposition based on the performance evaluation and computational efficiency. The decomposition process produces real and imaginary parts of complex wavelet coefficients, and we selected the magnitude of the complex coefficients. After the decomposition, we computed a set of statistical features from each of the coefficients. Although these coefficients may be directly fed into the machine learning algorithm, it has been observed that these decomposed signals are very sensitive to noise [40]. Therefore, we computed a set of statistical features from each of the coefficients as suggested in [1,19]. Those sets of features were successfully applied in EEG research to discriminate between the signals, such as in abnormality detection [20]. The computed features and their corresponding mathematical representations are presented below. For mathematical representations, M is the length of the signal in each sub-band, which is taken as 500 in this study, while $Y\{y_1, y_2, \ldots y_M\}$ and $Z\{z_1, z_2, \ldots z_M\}$ are two adjacent sub-bands [19,20,41].

1. Mean absolute values (MAV) of the coefficients in each sub-band, μ.

$$\mu = \frac{1}{M} \sum_{j=1}^{M} |y_j| \qquad (3)$$

2. Average power (AVP) of the coefficients in each sub-band, λ.

$$\lambda = \sqrt{\frac{1}{M} \sum_{j=1}^{M} y_j^2} \qquad (4)$$

3. Standard deviation (SD) of the coefficients in each sub-band, σ.

$$\sigma = \sqrt{\frac{1}{M}\sum_{j=1}^{M}(y_j - \mu)^2} \quad (5)$$

4. Ratio of the absolute mean values (RMAV) of adjacent sub-bands, χ.

$$\chi = \frac{\sum_{j=1}^{M}|y_j|}{\sum_{j=1}^{M}|z_j|} \quad (6)$$

5. Skewness (skew) of the coefficients in each sub-band, ϕ.

$$\phi = \sqrt{\frac{1}{M}\sum_{j=1}^{M}\frac{(y_j - \mu)^3}{\sigma^3}} \quad (7)$$

6. Kurtosis (Kurt) of the coefficients in each sub-band, ϕ_k.

$$\phi_k = \sqrt{\frac{1}{M}\sum_{j=1}^{M}\frac{(y_j - \mu)^4}{\sigma^4}} \quad (8)$$

The features across all the statistical coefficients corresponding to this interval signal are stacked together, which forms a 6 × 5 (statistical feature × DTCWT coefficients) feature matrix. We have 20 channels in TCP montage as mentioned in Table 4. Therefore, our resulting feature matrix is of size 20 × 6 × 5 (number of channels × statistical features × DTCWT coefficients), which is flattened to 1 × 600 vector for classification.

2.2.3. Feature Analysis

We analyzed the involved features to understand the importance of the features extracted by DTCWT. We used two feature analysis methods: filtering using ANOVA (analysis of variance) and LightGBM feature importance scores. In both techniques, selecting the top important features, such as 5, 10, or 20 top features, always led to a decrease in classification results regardless of the choice of the number of selected features. Figures 4 and 5 present the results of features obtained by DTCWT using ANOVA and LightGBM feature importance, respectively. We analyzed the extracted features channel-wise and presented the average of all those channels' features. It can be observed from the presented results that all extracted features by DTCWT in our technique play important roles in improving the performance of classification. Therefore, we used all the features, as we believe all features contribute to improving the classification results.

2.2.4. Classification

As mentioned in Table 1, we defined our problem in two classification problems: (1) classification of seven seizure types, including both specific and non-specific seizure types in TUSZ (see Table 1), and (2) classification of five seizure types, including only specific seizure types (see Table 1).

For both problems, we used Light Gradient Boosting Machine (LightGBM ver. 3.2.1) for classification. LightGBM is a gradient boosting decision tree framework which utilizes a tree-based learning algorithm. It is a proven to be an optimal choice to handle a large amount of data, as it is memory efficient, trains faster, and provides high accuracy [42]. The key characteristic of LightGBM is that it uses Gradient-based One-side Sampling (GOSS) in order to find the best split value. In addition, the exclusive feature bundling (EFB) technique is used in LightGBM to reduce the feature space complexity, and the tree growth in LightGBM is leaf-wise growth that leads to faster training [42].

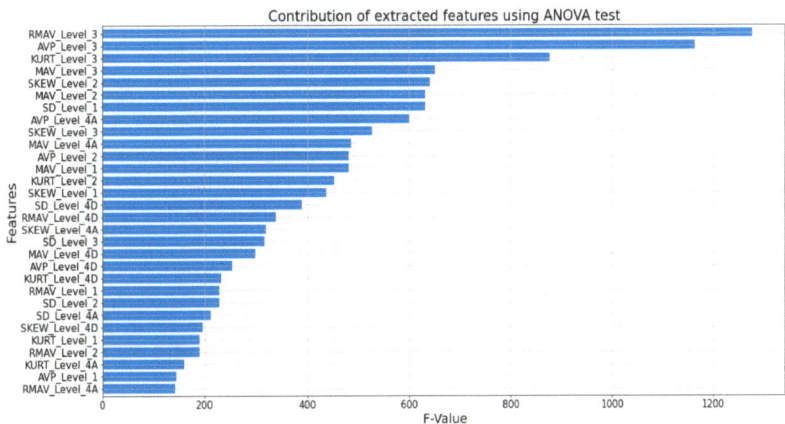

Figure 4. The obtained F-values of the features, using one-way ANOVA test.

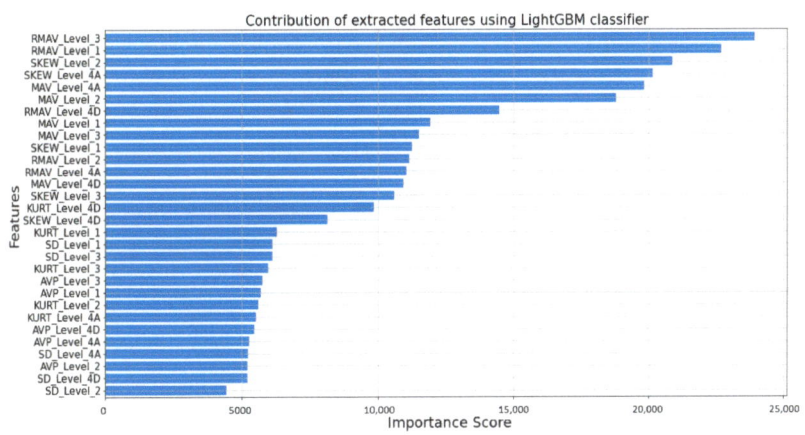

Figure 5. Importance scores for each feature obtained using LightGBM classifier.

In a recent study of EEG binary classification for abnormality detection [20], different classifiers were tested, and LightGBM was found to be one of the most effective classifiers in terms of results and training speed [42]. Therefore, we selected LightGBM for classification. Hyperopt [43] was used to discover the best hyperparameters for our LightGBM. The optimized hyperparameters are presented in Table 5.

3. Results

In this paper, we used TUSZ EEG Corpus ver.1.5.2 to test our proposed technique for seizure type classification. Firstly, we applied some preprocessing methods to remove noise and to accentuate spike activity. Afterwards, the DTCWT feature extraction method was applied, and finally, the LightGBM machine learning method was used for classification.

3.1. Experimental Settings

A desktop computer with 16 GB main memory (RAM), a 255 GB solid-state disk (SDD), a 3.6 GHz microprocessor (CPU), and the Windows 10 operating system was used for the experiments. The technique was developed in Python 3.7. using the DTCWT Python package library ver. 0.12.0.

3.2. Performance Evaluation

It can be observed from Table 2 that the TUSZ multi-class dataset suffers from the problem of class imbalance and the class distribution varies significantly. FNSZ, GNSZ, and CPSZ classes have a higher number of instances in the data as compared to the remaining classes. With this uneven class distribution, the accuracy alone cannot represent the performance of the proposed technique. Therefore, the average weighted F1-score is used to evaluate the performance of our proposed technique. Indeed, we report the average weighted sensitivity, specificity, and Cohen's Kappa scores.

As mentioned earlier, we applied our technique to two different classification problems: seven-class and five-class classification. Moreover, we also tested our technique for both seizure-wise and patient-wise cross-validation classification. In seizure-wise cross-validation, we used a stratified five-fold cross-validation to robustly evaluate and validate the performance of the proposed technique, which is inspired by a state-of-the-art technique [8]. The dataset is split into five folds, where in each fold, the proportional distribution of classes in the entire dataset is randomly allocated to five folds. The class distribution after this split is approximately equal in each fold. During classification, the model is estimated on four folds and tested on the remaining fifth fold (test set); this process is repeated until all folds have been used as a test set. For patient-wise cross-validation, we adopted the validation technique of Asif et al. [11], in which they applied three-fold cross-validation across patients. In this scenario, the data presented in Table 2 are split into three folds, as the selected classes of seizures include data from a minimum of three patients. Therefore, this ensures that data used for testing are always from distinct patients whose data have never been used in the training phase.

3.3. Experimental Results

In this section, we compare the obtained results for both evaluation scenarios. We present our proposed technique's performance for the seven-class problem followed by the five-class problem for each seizure-wise and patient-wise validation.

3.3.1. Seizure-Level Cross-Validation

For both classification problems, we performed a five-fold cross-validation. For a seven-class problem, our proposed technique achieved a weighted average F1-score of 96.04%. Figure 6 presents the classification performance in terms of F1-score for each class in the dataset for all five folds, while Figure 7 presents the confusion matrix for our proposed technique's performance on the seven-class classification problem.

For the five-class problem, when only the specific types of seizures are included (see Table 2), our method achieved a weighted average F1-score of 99.1%. This means that the non-specific seizures in the dataset have a big impact on the performance of the machine learning algorithm, as the results improved by more than 2%; we discuss this in more detail in later sections. Figures 8 and 9 present the classification performance in terms of F1-score for each class in the dataset for all five folds and the confusion matrix for the five-class classification problem, respectively.

Moreover, the performance results of the proposed technique in terms of F1-score, sensitivity, specificity, and Cohen's Kappa for each fold and for both classification problems are presented in Table 6.

3.3.2. Patient-Wise Cross-Validation

For patient-wise cross-validation, three-fold cross-validation was performed. We first evaluated our method for a seven-class problem, and our proposed technique achieved the weighted average F1-score of 56.22%. Similarly, for five-class classification problem, the performance of our proposed technique significantly improved, and the proposed technique achieved a 75.97% weighted average F1-score.

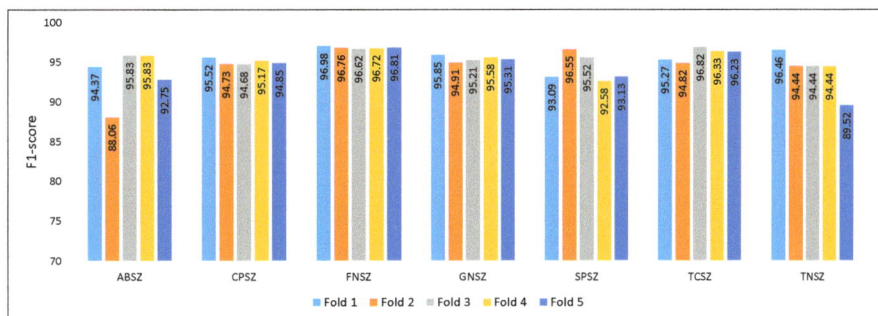

Figure 6. Performance of proposed technique on 7-class classification problem for each class having 5-fold cross-validation.

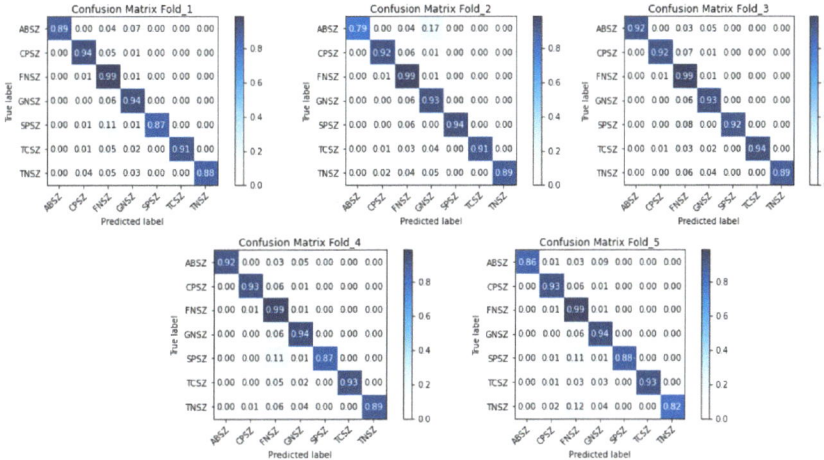

Figure 7. Confusion matrix for 7-class classification problem having 5-fold cross-validation: 1st to 5th fold (Left to right, top to bottom).

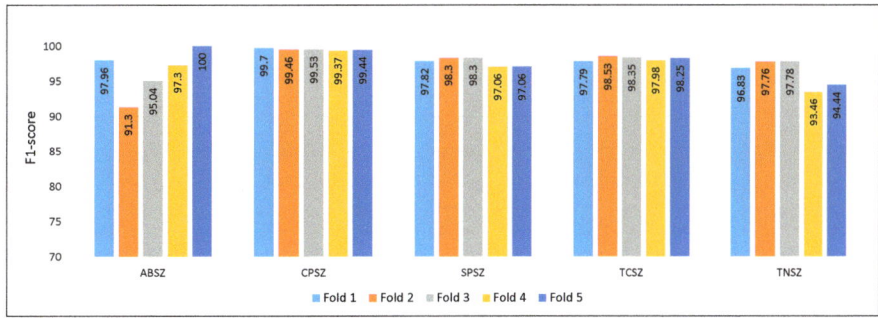

Figure 8. Performance of proposed technique on 5-class classification problem for each class having 5-fold cross-validation.

Table 6. Weighted average specificity, sensitivity, Cohen's Kappa, and F1 score of the proposed method for 7-class and 5-class problems, having five folds each.

		Specificity (%)	Sensitivity (%)	Cohen's Kappa (%)	F1 Score (%)
7-class problem	Fold 1	96.7	96.3	93.9	96.3
	Fold 2	96.4	95.9	93.3	95.9
	Fold 3	96.3	95.9	93.3	95.9
	Fold 4	96.4	96.1	93.6	96.1
	Fold 5	96.5	96.0	93.5	96.0
	Average	96.5	96.0	93.5	96.0
5-class problem	Fold 1	97.2	99.1	97.5	99.1
	Fold 2	96.8	99.1	97.5	99.1
	Fold 3	97.4	99.2	97.8	99.2
	Fold 4	96.6	98.9	97.0	98.9
	Fold 5	96.8	99.1	97.4	99.1
	Average	97.0	99.1	97.4	99.1

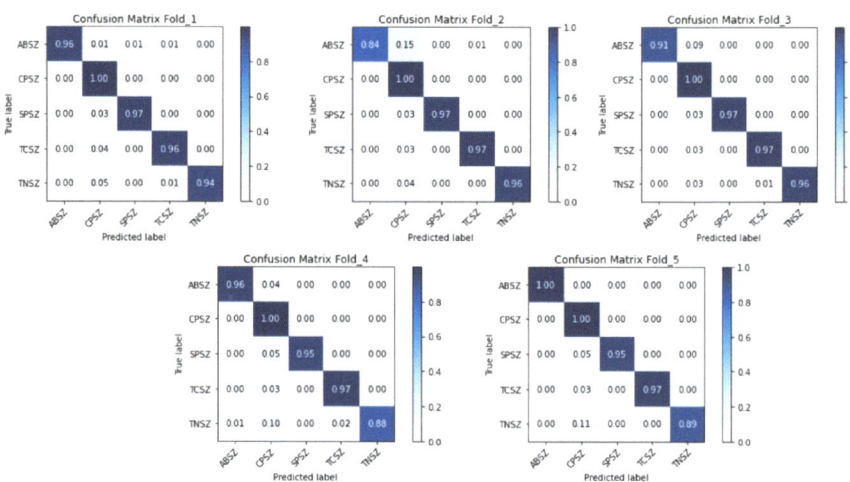

Figure 9. Confusion matrix for 5-class classification problem having 5-fold cross-validation: 1st to 5th fold (left to right, top to bottom).

4. Discussion

Table 3 presents state-of-the-art techniques applied to the problem of seizure-type classification. It is difficult to compare the performance of our proposed technique with the existing studies in the literature, as each of the studies chose a different number of seizure classes. Therefore, we selected all the state-of-the-art studies considering more than three classes and compared our technique's performance with them, as presented in Table 7.

It can be observed from Table 7 that our proposed technique's performance for specific seizure type classification is the best among all the techniques at both seizure-level classification as well as patient-level classification, achieving F1-scores of 99.1% and 74.7%, respectively. For the seven-class problem, ref. [11] reported an F1-score of 96.0% using an ensemble architecture of three DenseNets. Similarly, F1-scores of 94.5% and 90.1% were reported in [9,26], respectively. Our proposed technique outperformed all existing studies considering the same seven classes of seizures.

Table 7. Patient-wise cross-validation and seizure-wise cross-validation results for studies found in the literature.

Method	No. of Seizure Types	Seizure_Wise CV (%)	Patient_Wise CV (%)
SVM [29]	4	95.00 Acc	—
SVM [30]	4 *	95.14 Acc	—
SeizureNet [11]	7	95.00 F1	62 F1
KNN [9]	7	90.1 F1	40.1 F1
XGBoost [9]	7	85.1 F1	54.2 F1
SGD [9]	7	80.7 F1	46.9 F1
CNN [9]	7	71.8 F1	52.5 F1
NMN [26]	7	94.5 F1	—
Inceptionv3 [27]	8 *	88.3 Acc	—
AlexNet [28]	8 *	84.06 Acc	—
CNN+LSTM [8]	8	97.40 F1	—
This Work	5	99.1 F1	74.7 F1
This Work	7	96.04 F1	56.22 F1

* Including non-seizure EEG class.

For an eight-class problem, refs. [27,28] proposed CNN-based solutions and reported an accuracy of 84.06% and 88.3%, respectively. Similarly, Liu et al. [8] reported a high F1-score of 97.4% obtained by a symmetric bi-linear deep learning model consisting of two feature extractor models, CNN and LSTM. The study demonstrates a limitation in testing. The 1-second segments considered in the dataset have a 50% overlap, which always has the potential of data leaking, as mentioned by [8]. Most of the work mentioned in Table 7 is based on fast Fourier transform (FFT) [9,11,26], which has a high resolution in the frequency-domain but zero resolution in the time-domain, which is very essential for EEG signal processing [44]. The other approaches [8,27,28] were based on short-time Fourier transform (STFT), which is the known solution to overcome the limitation of FFT. STFT analyzes the frequency of the signals at a particular short time period to avoid losing temporal information. However, STFT cannot catch sharp signal events because of the use of a fixed window length and fixed basis function [44]. On the other hand, our proposed technique overcomes these shortcomings by providing a smooth representation of EEG signals. It enables generating detailed features that have strong correlations with the latent structure of seizure types in EEG signals. Additionally, our proposed method also demonstrated very high classification results compared to other classical machine learning techniques [9,29–31]. Moreover, all of the research studies mentioned in Table 7 utilized an older version of the TUSZ, which is ver.1.4.0, whereas the number of seizure events in the current version is much larger as compared to the previous version, as shown in Table 2. With a more challenging new version of TUSZ ver.1.5.2, which contains 1000 additional seizure events, our proposed method achieves better results for the seven-class classification problem as compared to existing techniques.

Most of the research studies in the literature chose to evaluate their methods only at the seizure level. Out of the eight studies presented in Table 3, only [9,11] considered the generalization of their models over different patients, or, in other words, the model is trained and evaluated on data from different patients. This ensures that the performance of the model is general and can be adopted for different patients. The performance of both studies [9,11] sharply decreases when evaluated using the patient-wise cross-validation technique, as shown in Table 7. Comparatively, our proposed method showed a competitive result for the seven-class problem, and it demonstrated more stable performance when evaluated across different patients for the five-class problem. Our proposed technique achieved F1-scores of 56.22% and 74.7% for the classification of the seven-class and five-class problems, respectively.

Regarding [23], DTCWT was employed to extract features from EEGs to classify epileptic vs. non-epileptic patients; however, in this study, we are using the DTCWT with a different set of features for a more complex problem, which is the identification of the seizure types, including specific and non-specific focal and generalized seizures. Moreover, we evaluated this technique seizure-wise and patient-wise on the most extensive available EEG dataset, TUSZ ver.1.5.2 [25], containing data from more than 300 patients. In [23], the data used for evaluation were obtained from only 21 subjects. This study has explored the generalization of our technique to be evaluated across different patients for better generalization capability.

In addition to the evaluation, we speculated that when only considering the EEG signals, it is not appropriate to treat the main seizure categories as unique seizure types against any of their sub-types. Since both SPSZ and CPSZ are sub-categories of focal seizures, it is unreasonable to train the machine learning algorithm to differentiate between them. According to the dataset, the reason for labeling an event as a focal non-specific seizure is the lack of information to make a decision [25]. After excluding the non-specific seizure labels, the experimental results demonstrated the stability of the classifier across different patients, as shown in Table 7. The performance of our proposed technique for specific seizure type classification showed nearly perfect results when evaluated at the seizure level, and it has the ability to generalize itself better on signals recorded from new patients, as compared to [9,11]. Therefore, considering the non-specific seizure labels in TUSZ as unique types of seizures, when only utilizing EEG data, does not reveal meaningful results. Instead, one must include clinical features that neurologists look for when making a diagnosis (i.e., video EEG monitoring). By doing so, the machine learning algorithm knows the reason for labeling an event as non-specific, as there is not enough information to make a decision. This is beyond the scope of this paper, as we focused solely on utilizing the EEG data. By knowing that the other specific seizure types in the corpus medically must be either focal or generalized at some point, we excluded the non-specific labels from our experiment, and, in turn, the results demonstrated high and stable performance.

During patient-wise cross-validation, we noticed that the majority of the seizure events of type SPSZ were classified as CPSZ. Again, and from medical perspectives, we can relate this misclassification to the fact that the difference between focal CPSZ and focal SPSZ can mainly be determined by clinical characteristics, as described in [4,25]. Therefore, considering the neurologist report in this situation might help in distinguishing between the two types. Moreover, as most epileptic conditions are age-determined [2], we suggest that one could also include age, gender, and medical history as extra input features to the machine learning model to obtain more accurate results, which is beyond our scope in this paper.

5. Conclusions

Epilepsy is one of the most common neurological diseases that affects people of all ages. It is characterized by sudden and recurrent seizure attacks that appear in different forms and are treated in different ways. Correct assessment of epileptic seizures is vital in overcoming the complications of the disease, and it provides accurate information to the affected person. This paper presents a novel technique utilizing DTCWT and machine learning for automatic seizure type classification in EEGs. The proposed method demonstrates a significant improvement in classification, achieving 96.04% and 99.1% for seven-class and five-class classification problems, respectively. We evaluated our proposed technique across different subjects, which is a very challenging task due to the limited amount of training data that are generalized to unseen test patients' EEG data. The achieved results show that our proposed technique performs significantly better as compared to the existing methods in the literature and is more general. The findings in this study enhance the applicability of artificial intelligence applications in assisting neurologists' decisions. In future research, we plan to investigate the use of different methods for feature extraction that can finely detect the differences between the seizures in an EEG.

Author Contributions: Conceptualization, H.A., G.M.H. and A.D.; methodology, H.A.; software, H.A.; validation, H.A., G.M.H. and A.D.; formal analysis, H.A.; investigation, H.A.; resources, A.D.; data curation, H.A.; writing—original draft preparation, H.A.; writing—review and editing, H.A.,

G.M.H. and A.D.; visualization, H.A.; supervision, G.M.H. and A.D. All authors have read and agreed to the published version of the manuscript.

Funding: This work was supported by a scholarship from University of Jeddah, Jeddah, Saudi Arabia.

Institutional Review Board Statement: Not applicable.

Informed Consent Statement: Not applicable.

Data Availability Statement: The dataset used in this research is publicly available as open source from the TUH EEG seizure corpus (TUSZ) v1.5.2. The TUSZ v1.5.2 was released in early 2020. It can be downloaded using the following link: https://www.isip.piconepress.com/projects/tuh_eeg/downloads.html, accessed on 1 May 2021.

Conflicts of Interest: The authors declare no conflict of interest.

References

1. Subasi, A.; Kevric, J.; Abdullah Canbaz, M. Epileptic seizure detection using hybrid machine learning methods. *Neural Comput. Appl.* **2019**, *31*, 317–325. [CrossRef]
2. Sanei, S.; Chambers, J.A. *EEG Signal Processing*; John Wiley & Sons: Hoboken, NJ, USA, 2013.
3. Scheffer, I.E.; Berkovic, S.; Capovilla, G.; Connolly, M.B.; French, J.; Guilhoto, L.; Hirsch, E.; Jain, S.; Mathern, G.W.; Moshé, S.L.; et al. ILAE classification of the epilepsies: Position paper of the ILAE Commission for Classification and Terminology. *Epilepsia* **2017**, *58*, 512–521. [CrossRef] [PubMed]
4. Fisher, R.S.; Cross, J.H.; French, J.A.; Higurashi, N.; Hirsch, E.; Jansen, F.E.; Lagae, L.; Moshé, S.L.; Peltola, J.; Roulet Perez, E.; et al. Operational classification of seizure types by the International League Against Epilepsy: Position Paper of the ILAE Commission for Classification and Terminology. *Epilepsia* **2017**, *58*, 522–530. [CrossRef] [PubMed]
5. Alotaiby, T.N.; Alshebeili, S.A.; Alotaibi, F.M.; Alrshoud, S.R. Epileptic seizure prediction using CSP and LDA for scalp EEG signals. *Comput. Intell. Neurosci.* **2017**, *2017*, 1240323. [CrossRef]
6. Goldenberg, M.M. Overview of drugs used for epilepsy and seizures: Etiology, diagnosis, and treatment. *Pharm. Ther.* **2010**, *35*, 392.
7. Obeid, I.; Picone, J. Machine learning approaches to automatic interpretation of EEGs. *Signal Process. Mach. Learn. Biomed. Big Data* **2018**, *70*, 271–300.
8. Liu, T.; Truong, N.D.; Nikpour, A.; Zhou, L.; Kavehei, O. Epileptic Seizure Classification with Symmetric and Hybrid Bilinear Models. *IEEE J. Biomed. Health Inform.* **2020**, *24*, 2844–2851. [CrossRef]
9. Roy, S.; Asif, U.; Tang, J.; Harrer, S. Seizure type classification using EEG signals and machine learning: Setting a benchmark. In Proceedings of the IEEE Signal Processing in Medicine and Biology Symposium, New York, NY, USA, 10 December 2020; pp. 1–6.
10. Panayiotopoulos, C. Optimal use of the EEG in the diagnosis and management of epilepsies. In *The Epilepsies: Seizures, Syndromes and Management*; Bladon Medical Publishing: Oxfordshire, UK, 2005.
11. Asif, U.; Roy, S.; Tang, J.; Harrer, S. SeizureNet: Multi-spectral deep feature learning for seizure type classification. In *Machine Learning in Clinical Neuroimaging and Radiogenomics in Neuro-Oncology*; Springer: Cham, Switzerland, 2020; pp. 77–87.
12. Shoeibi, A.; Ghassemi, N.; Alizadehsani, R.; Rouhani, M.; Hosseini-Nejad, H.; Khosravi, A.; Panahiazar, M.; Nahavandi, S. A comprehensive comparison of handcrafted features and convolutional autoencoders for epileptic seizures detection in EEG signals. *Expert Syst. Appl.* **2021**, *163*, 113788. [CrossRef]
13. Varone, G.; Gasparini, S.; Ferlazzo, E.; Ascoli, M.; Tripodi, G.G.; Zucco, C.; Calabrese, B.; Cannataro, M.; Aguglia, U. A comprehensive machine-learning-based software pipeline to classify EEG signals: A case study on PNES vs. control subjects. *Sensors* **2020**, *20*, 1235. [CrossRef]
14. Altunay, S.; Telatar, Z.; Erogul, O. Epileptic EEG detection using the linear prediction error energy. *Expert Syst. Appl.* **2010**, *37*, 5661–5665. [CrossRef]
15. Polat, K.; Güneş, S. Classification of epileptiform EEG using a hybrid system based on decision tree classifier and fast Fourier transform. *Appl. Math. Comput.* **2007**, *187*, 1017–1026. [CrossRef]
16. Subasi, A. EEG signal classification using wavelet feature extraction and a mixture of expert model. *Expert Syst. Appl.* **2007**, *32*, 1084–1093. [CrossRef]
17. Faust, O.; Acharya, U.R.; Adeli, H.; Adeli, A. Wavelet-based EEG processing for computer-aided seizure detection and epilepsy diagnosis. *Seizure* **2015**, *26*, 56–64. [CrossRef]
18. Acharya, U.; Sree, V.; Martis, R.; Suri, J. Automated diagnosis of epileptic electroencephalogram using independent component analysis and discrete wavelet transform for different electroencephalogram durations. *Proc. Inst. Mech. Eng. Part J. Eng. Med.* **2013**, *227*, 234–244.
19. Subasi, A.; Jukic, S.; Kevric, J. Comparison of EMD, DWT and WPD for the localization of epileptogenic foci using random forest classifier. *Measurement* **2019**, *146*, 846–855. [CrossRef]
20. Albaqami, H.; Hassan, G.M.; Subasi, A.; Datta, A. Automatic detection of abnormal EEG signals using wavelet feature extraction and gradient boosting decision tree. *Biomed. Signal Process. Control* **2021**, *70*, 102957. [CrossRef]

21. Hassan, A.R.; Siuly, S.; Zhang, Y. Epileptic seizure detection in EEG signals using tunable-Q factor wavelet transform and bootstrap aggregating. *Comput. Methods Programs Biomed.* **2016**, *137*, 247–259. [CrossRef]
22. Bhattacharyya, A.; Pachori, R.B.; Upadhyay, A.; Acharya, U.R. Tunable-Q wavelet transform based multiscale entropy measure for automated classification of epileptic EEG signals. *Appl. Sci.* **2017**, *7*, 385. [CrossRef]
23. Swami, P.; Gandhi, T.K.; Panigrahi, B.K.; Tripathi, M.; Anand, S. A novel robust diagnostic model to detect seizures in electroencephalography. *Expert Syst. Appl.* **2016**, *56*, 116–130. [CrossRef]
24. Obeid, I.; Picone, J. The temple university hospital EEG data corpus. *Front. Neurosci.* **2016**, *10*, 196. [CrossRef]
25. Shah, V.; Von Weltin, E.; Lopez, S.; McHugh, J.R.; Veloso, L.; Golmohammadi, M.; Obeid, I.; Picone, J. The temple university hospital seizure detection corpus. *Front. Neuroinform.* **2018**, *12*, 83. [CrossRef] [PubMed]
26. Ahmedt-Aristizabal, D.; Fernando, T.; Denman, S.; Petersson, L.; Aburn, M.J.; Fookes, C. Neural memory networks for seizure type classification. In Proceedings of the 42nd Annual International Conference of the IEEE Engineering in Medicine & Biology Society (EMBC), Montreal, QC, Canada, 20–24 July 2020.
27. Raghu, S.; Sriraam, N.; Temel, Y.; Rao, S.V.; Kubben, P.L. EEG based multi-class seizure type classification using convolutional neural network and transfer learning. *Neural Netw.* **2020**, *124*, 202–212. [CrossRef] [PubMed]
28. Sriraam, N.; Temel, Y.; Rao, S.V.; Kubben, P.L. A convolutional neural network based framework for classification of seizure types. In Proceedings of the 2019 41st Annual International Conference of the IEEE Engineering in Medicine and Biology Society (EMBC), Berlin, Germany, 23–27 July 2019; pp. 2547–2550.
29. Wijayanto, I.; Hartanto, R.; Nugroho, H.A.; Winduratna, B. Seizure Type Detection in Epileptic EEG Signal using Empirical Mode Decomposition and Support Vector Machine. In Proceedings of the 2019 International Seminar on Intelligent Technology and Its Application, ISITIA, Yogyakarta, Indonesia, 20–21 July 2019; pp. 314–319.
30. Saputro, I.R.D.; Maryati, N.D.; Solihati, S.R.; Wijayanto, I.; Hadiyoso, S.; Patmasari, R. Seizure type classification on EEG signal using support vector machine. In Proceedings of the Journal of Physics: Conference Series, Surabaya, Indonesia, 29–30 January 2019; Volume 1201, p. 012065.
31. Sarić, R.; Jokić, D.; Beganović, N.; Pokvić, L.G.; Badnjević, A. FPGA-based real-time epileptic seizure classification using artificial neural network. *Biomed. Signal Process. Control* **2020**, *62*, 102106. [CrossRef]
32. López, S.; Golmohammadi, M.; Obeid, I.; Picone, J. An analysis of two common reference points for EEGs. In Proceedings of the IEEE Signal Processing in Medicine and Biology Symposium, Philadelphia, PA, USA, 2 December 2016; pp. 1–4.
33. Ferrell, S.; Mathew, V.; Refford, M.; Tchiong, V.; Ahsan, T.; Obeid, I.; Picone, J. The temple university hospital eeg corpus: Electrode location and channel labels. *Inst. Signal Inf. Process. Rep.* **2020**, *1*, 1.
34. Shah, V.; Golmohammadi, M.; Ziyabari, S.; Von Weltin, E.; Obeid, I.; Picone, J. Optimizing channel selection for seizure detection. In Proceedings of the 2017 IEEE Signal Processing in Medicine and Biology Symposium (SPMB), Philadelphia, PA, USA, 2 December 2017; pp. 1–5.
35. Tuncer, T.; Dogan, S.; Subasi, A. Surface EMG signal classification using ternary pattern and discrete wavelet transform based feature extraction for hand movement recognition. *Biomed. Signal Process. Control* **2020**, *58*, 101872. [CrossRef]
36. Georgieva-Tsaneva, G. Wavelet based interval varying algorithm for optimal non-stationary signal denoising. In Proceedings of the 20th International Conference on Computer Systems and Technologies, Ruse, Bulgaria, 21–22 June 2019; pp. 200–206.
37. Kingsbury, N. Complex wavelets for shift invariant analysis and filtering of signals. *Appl. Comput. Harmon. Anal.* **2001**, *10*, 234–253. [CrossRef]
38. Selesnick, I.W.; Baraniuk, R.G.; Kingsbury, N.C. The dual-tree complex wavelet transform. *IEEE Signal Process. Mag.* **2005**, *22*, 123–151. [CrossRef]
39. Wareham, R.; Forshaw, S.; Roberts, T. DTCWT: A Python Dual Tree Complex Wavelet Transform Library. 2014. Available online: https://zenodo.org/record/9862#.Ypl5O-xBxPY (accessed on 3 May 2022).
40. Shoeb, A.H. Application of Machine Learning to Epileptic Seizure Onset Detection and Treatment. Ph.D. Thesis, Massachusetts Institute of Technology, Cambridge, MA, USA, 2009.
41. Kevric, J.; Subasi, A. Comparison of signal decomposition methods in classification of EEG signals for motor-imagery BCI system. *Biomed. Signal Process. Control* **2017**, *31*, 398–406. [CrossRef]
42. Ke, G.; Meng, Q.; Finley, T.; Wang, T.; Chen, W.; Ma, W.; Ye, Q.; Liu, T.Y. Lightgbm: A highly efficient gradient boosting decision tree. *Adv. Neural Inf. Process. Syst.* **2017**, *30*, 3146–3154.
43. Bergstra, J.; Komer, B.; Eliasmith, C.; Yamins, D.; Cox, D.D. Hyperopt: A python library for model selection and hyperparameter optimization. *Comput. Sci. Discov.* **2015**, *8*, 014008. [CrossRef]
44. Rajoub, B. Characterization of biomedical signals: Feature engineering and extraction. In *Biomedical Signal Processing and Artificial Intelligence in Healthcare*; Elsevier: Amsterdam, The Netherlands, 2020; pp. 29–50.

Article

EEG Signal Processing and Supervised Machine Learning to Early Diagnose Alzheimer's Disease

Daniele Pirrone [1], Emanuel Weitschek [1], Primiano Di Paolo [1], Simona De Salvo [2] and Maria Cristina De Cola [2,*]

[1] Department of Engineering, Uninettuno University, Corso Vittorio Emanuele II 139, 00186 Rome, Italy; daniele.pirrone@uninettunouniversity.net (D.P.); emanuel.weitschek@uninettunouniversity.net (E.W.); p.dipaolo@students.uninettunouniversity.net (P.D.P.)
[2] IRCCS Centro Neurolesi "Bonino-Pulejo", Contrada Casazza, SS113, 98124 Messina, Italy; simona.desalvo@irccsme.it
* Correspondence: mariacristina.decola@irccsme.it; Tel.: +39-090-60128141

Abstract: Electroencephalography (EEG) signal analysis is a fast, inexpensive, and accessible technique to detect the early stages of dementia, such as Mild Cognitive Impairment (MCI) and Alzheimer's disease (AD). In the last years, EEG signal analysis has become an important topic of research to extract suitable biomarkers to determine the subject's cognitive impairment. In this work, we propose a novel simple and efficient method able to extract features with a finite response filter (FIR) in the double time domain in order to discriminate among patients affected by AD, MCI, and healthy controls (HC). Notably, we compute the power intensity for each high- and low-frequency band, using their absolute differences to distinguish among the three classes of subjects by means of different supervised machine learning methods. We use EEG recordings from a cohort of 105 subjects (48 AD, 37 MCI, and 20 HC) referred for dementia to the IRCCS Centro Neurolesi "Bonino-Pulejo" of Messina, Italy. The findings show that this method reaches 97%, 95%, and 83% accuracy when considering binary classifications (HC vs. AD, HC vs. MCI, and MCI vs. AD) and an accuracy of 75% when dealing with the three classes (HC vs. AD vs. MCI). These results improve upon those obtained in previous studies and demonstrate the validity of our approach. Finally, the efficiency of the proposed method might allow its future development on embedded devices for low-cost real-time diagnosis.

Keywords: Alzheimer's disease; EEG signals; power spectrum; FIR filtering; supervised machine learning

1. Introduction

Alzheimer's diseases (AD) belong to the class of dementia, a neurodegenerative disease characterized by a range of impairments in brain functions, especially memory and learning, as well as executive and motor functions, complex attention, social cognition, and language [1]. The estimated proportion of the general population with dementia is around 50 million of people worldwide, and 60% of those cases correspond to AD [2]. It begins with a symptomatic stage of cognitive decline, called Mild Cognitive Impairment (MCI), characterized by an impairment in cognition that is not severe enough to compromise social and/or occupational functioning [3]. As the progression of this disease lasts for decades, from the appearance of the first sign to the onset of severe clinical symptoms, the clinician's first challenge is to identify the first significant cognitive changes [4]. Indeed, the diagnosis of dementia is usually made when the patient is at least partially dependent on his/her family members [5]. However, a timely diagnosis can facilitate care and support patients and their families in managing this very disabling disease [6].

Conventional techniques to detect AD are costly and distressing. However, electroencephalography (EEG) is a fast, inexpensive, and noninvasive technique to gather brain data, but its interpretation requires a visual inspection, which is often time-consuming and varies with the expertise experience. Moreover, when the EEG recording is long, then its manual review requires a lot of time with the risk of errors because of the presence of artifacts in the signal. Thus, automated methods based on EEG signal analysis in combination with supervised machine learning have become an important topic of research to assist clinicians in the challenging task of early AD detection [7].

The technique for the sampling of EEG signal consists of placing electrodes on the scalp according to a certain configuration, and the most commonly used is the international 10–20 system [8]. The electrodes record the postsynaptic biopotentials of all the neurons with the same spatial orientation in order to map the electrical activity of the cerebral cortex. The biopotentials, which are sampled in bipolar mode by different electrode pairs or in monopolar mode with a reference electrode, constitute the raw signals. Subsequently, preprocessing procedures clean the raw signals from artefacts and apply band-pass filtering to reject out-of-band noise. Basically, all preprocessing steps convert the raw signals into EEG signals [9]. The composition of the EEG signal is complex, but it can be divided into five frequency bands: delta (1–4 Hz, δ), theta (4–8 Hz, θ), alpha (8–13 Hz, α), beta (13–30 Hz, β), and gamma (30–40 Hz, γ) [10].

Features are extracted from EEG data through a procedure denominated *feature extraction*. Features should be independent and discriminative to facilitate the classification of the subjects. Usually, features such as complexity, coherence, or spectral power are extracted from the time and frequency domains [7]. Fourier transform (FT) is the main technique used to extract frequency domain features for AD detection. Nonetheless, EEG signals are nonstationary and nonlinear in nature. The wavelet transform (WT), which decomposes a signal into the combination of functions (wavelets) of finite length and different frequencies, represents a suitable alternative to address this issue [11]. On the contrary, the frequency domain represents the principal source of the EEG features for AD detection. Indeed, different changes in the frequency patterns of the brain waves have been found in MCI and AD patients compared to healthy aged subjects [12]. All feature extraction methods extrapolate EEG signal features from different domains (e.g., frequency and time) [13]. Then, statistical or machine learning analyses [14] can use these features to develop and validate models based on linear or nonlinear systems [15] to distinguish AD from MCI or normal aging. In particular, supervised machine learning (SML) permits the development of robust classification models for recognizing AD [16], frontotemporal dementia [17], and other pathologies [18]. Complex and heterogeneous symptoms complicate the diagnosis, because more often than not, biomarkers are intrinsically hidden in the EEG signal.

Rhythms are often used to analyze the EEG in a particular sub-band through filtering, due to the different activities between the frequency bands [19,20]. In fact, previous studies have shown that the relative power in fast rhythms (α and β) decreases, while, in slow rhythms (δ and θ), it increases [21,22]. This effect shifts the peak power towards lower frequencies, which is why it is also called "shift-to-the-left" (STTL) [23]. The method presented in this paper exploits the power intensity of EEG signals, filtered in the time domain by using both high-pass and a low-pass filters in order to analyze the STTL phenomenon. Our idea is to classify the absolute difference in power between fast and slow rhythms for each individual, using it as a biomarker. For this purpose, we also use the power spectrum density (PSD) calculated with the help of the spectrogram and SML to choose the best filter cutoffs and improve the classification performance.

2. Materials and Methods

We used an EEG dataset composed of 109 EEG recordings (49 AD, 37 MCI, and 23 HC) collected in resting condition and with closed eyes at the IRCCS Centro Neurolesi "Bonino-Pulejo" in Messina (Italy). A diagnosis of AD or MCI was formulated following the guidelines of the Diagnostic and Statistical Manual of Mental Disorders (fifth edition, DSM-5).

2.1. Data Acquisition and Preprocessing

Multi-channel EEG signals were recorded by using 19 electrodes placed according to the 10–20 system [8] in monopolar connection with the earlobe electrode as a reference.

Raw electrical brain activity (μV) recordings last about 300 s. For more details on data collection, the reader can refer to the previous study [24].

In the preprocessing step, the sampling rate is normalized to 256 Hz, and EEG are filtered at a 1-Hz low cutoff (high-pass) and at a 30-Hz high cutoff (low-pass). After filtering, artifacts are detected by visual inspection and rejected. One hundred and fifty seconds of cleaned EEG are considered for each subject, extracted from the central part of the EEG signal in order to maintain the maximum signal information to train classifiers on the same length signals but without losing too many instances. Thus, four subjects were excluded (i.e., 1 AD and 3 HC) due to an excessive number of artifacts, and the dataset dropped to 105 EEG recordings.

2.2. Feature Extraction

The feature extraction procedure includes two main steps: (i) data exploration in the time–frequency domains by means of PSD computed in the spectrogram and (ii) construction of the double digital filter and its application.

2.2.1. Data Exploration in the Time–Frequency Domains

For each subject, we generated a unique signal by concatenating the 19 biopotentials signals (i.e., one for each electrode). This concatenated signal provides a complete view of the whole subject's signal and allows to know the electrodes more involved in the STTL phenomenon. Therefore, the 3 classes (AD, MCI, and HC) contain as many concatenated signals as there are subjects in each of them. Then, the average of each class is calculated, resulting in 3 average signals. Therefore, we apply the MATLAB *pspectrum* function (it is included in the Signal Processing Toolbox introduced in version R2017b), setting the spectrogram mode and providing input in the sampling frequency (f_s) and the signal in the time domain. The power spectrum density of the signal is computed, also performing the Short-Time Fourier Transform (STFT) of the signal and evaluates its power [25,26]. This step allows to find the best cutoff frequencies (f_{cut}) that can separate the classes. For the sake of clarity, recall that the spectrogram is a function used to plot the STFT of the signal, determining both the sinusoidal frequency and phase contained in different time frames and composing the entire signal.

2.2.2. Double Digital Filter Construction

The second step includes the construction of two Finite Impulse Response (FIR) digital filters, i.e., a second-order Butterworth filter for high-pass (FIR-H) and low-pass (FIR-L) frequencies by using the cutoff frequencies (f_{cut}) previously identified [27]. Thus, each EEG signals provided by an electrode is double-filtered, and two signals are generated in the time domain, called $EEG^{(L)}$ and $EEG^{(H)}$, where the first value is the EEG filtered with FIR-L, while the second value is the EEG filtered with FIR-H. Subsequently, we compute the power (see Appendix A) of these two signals, $P_{xx}^{(L)}$ and $P_{xx}^{(H)}$, with the aim to calculate the square of their absolute difference:

$$P_{(L-H)}^2 = \left\| P_{xx}^{(L)} - P_{xx}^{(H)} \right\|^2 \quad (1)$$

This value is the extracted feature corresponding to our biomarker. Figure 1 shows a schematic representation of the double digital filter construction.

In this way, we are able to represent each initial EEG signal as an array of values sampled in a single feature, $P_{(L-H)}^2$, reducing the input size. This procedure is iterated for all subjects, as shown in Table 1. To ensure that the cutoff frequencies identified in the first stage of the procedure are really the best for class separation, we vary f_{cut} from 1 Hz to 18 Hz with a step size of 1 Hz, in order to achieve the best class separation.

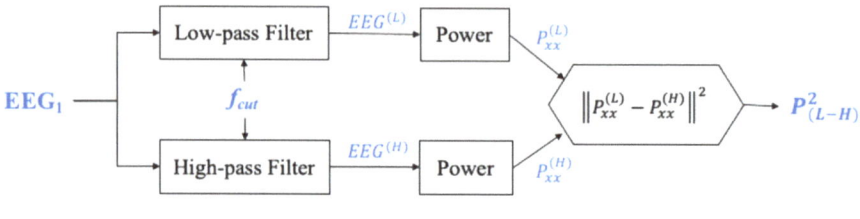

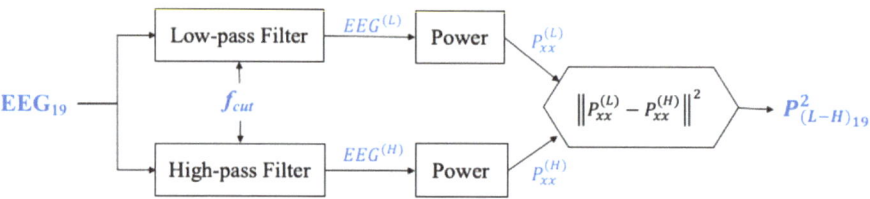

Figure 1. Schematic representation of the double filtering method. The initial input is the subsequence of the main EEG signal, which is divided and filtered according to the two main branches. The final result is the square of the power absolute difference. The block "Power" is explained in Appendix A.

Table 1. Schema of the feature extraction procedure: features extracted from an EEG recording are shown in the last column.

N-Subjects	EEG Signals	Label	Extracted Features
1	$EEG_1, EEG_2, \ldots, EEG_{19}$	AD	$P^2_{(L-H)_1}, P^2_{(L-H)_2}, \ldots, P^2_{(L-H)_{19}}$
2	$EEG_1, EEG_2, \ldots, EEG_{19}$	MCI	$P^2_{(L-H)_1}, P^2_{(L-H)_2}, \ldots, P^2_{(L-H)_{19}}$
…	… ..	…	… ..
109	$EEG_1, EEG_2, \ldots, EEG_{19}$	HC	$P^2_{(L-H)_1}, P^2_{(L-H)_2}, \ldots, P^2_{(L-H)_{19}}$

The column Label shows the initial labeling used in supervised machine learning.

2.3. Classification

The signal classification is performed by an SML analysis through three classification methods: decision trees (DT), support vector machines (SVM), and k-nearest neighbor (KNN) [28]. The algorithms are implemented in Python (version 3.7.21) by means of the scikit-learn toolkit [29]. Python is an open-source programming language, and its choice stems from the availability of many external libraries, frameworks, and tools from a huge community distributed all over the world.

In this study, we define five classification problems: (i) AD vs. HC, (ii) AD vs. MCI, (iii) MCI vs. HC, (iv) AD + MCI vs. HC, and (v) AD vs. MCI vs. HC. The first four problems were addressed in our previous study [24], whereas the last one defines the three class classification problem [30]. Then, for each problem, we perform 10 runs where the extracted features are randomly sampled. Finally, we adopt the following procedure:

1. Dataset splitting in training 70% and data tests 30%, except for the (v) case where the data has been divided into 80% training and 20% data tests;
2. Dataset size reduction with the Linear Discriminant Analysis (LDA) [31];
3. Application of the three aforementioned supervised machine learning methods;
4. Tuning of the hyperparameters of the machine learning algorithms combined with k-fold cross validation [32];
5. Data validation and performance evaluation through the confusion matrices.

Regarding point 1, data splitting can affect the performance of evaluators, so making appropriate decisions during this step is very challenging, as highlighted in [33]. Here, the authors summarize the challenges of data splitting into three main points: (i) Data imbalance, (ii) Data loss, and (iii) Concept drift. Taking these points into consideration, we divide the initial datasets into 70–30 and 80–20. The first choice shows very high levels of accuracy in the first four cases of the classification; on the contrary, 80–20 has an accuracy increase of 5% in the three-class classification problem. In order to reduce the less meaningful features, we applied the LDA to the extracted features. The LDA, cited at point 2, is a well-known data mining algorithm [34] and is suitable in those cases where classes are unbalanced or nonlinearly separable, such as the EEG signal. LDA automatically defines a separation hyperplane between the points that belongs to a class by generating two subclasses from the main one. Consequently, the Fisher criterion [35], also called Fisher's linear discriminant, maximizes the ratio of the between-class variance to the within-class variance in any particular dataset, ensuring the maximum separation. In this way, it is possible to discard those values of the extracted features that do not affect the variance of the main class, decreasing the sample dimensions.

With the purpose of improving the performance of the classification algorithms, we automatically introduce additional parameters, namely hyperparameters, provided by an external constructor. However, a wrong choice of the hyperparameters can lead to incorrect results and to obtaining a poor performance model [36,37]. In this work, we chose the grid search algorithm (GScv) [38], implemented through the python function GridSearchCV. GScv is the simplest algorithm for hyperparameter optimization [39]. However, it is time-consuming, since it considers all combinations to find the best point (Figure 2a), and each grid point needs cross-validation in the training phase.

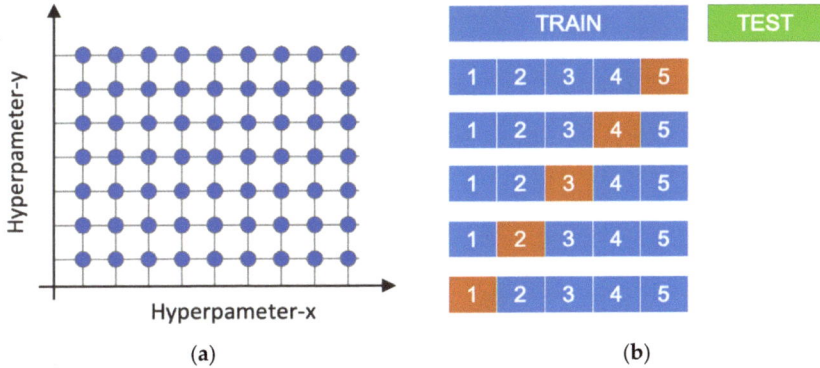

Figure 2. (**a**) Example of a grid search domain, where the hyperparameters are distributed into a matrix. (**b**) A schematic representation of 5-fold cross-validation. In (**b**), the data train (the blue boxes) is split in "k" subsegments, and one (the orange box) of them is used as the validation in each iteration.

In addition, the cross-validation procedure resamples the data randomly for the better evaluation of machine learning models. To improve the model validation, the procedure of cross-validation is iterated k times. Consequently, the data training is segmented into k subgroups [40]. For this reason, the procedure is often called k-fold cross-validation. Here, we split the training datasets into k = 10-fold without reinsertion, where 9-folds are used to train the model and 1-fold for the performance evaluation [41], as represented in Figure 2b. The estimator evaluation (e.g., accuracy) is the average of each estimator computed over the kth iteration.

3. Results

In this section, we report the results obtained after any steps of our method, e.g., from the feature extraction to the classification process, providing the accuracy measures of the classification algorithms.

First, we show in Figure 3 the allocation of the power spectrum for each patient. Looking closely at Figure 3, we can enhance that the PSD in AD and MCI is restricted in low frequencies (<7 Hz), while, in HC, the power is spread up to about 14 Hz. Thus, we expect that the next part of the feature extraction procedure also identifies that these frequencies are the best cutoffs for a good separation between classes.

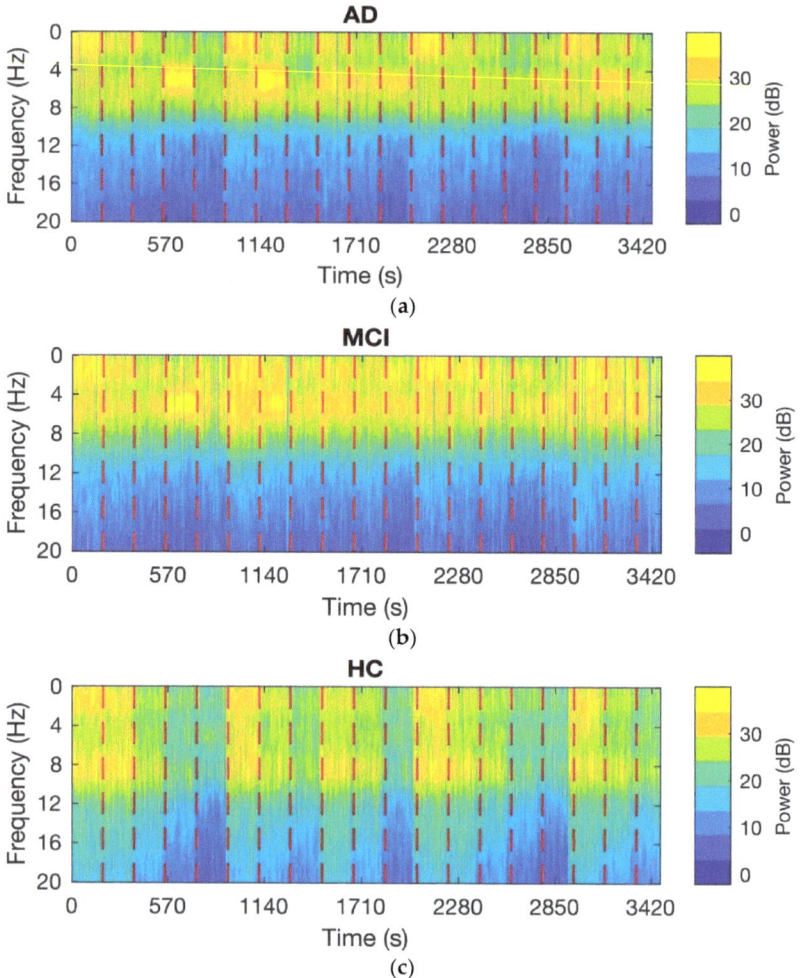

Figure 3. Computed pspectrum from 19 biopotentials (EEG signals) in a frequency range of 0–20 Hz for AD (**a**), MCI (**b**), and HC (**c**). The dashed red line separates each individual signal from the 19 biopotentials that make up the preprocessed EEG signal.

As viewable in Figure 4a, the statistical test carried out on the characteristics shows a good separation between classes in these f_{cut} values. Indeed, we found an excellent class separation between HC and AD-MCI for the value of f_{cut} equal to 7 Hz, as can be seen in Figure 4b, and an excellent separation between AD and MCI with f_{cut} = 16 Hz (Figure 4c).

These interesting results suggest applying the first filtering at 7 Hz to exclude controls and the subsequent filtering at 16 Hz for a better classification between MDI and AD.

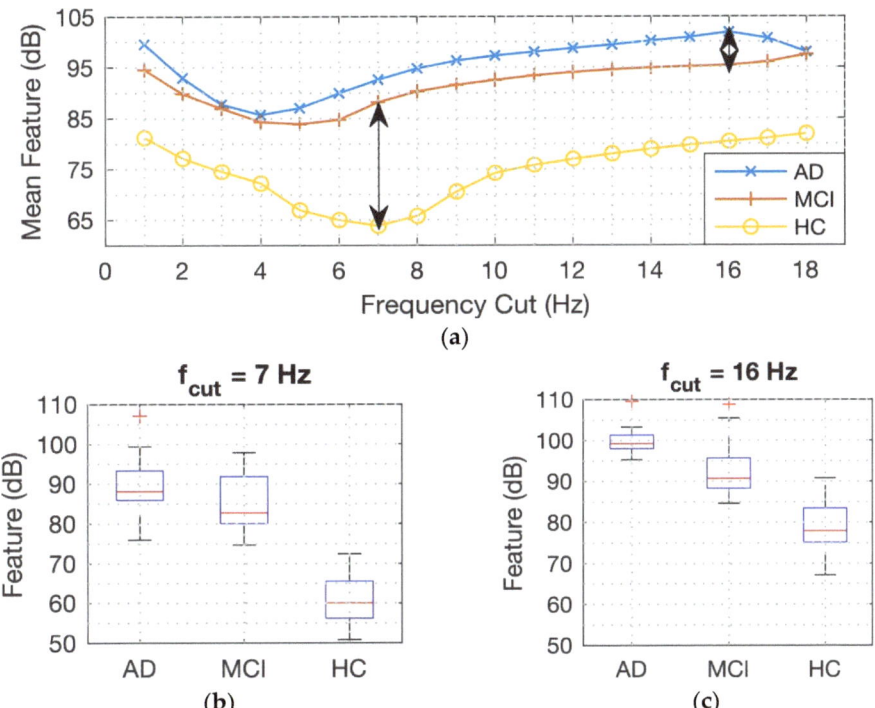

Figure 4. (**a**) The mean value of the feature extraction, the black double arrows, show the maximum distance between HC and AD + MCI for f_{cut} = 7 Hz and between AD and MCI for f_{cut} = 16 Hz. The statistical results are shown in (**b**) for 7 Hz and (**c**) for 16 Hz. In both of the last two graphs, the main box represents the data distribution, while the red line represents the median value, and the whisker stretches from the box show the range of the data, except for the outlays that are represented, such as the floating point (+).

The classification of the samples achieves a high level of accuracy, especially in distinguishing AD vs. HC and MDI vs. HC cases. In fact, the three classification procedures achieve an accuracy value of more than 87% and up to 97%, as shown in Figure 5a,b. Moreover, when we consider AD + MCI vs. HC, the accuracy reaches a value between 84% and 89% (Figure 5c). The effectiveness of the proposed method consists of tunneling f_{cut}, as shown in Figure 5d for the AD vs. MCI case. Here, the accuracy of the classification methods improves from 49–60% to 80–83% when f_{cut} is increased from 7 Hz to 16 Hz. Finally, Figure 5e shows the comparison between the three different cases and reinforces the hypothesis of the effectiveness of the proposed method for feature extraction. In the latter case, the accuracy value is between 73% and 86%.

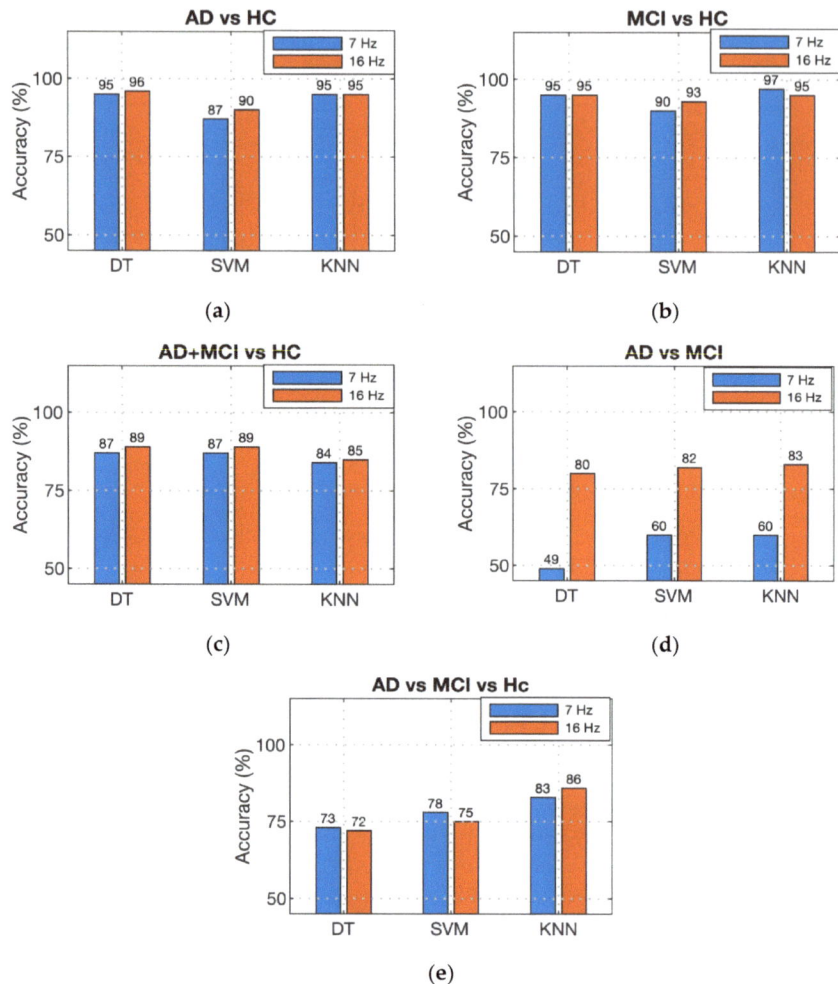

Figure 5. Each subfigure displays the comparison between the three classification methods: decision tree (DT), support vector machine (SVM), K-nearest neighbors (KNN) in the different problems: (**a**) AD vs. HC, (**b**) AD vs. MCI, (**c**) MCI vs. HC, (**d**) AD + MCI vs. HC, and (**e**) AD vs. MCI vs. HC. Furthermore, all the subfigures show the accuracy results achieved in percentages at f_{cut} = 7 Hz in the blue column and f_{cut} = 16 Hz in the red column.

In Table 2, we present the computation times of the classification procedure. The costs of feature extraction are low, and the computation time is about 0.1 s for a single subject. On the contrary, in the classification process, the computation time is higher because of the search for the best hyperparameters improving the accuracy. In the binary classification problems, the total execution time is, on average, 39.5 s, of which the DT takes about 21 s, the SVM takes about 15 s, and the K-NN takes about 3 s, except in the last case, when the execution time increases by two seconds. K-NN is the algorithm with the lowest runtime, while DT is the slowest. This is because KNN stores the training data in an n-dimensional space defining the pattern's spaces, and for each unknown sample, it assigns the pattern's space with a minor distance function [42] while DT extracts a classification model composed of features and value assignments requiring a longer runtime, although DT provides a better interpretation of the classification solution.

Table 2. Average execution time of the classification algorithm, including tuning of the hyperparameters.

Case	f_{cut} (Hz)	Time (s) DT/SVM/K-NN	Tot. Time (s)
AD vs. HC	7	21.8/14.0/3.0	38.8
	16	20.9/14.0/3.0	38.0
AD vs. MCI	7	22.3/14.8/3.0	40.1
	16	21.1/15.0/3.0	39.1
MCI vs. HC	7	20.7/13.3/3.0	37.0
	16	21.1/13.6/3.0	37.7
AD + MCI vs. HC	7	20.6/16.4/3.0	40.0
	16	21.1/18.5/3.0	42.6
AD vs. MCI vs. HC	7	25.0/22.6/4.0	51.6
	16	25.9/24.1/4.8	54.8

The Time column shows the execution time of each classifier, where DT = decision tree, SVM = support vector machine, and K-NN = K-nearest neighbors. The last column shows the sum of the execution times.

4. Discussion

In this paper, we proposed a new method for the feature extraction in AD recognition from EEG signals. Our findings confirm that AD affects the power spectrum of the patient, according to previous studies [43,44]. The proposed method is carried out in the time domain, exploiting the knowledge a priori of the EEG signals, i.e., power spectrum, spectral entropy, and phase synchronization. We used three different classification algorithms to validate our method, obtaining promising results. Indeed, the accuracy ranged between 73% and 97%, overcoming previous studies [24,45]. In particular, in [45], the best accuracy rate was 94%, obtained by using discrete wavelet transform to extract features, whereas, in [24], we used a DT classifier (i.e., the C4.5 algorithm), reaching the following levels of accuracy: 83%, 92%, 79%, and 73% in HC vs. AD, HC vs. MCI, MCI vs. AD, and HC vs. MCI + AD classification problems, respectively. Furthermore, in the last two cases, we increased the level of accuracy, with values higher than 80% combining in the information gain filter. Here, K-NN was the best classification algorithm as concerning the accuracy (almost always greater than 80%), as shown in Figure 5, with a runtime of about 3 s (Table 2). On the contrary, the accuracy of DT was the lowest, and its running times the highest because of the computation complexity in the construction of the solution in a human interpretable format.

The recent literature has reported several techniques using EEG signals for the early diagnosis of AD, differing in how features are extracted. Some of these, such as event-related potentials, signal a complexity analysis and relative power, involving a time domain signal analysis. Other techniques, instead, work in the Fourier domain, such as the coherence metric that evaluates the synchrony between two signals. In addition, there are techniques exploiting the analysis in the frequency domain, such as the continuous or discrete wavelet transform [7]. Cejnek et al. [46] employed a linear neural unit with gradient descent adaptation as the filter to predict AD, achieving a specificity of 85% and a sensitivity between 85% and 94%, depending on the classifier. In Reference [47], the authors developed an algorithm that consists of three cascade methods for analysis: discrete WT, PSD, and coherence. They tested this method on 35 subjects by means of the bagged trees classifier trained with five-fold cross-validation, obtaining a 96.5% accuracy.

There are only a limited number of works that have exploited time–frequency or bispectrum-based features, such as discriminating coefficients. The multimodal machine learning approach of Ieracitano et al. [48], where EEG signals are projected into the time–frequency domains by means of the continuous WT to extract a set of features from EEG sub-bands, while the nonlinear the phase-coupling information of EEG signals is also used to generate a second set of features from the bispectrum representation. This method provides high levels of accuracy with different classifiers in all considered problems: AD vs. HC, Ad vs. MCI, MCI vs. HC, and AD vs. MCI vs. HC and on a large cohort of subjects (i.e., 189 subjects: 63 AD, 63 MCI, and 63 HC). Similar to our method for the early detection of AD, the Lacosogram tool [21] performs a statistical analysis to measure distances between EEG sub-bands, obtaining an accuracy of 98.06% for HC vs. MCI, 95.99% for HC vs. AD, and 93.85% for MCI vs. AD. Kulkarni and Bairagi in [49] decomposed EEG by using the WT to decompose the EEG signal into its five sub-bands. The means and variances of the wavelet coefficients were evaluated and used as input to a SVM classifier, achieving an accuracy value of 88% in AD vs. HC classification.

Although, in our previous work [24], the experimental results showed that wavelet coefficients evaluated by applying the discrete wavelet transform achieved the highest accuracy rates (i.e., 83.3% for AD vs. HC, 91.7% for MCI vs. HC, and 79.1% for AD vs. MCI), the three-class classification did not achieve good results when we used only Fourier or Wavelet transform. On the contrary, here, the results of the three-class classification problem (AD vs. MCI vs. HC) were also reported, showing an average accuracy of 78% for the three classifiers and longer running times. This further proved the validity of this feature extraction method, which plays a key role in the analysis. Indeed, the method separates the high frequencies and the low frequencies of the EEG signal, and then, it computes their powers. The comparison of these powers shows an imbalance of energies in the frequency range, demonstrating the phenomenon of the STTL described in [15].

Our findings showed that a correct choice of f_{cut} increases the accuracy of discrimination between AD, MCI, and HC subjects. In addition, the proposed method for feature extraction is simple and fast in running time, and therefore, it is easily replicable in different development environments. This is undoubtedly its greatest strength, making its implementation and understanding very easy. The model, tested on a larger sample, could lead to the identification of biomarkers capable of determining features that discriminate electrical signals between different AD cohorts at specific electrodes.

As a future work, we plan to improve the classification method to the point of removing all the HC subjects, applying double filtering with f_{cut} = 7 Hz from the main sample, and distinguish AD from MCI by applying double filtering with f_{cut} = 16 Hz. However, given the heterogeneity of the disease, a larger cohort is necessary to confirm the results of this study. We also plan to test the method in different EEG recording protocols, maybe while the subject performs a cognitive task, in order to provide insights on how AD affects certain cognitive areas.

Since AD is expected to affect a large part of the worldwide population in the following years, EEG represents a suitable technique to assist clinicians in an early diagnosis. From this perspective, the method could be implemented on embedded devices and used in real time during EEG signal acquisition due to its low computational resource requirements. Considering the simplicity and robustness of the double filtering, we could promote it as an inexpensive and portable software suite by programming the current embedded electronic microprocessor, such as a Dev Board [50].

Author Contributions: Conceptualization, E.W. and D.P.; methodology, D.P.; software, D.P. and P.D.P.; validation, E.W., M.C.D.C. and S.D.S.; formal analysis, E.W. and D.P.; investigation, M.C.D.C.; resources, M.C.D.C.; data curation, S.D.S.; writing—original draft preparation, D.P., M.C.D.C. and E.W.; writing—review and editing, E.W. and M.C.D.C.; supervision, E.W. and M.C.D.C.; project administration, E.W. and M.C.D.C.; and funding acquisition, M.C.D.C. All authors have read and agreed to the published version of the manuscript.

Funding: This study was supported by Current Research Funds 2021, Ministry of Health, Italy.

Institutional Review Board Statement: The Ethical Committee of the IRCCS Centro Neurolesi "Bonino-Pulejo" approved the study after informed consent to participate in the study was signed by the enrolled subjects (reference number 40/2013).

Informed Consent Statement: Informed consent was obtained from all subjects involved in the study.

Data Availability Statement: The datasets analyzed during the current study are available upon request from the IRCCS Centro Neurolesi "Bonino-Pulejo" (M.C.D.C.).

Conflicts of Interest: The authors declare no conflict of interest.

Appendix A

In this section, we explain how we calculate the power of the EEG signal after splitting the signal into the high- and low-frequency components. This operation describes the function of the block "Power" in Figure 1. The power of a generical signal $s(t)$ in a time interval or period T is calculated as [51]:

$$P_{xx} = \frac{1}{T} * \int_{t_0}^{t_0+T} \|s(t)\|^2 \, dt \tag{A1}$$

In our work, we considered $t_0 = 0$, and the EEG is a sampled signal, i.e., a discrete-time signal, so that (A1) can be rewritten as:

$$P_{xx} = \frac{1}{N} * \sum_{n=0}^{N} EEG(n)^2 \tag{A2}$$

where n is the index of the nth sample that compose the EEG, and N is the total number of samples. As well-known, the ratio between N and the sampling rate returns T, and this indicates the linear dependence between N and T.

References

1. Finkel, S.I.; Costa, E.; Silva, J.; Cohen, G.; Miller, S.; Sartorius, N. Behavioral and psychological signs and symptoms of dementia: A consensus statement on current knowledge and implications for research and treatment. *Int. Psychogeriatr.* **1996**, *8*, 497–500. [CrossRef] [PubMed]
2. Patterson, C. *World Alzheimer Report 2018. The State of the Art of Dementia Research: New Frontiers*; Alzheimer's Disease International: London, UK, 2018.
3. Benbow, S.M.; Jolley, D. Dementia: Stigma and his effects. *Neurodegener. Dis. Manag.* **2012**, *2*, 165–172. [CrossRef]
4. Hugo, J.; Ganguli, M. Dementia and cognitive impairment: Epidemiology, diagnosis, and treatment. *Clin. Geriatr. Med.* **2014**, *30*, 421–442. [CrossRef] [PubMed]
5. Brooker, D.; La Fontaine, J.; Evans, S.; Bray, J.; Saad, K. Public health guidance to facilitate timely diagnosis of dementia: Alzheimer's Cooperative Valuation in Europe recommendations. *Int J. Geriatr. Psychiatry* **2014**, *29*, 682–693. [CrossRef]
6. Verbeek, H.; Meyer, G.; Leino-Kilpi, H.; Zabalegui, A.; Hallberg, I.R.; Saks, K.; Soto, M.E.; Challis, D.; Sauerland, D.; Hamers, J.P. RightTimePlaceCare Consortium. A European study investigating patterns of transition from home care towards institutional dementia care: The protocol of a RightTimePlaceCare study. *BMC Public Health* **2012**, *12*, 68. [CrossRef]
7. Perez-Valero, E.; Lopez-Gordo, M.A.; Morillas, C.; Pelayo, F.; Vaquero-Blasco, M.A. A Review of Automated Techniques for Assisting the Early Detection of Alzheimer's Disease with a Focus on EEG. *J. Alzheimers Dis.* **2021**, *80*, 1363–1376. [CrossRef]
8. Thatcher, R.W. Maturation of the human frontal lobes: Physiological evidence for staging. *Dev. Neuropsychol.* **1991**, *7*, 397–419. [CrossRef]
9. Sharma, N.; Kolekar, M.H.; Jha, K.; Kumar, Y. EEG and Cognitive Biomarkers Based Mild Cognitive Impairment Diagnosis. *Irbm* **2019**, *40*, 113–121. [CrossRef]
10. Abasolo, D.; Hornero, R.; Escudero, J.; Espino, P. A Study on the Possible Usefulness of Detrended Fluctuation Analysis of the Electroencephalogram Background Activity in Alzheimer's Disease. *IEEE Trans. Biomed. Eng.* **2008**, *55*, 2171–2179. [CrossRef]
11. Amini, M.; Pedram, M.; Moradi, A.; Ouchani, M. Diagnosis of Alzheimer's Disease by Time-Dependent Power Spectrum Descriptors and Convolutional Neural Network Using EEG Signal. *Comput. Math. Methods Med.* **2021**, *2021*, 5511922. [CrossRef]
12. Benz, N.; Hatz, F.; Bousleiman, H.; Ehrensperger, M.M.; Gschwandtner, U.; Hardmeier, M.; Ruegg, S.; Schindler, C.; Zimmermann, R.; Monch, A.U.; et al. Slowing of EEG background activity in Parkinson's and Alzheimer's disease with early cognitive dysfunction. *Front. Aging Neurosci.* **2014**, *6*, 314. [CrossRef] [PubMed]
13. Courtney, S.M.; Hinault, T. When the time is right: Temporal dynamics of brain activity in healthy aging and dementia. *Prog. Neurobiol.* **2021**, *203*, 102076. [CrossRef] [PubMed]
14. Susana, C.F.; Mónica, L.; Fernando, D. Event-related brain potential indexes provide evidence for some decline in healthy people with subjective memory complaints during target evaluation and response inhibition processing. *Neurobiol. Learn. Mem.* **2021**, *182*, 107450. [CrossRef] [PubMed]

15. Zhao, Y.; Zhao, Y.; Durongbhan, P.; Chen, L.; Liu, J.; Billings, S.A.; Zis, P.; Unwin, Z.C.; De Marco, M.; Venneri, A.; et al. Imaging of Nonlinear and Dynamic Functional Brain Connectivity Based on EEG Recordings With the Application on the Diagnosis of Alzheimer's Disease. *IEEE Trans. Med. Imaging* **2020**, *39*, 1571–1581. [CrossRef]
16. Miltiadous, A.; Tzimourta, K.D.; Giannakeas, N.; Tsipouras, M.G.; Afrantou, T.; Ioannidis, P.; Tzallas, A.T. Alzheimer's Disease and Frontotemporal Dementia: A Robust Classification Method of EEG Signals and a Comparison of Validation Methods. *Diagnostics* **2021**, *11*, 1437. [CrossRef]
17. Musaeus, C.S.; Salem, L.C.; Sabers, A.; Kjaer, T.W.; Waldemar, G. Associations between electroencephalography power and Alzheimer's disease in persons with Down syndrome. *J. Intellect. Disabil. Res.* **2019**, *63*, 1151–1157. [CrossRef]
18. Ahmedt-Aristizabal, D.; Fernando, T.; Denman, S.; Robinson, J.E.; Sridharan, S.; Johnston, P.J.; Fookes, C. Identification of Children at Risk of Schizophrenia via Deep Learning and EEG Responses. *IEEE J. Biomed. Health Inform.* **2021**, *25*, 69–76. [CrossRef]
19. Dauwels, J.; Vialatte, F.; Cichocki, A. Diagnosis of Alzheimer's disease from EEG signals: Where are we standing? *Curr. Alzheimer Res.* **2010**, *7*, 487–505. [CrossRef]
20. Moretti, D.V. Theta and alpha EEG frequency interplay in subjects with mild cognitive impairment: Evidence from EEG, MRI, and SPECT brain modifications. *Front. Aging Neurosci.* **2015**, *7*, 31. [CrossRef]
21. Rodrigues, P.M.; Bispo, B.C.; Garrett, C.; Alves, D.; Teixeira, J.T.; Freitas, D. Lacsogram: A New EEG Tool to Diagnose Alzheimer's Disease. *IEEE J. Biomed. Health Inform.* **2021**, *25*, 3384–3395. [CrossRef]
22. Kanda, P.A.M.; Oliveira, E.F.; Fraga, F.J. EEG epochs with less alpha rhythm improve discrimination of mild Alzheimer's. *Comput. Methods Programs Biomed.* **2017**, *138*, 13–22. [CrossRef] [PubMed]
23. Zhang, H.; Geng, X.; Wang, Y.; Guo, Y.; Gao, Y.; Zhang, S.; Du, W.; Liu, L.; Sun, M.; Jiao, F.; et al. The Significance of EEG Alpha Oscillation Spectral Power and Beta Oscillation Phase Synchronization for Diagnosing Probable Alzheimer Disease. *Front. Aging Neurosci.* **2021**, *13*, 291. [CrossRef] [PubMed]
24. Fiscon, G.; Weitschek, E.; Cialini, A.; Felici, G.; Bertolazzi, P.; De Salvo, S.; Bramanti, A.; Bramanti, P.; De Cola, M.C. Combining EEG signal processing with supervised methods for Alzheimer's patients classification. *BMC Med. Inform. Decis. Mak.* **2018**, *18*, 35. [CrossRef] [PubMed]
25. Kuo, C.E.; Chen, G.T.; Liao, P.Y. An EEG spectrogram-based automatic sleep stage scoring method via data augmentation, ensemble convolution neural network, and expert knowledge. *Biomed. Signal Process. Control* **2021**, *70*, 102981. [CrossRef]
26. Fang, W.; Wang, K.; Fahier, N.; Ho, Y.; Huang, Y. Development and Validation of an EEG-Based Real-Time Emotion Recognition System Using Edge AI Computing Platform with Convolutional Neural Network System-on-Chip Design. *IEEE J. Emerg. Sel. Top. Circuits Syst.* **2019**, *9*, 645–657. [CrossRef]
27. Sumaiyah, I.K.; Maha, S.D.; Soliman, A.M. Design of low power Teager Energy Operator circuit for Sleep Spindle and K-Complex extraction. *Microelectron. J.* **2020**, *100*, 104785.
28. Tzimourta, K.D.; Christou, V.; Tzallas, A.T.; Giannakeas, N.; Astrakas, L.G.; Angelidis, P.; Dimitrios, T.; Tsipouras, M.G. Machine Learning Algorithms and Statistical Approaches for Alzheimer's Disease Analysis Based on Resting-State EEG Recordings: A Systematic Review. *Int. J. Neural Syst.* **2021**, *31*, 2130002. [CrossRef]
29. Pedregosa, F.; Varoquaux, G.; Gramfort, A.; Michel, V.; Thirion, B.; Grisel, O.; Blondel, M.; Prettenhofer, P.; Weiss, R.; Dubourg, V.; et al. Scikit-learn: Machine Learning in Python. *J. Mach. Learn. Res.* **2011**, *12*, 2825–2830.
30. Chato, L.; Latifi, S. Machine learning and deep learning techniques to predict overall survival of brain tumor patients using MRI images. In Proceedings of the 2017 IEEE 17th International Conference on Bioinformatics and Bioengineering, Washington, DC, USA, 23–25 October 2017; pp. 9–14.
31. Sosulski, J.; Kemmer, J.P.; Tangermann, M. Improving Covariance Matrices Derived from Tiny Training Datasets for the Classification of Event-Related Potentials with Linear Discriminant Analysis. *Neuroinformatics* **2021**, *19*, 461–476. [CrossRef]
32. Fushiki, T. Estimation of prediction error by using K-fold cross-validation. *Stat. Comput.* **2011**, *21*, 137–146. [CrossRef]
33. Lyu, Y.; Li, H.; Sayagh, M.; Jiang, Z.M.; Hassan, A.E. An empirical study of the impact of data splitting decisions on the performance of AIOps solutions. *ACM Trans. Softw. Eng. Methodol.* **2021**, *30*, 54. [CrossRef]
34. Xanthopoulos, P.; Pardalos, P.M.; Trafalis, T.B. Principal Component Analysis. In *Robust Data Mining*; Springer: New York, NY, USA, 2013; pp. 27–33.
35. Fu, R.; Tian, Y.; Shi, P.; Bao, T. Automatic Detection of Epileptic Seizures in EEG Using Sparse CSP and Fisher Linear Discrimination Analysis Algorithm. *J. Med. Syst.* **2020**, *44*, 43. [CrossRef] [PubMed]
36. Kandasamy, K.; Vysyaraju, K.R.; Neiswanger, W.; Paria, B.; Collins, C.R.; Schneider, J.; P'oczos, B.; Xing, E.P. Tuning Hyperparameters without Grad Students: Scalable and Robust Bayesian Optimisation with Dragonfly. *J. Mach. Learn. Res.* **2020**, *21*, 1–27.
37. Weerts, H.J.; Mueller, A.C.; Vanschoren, J. Importance of tuning hyperparameters of machine learning algorithms. *arXiv* **2020**, arXiv:2007.07588.
38. Shekar, B.H.; Dagnew, G. Grid Search-Based Hyperparameter Tuning and Classification of Microarray Cancer Data. Proceeedings of the 2019 Second International Conference on Advanced Computational and Communication Paradigms (ICACCP), Gangtok, India, 25–28 February 2019; pp. 1–8.
39. Friedrichs, F.; Igel, C. Evolutionary tuning of multiple SVM parameters. *Neurocomputing* **2005**, *64*, 107–117. [CrossRef]
40. Wong, T.T.; Yeh, P.Y. Reliable accuracy estimates from k-fold cross validation. *IEEE Trans. Knowl. Data Eng.* **2019**, *32*, 1586–1594. [CrossRef]

41. James, G.; Witten, D.; Hastie, T.; Tibshirani, R. K-fold Cross validation. In *An Introduction to Statistical Learning with Application in R*; Casella, G., Fienberg, S., Olkin, I., Eds.; Springer: New York, NY, USA, 2013; Volume 112.
42. Phyu, T.N. Survey of classification techniques in data mining. In Proceedings of the International Multiconference of Engineers and Computer Scientists, Hong Kong, China, 18–20 March 2009.
43. Kulkarni, N.; Bairagi, V. *EEG-based Diagnosis of Alzheimer Disease: A Review and Novel Approaches for Feature Extraction and Classification Techniques*; Elsevier Academic Press: Cambridge, MA, USA, 2018.
44. Elgandelwar, S.M.; Bairagi, V.K. Power analysis of EEG bands for diagnosis of Alzheimer dis-ease. *Int. J. Med. Eng. Inform.* **2021**, *13*, 376–385.
45. Bairagi, V. EEG signal analysis for early diagnosis of Alzheimer disease using spectral and wavelet based features. *Int. J. Inf. Technol.* **2018**, *10*, 403–412. [CrossRef]
46. Cejnek, M.; Vysata, O.; Valis, M.; Bukovsky, I. Novelty detection-based approach for Alzheimer's disease and mild cognitive impairment diagnosis from EEG. *Med. Biol. Eng. Comput.* **2021**, *59*, 2287–2296. [CrossRef]
47. Oltu, B.; Akşahin, M.F.; Kibaroğlu, S. A novel electroencephalography based approach for Alzheimer's disease and mild cognitive impairment detection. *Biomed. Signal. Process. Control* **2021**, *63*, 102223. [CrossRef]
48. Ieracitano, C.; Mammone, N.; Hussain, A.; Morabito, F.C. A novel multi-modal machine learning based approach for automatic classification of EEG recordings in dementia. *Neural Netw.* **2021**, *121*, 176–190. [CrossRef] [PubMed]
49. Kulkarni, N.N.; Bairagi, V.K. Extracting salient features for EEG-based diagnosis of Alzheimer's disease using support vector machine classifier. *IETE J. Res.* **2017**, *63*, 11–22. [CrossRef]
50. Available online: https://coral.ai/docs/dev-board/datasheet/#features (accessed on 18 March 2022).
51. Cariolaro, G. Classical Signal Theory. In *Unified Signal Theory*; Springer: London, UK, 2011; p. 23.

Article

qEEG Analysis in the Diagnosis of Alzheimer's Disease: A Comparison of Functional Connectivity and Spectral Analysis

Maria Semeli Frangopoulou [1,*] and Maryam Alimardani [2]

1. Department of Communication & Cognition, Tilburg University, 5037 AB Tilburg, The Netherlands
2. Department of Cognitive Science & Artificial Intelligence, Tilburg University, 5037 AB Tilburg, The Netherlands; m.alimardani@tilburguniversity.edu
* Correspondence: m.s.frangopoulou@tilburguniversity.edu

Abstract: Alzheimer's disease (AD) is a brain disorder that is mainly characterized by a progressive degeneration of neurons in the brain and decline of cognitive abilities. This study compared an FFT-based spectral analysis against a functional connectivity analysis for the diagnosis of AD. Both quantitative methods were applied on an EEG dataset including 20 diagnosed AD patients and 20 age-matched healthy controls (HC). The obtained results showed an advantage of the functional connectivity analysis when compared to the spectral analysis; while the latter could not find any significant differences between the AD and HC groups, the functional connectivity analysis showed statistically higher synchronization levels in the AD group in the lower frequency bands (delta and theta), suggesting a 'phase-locked' state in AD-affected brains. Further comparison of functional connectivity between the homotopic regions confirmed that the traits of AD were localized to the centro-parietal and centro-temporal areas in the theta frequency band (4–8 Hz). This study applies a neural metric for Alzheimer's detection from a data science perspective rather than from a neuroscience one and shows that the combination of bipolar derivations with phase synchronization yields similar results to comparable studies employing alternative analysis methods.

Keywords: Alzheimer's disease (AD); EEG; fast Fourier transformation (FFT); functional connectivity; phase synchronization

1. Introduction

Alzheimer's disease (AD) is a brain disorder that is mainly characterized by a progressive degeneration of neurons in the brain. As the disease progresses, a cortical disconnection occurs, causing a deficit in memory and a decline in other cognitive capabilities [1]. AD-related effects on the patient's brain can be identified with various tools, one option being the electroencephalogram (EEG), which measures the electrical activity of the brain. EEG is a fast and non-invasive method that provides a high temporal resolution. However, it lacks in spatial resolution, meaning that it is not the most precise method for the diagnosis of a brain disorder.

Quantitative EEG (qEEG) analysis takes EEG recordings, commonly interpreted by clinicians using visualization tools, one step further, giving the possibility of digitally processing and presenting the signal characteristics in spectral and spatial domains [2]. In a spectral analysis, a given signal is broken down and examined in the frequency domain. This type of analysis is useful when finding differences between patients who are diagnosed with a disorder and healthy individuals, by examining the relevant frequency bands to identify a noticeable change in the activity within a particular frequency band [3]. A very common yet powerful tool used in spectral analysis is the fast Fourier transformation (FFT) [4]. This algorithm can be used to find band-specific differences by calculating the power of each band separately.

When conducting a spectral analysis, the power spectral density (PSD) is often used to determine differences in brain activity between frequency bands. Previous studies have shown that when compared to healthy controls (HC), AD patients show an increase of PSD in the theta band and a decrease in the alpha band [3,5,6]. In AD diagnosis specifically, a spectral analysis can show discrepancies between AD and other types of dementia, such as vascular dementia (VaD) [1].

However, while these studies suggest that EEG spectral analysis may differentiate AD patients from HC [7], several other studies that have examined the process of AD have concluded that this brain disorder is involved with changes in the distributed networks related to memory [8] and that the changes observed in the frequency bands may not sufficiently reflect this. In addition, it is known that the FFT is not the most adequate tool for non-stationary signals such as the EEG [9,10]. Moreover, as mentioned above, patients suffering from AD experience a cortical disconnection. It is therefore important to examine various regions of interest (ROIs) that are affected by the disease. Hence, more reliable signal processing methods are required to capture the complexity of this disorder and investigate the processes that underlie the occurring symptoms [11,12]. An alternative to a spectral analysis is the connectivity analysis; a method which allows one to study the communications between different regions of the brain [12].

Functional connectivity analysis measures the degree of synchronization between two EEG signals; a higher connectivity indicates more effective communication between the examined brain regions [13]. There are several ways of conducting a functional connectivity analysis. For instance, **Coherence analysis** has been used exhaustively in detecting differences between AD patients and HC. Recent studies indicated a decrease in the coherence levels between ROIs for AD [3,14]. Although coherence has brought some novelty in studies involving AD patients, it is worth mentioning that it solely accounts for linear correlations, thus not considering nonlinear interactions.

Nonlinear correlations, on the other hand, can give crucial information in a functional connectivity analysis. A widely used method for this is the **phase synchronization (PS)** analysis. PS looks at the oscillatory activity in two brain regions in terms of their phases [15]. The oscillations are therefore said to be synchronized if their phases are similar. PS excels over coherence analysis in terms of being able to account for nonlinearity [16]. Moreover, a study has shown that differences have been found in terms of the synchronization between within-band connections and between-band connections (e.g., within the delta band; between the delta and theta bands) [17]. This study in particular also discovered that AD patients showed much a lower strength of synchronization for between-frequency band analysis when compared to HC.

PS has several indices of measurement, with the **phase-lag index (PLI)** and **phase-locking value (PLV)** being the most-used measures [18]. The PLI takes a time-series of phase differences and computes the asymmetry corresponding to the distribution of these phase differences [19]. In a recent PS study using the PLI as the index of choice, results showed that in AD patients, the lower alpha band presented a decrease in functional connectivity situated in the posterior region [20]. On the other hand, PLV looks at the consistency in phase difference. The PLV value ranges from zero, indicating random phase differences, to one, indicating a fixed phase difference [21]. For example, a study performing cross-frequency coupling (CFC), using PLV on AD patients, reached the conclusion that oscillations in the alpha band, and more specifically around the dominant peak, are phase-locked with the gamma band power [22]. Results were observable in the posterior region of the brain, suggesting that AD elicits a region-specific change in functional connectivity.

It is noteworthy that functional connectivity entails its own drawbacks too. In a comprehensive review of different methods for functional connectivity analysis and their caveats, Bastos and Schoffelen indicated that many of the connectivity measures are prone to volume conduction and hence can yield false positive results [16]. This is particularly the case for methods using unipolar derivations with a common reference. However, as it will be further explained in this paper, this study employed bipolar EEG montages, thus greatly diminishing the issue. The same review study [16] highlighted another disadvantage of functional connectivity, namely the signal-to-noise ratio (SNR) problem; an excessive amount of noise in the signal of interest during the EEG signal retrieval may cause false connectivity results.

In sum, the current state-of-the-art calls for a comparison between computational methods that are used for diagnosis of Alzheimer's disease. The reviewed literature focusing on spectral analysis and functional connectivity demonstrates a scarcity of studies employing both methods in diagnosis of AD. It should be mentioned that although spectral analysis and functional connectivity analysis have different applications from a physiological perspective, applying both methods on a single dataset can be useful in determining the more accurate method in the diagnosis of AD based on brain activity. This is the gap that this study aimed to address. So far, several studies have reported the outcomes of either of the two methods [23,24]. However, conducting a connectivity analysis and comparing it with a spectral analysis using the same dataset presents two advantages; (1) it shows which method can yield the most accurate and complete information in AD diagnosis [3], and (2) it can identify the affected ROIs instead of solely looking at whether the patient suffers from AD, which is believed to help predicting AD in its early stages of development [12].

The proposed study serves as a comparison between two methods in qEEG, namely the spectral and functional connectivity analyses. Having highlighted the advantages of PS over other connectivity measures, this study aimed to investigate the performance of PS in differentiating AD patients from HCs as compared to conventional FFT. This is done with the goal of identifying the most efficient method in implementation of machine-assisted diagnosis tools in the future clinical assessment of AD. The two types of analysis were conducted on a set of EEG recordings obtained from patients suffering from AD and from their respective healthy controls (HC), in an attempt to address the following research question:

RQ1. *How does a functional connectivity analysis perform against a spectral analysis in finding differences between patients diagnosed with Alzheimer's disease (AD) and healthy controls (HC)?*

Moreover, this study attempted to answer a secondary research question:

RQ2. *Can a functional connectivity analysis localize the differences identified in the brain activity of AD subjects when compared to that of the HCs?*

To answer this question, a series of statistical tests were performed using the results provided by the connectivity analysis.

2. Materials and Methods

2.1. Dataset and Preprocessing

The EEG dataset was provided by the University of Sheffield, from an open study and under a relevant NDA. All subjects were informed about the experiment and signed an informed consent form. The dataset consists of 12 s, eyes-open recordings of 20 AD-diagnosed patients and 20 age-matched HC, younger than 70 years of age (Table 1).

Table 1. General information of the AD and HC groups including sample size, age mean with standard deviation, and gender ratio per group.

	AD	HC
Size	$N = 20$	$N = 20$
Age	60 ($SD = 4.40$)	61 ($SD = 6.67$)
Gender (F/M)	8/12	12/8

The participants' EEGs were recorded using the International 10–20 system [25]. To reduce volume conduction effects from a common reference [26], 23 bipolar derivations were used in this study. Figure 1 gives an overview of the electrodes and bipolar channels. More specifically, the following bipolar channels were used: F8-F4, F7-F3, F4-C4, F3-C3, F4-FZ, FZ-CZ, F3- FZ, T4-C4, T3-C3, C4-CZ, C3-CZ, CZ-PZ, C4-P4, C3-P3, T4-T6, T3-T5,

P4-PZ, P3-PZ, T6-O2, T5-O1, P4-O2, P3-O1, and O2-O1. These bipolar channels are the most commonly used in clinical practice [27]. During the recording, the participants were instructed to reduce their movements and not to think of anything in particular (i.e., resting state EEG).

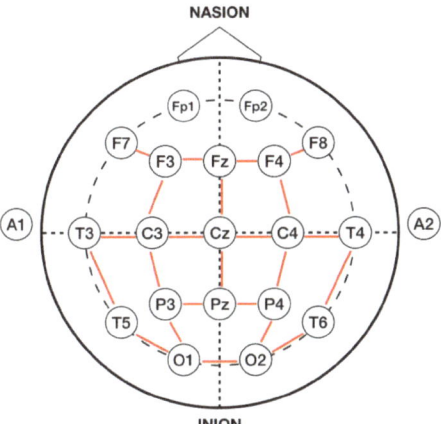

Figure 1. EEG signals were collected from 23 bipolar channels based on the 10–20 international system.

The raw EEG signals were preprocessed in EEGLAB (v.2021.0), MATLAB. First, the signals were downsampled to 500 Hz. Next, a band-pass filter was applied between 0.1 and 100 Hz using EEGLAB functions. This was done following the requirements used for the phase synchronization (see Section 2.3. 'Functional Connectivity Analysis') to avoid phase distortion. Additionally, a notch filter was used to attenuate signals in the 48–52 Hz range.

2.2. Spectral Analysis

The power spectral density (PSD) of the entire signal for each of the bipolar montages was calculated using EEGLAB's spectopo() function. This function makes use of the FFT algorithm to extract and plot the PSD. The signal was subsequently divided into five frequency bands: delta (1–4 Hz), theta (4–8 Hz), alpha (8–13 Hz), beta (13–30 Hz), and gamma (36–44 Hz) and the mean power in each band was computed. These ranges were selected according to [28] and were also used in the connectivity analysis. A Shapiro–Wilk test was applied to the data to check for normality and subsequently, a Mann–Whitney U-test was used to compare band power medians between the AD and HC groups.

2.3. Functional Connectivity Analysis

Functional connectivity analysis was carried out using the PLV index [28]. First, a continuous wavelet transform was applied (i.e., the complex Morlet wavelet), with this wavelet being used as a kernel to compute the PLV, which is defined by Equation (1):

$$\text{PLV}(t) = \frac{1}{N} \left| \sum_{n=1}^{N} e^{i\theta(t,\,n)} \right| \qquad (1)$$

where n is an index for the trial number and θ indicates the phase difference. The phase-locking value yielded by PLV ranges from 0 to 1, with 1 indicating that two signals have an identical relative phase across N trials. Conversely, values that approach 0 indicate little to no phase synchrony between the signals. For every subject, the PLV was calculated for all possible 253 bipolar channel combinations in five frequency bands as defined above. Next, inspired by [29], 'Global Connectivity' and 'Homotopic Pair Connectivity' were computed using the extracted PLV values and were compared between the groups.

2.3.1. Global Connectivity

Global connectivity was computed by averaging all 253 PLV values that were obtained per frequency band. This led to a total of five PLV_{mean} values per subject (i.e., one PLV_{mean} per frequency band). Following the Shapiro–Wilk test, a Mann–Whitney U-test was used to compare the mean PLVs between the AD and HC groups. The aim of this evaluation was to determine whether band-specific differences could be found in the global functional connectivity of the AD subjects against the HCs.

2.3.2. Homotopic Pair Connectivity

Homotopic Pair Connectivity was computed by focusing on certain pairs of bipolar derivations that were homotopic in the left and right brain hemispheres (mirror areas of the brain hemispheres). Based on previous classifications [30,31], four pairs that have been, in part, shown to be most affected by Alzheimer's disease were selected. These pairs are depicted in Figure 2. Pair A consisted of the homotopic pair located in the centro-parietal region of the brain (C3-P3 and C4-P4). Pair B corresponded to the pair in the fronto-central area (F3-C3 and F4-C4), Pair C consisted of electrodes located in the parieto-occipital region (P3-O1 and P4-O2), and Pair D consisted of electrodes placed in the centro-temporal area (C3-T3 and C4-T4). For each pair, the PLV was computed in the five frequency bands and a Mann Whitney U–test was carried out to compare the band-specific PLVs between the two AD and HC groups.

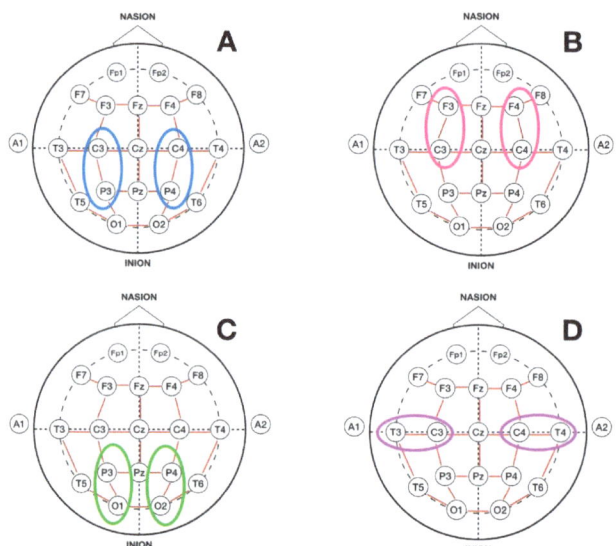

Figure 2. Homotopic pair connectivity was examined in four mirror regions in the left and right hemispheres including (**A**) centro-parietal (C3-P3 and C4-P4), (**B**) fronto-central area (F3-C3 and F4-C4), (**C**) parieto-occipital (P3-O1 and P4-O2) and (**D**) centro-temporal (C3-T3 and C4-T4) connections.

2.3.3. Localization of AD Using Homotopic Pair Connectivity

To answer the secondary RQ, the four homotopic pairs were compared against each other to ascertain as to which areas displayed a significant connectivity difference between the two groups. To do this, the PLV values obtained from both subject groups in each of the above-mentioned homotopic pairs were compared using linear mixed effects (LME) regression models. LME was fitted in RStudio (v.2021.09.10) using the lme4() package [32] and was chosen for this analysis because the repeated measure from the homotopic pairs was correlated, violating the assumptions of other tests, such as ANOVA.

The analysis included two steps; first, the LME model was fitted with PLVs as the response variable and pair and group as predictors. Participants were included as a random factor in the model. The interaction term was included to prevent the overly enthusiastic outcome that there is a difference in connectivity between HC and AD for all pairs. Next, following verification of the main effects, post-hoc comparisons were conducted between pairs to examine which brain regions showed significant differences between the two groups. These steps were only applied to the frequency bands that showed a statistically significant difference between the AD and HC groups in at least one of the homotopic pairs in the 'Homotopic Pair Connectivity' analysis.

3. Results

3.1. Spectral Analysis

The Shapiro–Wilk test applied to the band power data rejected the null hypothesis of normal populations distributions ($p < 0.05$). Therefore, the non-parametric Mann–Whitney U-test was used to compare the groups in each frequency band. The test did not find any significantly different delta power for the AD subjects ($Mdn = 4.23$) than the healthy controls ($Mdn = 4.07$), $U = 174$, $p = 0.45$. Similar results were observed for the theta ($Mdn = 2.42$ vs. $Mdn = 4.30$, $U = 146$, $p = 0.15$), alpha ($Mdn = 1.88$ vs. $Mdn = 2.29$, $U = 162$, $p = 0.31$), beta ($Mdn = 1.67$ vs. $Mdn = 1.76$, $U = 152$, $p = 0.2$), and gamma bands ($Mdn = 0.67$ vs. $Mdn = 0.90$, $U = 158$, $p = 0.26$). Therefore, it can be concluded that the spectral analysis yielded no significant differences between the AD subjects versus HC in any of the five frequency bands.

3.2. Functional Connectivity Analysis

3.2.1. Global Connectivity

Figure 3 illustrates the distribution of PLV_{mean} from all subjects in the AD and HC groups in all five frequency bands. The result of the Mann–Whitney U-test indicated that the average PLVs from all channel combinations were significantly higher in the theta band for the AD participants ($Mdn = 0.31$) when compared to the HCs ($Mdn = 0.26$), $U = 326$, $p = 0.0004$. This was not the case for the delta ($Mdn = 0.30$ vs. $Mdn = 0.28$, $U = 258$, $p = 0.12$), alpha ($Mdn = 0.26$ vs. $Mdn = 0.24$, $U = 264$, $p = 0.09$), beta ($Mdn = 0.18$ vs. $Mdn = 0.18$, $U = 226$, $p = 0.50$), and gamma bands ($Mdn = 0.18$ vs. $Mdn = 0.18$, $U = 181$, $p = 0.62$).

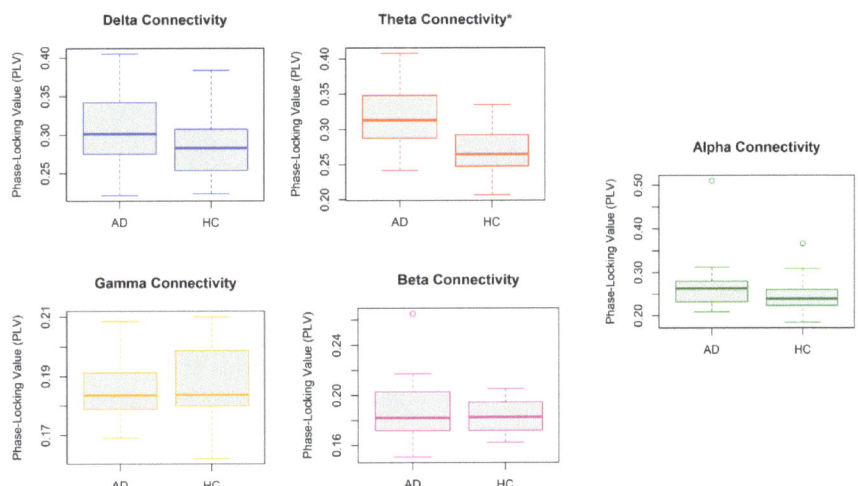

Figure 3. The average PLVs obtained from all connectivity pairs for the five frequency bands (global connectivity). Plots marked with * indicate a statistically significant difference ($p < 0.05$) between AD patients and HCs.

3.2.2. Homotopic Pair Connectivity

Figure 4 illustrates the PLV values obtained from the homotopic pair in the centro-parietal area (Pair A) of the AD and HC groups in the five frequency bands. The Mann–Whitney U-test displayed a significantly higher PLV in the theta band for AD participants (Mdn = 0.64) as compared to HCs (Mdn = 0.52), U = 292, p = 0.01. No significant results were found for the other four frequency bands.

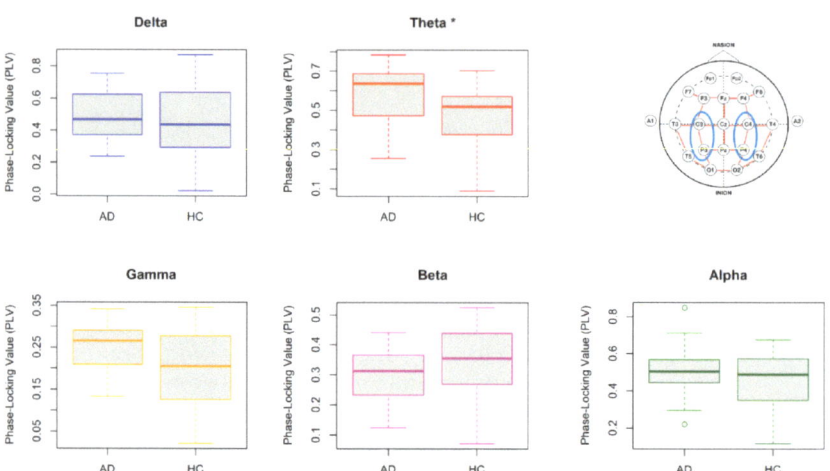

Figure 4. The PLVs obtained from the homotopic pair in the centro-parietal region (Pair A). Plots marked with * indicate a statistically significant difference (p < 0.05) between AD patients and HCs.

Figure 5 illustrates the PLV values obtained from the homotopic pair in the fronto-central region (Pair B) of the AD and HC groups in the five frequency bands. The Mann–Whitney U-test displayed a significantly higher PLV for AD participants in both the delta (AD Mdn = 0.57, HC Mdn = 0.45, U = 275, p = 0.04) and theta bands (AD Mdn = 0.65, HC Mdn = 0.50, U = 282, p = 0.03). No significant results were found for the other three frequency bands.

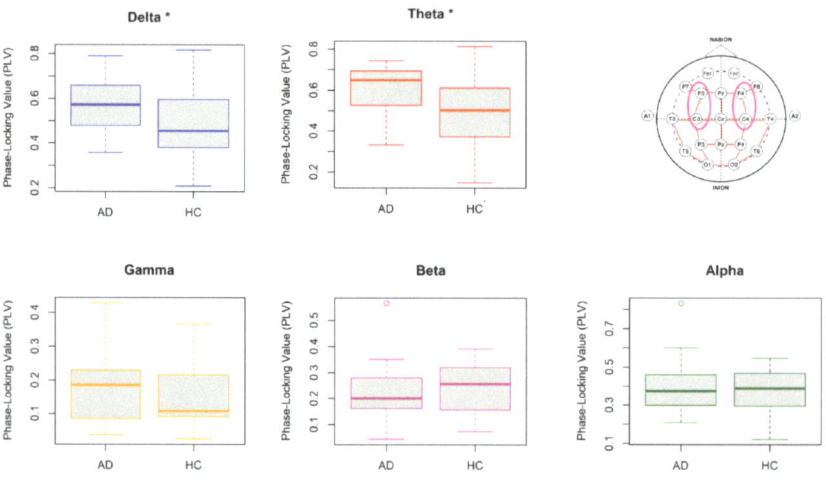

Figure 5. The PLVs obtained from the homotopic pair in the fronto-central region (Pair B). Plots marked with * indicate a statistically significant difference (p < 0.05) between AD patients and HCs.

Figure 6 illustrates the PLV values obtained from homotopic pairs in the parieto-occipital region (Pair C) of the AD and HC groups in the five frequency bands. The Mann–Whitney U-test indicated a significantly higher PLV for the AD group ($Mdn = 0.64$) as compared to the HCs ($Mdn = 0.49$), solely in the delta band ($U = 293$, $p = 0.01$). The test resulted in an insignificant outcome for the other four frequency bands.

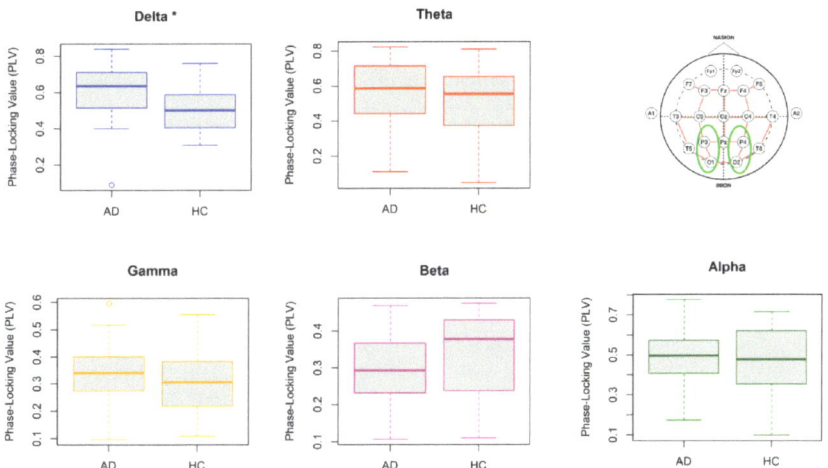

Figure 6. The PLVs obtained from the homotopic pair in the parieto-occipital region (Pair C). Plots marked with * indicate a statistically significant difference ($p < 0.05$) between AD patients and HCs.

Lastly, Figure 7 shows the PLV values obtained from homotopic pairs in the centro-temporal region (Pair D) of the AD and HC groups in the five frequency bands. The Mann–Whitney U-test indicated a significantly higher PLV solely in the theta band of AD participants ($Mdn = 0.48$) as compared to the HCs ($Mdn = 0.40$, $U = 280$, $p = 0.03$). The results of group comparisons in the other four frequency bands remained insignificant.

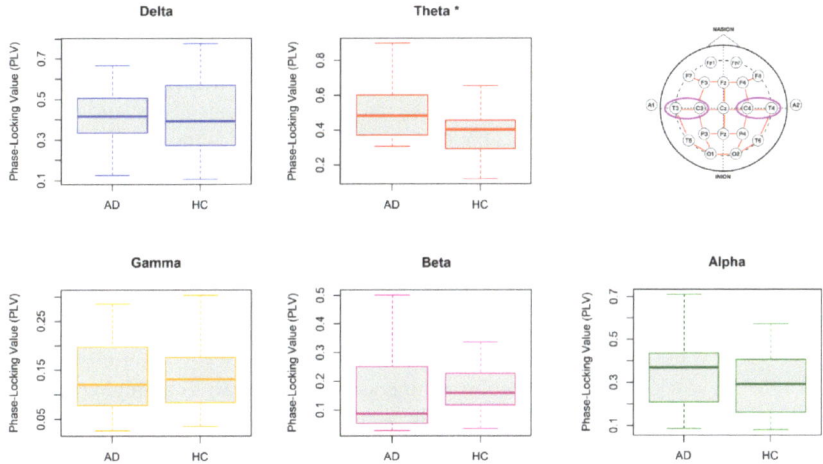

Figure 7. The PLVs obtained from the homotopic pair in the centro-temporal region (Pair D). Plots marked with * indicate a statistically significant difference ($p < 0.05$) between AD patients and HCs.

Summing up, the comparison of PLV in the selected homotopic pairs resulted in observing the main differences in the low frequency bands of delta and theta.

3.2.3. Localization of AD Using Homotopic Pair Connectivity

To compare the connectivity across homotopic pairs and to identify the most relevant brain region affected by AD, LME regression models were applied to the homotopic PLVs in the theta and delta frequency bands. The model confirming main effects for both pair and group was selected, and a post-hoc analysis using the Tukey's adjusted pairwise comparisons of least-squares means were conducted. Table 2 summarizes the outcome of the post-hoc comparisons.

Table 2. Summary of the results of the post-hoc analysis of the LME regression.

Homotopic Pair	Delta Band (1–4 Hz)	Theta Band (4–8 Hz)
A	✗	✓
B	✗	✗
C	✗	✗
D	✗	✓

In the theta band, the pairwise difference between AD patients and HCs reached significance for Pair A (*LSM difference* = 0.115, *SE* = 0.0517, p = 0.028) and Pair D (*LSM difference* = 0.101, *SE* = 0.0517, p = 0.038). While not statistically significant, trends were observed for Pair B (*LSM difference* = 0.097, *SE* = 0.0517, p = 0.064), whereas the difference between the AD and the HC group did not reach significance for Pair C. No significance was observed in the delta band.

4. Discussion

The current study explored differences in the brain activity of patients afflicted with Alzheimer's disease compared to a healthy control cohort, using the quantitative analyses of EEG signals. In particular, two types of analyses were conducted and compared. First, a conventional spectral analysis was conducted to find spectral band power differences between AD subjects and healthy controls. The second approach employed a functional connectivity analysis using phase synchronization across five frequency bands to compare the intra-brain connectivity (global or local) between healthy brains and the ones affected by AD-induced dementia. The results indicated that the spectral analysis did not yield any significant differences between the AD and HC groups, suggesting that it is not an ideal method for the diagnosis of AD based on the EEG. On the other hand, the functional connectivity analysis using the PLV measure showed significant differences between the groups, both in terms of global connectivity and homotopic connectivity. The further analysis of homotopic pairs revealed significantly higher theta-band connectivity localized in the centro-parietal and centro-temporal regions.

The dataset used in this study consisted of bipolar derivations, instead of unipolar channel values that are more commonly used in the qEEG analysis [22,29]. The use of bipolar derivations is seen as a more advantageous method when compared to unipolar or average referencing methods [33], as it can mitigate the issues associated with common active referencing such as volume conduction [34]. Volume conduction, which refers to the leakage of electrical potentials to the neighboring electrodes, can complicate the interpretation of connectivity metrics. Therefore, the use of bipolar derivations in the computation of functional connectivity is highly recommended, as was demonstrated in a recent study in the field of AD detection [30].

The spectral analysis, although being a powerful tool for analyzing specific types of signals, could not confirm any differences between the Alzheimer's patients recruited in this study and their age-matched healthy controls. This is inconsistent with previous reports in which the development of AD was associated with an increase of delta and theta activity as well as a decrease in alpha and beta activity [5,6,17]. An explanation for the lack of evidence in the current study could be that the AD subjects included in the sample were only moderately affected by this disorder. In addition, the use of bipolar derivations could have affected the results observed, since past studies made use of unipolar derivations to obtain the spectral powers of the specific EEG channels [3,5]. Another reason could be due to the number of participants in the selected dataset; had it been higher, different results might have emerged. For example, the study by Fraga et al., [5] made use of 75 participants, split into three categories (i.e., 27 with mild AD, 22 with moderate AD, and 27 HCs). While this leaves room for future research to confirm the most suitable computational approach for the detection of severe cases, this study proposes functional connectivity as being a promising tool in detection of early signs of AD from EEG signals [35].

The two connectivity analyses that were subsequently carried out, namely 'Global Connectivity' and 'Homotopic Pair Connectivity', displayed increased communication between brain networks in the AD subject group when compared to the HCs. These findings were first identified in the global connectivity analysis and subsequently confirmed in the homotopic pair connectivity analysis. The global connectivity analysis gave an overview of the AD process in the brain. Although it resulted in identifying higher connectivity that was distributed in the brains of the AD group, it could not localize the effect. Indeed, the effect of Alzheimer's disease tends to be more prominent in some areas of the brain than others [30,31], hence justifying a motive to pursue a further analysis with the homotopic pair connectivity. Similar to the band division performed to retrieve the PSD in a spectral analysis, in functional connectivity studies involving AD subjects, electrode pairs can be singled out and evaluated separately, instead of combining them all together [36]. The analysis of homotopic pairs revealed a significant difference of connectivity in the delta band of the pairs in the fronto-central and the parieto-occipital regions, whereas these effects were diminished in the global connectivity analysis, which only found a significant difference between the groups in the theta band.

The result indicating a higher functional connectivity of the AD brains conflicts with the study of Hata et al. [12], who reported a lower lagged phase synchronization in the delta and theta bands of AD patients. Indeed, a decreased connectivity between brain regions can be expected, as AD is known to cause neuronal loss and the damage of neural pathways [1,8,22]. However, other studies suggest that the impact of such damage is only reflected in fast signals, as healthy participants have a higher brain connectivity in alpha and beta bands [19,20] but not in the lower frequency bands. On the other hand, it has been shown in the past that patients suffering from neuropsychiatric disorders such as schizophrenia and epilepsy display an increased functional activity between brain networks as a sign of anomaly in information communication [37,38]. In the study of Cai et al. [17], similar patterns were reported for AD patients, where the connectivity within the same frequency band (intra-band connectivity) was stronger in AD brains than in the healthy brain, whereas the connectivity between the frequency bands (inter-band connectivity) was significantly weaker. Observing higher synchronization values in the lower frequency bands for AD subjects can therefore be interpreted as a sign of brain dysfunction [17,20]. More specifically, this study demonstrated that the brains affected by Alzheimer's disease seemed to be in a 'phase-lock' state, causing a high connectivity in the low frequency bands; this is an observation that is well in line with the existing literature [6,17,19,30,36].

The localization of AD analysis reached the conclusion that there was a significant difference in the connectivity between the AD and HC groups in the theta band for two out of four homotopic pairs. The answer to the secondary research question (RQ2) is therefore positive; it is possible to localize to some extent the differences between a healthy brain and one suffering from AD-induced dementia. The findings of this study therefore provide further evidence for damaged neural connections and, consequently, abnormal network dynamics in AD-affected brains, particularly in the centro-parietal and centro-temporal regions. While older studies such as [39] suggested that the effects of AD are not situated in one specific area of the brain, the regions identified by this study are in line with the report of more recent studies such as Deng et al. [40] which observed a significant decrease in signal complexity of the AD group in the occipito-parietal and temporal regions of the brain using 'multivariate multi-scale weighted permutation entropy' (MMSWPE) [40]. It is important to note that the findings of this study do not suggest that other brain regions should be excluded in future comparisons between the AD and HC groups. In fact, several studies have indicated that patients suffering from AD tend to have a disruption in brain connectivity in various areas of the brain, as well as in various stages of the disease [41]. The results of the proposed study, although solely being able to identify AD in the areas mentioned above, do not reject the hypothesis that AD can affect connectivity in other regions of the brain and hence this provides opportunity for future studies to confirm.

Clearly, this study is not without limitations. A first limitation arises from the duration of the epochs that were available in the dataset (12 s per subject). Longer epochs would have provided more EEG samples for phase synchronization analysis, as well as an opportunity to evaluate the dynamic changes of connectivity over time, as had been previously done in Zhao et al. [30]. Another limitation involved the number of participants, as previously mentioned. The dataset used in this research consisted of 20 AD participants and 20 age-matched HCs. Given the individual differences inherent to the progress of AD, a larger dataset would have been optimal to yield more reliable results. Moreover, this study made use of the phase-locking value as an index for phase synchronization, as the data was recorded in a bipolar manner, and therefore the analysis was considered to be robust to the common source effects [27,30]. Future research could use other indices of functional connectivity, such as coherence and phase-lag index (PLI), to investigate their efficacy in detecting AD impacts on the brain activity.

Finally, it shall be noted that this study applies a neural metric for Alzheimer's Disease detection from a data science perspective rather than a neuroscience one. This implies that the methodology employed in this study strived to find an accurate tool for the detection of AD from EEG brain activity, rather than attempting to explain the cognitive and neural mechanisms that underlie the observed differences between AD patients and healthy controls. In this case, the findings of this research are well in line with the existing literature regarding AD detection and brain connectivity and show that the combination of bipolar derivations with phase synchronization can yield comparable results to studies that used other connectivity methods. This qEEG analysis could therefore be considered as secondary tool, to be used alongside the visual EEG analysis that is employed by clinicians. Furthermore, as solely post-mortem studies can identify AD with perfect accuracy [6], the authors are aware that the proposed methods in this study may not yield the most accurate results in the prediction of AD from EEG. However, the data science perspective proposed in this study leaves room for more advanced techniques, such as machine learning (ML), which has been growing in the field of AD detection over the past years. Promising results have been observed in regular support vector machine (SVM) models [42], as well as in new, innovative ML methods [43], leaving fertile ground for future research in this domain.

5. Conclusions

This research served to find a promising tool for the diagnosis of early signs of Alzheimer's Disease (AD), through the interpretation of brain activity. This was done by comparing two quantitative EEG methods, namely spectral analysis and functional connectivity analysis, in two groups of AD patients and age-matched healthy controls. The results indicated that the old-school spectral analysis failed to yield any statistically significant results that could help differentiate a brain that is affected by AD from a healthy one, whereas the functional connectivity analysis using phase synchronization found a significantly stronger global 'phase-locked' state in the theta activity of AD-affected brains. Moreover, by extracting the functional connectivity metrics in four homotopic pairs of electrodes, it was possible to localize significant differences concerning the theta band in the centro-parietal and centro-temporal areas of the brain. To conclude, the findings of this research show that functional connectivity analysis using phase synchronization offers a promising quantitative method for future research in the area of AD detection. This method in combination with the standard cognitive tests that are commonly employed in dementia screening can put forward a more accurate diagnosis for patients who suffer from early symptoms of AD.

Author Contributions: Conceptualization, M.S.F. and M.A.; methodology, M.S.F. and M.A.; formal analysis, M.S.F.; writing—original draft preparation, M.S.F.; writing—review and editing, M.S.F. and M.A.; visualization, M.S.F.; supervision, M.A.; funding acquisition, M.A. All authors have read and agreed to the published version of the manuscript.

Funding: This research was funded by Department of Cognitive Science and AI, Tilburg University.

Institutional Review Board Statement: The study was conducted in accordance with the Declaration of Helsinki and approved by the Yorkshire and The Humber (Leeds West) Research Ethics Committee (reference number 14/YH/1070).

Informed Consent Statement: Informed consent was obtained from all subjects involved in the study.

Data Availability Statement: The raw dataset used for this study is under a Non-Disclosure Agreement (NDA) and is therefore not available to the public. The code used to support the findings of this study has been deposited in the GitHub repository, and can be accessed at https://github.com/SemeliF/AD_paper (accessed on 24 February 2022).

Acknowledgments: Authors would like to thank Daniel J. Blackburn and Ptolemaios G. Sarrigiannis from University of Sheffield for providing the EEG dataset used is this research, Sue Yoon from Eindhoven University of Technology for sharing her experience with the functional connectivity analysis and Peter Hendrix from Tilburg University for his guidance regarding the statistical analysis.

Conflicts of Interest: The authors declare no conflict of interest.

References

1. Neto, E.; Allen, E.A.; Aurlien, H.; Nordby, H.; Eichele, T. EEG Spectral Features Discriminate between Alzheimer's and Vascular Dementia. *Front. Neurol.* **2015**, *6*, 25. [CrossRef] [PubMed]
2. Smailovic, U.; Jelic, V. Neurophysiological Markers of Alzheimer's Disease: Quantitative EEG Approach. *Neurol. Ther.* **2019**, *8* (Suppl. 2), 37–55. [CrossRef] [PubMed]
3. Wang, R.; Wang, J.; Yu, H.; Wei, X.; Yang, C.; Deng, B. Power Spectral Density and Coherence Analysis of Alzheimer's EEG. *Cogn. Neurodyn.* **2015**, *9*, 291–304. [CrossRef]
4. Heideman, M.; Johnson, D.; Burrus, C. Gauss and the History of the Fast Fourier Transform. *IEEE ASSP Mag.* **1984**, *1*, 14–21. [CrossRef]
5. Fraga, F.J.; Falk, T.H.; Kanda, P.A.M.; Anghinah, R. Characterizing Alzheimer's Disease Severity via Resting-Awake EEG Amplitude Modulation Analysis. *PLoS ONE* **2013**, *8*, e72240. [CrossRef]
6. Cassani, R.; Estarellas, M.; San-Martin, R.; Fraga, F.J.; Falk, T.H. Systematic Review on Resting-State EEG for Alzheimer's Disease Diagnosis and Progression Assessment. *Dis. Markers* **2018**, *2018*, e5174815. [CrossRef]
7. Hampel, H.; Lista, S.; Teipel, S.; Garaci, F.; Nisticò, R.; Blennow, K.; Zetterberg, H.; Bertram, L.; Duyckaerts, C.; Bakardjian, H.; et al. Perspective on Future Role of Biological Markers in Clinical Therapy Trials of Alzheimer's Disease: A Long-Range Point of View beyond 2020. *Biochem. Pharmacol.* **2014**, *88*, 426–429. [CrossRef]
8. Sperling, R.A.; Dickerson, B.C.; Pihlajamaki, M.; Vannini, P.; LaViolette, P.S.; Vitolo, O.V.; Hedden, T.; Becker, J.A.; Rentz, D.M.; Selkoe, D.J.; et al. Functional Alterations in Memory Networks in Early Alzheimer's Disease. *Neuromol. Med.* **2010**, *12*, 27–43. [CrossRef]
9. Al-Fahoum, A.S.; Al-Fraihat, A.A. Methods of EEG Signal Features Extraction Using Linear Analysis in Frequency and Time-Frequency Domains. *ISRN Neurosci.* **2014**, *2014*, 730218. [CrossRef]
10. Arrufat-Pié, E.; Estévez-Báez, M.; Estévez-Carreras, J.M.; Machado-Curbelo, C.; Leisman, G.; Beltrán, C. Comparison between Traditional Fast Fourier Transform and Marginal Spectra Using the Hilbert–Huang Transform Method for the Broadband Spectral Analysis of the Electroencephalogram in Healthy Humans. *Eng. Rep.* **2021**, *3*, e12367. [CrossRef]
11. Leuchter, A.F.; Cook, I.A.; Newton, T.F.; Dunkin, J.; Walter, D.O.; Rosenberg-Thompson, S.; Lachenbruch, P.A.; Weiner, H. Regional Differences in Brain Electrical Activity in Dementia: Use of Spectral Power and Spectral Ratio Measures. *Electroencephalogr. Clin. Neurophysiol.* **1993**, *87*, 385–393. [CrossRef]
12. Hata, M.; Kazui, H.; Tanaka, T.; Ishii, R.; Canuet, L.; Pascual-Marqui, R.D.; Aoki, Y.; Ikeda, S.; Kanemoto, H.; Yoshiyama, K.; et al. Functional Connectivity Assessed by Resting State EEG Correlates with Cognitive Decline of Alzheimer's Disease—An ELORETA Study. *Clin. Neurophysiol.* **2016**, *127*, 1269–1278. [CrossRef]
13. Lombardi, A.; Tangaro, S.; Bellotti, R.; Bertolino, A.; Blasi, G.; Pergola, G.; Taurisano, P.; Guaragnella, C. A Novel Synchronization-Based Approach for Functional Connectivity Analysis. *Complexity* **2017**, *2017*, e7190758. [CrossRef]

14. Babiloni, C.; Lizio, R.; Marzano, N.; Capotosto, P.; Soricelli, A.; Triggiani, A.I.; Cordone, S.; Gesualdo, L.; Del Percio, C. Brain Neural Synchronization and Functional Coupling in Alzheimer's Disease as Revealed by Resting State EEG Rhythms. *Int. J. Psychophysiol.* **2016**, *103*, 88–102. [CrossRef]
15. Fell, J.; Axmacher, N. The Role of Phase Synchronization in Memory Processes. *Nat. Rev. Neurosci.* **2011**, *12*, 105–118. [CrossRef]
16. Bastos, A.M.; Schoffelen, J.-M. A Tutorial Review of Functional Connectivity Analysis Methods and Their Interpretational Pitfalls. *Front. Syst. Neurosci.* **2015**, *9*, 175. [CrossRef]
17. Cai, L.; Wei, X.; Wang, J.; Yu, H.; Deng, B.; Wang, R. Reconstruction of Functional Brain Network in Alzheimer's Disease via Cross-Frequency Phase Synchronization. *Neurocomputing* **2018**, *314*, 490–500. [CrossRef]
18. Lachaux, J.P.; Rodriguez, E.; Martinerie, J.; Varela, F.J. Measuring Phase Synchrony in Brain Signals. *Hum. Brain Mapp.* **1999**, *8*, 194–208. [CrossRef]
19. Yu, M.; Gouw, A.A.; Hillebrand, A.; Tijms, B.M.; Stam, C.J.; van Straaten, E.C.W.; Pijnenburg, Y.A.L. Different Functional Connectivity and Network Topology in Behavioral Variant of Frontotemporal Dementia and Alzheimer's Disease: An EEG Study. *Neurobiol. Aging* **2016**, *42*, 150–162. [CrossRef]
20. Engels, M.M.A.; Stam, C.J.; van der Flier, W.M.; Scheltens, P.; de Waal, H.; van Straaten, E.C.W. Declining Functional Connectivity and Changing Hub Locations in Alzheimer's Disease: An EEG Study. *BMC Neurol.* **2015**, *15*, 145. [CrossRef]
21. Bruña, R.; Maestú, F.; Pereda, E. Phase Locking Value Revisited: Teaching New Tricks to an Old Dog. *J. Neural Eng.* **2018**, *15*, 056011. [CrossRef]
22. Poza, J.; Bachiller, A.; Gomez, C.; Garcia, M.; Nunez, P.; Gomez-Pilar, J.; Tola-Arribas, M.A.; Cano, M.; Hornero, R. Phase-Amplitude Coupling Analysis of Spontaneous EEG Activity in Alzheimer's Disease. In Proceedings of the 2017 39th Annual International Conference of the IEEE Engineering in Medicine and Biology Society (EMBC), Jeju, Korea, 11–15 July 2017; Volume 2017, pp. 2259–2262. [CrossRef]
23. Gallego-Jutgla, E.; Elgendi, M.; Vialatte, F.; Sole-Casals, J.; Cichocki, A.; Latchoumane, C.; Jeong, J.; Dauwels, J. Diagnosis of Alzheimer's Disease from EEG by Means of Synchrony Measures in Optimized Frequency Bands. In Proceedings of the 2012 Annual International Conference of the IEEE Engineering in Medicine and Biology Society, San Diego, CA, USA, 28 August–1 September 2012; pp. 4266–4270. [CrossRef]
24. Elgendi, M.; Vialatte, F.; Cichocki, A.; Latchoumane, C.; Jeong, J.; Dauwels, J. Optimization of EEG Frequency Bands for Improved Diagnosis of Alzheimer Disease. In Proceedings of the 2011 Annual International Conference of the IEEE Engineering in Medicine and Biology Society, Boston, MA, USA, 30 August–3 September 2011; pp. 6087–6091. [CrossRef]
25. Jasper, H. The Ten-Twenty Electrode System of the International Federation. *Electroencephalogr. Clin. Neurophysiol.* **1958**, *10*, 370–375.
26. Rutkove, S.B. Introduction to Volume Conduction. In *The Clinical Neurophysiology Primer*; Humana Press: Totowa, NJ, USA, 2007; pp. 44–53.
27. Blackburn, D.; Zhao, Y.; De Marco, M.; Bell, S.; He, F.; Wei, H.-L.; Lawrence, S.; Unwin, Z.; Blyth, M.; Angel, J.; et al. A Pilot Study Investigating a Novel Non-Linear Measure of Eyes Open versus Eyes Closed EEG Synchronization in People with Alzheimer's Disease and Healthy Controls. *Brain Sci.* **2018**, *8*, 134. [CrossRef]
28. Yoon, S.; Alimardani, M.; Hiraki, K. The Effect of Robot-Guided Meditation on Intra-Brain EEG Phase Synchronization. In Proceedings of the Companion of the 2021 ACM/IEEE International Conference on Human-Robot Interaction, Boulder, CO, USA, 8–11 March 2021; ACM: New York, NY, USA, 2021; pp. 318–322. [CrossRef]
29. Leeuwis, N.; Yoon, S.; Alimardani, M. Functional Connectivity Analysis in Motor-Imagery Brain Computer Interfaces. *Front. Hum. Neurosci.* **2021**, *15*, 732946. [CrossRef]
30. Zhao, Y.; Zhao, Y.; Durongbhan, P.; Chen, L.; Liu, J.; Billings, S.A.; Zis, P.; Unwin, Z.C.; De Marco, M.; Venneri, A.; et al. Imaging of Nonlinear and Dynamic Functional Brain Connectivity Based on EEG Recordings With the Application on the Diagnosis of Alzheimer's Disease. *IEEE Trans. Med. Imaging* **2020**, *39*, 1571–1581. [CrossRef]
31. Durongbhan, P.; Zhao, Y.; Chen, L.; Zis, P.; De Marco, M.; Unwin, Z.C.; Venneri, A.; He, X.; Li, S.; Zhao, Y.; et al. A Dementia Classification Framework Using Frequency and Time-Frequency Features Based on EEG Signals. *IEEE Trans. Neural Syst. Rehabil. Eng.* **2019**, *27*, 826–835. [CrossRef]
32. Bates, D.; Mächler, M.; Bolker, B.; Walker, S. Fitting Linear Mixed-Effects Models Using Lme4. *J. Stat. Soft.* **2015**, *67*, 1–48. [CrossRef]
33. Acharya, J.N.; Acharya, V.J. Overview of EEG Montages and Principles of Localization. *J. Clin. Neurophysiol.* **2019**, *36*, 325–329. [CrossRef]
34. Trongnetrpunya, A.; Nandi, B.; Kang, D.; Kocsis, B.; Schroeder, C.E.; Ding, M. Assessing Granger Causality in Electrophysiological Data: Removing the Adverse Effects of Common Signals via Bipolar Derivations. *Front. Syst. Neurosci.* **2016**, *9*, 189. [CrossRef]
35. Rossini, P.M.; Di Iorio, R.; Vecchio, F.; Anfossi, M.; Babiloni, C.; Bozzali, M.; Bruni, A.C.; Cappa, S.F.; Escudero, J.; Fraga, F.J.; et al. Early Diagnosis of Alzheimer's Disease: The Role of Biomarkers Including Advanced EEG Signal Analysis. Report from the IFCN-Sponsored Panel of Experts. *Clin. Neurophysiol.* **2020**, *131*, 1287–1310. [CrossRef]
36. Sankari, Z.; Adeli, H.; Adeli, A. Intrahemispheric, Interhemispheric, and Distal EEG Coherence in Alzheimer's Disease. *Clin. Neurophysiol.* **2011**, *122*, 897–906. [CrossRef]

37. Yin, Z.; Li, J.; Zhang, Y.; Ren, A.; Von Meneen, K.M.; Huang, L. Functional Brain Network Analysis of Schizophrenic Patients with Positive and Negative Syndrome Based on Mutual Information of EEG Time Series. *Biomed. Signal Process. Control* **2017**, *31*, 331–338. [CrossRef]
38. Quraan, M.A.; McCormick, C.; Cohn, M.; Valiante, T.A.; McAndrews, M.P. Altered Resting State Brain Dynamics in Temporal Lobe Epilepsy Can Be Observed in Spectral Power, Functional Connectivity and Graph Theory Metrics. *PLoS ONE* **2013**, *8*, e68609. [CrossRef]
39. Adeli, H.; Ghosh-Dastidar, S.; Dadmehr, N. A Spatio-Temporal Wavelet-Chaos Methodology for EEG-Based Diagnosis of Alzheimer's Disease. *Neurosci. Lett.* **2008**, *444*, 190–194. [CrossRef]
40. Deng, B.; Cai, L.; Li, S.; Wang, R.; Yu, H.; Chen, Y.; Wang, J. Multivariate Multi-Scale Weighted Permutation Entropy Analysis of EEG Complexity for Alzheimer's Disease. *Cogn. Neurodyn.* **2017**, *11*, 217–231. [CrossRef]
41. Krajcovicova, L.; Marecek, R.; Mikl, M.; Rektorova, I. Disruption of Resting Functional Connectivity in Alzheimer's Patients and At-Risk Subjects. *Curr. Neurol. Neurosci. Rep.* **2014**, *14*, 491. [CrossRef]
42. Bari Antor, M.; Jamil, A.H.M.S.; Mamtaz, M.; Monirujjaman Khan, M.; Aljahdali, S.; Kaur, M.; Singh, P.; Masud, M. A Comparative Analysis of Machine Learning Algorithms to Predict Alzheimer's Disease. *J. Healthc. Eng.* **2021**, *2021*, e9917919. [CrossRef]
43. Liu, L.; Zhao, S.; Chen, H.; Wang, A. A New Machine Learning Method for Identifying Alzheimer's Disease. *Simul. Model. Pract. Theory* **2020**, *99*, 102023. [CrossRef]

Article

Use Electroencephalogram Entropy as an Indicator to Detect Stress-Induced Sleep Alteration

Yun Lo [1], Yi-Tse Hsiao [1,*] and Fang-Chia Chang [1,2,3,4,*]

[1] Department of Veterinary Medicine, School of Veterinary Medicine, National Taiwan University, Taipei 10617, Taiwan; f05629004@ntu.edu.tw
[2] Graduate Institute of Brian and Mind Sciences, College of Medicine, National Taiwan University, Taipei 10617, Taiwan
[3] Graduate Institute of Acupuncture Science, College of Chinese Medicine, China Medical University, Taichung 406040, Taiwan
[4] Department of Medicine, College of Medicine, China Medical University, Taichung 406040, Taiwan
* Correspondence: ythsiao@ntu.edu.tw (Y.-T.H.); fchang@ntu.edu.tw (F.-C.C.)

Abstract: An acute stressor can cause sleep disruptions. Electroencephalography (EEG) is one of the major tools to measure sleep. In rats, sleep stages are classified as rapid-eye movement (REM) sleep and non-rapid-eye movement (NREM) sleep, by different characteristics of EEGs. Sleep alterations after exposure to an acute stress are regularly determined by the power spectra of brain waves and the changes of vigilance stages, and they all depend on EEG analysis. Herein, we hypothesized that the Shannon entropy can be employed as an indicator to detect stress-induced sleep alterations, since we noticed that an acute stressor, the footshock stimulation, causes certain uniformity changes of the spectrograms during NREM and REM sleep in rats. The present study applied the Shannon entropy on three features of brain waves, including the amplitude, frequency, and oscillation phases, to measure the uniformities in the footshock-induced alterations of sleep EEGs. Our result suggests that the footshock stimuli resulted in a smoother and uniform amplitude as well as varied frequencies of EEG waveforms during REM sleep. In contrast, the EEGs during NREM sleep exhibited a smoother, but less uniform, amplitude after the footshock stimuli. The result depicts the change property of brain waves after exposure to an acute stressor and, also, demonstrates that the Shannon entropy could be used to detect EEG alteration in sleep disorders.

Keywords: acute stress disorder; electroencephalogram (EEG); footshock; information theory; sleep

1. Introduction

Electroencephalography (EEG) corresponds to several vigilance states of the brain. These states, such as wakefulness and sleep, are demonstrated by distinct brain waves [1,2]. Nevertheless, stress influences brain rhythms in both sleep and wakefulness in humans and rodents [3–6]. For instance, theta waves are predominant during an anxious condition in rodents [7–11]. Moreover, the literature, also, demonstrates that theta waves are dominant during learning and navigation in rodents [1,10,12]. In comparison to wakefulness, the function of electrical activities during sleep is much less known. There are two distinct states of sleep, rapid eye movement (REM) sleep and non-rapid-eye movement (NREM) sleep, and each sleep state has a unique EEG pattern. The delta powers of EEGs during NREM sleep are positively correlated with the level of sleep depth, implying the sleep quality. Stages 3 and 4 of NREM sleep generate large slow waves within the delta frequency band (0.5–4 Hz). Theta waves can be observed during REM sleep in rodents [1,2,10,13].

Diagnostic criteria for acute stress disorder (ASD) or post-traumatic stress disorder (PTSD) include hyperarousal conditions, such as insomnia or hypervigilance [14]. Stress is frequently associated with sleep disturbance (approximate 78%) in insomnia patients [15]. The studies for measuring stress-related sleep problem mainly count on the parameters of polysomnography, e.g., the duration and transitions between different states [5] and the powers of different frequency spectra [4,16,17]. EEG is a composition of distinct brain waves, which contain certain basic features, including amplitude, frequency, and oscillation phases. Some researchers link these features to potential physiological meanings. These features might be correlated and associated with physiological functions. For example, the amplitudes extracted from frequencies between 0.5 and 4 Hz are greater during NREM sleep, during which the amplitudes are positively correlated with sleep intensity [18,19]. The coupling of oscillation phases and frequencies in the hippocampus are elevated when rodents learn a new behavioral task [20,21]. These EEG features can also, solely, correlate with a physiological meaning. For example, instantaneous running speed is positively correlated with the frequencies of theta [11,22] and gamma waves in rats [23,24]. Therefore, the aim of the present study is investigating the features of EEGs during sleep after exposure to an acute stressor and revealing the stressor's potential effects on EEG features.

Measuring the spectrum of a period of EEGs, or counting densities of a unique waveform, is a regular method to study the alterations of EEGs [4,5,16,17]. However, EEG has a unique property: EEG amplitude decreases as the frequency increases. The inverse relationship between amplitude and frequency is due to the slower oscillation of EEGs, which involves larger numbers of neurons and generates stronger brain waves [1]. In other words, the distributions between the amplitude and frequency of EEG are non-linear [1]. Moreover, the features of EEGs, e.g., amplitude, frequency, and phase, are, also, non-linear or circular. For instance, the amplitude of EEGs as a function of time can be regard as a sine wave but not a straight line. Therefore, it is important to develop a method to describe these non-linear changes of EEGs, when animals come across a certain stimulation. In the present study, we focus on stress-induced EEG alterations. The Shannon entropy (information theory) can measure the uncertainty, by calculating the probability of a variable appearing [25]. It has the potential to detect non-linear changes of EEGs, since the Shannon entropy is related to probabilities and distributions but is not affected by the distribution of data. If the uncertainty level is high, which means the probability of appearance is even, the entropy would be high. This theory has been modified into several mathematical models in the fields of neuroscience, such as consciousness investigation and reconstruction of input signals (e.g., sensory input) to form action potentials of neurons [1]. Since the Shannon entropy depends on the distributions of occurrent rates for certain measuring variables, it can detect not only linear, but also non-linear, or even circular, changes. The aim of this study is to detect the changes of amplitude, frequency, and phase of EEGs. We prospected to apply the information theory to detecting the alterations of EEG features. Although rodent study has demonstrated that stress increases theta powers during a vigilance state [8], there are still few studies investigating the post-effect of stress on the features of EEGs. In the present study, we calculated the entropies of the amplitudes, frequencies, and oscillation phases of sleep EEGs and compared these entropies between naïve and footshock-stimulated rats. The probabilities of the appearance of amplitude, frequency, and oscillation phases, between the control and footshock-stimulated rats, are quantified by Shannon entropy. The analysis methods and results may provide a distinct way for detecting EEG alterations after exposure to a stressor.

2. Materials and Methods

2.1. Animals

This study included 34 male Sprague-Dawley rats (250–300 g; BioLASCO Co., Ltd., Taipei, Taiwan). These rats were housed individually in their home cages, with a consistent temperature of 23 ± 1 °C. The circadian rhythm was controlled in a 12:12 h light:dark cycle (with 40 watts $\times$ 4 tubes illumination). For minimizing interference of the circadian rhythm, all the experimental procedures and daily care were manipulated 30 min prior to the light period. Food and water were available ad libitum. All procedures performed in this study were approved by the National Taiwan University Animal Care and Use Committee.

2.2. Surgery

The subjects were anesthetized with 50 mg/kg ZoletilTM (Tiletamine:Zolazepam = 1:1; Virbac, Carros, France) and surgically implanted with two EEG screw electrodes (Plastics One, Roanoke, VA, USA). The EEG screw electrodes were implanted on the skull, and the tips of screws were placed on the surface of the cortex for recording. The coordinates for EEG electrode implantation are as follows: one electrode is on ML: 2.5 mm and AP: 2.0 mm relative to bregma, and the other one is on ML: -3.0 mm and AP: -11.5 mm relative to bregma. Insulated leads from EEG electrodes were routed to a Teflon pedestal (Plastics One), and, then, cemented to the skull with dental acrylic (Tempron, GC Co., Tokyo, Japan). After surgery, penicillin G (5000 IU; Sigma-Aldrich, St. Louis, MO, USA) was administered systemically, and the incision was treated topically with polymyxin B sulphate-bacitracin zinc. Ibuprofen was dissolved in drinking water (140 mg/250 mL) for rats to relieve pain. Animals were allowed to recover for seven days before proceeding with the experiments. The antibiotic and analgesic were also dissolved in the drinking water for seven days during the recovery. In order to reduce the influence of sleep by the tether, the tether was plugged into the pedestal four days before the experiment and was only unplugged when performing footshocks.

2.3. Apparatus

Foot electrical stimuli were given by a custom-made footshock stimulation box (40 cm × 22 cm × 29 cm). The box randomly generated electrical stimuli 12 times within 10 min. The intensity of each stimulus current was 0.5 mA, and the stimulation duration was 50 ms. No extra cues or escapable places were provided in the footshock stimulation box.

Signals from the EEG electrodes were amplified at a factor of 5000 and analogue filtered between 0.1 and 40 Hz (frequency response: ±3 dB; filter frequency roll off: 12 dB/octave) by an amplifier (model V75-01; Colbourn Instruments, Lehigh Valley, PA, USA). Gross body movements were detected by infrared-based motion detectors (Biobserve GmbH, Bonn, Germany), and the movement activity was converted to a voltage output that was digitized and integrated into 1-s bins. The EEGs and gross body movements were subjected to analogue-to-digital conversion, with 16-bit precision at a sampling rate of 128 Hz (NI PCI-6033E; National Instruments, Austin, TX, USA).

2.4. Experimental Procedure

After recovery, the animals were randomly divided into the control ($n = 18$) and footshock ($n = 16$) groups. Rats in the footshock group received footshocks 12 times within 10 min, as described in Section 2.3. Upon completion of the footshock stimulation, each rat was moved back to its recording cage, and the tether was plugged into the pedestal again. We executed the footshock protocol during the last 10 min of the dark period; thus, the sleep EEGs were acquired from the subsequent resting (light) period. The rats in the control group stayed in their recording cages, while their sleep EEGs were recorded during the light period. The digitized EEG waveforms and integrated values of body movement were stored as binary files, pending for subsequent analyses.

2.5. Data Analysis
2.5.1. Analysis of the Vigilance States

The distinct vigilance states were categorized by visually scoring 12 s epochs of EEGs using custom software (ICELUS, M. R. Opp) written in LabView (National Instruments, Austin, TX, USA). The sleep–wake state was classified as either NREM sleep, REM sleep, or wakefulness, based on previously defined criteria (Chang and Opp, 1998). NREM sleep is characterized by a large amplitude of EEG slow waves, high-power-density values in the delta frequency band, and lack of gross body movements. During REM sleep, the amplitude of EEG is reduced, the predominant EEG power density occurs within the theta frequency, and there are phasic body twitches. During wakefulness, the rats are, generally, active, with protracted body movements. The amplitude of the wakefulness EEG is similar to that observed during REM sleep, but mixed with frequencies that were high than the theta band.

2.5.2. EEG Parameter Investigation

The program for analyzing EEGs in the aspects of spectrograms for the amplitude, frequency, and oscillation phases was designed by using custom-written scripts in MATLAB R2016b (MathWorks, Natick, MA, USA) or combining with open-sourced codes. To match up the epochs of EEG analyses and vigilance states, the raw EEGs were also cut into non-overlapping 12 s epochs and, then, the spectrum and the entropies of the amplitude, frequency and phase were calculated. The analyzed data were subsequently classified into NREM sleep, REM sleep, or wakefulness, based on the criteria of the vigilance states (Section 2.5.1). A 12 s resolution is not appropriate for representing 12 h sleep–wake states, given that certain vigilance states, normally, do not occur during certain times (e.g., REM sleep at the very first 12 s epoch). Therefore, we collected 75 epochs of 12 s bins and averaged them to generate the time resolution of 15 min. The analyzed data were excluded, if no certain vigilance state was found. The methods for measuring the spectrum, Shannon entropy, instantaneous amplitude, instantaneous frequency, and instantaneous phase were specified as follows.

Spectrum

The spectrogram was computed using the multitaper method from the open-source MATLAB toolbox Chronux (version 2.10) [26]. The time–bandwidth product parameter was set at 3, and the number of tapers was set at 5. The power for each frequency was subsequently z-scored, to minimize the difference of absolute power between different subjects.

Instantaneous Amplitude, Frequency, and Phase

Before analyzing the entropies of amplitude, frequency, and phase, the instantaneous amplitude, frequency, and phase of every sampling point were calculated first. To be specific, the raw EEGs were converted by Hilbert transformation. This transformation gave complex numbers for every sampling point. The instantaneous amplitudes were obtained by calculating the absolute values of the complex numbers [27]. Similarly, the instantaneous phases were obtained by calculating the angles of the complex numbers. For the instantaneous frequencies, the raw EEG data were filtered between 0.5 to 30 Hz from a Hamming window and, then, converted by Hilbert transformation. Subsequently, the differences in the phases between sampling points were then computed, by measuring the angles of complex numbers. By determining the differences of the phases within the delta time (in our setup, the sampling rate was 128 Hz), the instantaneous frequencies can be acquired [11]. For instance, a difference of pi/128 gives an instance frequency of 2 Hz.

Shannon Entropy

The Shannon entropy was used to compare the differences of EEG waveforms after being exposed to the footshock. The Shannon entropy equation is as follows [25].

$$H(X) = -\sum_{i=1}^{n} P(x_i) \log_2 P(x_i)$$

The entropy H(X) is the summation of $P(x_i)*\log_2 P(x_i)$, and the $P(x_i)$ is the probability of occurrence of event x under i circumstance. In our study, we applied the equation to measuring the entropies of amplitude, frequency, and phase. We, firstly, sorted the variables (i.e., amplitude, frequency, or phase) into n levels (nAmplitude = 15, nFrequency = 25, nPhase = 12). For amplitude entropy, the z-scored amplitudes between −0.4 and 1.4 were sorted into 15 levels, as we noticed most amplitudes were within this range. For frequency entropy, frequencies between 0.5 and 30 Hz (the range of predominant frequencies for sleep and wakefulness) were sorted into 25 levels. For phase entropy, the phases between 0 and

2*pi were sorted into 12 levels. The numbers or levels were related to data distribution and the number of data points, so we used the Freedman–Diaconis rule to determine them [28].

$$nlevels = \frac{\max(x) - \min(x)}{2Q_x n^{\frac{-1}{3}}}$$

Q_x is the 25th and the 75th percentiles of the data X, n is the total number of data points, and max(x) and min(x) are the maximum and minimum values of the data X. We, then, calculated the occurrence rate of each level during each 12 s epoch and obtained the values of PAmplitude(x_i), PFrequency(x_i), and PPhase(x_i), and, finally, calculated the entropies of the amplitude, frequency, and phase using the Shannon entropy equation.

2.6. Statistics

A statistically significant difference was indicated by one-way ANOVA, with an alpha level of $p < 0.05$, using the software SPSS (Version: 10.0.7, IBM, New York, NY, USA).

3. Results

3.1. Alterations of Sleep Duration and Spectrogram after Footshock

The aim of the present study is to investigate sleep EEGs, after exposure to an acute stressor. Electrical footshock stimulation is a common protocol, for creating an animal model to mimic stress-related diseases and the subsequent sleep alterations in humans [3]. Therefore, the sleep EEGs obtained from the control (n = 18) and footshock (n = 16) groups were analyzed and compared. In the footshock group, the rats received 12 footshock stimuli within 10 min before the light period. EEGs were acquired for 12 h, after the rats returned to their home cages. We determined whether the footshock protocol altered the sleep patterns of the subjects. By calculating the duration of the vigilance states (Figure 1), we found that the inescapable and random footshock stimuli only decreased the duration of NREM sleep for the first two hours of the light period (Figure 1A), but the mean duration during the light period did not demonstrate a significant difference between the footshock and control groups (Figure 1B, control vs. footshock: 7.19 ± 0.11 vs. 6.85 ± 0.14 (min), F(1,1393) = 3.48, p = 0.065). However, the durations of REM sleep (Figure 1C,D) and wakefulness (Figure 1E,F) were significantly altered; the mean duration during the light period was reduced in REM sleep (Figure 1D, control vs. footshock: 3.03 ± 0.08 vs. 2.54 ± 0.09 (min), F(1,1393) = 16.60, p < 0.01) and was enhanced in wakefulness (Figure 1F, control vs. footshock: 4.74 ± 0.15 vs. 5.58 ± 0.19 (min), F(1,1393) = 12.33, p < 0.01). The results indicated that the present footshock protocol affected the subsequent sleep behavior in the rats. We then analyzed the spectrum of EEG between different sleep–wake states.

Figure 2A,B represented the EEG spectrograms of NREM sleep for the control and footshock group, respectively. Figure 2A,B revealed, approximately, 5 h of strong delta EEG powers at the beginning of light period, which gradually decreased. This phenomenon demonstrated that both the control and footshock groups entered deeper sleep stages because the delta power of EEGs during NREM sleep reflects the depth of sleep [18,19]. We, further, determined the statistical differences between Figure 2A,B. The footshock significantly decreased the power of slow waves (1–3 Hz) during the first 2 h (the black-dashed-line area of Figure 2C). Although the averaged power of frequencies from 1 to 6 Hz during the first 5 h of light period were stronger in the footshock group, they still did not reach statistical significance (the red-dashed-line area of Figure 2C). Interestingly, the significances were mainly in frequencies between 6 and 12 Hz (the green-dashed-line area of Figure 2C), which are, predominantly, theta frequencies during paradoxical sleep in rats [13]. Theta rhythms mainly occur during REM sleep; thus, we, next, determined the spectrogram during REM sleep (Figure 2D,E). Similar to the spectrogram of NREM sleep, the footshock significantly enhanced the power of the theta band near the second hour of light period (the black-dashed-line area of Figure 2F). We were, also, interested in the EEG features during their wakefulness, since Figure 1E showed the increases of wakefulness, especially during the first 2 h. The spectrograms (Figure 2G,H) and panel of significance test (Figure 2I) depicted an enhancement of high frequencies (25–30 Hz) during the first 2 h after the footshock stimuli (the black-dashed-line area of Figure 2I). From the result of the spectrogram analysis, we postulated that the distinct EEG profiles would be clearer if we measured entropies of the EEG features because we noticed the consistent alterations of EEGs between the control and footshock groups. For instance, the control group had a wider theta band than those in the footshock group (Figure 2D vs. Figure 2E, the arrow area demonstrated the power intensity). Since entropy is only related to the probabilities and distributions of the data, it can detect both linear and non-linear changes. Moreover, entropy is insensitive to the effect of the power law (i.e., higher frequency shows lower power), so it may reveal some underlying changes of EEGs.

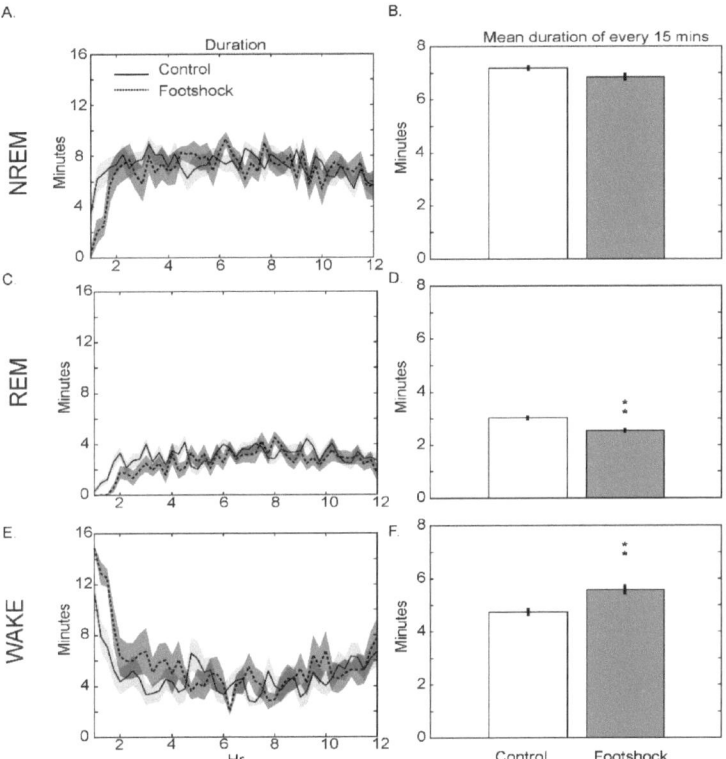

Figure 1. Comparison of the duration between distinct vigilance stages. (**A**) NREM sleep duration. The solid line and light gray shadow represent the values of means ± SEMs, for every 15 min during the light period in the control group. The dashed line and dark gray shadow are the values acquired from the footshock group. (**B**) Mean duration of 15 min bins across the total 12 h light period. The white bar is the value of mean ± SEM obtained from the control grou,p and the gray bar represents data acquired from the footshock group. (**C,D**) Represented the data from REM sleep. (**E,F**) Demonstrated the result of wakefulness. ** Indicates the difference reaches statistical significance of $p < 0.01$.

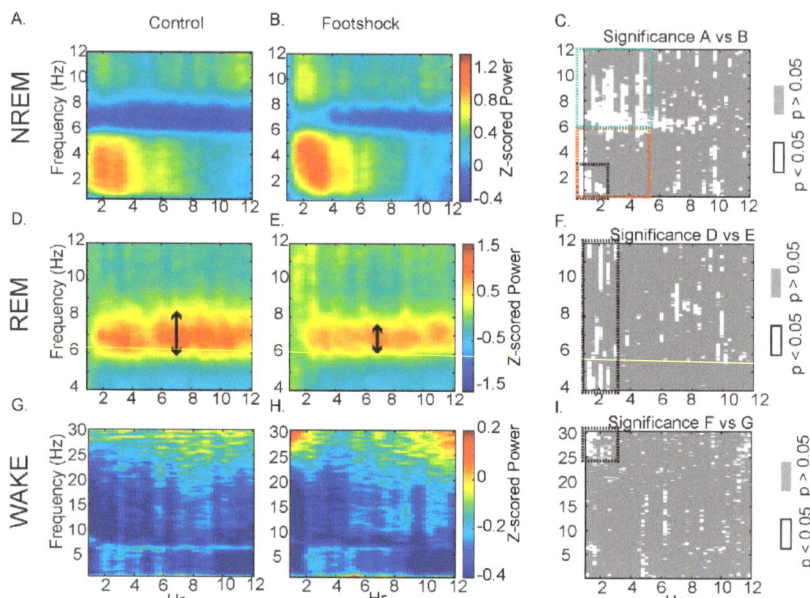

Figure 2. Alterations of spectrogram and statistical significance testing after the footshock stimuli. (**A**) Mean spectrogram of NREM sleep obtained from the control group. (**B**) Mean spectrogram of NREM sleep acquired from the footshock group. (**C**) Statistical significance testing for the difference between (**A**,**B**). The black-dashed-line area marks 1–3 Hz, the red-dashed-line area marks 1–6 Hz, the green-dashed-line area marks 6–12 Hz. (**D**) Mean spectrogram of REM sleep obtained from the control group. (**E**) Mean spectrogram of REM sleep observed from the footshock group. The arrows in (**D**,**E**) represent a wider predominant frequency band in the control than that obtained from the footshock. (**F**) Statistical significance testing for the difference between (**D**,**E**). The black-dashed-line area marks 4–12 Hz. (**G**) Mean spectrogram of wakefulness acquired from the control group. (**H**) Mean spectrogram of wakefulness recorded from the footshock group. (**I**) Statistical significance testing for the difference between (**G**,**H**). The black-dashed-line area marks 25–30 Hz. The *p*-value applied was $p = 0.05$; $p < 0.05$ was marked by the white color.

3.2. Amplitude Entropy

We, firstly, investigated the amplitude entropies between the control and the footshock groups (Figure 3). Figure 3A illustrates that a uniform amplitude of the wave results in a low entropy. The distributions of amplitude entropy in NREM sleep (Figure 3B) demonstrated that the footshock group (dashed line) had a higher entropy than the control group (solid line). We further analyzed the 15 min mean amplitude entropy and showed an increase in amplitude entropy after the footshock stimuli (Figure 3C, control vs. footshock: 2.491 ± 0.011 vs. 2.544 ± 0.010, $F(1,1394) = 137.60$, $p < 0.01$). In contrast, the footshock stimuli significantly decreased the amplitude entropy during REM sleep (Figure 3D,E, control vs. footshock: 2.63 ± 0.003 vs. 2.529 ± 0.003, $F(1,1226) = 47.46$, $p < 0.01$). The amplitude entropy demonstrated no significant change between the control and footshock groups during wakefulness (Figure 3F,G). This result suggests that the footshock stimuli potentiated the variation of EEGs during NREM sleep but attenuated the variation of EEGs in REM sleep.

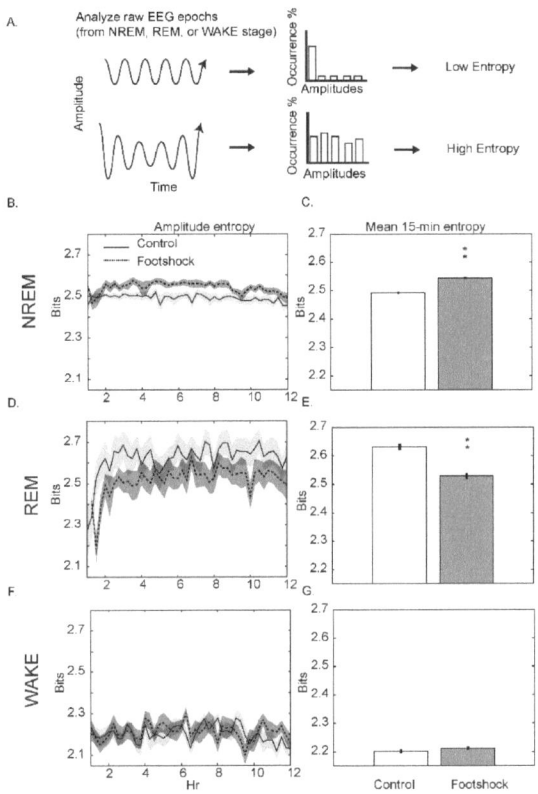

Figure 3. Comparison of the amplitude entropy after the footshock stimuli. (**A**) Uniform amplitudes of EEGs result in a low entropy, whereas the variety of amplitudes causes uncertainty of occurrence rate and results in a high entropy. (**B**) Distribution of amplitude entropy for NREM sleep. The solid line and light gray shadow are the values of means ± SEMs for every 15 min during the light period, obtained from the control group. The dashed line and dark gray shadow are results acquired from the footshock group. (**C**) Statistical significant test for the amplitude entropy. The white bar represents the value of mean ± SEM for 15 min bins across the total 12 h light period, obtained from the control group, and the gray bar depicts the amplitude entropy acquired from the footshock group. (**D**,**E**) Represented the data of REM sleep. (**F**,**G**) Demonstrated the results for the wakefulness. ** Indicates the difference reaches statistical significance of $p < 0.01$.

3.3. Frequency Entropy

We, further, calculated the frequency entropy of EEGs during sleep–wake states (Figure 4). The entropy elevated when the EEG contained various instantaneous frequencies (Figure 4A–G). During NREM sleep (Figure 4B,C), we noticed a rising frequency entropy as a function of time, no matter whether in the control or footshock groups. This finding suggests that the composition of frequencies in NREM sleep gradually became more complex as the resting time (light period) moved toward the active time (dark period). Regarding REM sleep (Figure 4D,E) and wakefulness (Figure 4F,G), the frequency entropies significantly increased after the footshock stimuli (Figure 4E, REM sleep: control vs. footshock: 3.623 ± 0.003 vs. 3.665 ± 0.003, $F(1,1199) = 109.71$, $p < 0.01$; Figure 4G, wakefulness: control vs. footshock: 3.667 ± 0.002 vs. 3.688 ± 0.002, $F(1,1513) = 66.91$, $p < 0.01$). These results implied that the footshock stimuli increased the complexity of instantaneous frequencies during REM sleep and wakefulness.

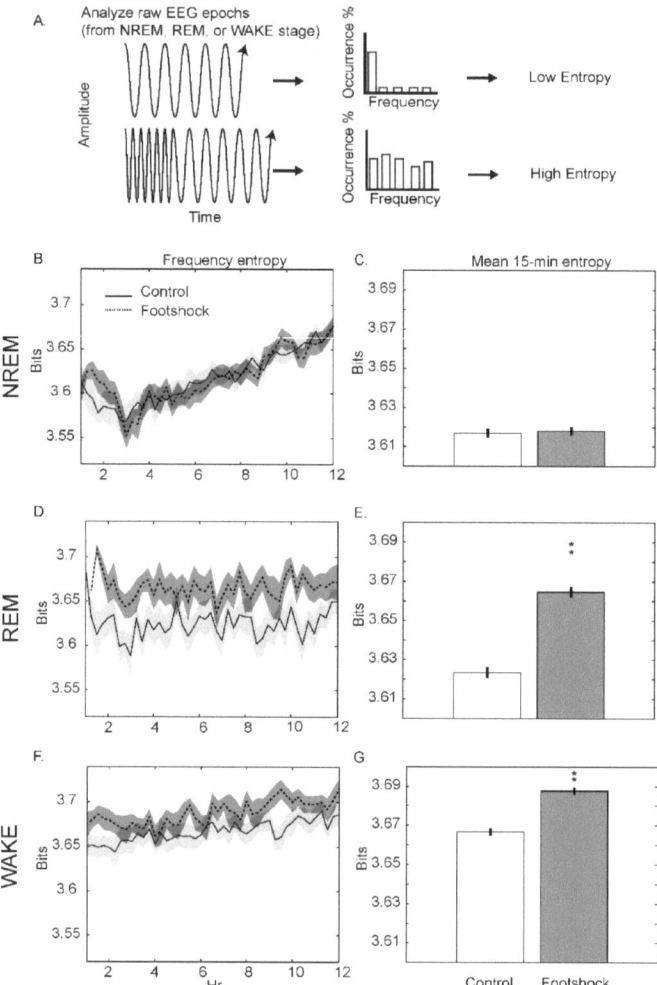

Figure 4. Comparison of the frequency entropy after the footshock stimuli. (**A**) Uniform frequencies of EEGs result in a low entropy, whereas the variety of frequencies causes an uncertainty of the occurrence rate and results in a high entropy. (**B**) Distribution of frequency entropy for NREM sleep. The solid line and light gray shadow represent the values of means ± SEMs for every 15 min during the light period, obtained from the control group. The dashed line and dark gray shadow are the results acquired from the footshock group. (**C**) Statistical significant test for the frequency entropy. The white bar is the value of mean ± SEM for 15 min bins across the total 12 h light period, obtained from the control group, and the gray bar represents the frequency entropy acquired from the footshock group. (**D,E**) Represented the data for REM sleep. (**F,G**) Demonstrated the results for wakefulness. ** Indicates the difference reaches statistical significance of $p < 0.01$.

3.4. Phase Entropy

The phase entropies between the control and footshock groups were compared in Figure 5. The phase entropy reflects the shape of EEG waveforms; that is, smooth sine-wave-like oscillations result in an even phase distribution and a higher phase entropy (Figure 5A). By analyzing the phase entropy, we noticed that the footshock enhanced the phase entropies during NREM sleep (Figure 5B,C), REM sleep (Figure 5D,E) and wakefulness (Figure 5F,G). The 15-min mean phase entropies were significantly higher after the footshock stimuli

(Figure 5C, NREM sleep: control vs. footshock: 3.5791 ± 0.0001 vs. 3.5801 ± 0.00003, $F(1,1394) = 86.21$, $p < 0.01$; Figure 5E, REM sleep: control vs. footshock: 3.5793 ± 0.0001 vs. 3.5807 ± 0.00004, $F(1,1226) = 169.02$, $p < 0.01$; Figure 5G, wakefulness: control vs. footshock: 3.5795 ± 0.0001 vs. 3.5805 ± 0.00003, $F(1,1473) = 77.72$, $p < 0.01$). These results suggest that the waveforms became smoother after receiving the footshock stimuli when compared with those obtained from the control group. In summary, the footshock stimuli profoundly affected the duration, EEG amplitude, frequency and oscillation phase during the distinct vigilance states.

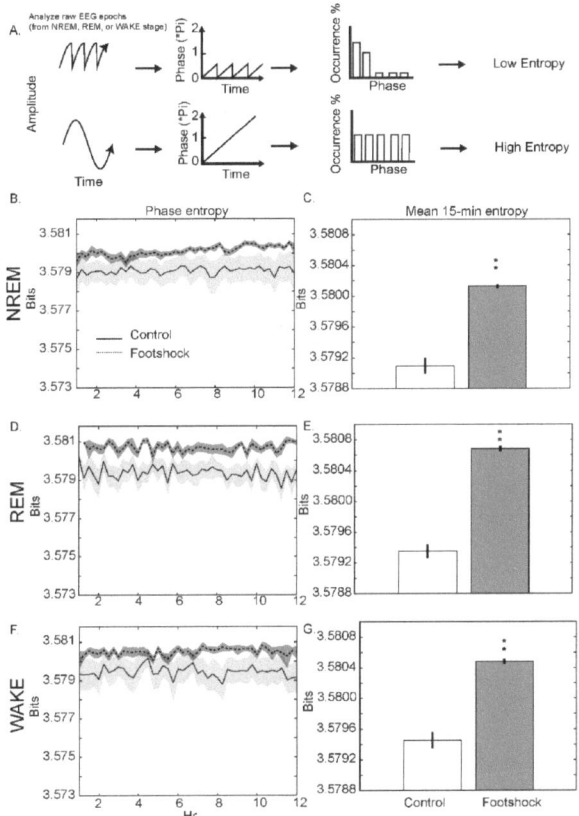

Figure 5. Comparison of the phase entropy after the footshock stimuli. (**A**) Abrupt and edge waveforms result in a low entropy whereas smooth sine-wave-like waveforms cause uncertainty of occurrence rate and results in a high entropy. (**B**) Distribution of phase entropy obtained from NREM sleep. Solid line and light gray shadow are the values of means ± SEMs for every 15 min of light period obtained from the control group. Dashed line and dark gray shadow are results acquired from the footshock group. (**C**) Statistical significant test for the phase entropy. White bar is the value of mean ± SEM for 15-min bins across the total of 12-h light period obtained from the control group and gray bar represents the phase entropy acquired from the footshock group. (**D,E**) represented the data of REM sleep. (**F,G**) demonstrated the results for wakefulness. ** Indicates the difference reaches statistical significance of $p < 0.01$.

4. Discussion

4.1. Footshock Reduces REM Sleep Duration

The present study demonstrated that an acute stressor not only altered the duration of sleep stages, but also affected the EEG features during the subsequent sleep–wake states.

We showed that the random and inescapable footshock stimuli reduced NREM sleep in the first hour and it changed the quantity of distinct vigilance states during a quarter hour; that is, REM sleep was decreased and wakefulness was increased. Further investigating the values, we found that the increased wakefulness mostly contributed to the changes in the composition of sleep–wake states. The wakefulness duration increased 0.84 min and REM sleep duration decreased approximately 0.5 min in a unit of 15 min. This finding is similar to other report which used the footshock to create a rodent model of stress-induced sleep disorder [29]. We postulated that our protocol employing the footshock stimuli partially simulates the stress-induced sleep problems in humans, such as acute stress disorder (ASD) or post-traumatic stress disorder (PTSD). Hyper-arousal is one of the main symptoms of ASD and PTSD [14]. Although most of the clinical studies report a reduction in NREM sleep in patients with PTSD, the alterations in REM sleep are still inconclusive [30]. A meta-analysis report states that the percentage of REM sleep decreases in PTSD patients with age below 30-year-old and affects less in patients older than 30 [30]. Even though the inescapable footshock is not specific as the PTSD model in rodents [3], our data suggest the hypothesis of decreased REM sleep in PTSD patients.

4.2. The EEG Spectrum after Footshock

Spectrum is a common way for analyzing the power as a function of frequency. The spectrum between the control and footshock group did not show extensive changes. If we focus on the x-axis (time), the spectrum demonstrates the alterations mostly occurred during the first 2 h of the light period; the power of delta band (0.5–4 Hz) was attenuated in NREM sleep, theta power (4–12 Hz) during REM sleep was increased, and gamma power (25–30 Hz) in wakefulness was also increased. If we focus on the y-axis (frequency) of each stage, the footshock enhanced theta power in NREM sleep, although delta power is the predominant frequency during NREM sleep. Since the type-1 (6–12 Hz) and type-2 (4–9 Hz) [10,31] theta bands contain an overlap frequency (6–9 Hz) in rats, we can hardly differentiate which type(s) of theta band(s) was (were) mainly affected without pharmacological approaches. However, we noticed the enhanced theta was very likely the type-1 since it contained 9 to 12 Hz frequencies as we demonstrated in Figure 2C. Regarding to REM sleep, Figure 2F demonstrated the increases of type-1 and type-2 theta rhythms, and they mainly occurred during the first 2 h of sleep. Studies depicts that type-1 theta rhythm was correlated to movement and exploration [32] and type-2 theta oscillation was observed during a predator presence [10,31]. Regarding to the wakefulness, slow gamma (25–30 Hz) power was increased after footshock. Slow gamma is known to be generated when rats are retrieving memories [33]. Taken the presence of theta and gamma oscillations together, we proposed that the aversive memories from footshock cause sleep disturbances during the subsequent light period. This phenomenon may link to the occurrence of nightmare in ASD and PTSD patients [14]. However, this hypothesis needs to be further investigated in the future.

4.3. Amplitude Entropy

Using Fourier transform to measure the EEG features is a standard method for analyzing polysomnography findings during sleep. Fourier transform is temporally dependent. In Figure 2 we observed a linear correlation with the time variable by using Fourier transform. For instance, the EEG power of NREM sleep gradually attenuated as the function of time. We further noticed some nonlinear correlations. For example, the power for frequencies of each temporal bin (15 min) is not linear and the width of theta frequency during REM sleep was narrower in the footshock group than that in the control group. Therefore, using the Shannon entropy can compromise this limitation. We thoroughly measured the entropies of instantaneous amplitudes, frequencies, and phases to depict the EEG profiles after the footshock stimuli. The footshock stimuli increased the amplitude entropy of NREM sleep, but decreased the amplitude entropy of REM sleep, which suggests that the footshock affected the uniformity of amplitude. During NREM sleep, the amplitudes of EEGs were changed more frequently after the footshock because their amplitude entropy increased. These results can be interpreted by two possible explanations. One possibility is the poor sleep quality affected by the footshock stimuli, given that the amplitude of EEGs during NREM sleep corresponds to sleep quality. The various amplitudes may reflect that the rats entered various NREM sleep stages after the footshock stimuli, although it is difficult to further differentiate the sleep stages of NREM sleep in rodents compared to humans. The other possible explanation is that there may be certain sleep compensation for the first hour of sleep deficit after the footshock. This sleep compensation might result in greater amplitudes, indicating the sleep quality, rather than the regular baseline and the greater amplitudes, contributed to the lower uniformity of the amplitude. In contrast, the footshock stimuli caused a low entropy of amplitudes during REM sleep, which may result from the weak power spectrum and the smaller amplitudes of REM EEGs, as indicated in Figure 2D,E. We believe the spectrogram of REM sleep in Figure 2D,E reflects this finding; that is, the power of theta waves between 6 to 8 Hz decreased after the footshock stimuli, although it did not reach significant meaning. Some human [34] and rodent studies [4], also, report that PTSD patients or susceptible animals have a reduction in theta powers during REM sleep. Weak amplitudes contributed to lower amplitude entropies because the difference of amplitudes at different levels is low and would be sorted into the same level, which would, subsequently, exhibit a low entropy.

4.4. Frequency Entropy

Although the classification of brain oscillations was defined by certain ranges of frequency, the dominant frequencies were not always the same in the animals' behaviors. For instance, the frequencies of the theta band are positively correlated with the running speed of rats [11,22–24]. This finding implies that the behavior not only affects the power of brain waves, but also affects the frequencies when a brain is processing the information. It, also, means that frequencies can be a measurable variable. EEGs are mixed with more than one frequency band and are dominated by one frequency band for a particular behavior or event. For instance, gamma oscillations could nest in the theta waves [35], and theta waves are predominant during REM sleep in rodents [1,2]. High entropy of instantaneous frequency reflects the oscillations of various frequency bands. We still do not know the physiological relevance of high entropy frequency, but we hypothesized that a complex composition of frequencies corresponds to an un-resting and busy brain. Evidence in Figure 4 supports our hypothesis. The first, the tendency of entropy matched spectral power of NREM sleep, represents the depth of sleep [18,19]. We found that the deeper the sleep, the lower the frequency entropy, especially during the first few hours of the light period. In humans, the deeper stage of sleep, called slow wave sleep, shows a uniform and synchronized slow frequency [2,18,19]. The second, the frequency entropies, were highest in wakefulness and lowest in NREM sleep, which suggested that frequency entropy is positively correlated with the level of vigilance states. Therefore, based on the aforementioned hypothesis, we demonstrated that the footshock stimuli reduced REM sleep intensity because the frequency entropy of REM sleep was higher after the footshock. Moreover, the brain activities during wakefulness were also increased after the footshock because higher frequency entropy was observed. We postulated that the footshock-induced changes in frequency entropy mimic the symptom of hyperarousal in ASD or PTSD patients.

4.5. Phase Entropy

We, next, analyzed the phase features of EEGs. An evenly distributed phase generates a sine-wave-like waveform and results in a high-phase entropy. Our data demonstrated that the footshock increased phase entropy in NREM sleep, REM sleep, and wakefulness, which suggests a smoother and uniform EEG waveform after the footshock stimuli. The physiological meaning of low/high phase entropy is still unclear. Sleep spindles (found in both rodents and humans) and sawtooth waves (mainly found in humans) are abrupt and edge EEG waveforms during sleep [4,36], which may cause lower phase entropy.

Therefore, we hypothesized that an acute stressor may disrupt spindles or sawtooth waves. In rodents, sleep spindles occur during the transition from NREM sleep to REM sleep [4]. Researchers propose that the sleep spindle facilitates memory consolidations and eliminates unnecessary memory [37]. A rodent PTSD model demonstrates that the decrease in sleep spindles occurs when rats are exposed to a single prolonged stress [4]. On the other hand, the sawtooth waves are observed during REM sleep in normal humans, and the density of the sawtooth waves increases after the first sleep cycle [36]. The function of sawtooth waves is also unclear. We hypothesized that a normal waveform of sleep EEGs should contain sharp and edge waves, and the footshock stimuli may decrease this kind of waveform, reflecting the ability of forgetting aversive memory. However, this hypothesis needs to be further confirmed in the future.

5. Conclusions

The alterations of EEGs after exposure to the footshock stimuli can be determined by the Shannon entropy. The entropies suggest that the footshock resulted in a smoother, lower amplitude variation, and more frequent variation of EEG waveforms in REM sleep. The EEGs during NREM sleep were smoother, and the amplitude variation was higher, after the footshock stimuli. The Shannon entropy (information theory) could be applied to detect the alterations of sleep EEGs in stress-related disorders such as ASD and PTSD.

Author Contributions: The manuscript was written by Y.L., Y.-T.H. and F.-C.C. Data were analyzed and collected by Y.L. and Y.-T.H. The experiments were designed by Y.-T.H. and F.-C.C. All authors have read and agreed to the published version of the manuscript.

Funding: This work was supported by the Ministry of Science and Technology, Taiwan (108-2320-B-002-074, 109-2320-B-002-023-MY2) and National Taiwan University (111L892302).

Institutional Review Board Statement: The animal study protocol was approved by the National Taiwan University Animal Care and Use Committee, ID: B201800191, June 2018.

Informed Consent Statement: Not applicable.

Data Availability Statement: Data are available based upon request, by contacting the corresponding author.

Acknowledgments: The authors thank the Ministry of Science and Technology (108-2320-B-002-074, 109-2320-B-002-023-MY2) and National Taiwan University (111L892301 and 111L892302) for their support.

Conflicts of Interest: The authors declare no conflict of interests.

Abbreviations

ASD	Acute stress disorder
EEG	Electroencephalography
NREM	Non-rapid-eye movement
PTSD	Post-traumatic stress disorder
REM	Rapid-eye movement

References

1. Buzsaki, G. *Rhythms of the Brain*; Oxford University Press: Oxford, UK; New York, NY, USA, 2006.
2. Jouvet, M. Neurophysiology of the states of sleep. *Physiol. Rev.* **1967**, *47*, 117–177. [CrossRef] [PubMed]
3. Vanderheyden, W.M.; Poe, G.R.; Liberzon, I. Trauma exposure and sleep: Using a rodent model to understand sleep function in PTSD. *Exp. Brain Res.* **2014**, *232*, 1575–1584. [CrossRef] [PubMed]
4. Vanderheyden, W.M.; George, S.A.; Urpa, L.; Kehoe, M.; Liberzon, I.; Poe, G.R. Sleep alterations following exposure to stress predict fear-associated memory impairments in a rodent model of PTSD. *Exp. Brain Res.* **2015**, *233*, 2335–2346. [CrossRef] [PubMed]
5. Kim, E.J.; Dimsdale, J.E. The effect of psychosocial stress on sleep: A review of polysomnographic evidence. *Behav. Sleep Med.* **2007**, *5*, 256–278. [CrossRef] [PubMed]

6. Thibodeau, R.; Jorgensen, R.S.; Kim, S. Depression, anxiety, and resting frontal EEG asymmetry: A meta-analytic review. *J. Abnorm. Psychol.* **2006**, *115*, 715–729. [CrossRef]
7. Vanderwolf, C.H. Hippocampal electrical activity and voluntary movement in the rat. *Electroencephalogr. Clin. Neurophysiol.* **1969**, *26*, 407–418. [CrossRef]
8. Hsiao, Y.T.; Yi, P.L.; Cheng, C.H.; Chang, F.C. Disruption of footshock-induced theta rhythms by stimulating median raphe nucleus reduces anxiety in rats. *Behav. Brain Res.* **2013**, *247*, 193–200. [CrossRef]
9. Seidenbecher, T.; Laxmi, T.R.; Stork, O.; Pape, H.C. Amygdalar and hippocampal theta rhythm synchronization during fear memory retrieval. *Science* **2003**, *301*, 846–850. [CrossRef]
10. Lever, C.; Kaplan, R.; Burgess, N. The Function of Oscillations in the Hippocampal Formation. In *Space, Time and Memory in the Hippocampal Formation*; Derdikman, D., Knierim, J.J., Eds.; Springer Vienna: Vienna, Austria, 2014; pp. 303–350.
11. Wells, C.E.; Amos, D.P.; Jeewajee, A.; Douchamps, V.; Rodgers, J.; O'Keefe, J.; Burgess, N.; Lever, C. Novelty and anxiolytic drugs dissociate two components of hippocampal theta in behaving rats. *J. Neurosci.* **2013**, *33*, 8650–8667. [CrossRef]
12. Winson, J. Loss of hippocampal theta rhythm results in spatial memory deficit in the rat. *Science* **1978**, *201*, 160–163. [CrossRef]
13. Corsi-Cabrera, M.; Perez-Garci, E.; Del Rio-Portilla, Y.; Ugalde, E.; Guevara, M.A. EEG bands during wakefulness, slow-wave, and paradoxical sleep as a result of principal component analysis in the rat. *Sleep* **2001**, *24*, 374–380. [CrossRef] [PubMed]
14. American Psychiatric Association. *DSM-V: Diagnostic and Statistical Manual of Mental Disorders: DSM-5*; American Psychiatric Association: Washington, DC, USA, 2013.
15. Bastien, C.H.; Vallieres, A.; Morin, C.M. Precipitating factors of insomnia. *Behav. Sleep Med.* **2004**, *2*, 50–62. [CrossRef] [PubMed]
16. Benca, R.M.; Obermeyer, W.H.; Larson, C.L.; Yun, B.; Dolski, I.; Kleist, K.D.; Weber, S.M.; Davidson, R.J. EEG alpha power and alpha power asymmetry in sleep and wakefulness. *Psychophysiology* **1999**, *36*, 430–436. [CrossRef] [PubMed]
17. Anderson, C.; Horne, J.A. Prefrontal cortex: Links between low frequency delta EEG in sleep and neuropsychological performance in healthy, older people. *Psychophysiology* **2003**, *40*, 349–357. [CrossRef]
18. Franken, P.; Dijk, D.J.; Tobler, I.; Borbely, A.A. Sleep deprivation in rats: Effects on EEG power spectra, vigilance states, and cortical temperature. *Am. J. Physiol.* **1991**, *261*, R198–R208. [CrossRef]
19. Borbely, A.A. A two process model of sleep regulation. *Hum. Neurobiol.* **1982**, *1*, 195–204.
20. Tort, A.B.; Kramer, M.A.; Thorn, C.; Gibson, D.J.; Kubota, Y.; Graybiel, A.M.; Kopell, N.J. Dynamic cross-frequency couplings of local field potential oscillations in rat striatum and hippocampus during performance of a T-maze task. *Proc. Natl. Acad. Sci. USA* **2008**, *105*, 20517–20522. [CrossRef]
21. Tort, A.B.; Komorowski, R.W.; Manns, J.R.; Kopell, N.J.; Eichenbaum, H. Theta-gamma coupling increases during the learning of item-context associations. *Proc. Natl. Acad. Sci. USA* **2009**, *106*, 20942–20947. [CrossRef]
22. Hinman, J.R.; Penley, S.C.; Long, L.L.; Escabi, M.A.; Chrobak, J.J. Septotemporal variation in dynamics of theta: Speed and habituation. *J. Neurophysiol.* **2011**, *105*, 2675–2686. [CrossRef]
23. Ahmed, O.J.; Mehta, M.R. Running speed alters the frequency of hippocampal gamma oscillations. *J. Neurosci.* **2012**, *32*, 7373–7383. [CrossRef]
24. Zheng, C.; Bieri, K.W.; Trettel, S.G.; Colgin, L.L. The relationship between gamma frequency and running speed differs for slow and fast gamma rhythms in freely behaving rats. *Hippocampus* **2015**, *25*, 924–938. [CrossRef] [PubMed]
25. Shannon, C.E. A mathematical theory of communication. *Bell Syst. Tech. J.* **1948**, *27*, 379–423. [CrossRef]
26. Bokil, H.; Andrews, P.; Kulkarni, J.E.; Mehta, S.; Mitra, P.P. Chronux: A platform for analyzing neural signals. *J. Neurosci. Meth.* **2010**, *192*, 146–151. [CrossRef] [PubMed]
27. Adhikari, A.; Sigurdsson, T.; Topiwala, M.A.; Gordon, J.A. Cross-correlation of instantaneous amplitudes of field potential oscillations: A straightforward method to estimate the directionality and lag between brain areas. *J. Neurosci. Methods* **2010**, *191*, 191–200. [CrossRef]
28. Freedman, D.; Diaconis, P. On the histogram as a density estimator: L2 theory. *Z. Wahrscheinlichkeitstheorie Verwandte Geb.* **1981**, *57*, 453–476. [CrossRef]
29. Jha, S.K.; Brennan, F.X.; Pawlyk, A.C.; Ross, R.J.; Morrison, A.R. REM sleep: A sensitive index of fear conditioning in rats. *Eur. J. Neurosci.* **2005**, *21*, 1077–1080. [CrossRef]
30. Zhang, Y.; Ren, R.; Sanford, L.D.; Yang, L.; Zhou, J.; Zhang, J.; Wing, Y.K.; Shi, J.; Lu, L.; Tang, X. Sleep in posttraumatic stress disorder: A systematic review and meta-analysis of polysomnographic findings. *Sleep Med. Rev.* **2019**, *48*, 101210. [CrossRef]
31. Sainsbury, R.S.; Heynen, A.; Montoya, C.P. Behavioral correlates of hippocampal type 2 theta in the rat. *Physiol. Behav.* **1987**, *39*, 513–519. [CrossRef]
32. O'Keefe, J. Hippocampus, theta, and spatial memory. *Curr. Opin. Neurobiol.* **1993**, *3*, 917–924. [CrossRef]
33. Bieri, K.W.; Bobbitt, K.N.; Colgin, L.L. Slow and fast gamma rhythms coordinate different spatial coding modes in hippocampal place cells. *Neuron* **2014**, *82*, 670–681. [CrossRef]
34. Cowdin, N.; Kobayashi, I.; Mellman, T.A. Theta frequency activity during rapid eye movement (REM) sleep is greater in people with resilience versus PTSD. *Exp. Brain Res.* **2014**, *232*, 1479–1485. [CrossRef] [PubMed]
35. Colgin, L.L.; Denninger, T.; Fyhn, M.; Hafting, T.; Bonnevie, T.; Jensen, O.; Moser, M.B.; Moser, E.I. Frequency of gamma oscillations routes flow of information in the hippocampus. *Nature* **2009**, *462*, 353–357. [CrossRef] [PubMed]

36. Pearl, P.L.; LaFleur, B.J.; Reigle, S.C.; Rich, A.S.; Freeman, A.A.; McCutchen, C.; Sato, S. Sawtooth wave density analysis during REM sleep in normal volunteers. *Sleep Med.* **2002**, *3*, 255–258. [CrossRef]
37. Poe, G.R. Sleep Is for Forgetting. *J. Neurosci.* **2017**, *37*, 464–473. [CrossRef] [PubMed]

Article

EEG Oscillatory Power and Complexity for Epileptic Seizure Detection

Lina Abou-Abbas [1,2,*], Imene Jemal [2,3], Khadidja Henni [1,2], Youssef Ouakrim [1,2], Amar Mitiche [2,3] and Neila Mezghani [1,2]

1. Imaging and Orthopedics' Research Laboratory, The CHUM Research Center, Montreal, QC H2X 0A9, Canada; khadidja.henni@teluq.ca (K.H.); oua.youssef@gmail.com (Y.O.); neila.mezghani@teluq.ca (N.M.)
2. Research Center LICEF, Teluq University, Montreal, QC H2S 3L4, Canada; imen.djmal@gmail.com (I.J.); amar.mitiche@inrs.ca (A.M.)
3. INRS-Centre Énergie, Matériaux et Télécommunications, Montreal, QC H5A 1K6, Canada
* Correspondence: lina.abou-abbas@teluq.ca

Abstract: Monitoring patients at risk of epileptic seizure is critical for optimal treatment and ensuing the reduction of seizure risk and complications. In general, seizure detection is done manually in hospitals and involves time-consuming visual inspection and interpretation by experts of electroencephalography (EEG) recordings. The purpose of this study is to investigate the pertinence of band-limited spectral power and signal complexity in order to discriminate between seizure and seizure-free EEG brain activity. The signal complexity and spectral power are evaluated in five frequency intervals, namely, the delta, theta, alpha, beta, and gamma bands, to be used as EEG signal feature representation. Classification of seizure and seizure-free data was performed by prevalent potent classifiers. Substantial comparative performance evaluation experiments were performed on a large EEG data record of 341 patients in the Temple University Hospital EEG seizure database. Based on statistically validated criteria, results show the efficiency of band-limited spectral power and signal complexity when using random forest and gradient-boosting decision tree classifiers (95% of the area under the curve (AUC) and 91% for both F-measure and accuracy). These results support the use of these automatic classification schemes to assist the practicing neurologist interpret EEG records more accurately and without tedious visual inspection.

Keywords: epileptic seizure; entropy; spectral power; random forest; gradient-boosting decision tree; support vector machine; k-nearest neighbors

1. Introduction

Epilepsy is a neurological disorder characterized by recurrent epileptic seizures. It affects as many as 50 million people of all ages worldwide. Uncontrolled seizures can lead to a disruption of the nervous system and physical risks such as injury and even death. It is generally accepted that better seizure control is essential for better brain health in the long term. Brain activity can be recorded by one of several means, such as electroencephalography (EEG), magnetoencephalography (MEG), and functional magnetic resonance imaging (fMRI). Of these, EEG is the most prevalent because it is simple to effect, low cost, and has high temporal resolution. Its interpretation is commonly used in the study of various brain disorders such as epilepsy, autism, and attention deficit hyperactivity. It is also the most common method in seizure detection and prediction [1–18].

Visual inspection and interpretation of EEG signals are tedious and complex tasks done by experts, which justify research and development of computer-processing methods of EEG that use EEG data representation and machine-learning technics.

Investigations of EEG patterns of seizure have shown that, in general, seizure events have been observed within a wide range of EEG bandwidths. Likewise, their amplitude and frequency in the delta (1–4 Hz), theta (4–8 Hz), alpha (8–14 Hz), beta (13–30 Hz), and gamma (30–70 Hz) bands are significantly different than normal [17,18]. In line with this observation, several studies have used EEG spectral power density (PSD) for seizure detection [3,7,8,19]. PSD is commonly used to evaluate the power of each observed frequency in different neurological brain states. According to previous research, PSDs during seizure events differ significantly from the PSD during rest or sleep states [20], and EEG PSD features have been shown to be able to distinguish between seizure and seizure-free events in the frequency domain [3,7,8,19]. Using the absolute and relative band power of the PSD in 13 frequency sub-bands [21], the authors achieved a 91.36% AUC in a patient-specific study. In another study [22], they found that relative spectral power features led to improved seizure prediction and an average fault rate per hour of 0.1. In a recent study, a comparison of power spectrum analysis and time-frequency analysis was performed. The results revealed that power spectrum features are likely to be seizure markers, and there was a significant difference between the distribution of the power spectrum in seizure segments versus segments in which there were no seizures [23]. Time and frequency domain signal complexity measures have also been used to describe EEG data and serve for the classification of normal and seizure EEG data patterns [2,5,7,11–13,17–19,24]. Other EEG data descriptions and properties have been used as well—for instance, permutation entropy, which decreases significantly in the transition from a seizure-free phase to a seizure phase [16]; Shannon and logarithmic energy entropy, which decrease during epileptic seizures [4]; spectral entropy, which increases for seizure segments [14]; and wavelet entropy [5,9]. Although some studies have considered single complexity features in their proposed approaches [5,16], others have combined several such measures [12,14].

The clinical utility of automated classification methods that involve neuroimaging techniques have become increasingly important in many areas of health care due to their ability to identify atypical neural activity, such as seizure episodes, without visual inspection. Accordingly, research has addressed the design of end-to-end machine learning (ML) systems to detect seizure segments and determine seizure type. ML offers the opportunity to automatically distinguish between patterns in seizure and seizure-free segments. Automatic methods that use ML classification have been investigated for their potential for classification in several healthcare domains, including epilepsy detection, seizure detection, seizure localization, and seizure type identification [25].

There have been several studies of low false alarm detection of epileptic seizure episodes in EEG recordings by ML [4,8,9,14,18]. Along this vein, various classifiers have been used, such as the k-nearest neighbors algorithm (k-NN) [14], support vector machines (SVM) [15,26,27], neural networks (NN) [1,28], and decision trees (DT) [7]. In [27], for neonatal seizure detection, the authors used 55 features with SVM and reported an overall performance of 89% with only one false seizure detection per hour. The system was trained and tested using EEG data collected from 17 newborns with seizures. The study [29] focused on EEG feature selection by relevance and redundancy analysis and used a back propagation neural network to evaluate the effectiveness of feature selection. Results showed an average seizure detection rate of 91% with a false detection rate of 1.17 per hour. Using the Children's Hospital Boston-Massachusetts Institute of Technology (CHB-MIT) dataset, the study [30] reported an accuracy of 96% with a false-positive rate of 0.08 per hour. Using the same database, which consists of only 23 subjects, a recent study [31] showed an overall accuracy of 91.8% by deploying linear discriminant analysis as a classifier validated on only five subjects. With the European Epilepsy Database, results achieved include an average sensitivity of 91.72% by neural network classification [32], 90.8% by SVM (false-positive rate of 0.094 per hour) [33], and 93.8% by a feed-forward back propagation artificial neural network [34]. These results all relate to patient-specific experimentation and have not been confirmed on across-patient data.

Although there are many studies that focus on seizure detection based on EEG, most of them have some limitations, mainly including experimentation on data from few patients and using of invasive intracranial EEG records. Furthermore, previous research has used a patient-specific problem formulation and experimental test in which the model is trained and tested on a single patient. Despite attempts to create algorithms to obtain values as accurate as possible, no single one has yet gained widespread acceptance in clinical practice, and the results of studies are far from substituting manual interpretation of the EEG [35,36]. Indeed, despite the numerous commercially available seizure detection devices, there are several issues that must be taken into consideration before such devices can be used in everyday clinical practice. According to a study published in 2021 [37], a comparative study was conducted between 23 devices on the market for seizure detection/alerting. This study showed that these devices rely largely on movement detectors, autonomic change detectors, heart rates, or eye tracking to detect seizures. According to the study's authors, almost no commercially available seizure detection methods use EEG except for Epihunter, a wearable headband device that is connected to a smartphone and is specifically designed for automatic absence seizure detection, which is a specific type of seizure. Thus, the authors concluded that commercially available seizure detection devices do not have the capability of detecting multiple seizure types. Another recent study [38] compared the sensitivity of three well-known commercial seizure detection softwares (Besa, Encevis, and Persyst) to determine whether they correctly detected seizures over long-term video-electroencephalography monitoring (VEM). Based on results from 81 unseen patients, the researchers found a sensitivity of 67.6% for Besa, 77.8% for Encevis, and 81.6% for Persyst on a patient-by-patient basis. They suggested that the false alarm rate needs to be improved. Furthermore, in another study published in 2020 [39], the authors concluded that most commercially available methods focus on using non-electroencephalography EEG signals, which is perceived as a major limitation given the inability of these devices to detect all seizures types. Therefore, it is crucial to develop an automated system for the detection of epileptic seizures based on EEG signals. To develop such an automated system, significant further research and experimentation are necessary so as to develop seizure detection methods that are clinically relevant, high-performing, and statistically validated. Therefore, testing the clinical relevance of state-of-the-art findings through the analysis of a large number of EEG data collated from an accurate representation of clinical situations and collected from multiple sites will be necessary.

The purpose of this study is to investigate a new data representation model that simultaneously exploits signal oscillatory power in frequency bands and signal complexity as EEG. This representation serves prevailing potent ML algorithms for seizure vs. seizure-free classification. We will rigorously evaluate the proposed EEG data representation features on the publicly available large dataset of Temple University Hospital (TUSZ) collected in clinical settings [40].

The main contribution of this paper can be summarized as follows: (1) An analytical framework for seizure and seizure-free EEG classification is proposed and validated on a new large EEG seizure corpus. (2) The feature significance level is investigated by a univariate data analysis. (3) The performance of complexity measures and oscillatory power is analyzed individually and then combined, as EEG signal representation features, to classify normal and seizure data. (4) The results of four supervised methods are compared: RF, GBDT, SVM, and k-NN. (5) The impact of adding a feature selection step on the performance of classifiers is investigated. (6) The contribution of each channel to the performance of classifiers is examined.

2. Materials and Methods

2.1. Dataset

EEG data were drawn from the TUH EEG seizure database (TUSZ), which is a part of Temple University Hospital EEG corpus, the largest open-source EEG corpus. The latter comprises more than 16,986 sessions of EEG recordings collected from 10,874 unique subjects [40,41]. The version of the database used in this work was v1.5.1, released in March 2020. In this study, the dataset consists of EEG signals collected from 341 patients, providing a substantial amount of data for seizure detection investigation. The EEG segments were labelled by a team of neuroscience and bioengineering students who underwent several months of intensive training to gain annotation skills. They have worked closely with a team of neurologists at Temple Hospital to understand their workflow and their clinical needs while labeling EEG segments. There have been numerous revisions of the corpus during the annotation process (two or three annotators per file system), each time applying more refinement and criteria to enhance the clarity and accuracy of the data. Annotations are created by viewing files with open-source software known as an EDF viewer or by using a customized annotation tool developed by the team, which provides time and frequency domain visualizations. The annotation was made available publicly, except for an evaluation corpus not released to the public, to be allocated for research competition purposes. The dataset used contains 886 sessions that were broken down into 7634 files, of which 1780 are seizure activities of different lengths, in seconds, for a total of 40.40 h. The data were collected in real-time clinical environments, including an intensive care unit, an epilepsy monitoring unit, emergency rooms, cardiac intensive care, and surgical and respiratory intensive care units [41]. Most of the EEG recordings have at least 19 electrodes corresponding to the international standard 10/20 system and range from one second to one hour in duration. The sampling rate varied between 250 Hz and 500 Hz. We resampled all data at 256 Hz. In this study, we discarded seizure events that lasted less than our sliding window of 20 s. Eight types of seizures were present in the TUSZ database: focal non-specific seizure, generalized non-specific seizure, complex partial seizure, simple partial seizure, tonic–clonic seizure, absence seizure, tonic seizure, and myoclonic seizure. Since our goal was not to detect the type of seizure but to determine the presence or absence of seizures, we combined all these types under one label, "seizure."

The dataset description is summarized in Table 1, and Figure 1 shows an example of seizure and seizure-free EEG epochs. Detailed information about the database annotation can be found in the following reference [42].

Table 1. Overview of the subset of the TUSZ EEG corpus used in this study for seizure detection.

	Number
Total patients (female)	341 (188 F)
Patients with seizure (female)	133 (72 F)
Sessions	886
Files	7634
Seizure files	1780
Seizure-free files	5854
Total duration in hours	655.36

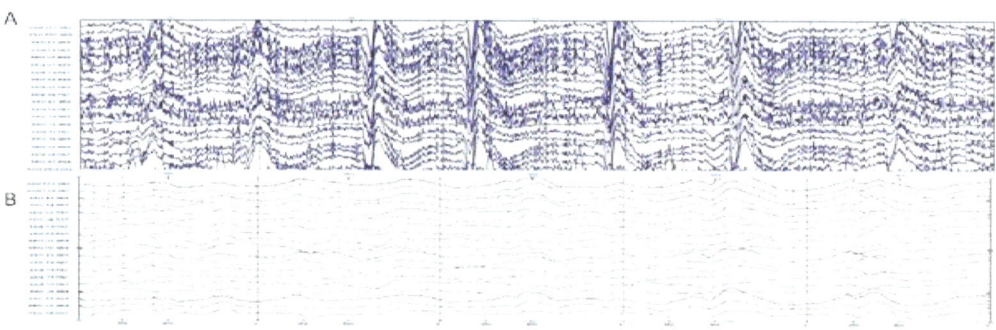

Figure 1. (**A**) An example of raw EEG including seizure epochs. (**B**) An example of normal EEG—seizure-free epochs.

2.2. Methodology

An overview of this study's generalized ML-processing strategy is shown in Figure 2. It gives the main steps for data pre-processing, feature extraction, and, finally, classification, which includes training and evaluation. The TUSZ EEG input data consist of pruned EEG recordings, and all uninteresting portions of the data (muscle artifacts, noise, electrode movements, and eye blinks) were purged from the cortical EEG signals. The steps involved in our proposed approach for the classification of epileptic EEG seizures are as follows:

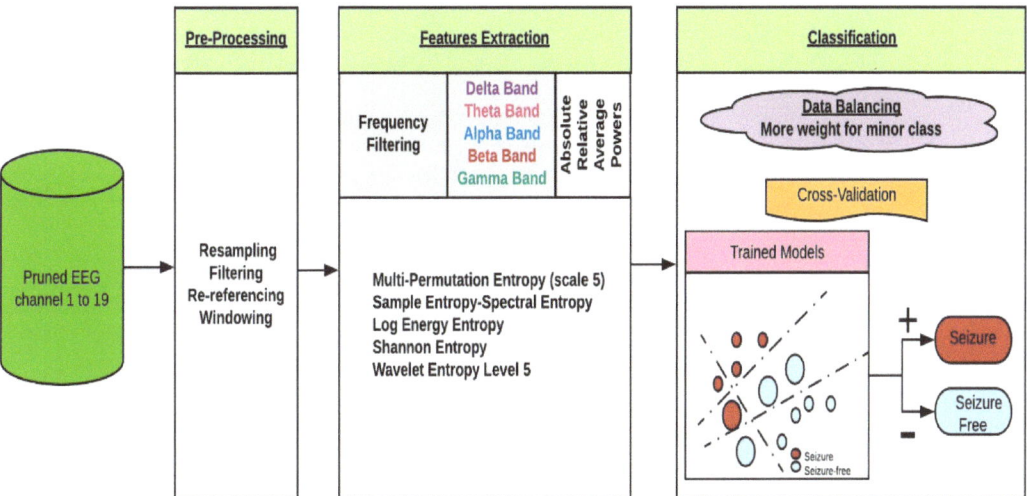

Figure 2. A block diagram of this study's automatic seizure detection method. A total of 19 pruned EEG channels from the TUSZ corpus are resampled, filtered, re-referenced, and segmented. Features are subsequently extracted and used as input for modelling and testing four classifiers separately (RF, GB, k-NN, and SVM).

2.2.1. Pre-Processing

In the staging phase, a 60 Hz infinite impulse response notch filter was applied to attenuate the power line, followed by a bandpass filter with lower and higher cut-off frequencies of 0.5 Hz and 75 Hz, respectively. This was followed by resampling to 256 Hz and re-referencing to the average of all electrodes. From the remaining EEG signal, 20 s fixed-length epochs were segmented.

2.2.2. Feature Extraction

Nine signal complexity measures were derived from the pre-processed EEG: multi-permutation entropy (4 levels), sample entropy, wavelet entropies, logarithmic entropy, Shannon entropy, and spectral entropy. The permutation entropy is a complexity measure with a low computational cost for time series based on comparing neighboring values using the distribution of order patterns [16]. The sample entropy quantifies the regularity in the EEG signal regardless of its length [14]. The discrete wavelet transform is used to compute the wavelet entropies [5], whereas the wavelet packet decomposition is used to compute log energy entropy and Shannon entropy to measure the degree of uncertainty in the signal and to evaluate the dynamical order of the signal [4]. The spectral entropy is calculated using the normalized power spectral distribution of the EEG signal [14]. For the frequency domain analysis, we performed the Welch method, which takes an average of the periodograms obtained using fast Fourier transform (FFT). We calculated the absolute power density and relative power density within each of the five frequency sub-bands—delta (1–4 Hz), theta (4–8 Hz), alpha (8–14 Hz), beta (13–30 Hz), and gamma (30–70 Hz)—to increase the feature vector size to 19 per channel. The absolute power of a band can be calculated as the sum of all of its power values in that frequency range, whereas the relative power (RP) index for each band was determined based on the absolute power in each frequency band, expressed as a percentage of the absolute power (AP) totaled across all frequency bands. Both absolute and relative PSD analysis are essential for achieving accurate brain analysis. Feature values were averaged using a scalar method before serving as input to the classifier. To investigate whether a subset of the extracted features had a greater significance compared to the entire set, we subdivided the features based on their category: frequency domain and information theory (see Table 2). A univariate feature selection method based on the one-way ANOVA test was also applied for comparison to investigate the effect of reducing the vector size. We evaluated the statistical difference of each feature between the two classes: seizure and seizure-free. In order to identify highly significant features, we sorted features based on their p-values, which correspond to the probability of the results being observed if the null hypothesis H0 is true. In this work, the null hypothesis H0 implies that there is no difference between the means of two groups (seizure and seizure-free records). In general, the lower the p-value, the higher the reliability of results. In medical applications, 0.05, 0.01, or 0.001 are often the recommended threshold values. The features' level of significance was implemented using Python (3.8.0), specifically the Sklearn library.

Table 2. Subset of features in correspondence to their category.

Type	Features	Category
Complexity measures	Sample entropy Logarithmic entropy Wavelet entropy Spectral entropy Shannon entropy Permutation entropy	Information theory
Oscillatory power	Absolute power relative power (delta, theta, alpha, beta, gamma)	Frequency domain

2.2.3. Training and Evaluation

Comparative supervised classification was carried out with four classifiers: random forest (RF), gradient-boosting decision tree (GBDT), support vector machine (SVM), and k-nearest neighbors (k-NN). The four classifiers were chosen based on findings from previous studies, which showed that they had superior classification compared to others [15,26,27,43,44] The random forest algorithm, an ensemble learning method, involves multiple un-pruned decision tree classifiers. Every tree is an individual classifier, built at the training stage using randomly selected attributes from the original data and at each node to determine the best split. Classification is determined based on majority voting [45]. The gradient-boosting decision tree classifier runs an iterative algorithm to build multiple trees sequentially, where each tree learns and updates its model from the errors of all preceding ones [46]. The support vector machine, an extensively adopted supervised learning classifier, has been widely and successfully used in EEG binary classification problems and specifically in the medical field (automatic seizure detection, mental task classification, epileptic EEG classification, emotion recognition, etc.). The k-nearest neighbors algorithm assigns an unlabeled sample to the most frequently occurring class among the k-nearest labeled training samples. The Euclidean distance is used as the distance metric for k-NN [47].

These proposed ML techniques are based on the most widely used algorithms in previous studies and in various classification problems, as these algorithms have better predictive power and are designed to perform better than linear algorithms, especially for complex non-linearly separable EEG data. Additionally, these techniques have the great advantage of having fewer parameters and hyper-parameters, which makes optimization easier. The model was developed with a view toward being sufficiently transferable to enable the integration in the connected objects for real-time detection.

To avoid overfitting, for each experiment, the dataset was partitioned randomly into training and testing portions, using 10-fold cross-validation repeated 100 times so that one-fold served for testing and the remaining k-1 folds were used for training. The parameter settings of the classifiers in all experiments were as follows: for RF, the number of trees was 100, for GBDT it was 100 as well, for SVM the kernel type was linear, and for k-NN the number of nearest neighbors was 5.

To handle imbalanced data, which may occur due to uneven data representation of classes in the training dataset, with a ratio of seizure to seizure-free of approximately 1:4, we corrected the weights of minority and majority classes according to the distribution of the classes in the entire training set. Furthermore, the weights were inversely proportional to the frequency of classes in the dataset. In a final step, the accuracy and the area under the curve (AUC), an effective metric that combines sensitivity and specificity for classification performance evaluation, were computed for each subset of features and also for the entire set. An F-measure representing the harmonic mean of precision and recall was used as well to study performance. For clarity, results are given per 20 s window size input vector and with respect to the three set of features: complexity, power, and a combination of the two. We compared the results with and without the feature selection step.

3. Results

3.1. Univariate Data Analysis

Each feature was subjected to a univariate data analysis to identify its level of significance with the studied classes, i.e., normal or seizure class, and to investigate its discrimination capability in characterizing seizure records. Tables 3 and 4 show the normalized values of the extracted measures (mean ± standard deviation) for seizure and seizure-free segments. Significant p-values (obtained using the ANOVA test) indicate a high discrimination capability in characterizing seizure segments. The values displayed in Tables 3 and 4 correspond to the highest p-value per feature. It is evident from Table 3 that seizure-free records had lower complexity values compared to epileptic seizure records. Results in Table 4 show that the relative power in the alpha band had the lowest p-value. The results also show that oscillatory power change features had higher levels of importance than complexity measures (Tables 3 and 4), which indicates that oscillatory power change in all channels can contribute to better classification. In addition, Shannon entropy and relative power in the alpha band had lower p-values, which indicates that they had higher levels of importance than other features of this category.

Table 3. Range of values of the complexity measures for the two classes of seizure and seizure-free. The lowest p-value of 4.130×10^{-98} for multi-permutation entropy at level 2 indicates the great discrimination capability of this feature in characterizing seizure records.

Features	Seizure-Free (Mean ± Std)	Seizure (Mean ± Std)	p-Value
Shannon entropy	0.242 ± 0.125	0.1556 ± 0.095	6.159×10^{-15}
Sample entropy	0.279 ± 0.132	0.199 ± 0.118	1.760×10^{-96}
Spectral entropy	0.288 ± 0.143	0.203 ± 0.123	2.193×10^{-95}
Log energy entropy	0.320 ± 0.152	0.217 ± 0.139	7.184×10^{-11}
Wavelet entropy	0.262 ± 0.125	0.181 ± 0.112	6.657×10^{-11}
Multi-permutation 1	0.267 ± 0.122	0.196 ± 0.117	1.212×10^{-80}
Multi-permutation 2	0.296 ± 0.145	0.207 ± 0.129	4.130×10^{-98}
Multi-permutation 3	0.275 ± 0.130	0.194 ± 0.125	3.050×10^{-90}
Multi-permutation 4	0.291 ± 0.141	0.196 ± 0.123	3.467×10^{-12}

Table 4. Range of values of the spectral power measures for the two classes of seizure and seizure-free. The lowest p-value of 7.173×10^{-145} for the relative power at the alpha band indicates the great discrimination capability of this feature in characterizing seizure records.

Features	Band	Seizure-Free (Mean ± Std in mV)	Seizure (Mean ± Std in mV)	p-Value
Absolute power	Delta	0.89 ± 0.02	0.87 ± 0.03	5.694×10^{-124}
	Theta	0.89 ± 0.02	0.87 ± 0.03	1.156×10^{-129}
	Alpha	0.89 ± 0.02	0.87 ± 0.03	7.125×10^{-133}
	Beta	0.90 ± 0.02	0.87 ± 0.04	9.067×10^{-133}
	Gamma	0.90 ± 0.02	0.87 ± 0.04	3.423×10^{-133}
Relative power	Delta	0.88 ± 0.02	0.86 ± 0.03	8.007×10^{-132}
	Theta	0.88 ± 0.02	0.53 ± 0.04	9.594×10^{-128}
	Alpha	0.25 ± 0.16	0.15 ± 0.11	7.173×10^{-145}
	Beta	0.26 ± 0.15	0.16 ± 0.11	1.367×10^{-129}
	Gamma	0.27 ± 0.14	0.19 ± 0.11	1.344×10^{-93}

3.2. Performance of Group of Features Extracted from All Channels without Feature Selection

We evaluated classification performance corresponding to the three sets of features: complexity measures (9 features per channel), oscillatory power (10 features per channel), and a combination of the two (19 features per channel). Table 5 displays the performance of the three sets of features. Classification of seizure versus seizure-free reached an accuracy and F-measure of 90.68% and 91.05%, respectively, using the complexity measures, and 90.95% and 91.33%, respectively, using relative and absolute power in the alpha, beta, gamma, theta, and delta bands. A high AUC value of 95%, an F-measure of 91.41%, and an accuracy of 91.07% were obtained using the RF classifier when the entire set of features was given as classification input. RF and GBDT outperformed SVM and k-NN in all experiments. There was only a slight difference in performance between RF and GBDT.

Table 5. Classification performance for the complexity measures, oscillatory power, and a combination of the two without the univariate feature selection.

Classifiers	RF			GBDT			SVM			K-NN		
Features	AUC%	F%	ACC%	AUC%	F%	ACC%	AUC%	F%	ACC%	AUC%	F%	ACC%
Complexity measures	95 ± 01	91.05	90.68	93 ± 01	89.86	89.48	79 ± 03	71.44	73.57	85 ± 02	88.92	87.90
Oscillatory power	95 ± 01	91.33	90.95	94 ± 01	90.68	90.40	88 ± 02	79.29	80.65	89 ± 02	89.67	88.73
Complexity and power	95 ± 01	91.41	91.07	94 ± 01	90.95	90.67	86 ± 02	77.24	78.72	88 ± 02	90.09	89.16

3.3. Performance of Group of Features Extracted from All Channels with Feature Selection

For this study, we executed three sets of experiments to investigate the effect of the recommended thresholds. Consistent with previous findings [48,49], we found that reducing the features by the criteria that their p-values be less than 0.001 led to improved performance. All features with a p-value higher than 0.001 were discarded from the feature vector as non-descriptive. Classification performance was evaluated using AUC, F-measure, and accuracy. Table 6 displays the performance of the three sets of features. RF was able to achieve an accuracy of 90.30% and an F-measure of 90.76%, with a decrease of 1%, compared to the experiments done without the feature selection step. There was no significant change in the performance of the three classifiers RF, GBDT, and k-NN between the experiments done with and without univariate feature selection. SVM accuracy increased from 78.72% to 81.92%.

Table 6. Classification performance for the complexity measures, oscillatory power, and a combination of the two with the univariate feature selection.

Classifiers	RF			GBDT			SVM			K-NN		
Features	AUC%	F%	ACC%	AUC%	F%	ACC%	AUC%	F%	ACC%	AUC%	F%	ACC%
Complexity measures	87 ± 02	88.65	86.00	83 ± 03	87.13	84.52	81 ± 03	73.09	75.07	84 ± 02	88.80	87.47
Oscillatory power	94 ± 01	90.07	89.60	93 ± 01	89.56	89.21	91 ± 01	81.73	82.96	88 ± 02	89.50	88.43
Complexity and power	94 ± 01	90.76	90.30	94 ± 01	89.76	89.43	90 ± 01	80.63	81.92	88 ± 02	90.04	89.07

3.4. Performance of Classifiers with Features Extracted from One Channel at a Time

Table 7 displays the results of RF- and GBDT-based classification using the combined complexity and power features for the 19 EEG channels. A total of 19 features was considered per run. The best AUC, F-measure, and accuracy of 94%, 89.96%, and 89.51% respectively, were obtained with the RF classifier in the Pz channel.

Table 7. Classification performance per channel using complexity measures and oscillatory power.

	RF			GBDT		
Channel	AUC%	F%	ACC%	AUC%	F%	ACC%
FP1	91 ± 01	87.27	86.41	88 ± 01	85.63	84.85
FP2	94 ± 01	89.78	89.38	92 ± 01	88.09	87.70
F3	79 ± 03	88.15	83.86	73 ± 03	87.50	82.78
F4	80 ± 02	87.82	83.39	73 ± 02	87.44	81.89
F7	90 ± 02	88.47	87.02	88 ± 02	87.07	85.55
F8	78 ± 03	87.86	81.70	73 ± 03	86.90	81.00
Fz	80 ± 02	88.05	83.78	73 ± 02	87.76	81.25
C3	92 ± 01	88.66	88.08	90 ± 01	87.15	86.76
C4	88 ± 01	86.43	85.55	87 ± 01	85.05	84.82
Cz	88 ± 01	85.96	85.11	87 ± 01	85.10	84.61
T3	92 ± 01	89.00	88.36	89 ± 01	86.34	85.66
T4	86 ± 02	86.64	84.14	85 ± 02	84.77	83.28
T5	91 ± 01	87.60	86.94	89 ± 01	86.25	85.71
T6	92 ± 01	88.20	87.52	89 ± 01	86.06	85.41
P3	90 ± 01	87.75	86.91	87 ± 01	85.43	84.69
P4	90 ± 01	88.22	87.00	87 ± 01	86.66	85.00
Pz	94 ± 01	89.96	89.51	92 ± 01	88.16	87.72
O1	80 ± 02	88.19	84.33	73 ± 02	87.83	81.63
O2	85 ± 02	88.13	85.25	80 ± 03	86.32	83.27

3.5. ROC Analysis of RF and GBDT Classifiers

The performance of this study's proposed approach was measured using the receiver operating curve (ROC) analysis and AUC metric. A higher AUC designates better performance. Figure 3 displays a comparison between ROC curves when the entire set of features is used. The AUC for each fold is shown in the caption of each figure, and the mean of the AUC was computed. Figure 3 illustrates the ranking performance of 100 iterations of 10-fold cross-validation. The ROC curve summarizes the results and shows that performance was good for all folds, giving an average AUC equal to 95% for RF and 94% for GBDT. The

RF classifier yielded better performance for each of the 10 folds compared to GBDT. It is remarkable for both classifiers that the fluctuations of the AUC were smaller among the different folds, indicating the stability of the feature set.

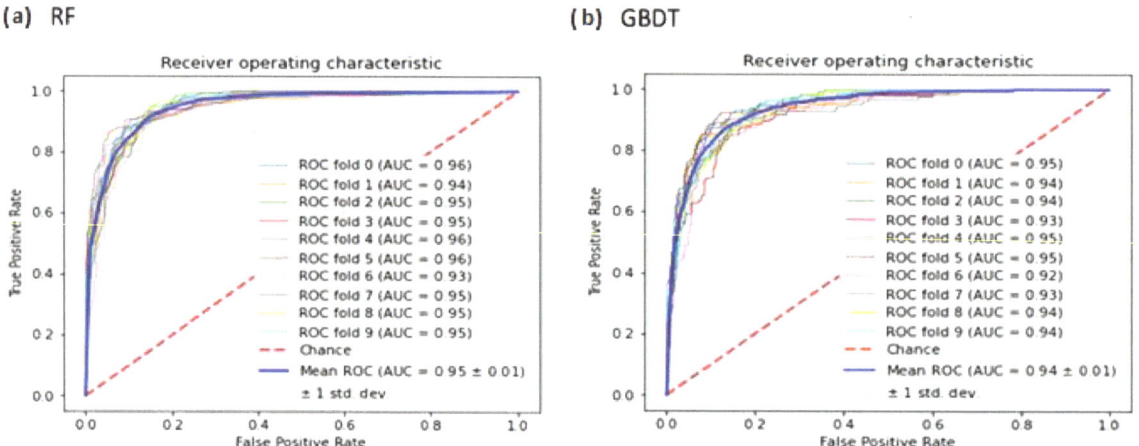

Figure 3. Comparison between ROC curves showing the best results achieved when the entire set of features is used. (**a**) The left ROC curve corresponds to the RF classifier, and (**b**) the right curve corresponds to the GBDT classifier. Each curve denotes one-fold of the 10-fold cross-validation, and the area under the curve (AUC) is displayed for each fold in the figure legend. The mean AUC is computed for the 10-fold.

3.6. Comparison of ROC Curves and Accuracies of Classifiers

A comparison between the ROC curves of the four classifiers is displayed in Figure 4. All four classifiers achieved a good AUC. RF produced the highest AUC of 96% and differed slightly from GDBT at 95%. RF outperformed k-NN and SVM in all experiments. The 10-fold cross-validation was performed and the mean and standard deviation of the cross-validation scores over 100 iterations were computed for each classifier. A comparison between accuracies achieved by the classifiers is shown in Figure 5. It is remarkable that the performance of RF classification was superior in both ROC and accuracy. An average accuracy of 90.90% was reached using RF, followed by 89.87%, 89.16%, and 78.72% with GBDT, k-NN, and SVM, respectively.

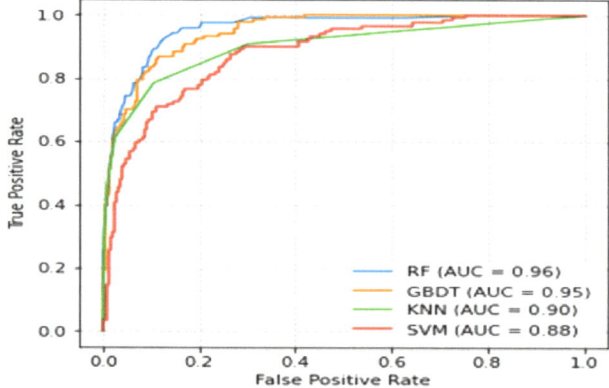

Figure 4. A comparison between ROC curves of the four classifiers: RF, GBDT, KNN and SVM. The AUC of each classifier is displayed.

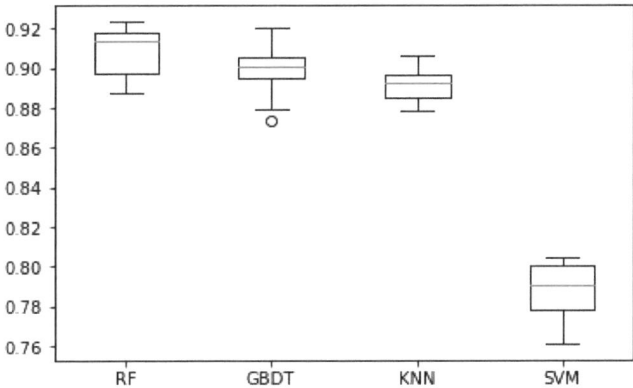

Figure 5. Box plots corresponding to the accuracy of the classification of the two classes of seizure and seizure-free. The orange line in the boxes marks the median. The final accuracy score is determined by averaging the 10 ROC curves and calculating the AUC of the mean ROC.

4. Discussion

This study addressed the problem of seizure vs. seizure-free EEG record classification. The classifier design and validation used data from 341 subjects in the TUSZ database, a larger amount of data than previously used.

Several studies have highlighted the advantage of complexity measures as EEG signal description in seizure detection [2,4,5,9–11,13,15,19,20], and others showed the importance, as well, of signal power in different frequency bands [3,7,8,19]. We inquired into the performance of these features with the larger TUSZ database to gain some understanding of the effect of the experimentation database size on seizure detection potency. We followed this with the study of a new method that exploits both types of features. We investigated different feature-grouping options, followed by machine-learning classification of selected features to identify seizure segments.

Two feature subsets of prevailing features, as well as their combination, were investigated to evaluate their seizure detection potency. In contrast to earlier studies that focused on patient-specific seizure detection, this study addressed patient-independent detection, using data from 341 subjects. Signal complexity and power both yielded good classification, with signal power providing slightly better results. When used jointly, signal complexity and power gave superior performance: 95% AUC with RF classification, 94% with GBDT, 86% with SVM, and 88% with k-NN. RF outperformed other classifiers systematically in all experiments. This may be explained by the fact that decisions made by the RF classifier correspond to unpruned and diverse trees that lead to high resolution in feature space. Random operations in the training and voting procedures of RF also contribute to better classification by addressing the issue of overfitting [50].

This study showed the classification significance of the features at the channel level. Unlike most others, which limited features to one channel (e.g., BONN dataset) or six channels (e.g., Freiburg dataset), the TUSZ database used in validating the method in this study has more than 19 channels corresponding to the 10–20 standard system. This allowed us to investigate the performance of the method using only one channel at each run. Results showed that a maximum distinction between seizure and seizure-free records was obtained from the Pz channel, with a corresponding AUC of 94%. In addition, the method results show that the parietal and frontal-parietal regions were effective in extracting features that can discriminate seizure from seizure-free records, wherein the performance for each channel obtained was above 90%. Additionally, this study addressed the effect of adding a feature selection step prior to classification. The analysis removed the less informative features using the *p*-value test: If the *p*-value was greater than 0.001, the feature extracted from a specified channel was discarded from the group considered. Only a slight decrease in performance occurred for RF and GBDT with feature selection, keeping in mind that the method is computationally more efficient when running on a smaller set of features.

The results are consistent with findings in a previous study [51] that compared random forest variable selection methods for classification. Indeed, the latter concluded that there was practically no difference in prediction error rates between all methods. However, it was obvious that SVM was more sensitive to the number of features, with a decrease in accuracy rate as the number of features increased.

It may be important to note that the good performance of all classifiers is an indirect indication that the feature set is stable in addition to being representative of EEG data for the problem of seizure detection. An analysis of the significance level of a set of entropy properties determined that seizure records have invariably lower signal complexity than seizure-free records. This finding agrees with previous studies that concluded that the EEG signal during seizure is less complex than when seizure free, and therefore, a reduction in signal information content and complexity can be inferred [11,12,14,15]. Additionally, an analysis of the significance level of relative and absolute signal power in frequency bands, typically, delta (1–4 Hz), theta (4–8 Hz), alpha (8–14 Hz), beta (13–30 Hz), and gamma (30–70 Hz), indicated that these power features can contribute to higher classification, explaining in part their successful use in [3,7,8,19]. A higher level of importance for oscillatory power-based feature vectors in comparison to complexity measures was apparent and could justify their outperformance in classification.

It is noteworthy to mention that one of the limitations of this study is related to the medical frequency sub-bands that were not considered in this work. This is the case of beta frequency Beta-1, 13–15 Hz; Beta-2, 15–18 Hz; Beta-3, 18–25 Hz; and Hi-Beta, 25–30 Hz. Recent reports [52,53] indicate that additional frequency band analysis is beneficial for epilepsy detection and shows the impact of the frequency sub-bands to the epileptic EEG classification accuracy, and the obtained results revealed several frequency sub-band combinations that achieved high classification accuracy, including the medical frequency sub-bands.

5. Conclusions

This study investigated the use of various ML methods for seizure vs. seizure-free EEG record classification. The classification performance of three sets of extracted features—complexity, oscillatory power, and a combination of the two—was evaluated. Our results showed that abnormalities in a seizure EEG record could be distinguished from a seizure-free EEG record by using its complexity measures and oscillatory power, collected as one feature vector fitted to a random forest classifier. This study's proposed approach could provide a prominent contribution to the development of a fully automated seizure detection system. In a future work, we will expand our study to include pre-ictal and ictal. As there are eight types of seizures present in the TUSZ database, we plan to classify seizures by type, incorporating gender as a predictor, since there are certain differences associated with epilepsy patterns between genders. In addition, we will include a cross-database evaluation to validate the effectiveness of our proposed method and to confirm its generalization ability in seizure classification.

Author Contributions: Methodology, L.A.-A.; software, L.A.-A. and Y.O.; validation, L.A.-A., Y.O. and I.J.; formal analysis, L.A.-A.; writing—original draft preparation, L.A.-A.; writing—review and editing, L.A.-A., I.J., K.H., A.M. and N.M.; supervision, N.M. and A.M. All authors have read and agreed to the published version of the manuscript.

Funding: This research was funded by Fonds de recherche du Quebec—Nature et technologies (L.A.-A.) and the Canada Research Chair on Biomedical Data Mining (950-231214).

Institutional Review Board Statement: Not applicable.

Informed Consent Statement: Not applicable.

Data Availability Statement: The dataset used for this study is available as open source from the TUH EEG seizure corpus (TUSZ) v1.5.1. This version of data was released in 20 March 2020. It can be downloaded using the following link: https://www.isip.piconepress.com/projects/tuh_eeg/html/downloads.shtml. Informed consent was obtained from all subjects involved in the study at Temple Hospital. The data has been properly anonymized before being released from Temple Hospital.

Conflicts of Interest: The authors declare no conflict of interest.

References

1. Alam, S.M.S.; Bhuiyan, M.I.H. Detection of Seizure and Epilepsy Using Higher Order Statistics in the EMD Domain. *IEEE J. Biomed. Health Inform.* **2013**, *17*, 312–318. [CrossRef] [PubMed]
2. Martis, R.J.; Acharya, U.R.; Tan, J.H.; Petznick, A.; Yanti, R.; Chua, C.K.; Ng, E.Y.; Tong, L. Application of Empirical Mode Decomposition (Emd) for Automated Detection of Epilepsy using Eeg Signals. *Int. J. Neural Syst.* **2012**, *22*, 1250027. [CrossRef]
3. Birjandtalab, J.; Heydarzadeh, M.; Nourani, M. Automated EEG-Based Epileptic Seizure Detection Using Deep Neural Networks. In Proceedings of the 2017 IEEE International Conference on Healthcare Informatics (ICHI 2017), Park City, UT, USA, 23–26 August 2017; pp. 552–555.
4. Harlalka, V.; Puntambekar, V.P.; Raviteja, K.; Mahalakshmi, P. Detection of Epileptic Seizure Using Wavelet Analysis based Shannon Entropy, Logarithmic Energy Entropy and Support Vector Machine. *Int. J. Eng. Technol.* **2018**, *7*, 935–939. [CrossRef]
5. Kumar, Y.; Dewal, M.L.; Anand, R.S. Wavelet entropy based EEG analysis for seizure detection. In Proceedings of the 2013 IEEE International Conference on Signal Processing, Computing and Control (ISPCC 2013), Solan, India, 26–28 September 2013.
6. Sharma, R.; Pachori, R.B. Classification of epileptic seizures in EEG signals based on phase space representation of intrinsic mode functions. *Expert Syst. Appl.* **2015**, *42*, 1106–1117. [CrossRef]
7. Tzallas, A.T.; Tsipouras, M.G.; Fotiadis, D.I. Epileptic Seizure Detection in EEGs Using Time–Frequency Analysis. *IEEE Trans. Inf. Technol. Biomed.* **2009**, *13*, 703–710. [CrossRef]
8. Tzallas, A.T.; Tsipouras, M.G.; Fotiadis, D.I. Automatic Seizure Detection Based on Time-Frequency Analysis and Artificial Neural Networks. *Comput. Intell. Neurosci.* **2007**, *2007*, 080510. [CrossRef] [PubMed]
9. Raghu, S.; Sriraam, N.; Kumar, G.P. Classification of epileptic seizures using wavelet packet log energy and norm entropies with recurrent Elman neural network classifier. *Cogn. Neurodynamics* **2017**, *11*, 51–66. [CrossRef] [PubMed]
10. Jemal, I.; Mitiche, A.; Mezghani, N. A Study of EEG Feature Complexity in Epileptic Seizure Prediction. *Appl. Sci.* **2021**, *11*, 1579. [CrossRef]
11. Kannathal, N.; Choo, M.L.; Acharya, U.R.; Sadasivan, P. Entropies for detection of epilepsy in EEG. *Comput. Methods Programs Biomed.* **2005**, *80*, 187–194. [CrossRef]
12. Acharya, U.R.; Fujita, H.; Sudarshan, V.K.; Bhat, S.; Koh, J.E. Application of entropies for automated diagnosis of epilepsy using EEG signals: A review. *Knowl. Based Syst.* **2015**, *88*, 85–96. [CrossRef]
13. Orosco, L.; Laciar, E.; Correa, A.G.; Torres, A.; Graffigna, J.P. An epileptic seizures detection algorithm based on the empirical mode decomposition of EEG. In Proceedings of the 31st Annual International Conference of the IEEE Engineering in Medicine and Biology Society: Engineering the Future of Biomedicine, EMBC 2009, Minneapolis, MN, USA, 3–6 September 2009; pp. 2651–2654.
14. Acharya, U.R.; Molinari, F.; Sree, S.V.; Chattopadhyay, S.; Ng, K.-H.; Suri, J.S. Automated diagnosis of epileptic EEG using entropies. *Biomed. Signal Process. Control* **2012**, *7*, 401–408. [CrossRef]
15. Sharma, R.; Pachori, R.B.; Acharya, U.R. Application of Entropy Measures on Intrinsic Mode Functions for the Automated Identification of Focal Electroencephalogram Signals. *Entropy* **2015**, *17*, 669–691. [CrossRef]
16. Li, J.; Yan, J.; Liu, X.; Ouyang, G. Using Permutation Entropy to Measure the Changes in EEG Signals during Absence Seizures. *Entropy* **2014**, *16*, 3049–3061. [CrossRef]
17. Wendling, F.; Bartolomei, F.; Bellanger, J.-J.; Bourien, J.; Chauvel, P. Epileptic fast intracerebral EEG activity: Evidence for spatial decorrelation at seizure onset. *Brain* **2003**, *126*, 1449–1459. [CrossRef]
18. Shantha Selva Kumari, R.; Prabin Jose, J. Seizure detection in EEG using time frequency analysis and SVM. In Proceedings of the 2011 International Conference on Emerging Trends in Electrical and Computer Technology, ICETECT 2011, Nagercoil, India, 23–24 March 2011; pp. 626–630.
19. Ridouh, A.; Boutana, D.; Bourennane, S. EEG Signals Classification Based on Time Frequency Analysis. *J. Circuits Syst. Comput.* **2017**, *26*, 1750198. [CrossRef]
20. Myers, M.H.; Jolly, E.; Li, Y.; de Jongh Curry, A.; Parfenova, H. Power Spectral Density Analysis of Electrocorticogram Recordings during Cerebral Hypothermia in Neonatal Seizures. *Ann. Neurosci.* **2017**, *24*, 12–19. [CrossRef] [PubMed]
21. Zhang, Z.; Parhi, K.K. Seizure detection using regression tree based feature selection and polynomial SVM classification. In Proceedings of the 2015 37th Annual International Conference of the IEEE Engineering in Medicine and Biology Society (EMBC), Milan, Italy, 25–29 August 2015; pp. 6578–6581.
22. Bandarabadi, M.; Teixeira, C.; Rasekhi, J.; Dourado, A. Epileptic seizure prediction using relative spectral power features. *Clin. Neurophysiol.* **2015**, *126*, 237–248. [CrossRef]

23. Chen, J.; Zhou, X.; Jin, L.; Lu, Q.; Sun, H.; Liu, Q.; Huang, Y. Can Spectral Power Be Used as a Candidate Seizure Marker of the Periodic Discharges Pattern? *Front. Neurol.* **2021**, *12*, 642669. [CrossRef]
24. Subha, D.P.; Joseph, P.K.; Acharya, R.; Lim, C.M. EEG Signal Analysis: A Survey. *J. Med. Syst.* **2010**, *34*, 195–212. [CrossRef]
25. Kotsiantis, S.B. Supervised machine learning: A review of classification techniques. *Informatica* **2007**, *31*, 249–268.
26. Fu, K.; Qu, J.; Chai, Y.; Dong, Y. Classification of seizure based on the time-frequency image of EEG signals using HHT and SVM. *Biomed. Signal Process. Control* **2014**, *13*, 15–22. [CrossRef]
27. Temko, A.; Thomas, E.; Marnane, W.; Lightbody, G.; Boylan, G. EEG-based neonatal seizure detection with Support Vector Machines. *Clin. Neurophysiol.* **2011**, *122*, 464–473. [CrossRef]
28. Gómez, C.; Arbeláez, P.; Navarrete, M.; Alvarado-Rojas, C.; Le Van Quyen, M.; Valderrama, M. Automatic seizure detection based on imaged-EEG signals through fully convolutional networks. *Sci. Rep.* **2020**, *10*, 21833. [CrossRef] [PubMed]
29. Aarabi, A.; Wallois, F.; Grebe, R. Automated neonatal seizure detection: A multistage classification system through feature selection based on relevance and redundancy analysis. *Clin. Neurophysiol.* **2006**, *117*, 328–340. [CrossRef]
30. Shoeb, A.H. Application of Machine Learning to Epileptic Seizure Onset Detection and Treatment. Ph.D. Thesis, Mass NSL of Technology, Massachusetts Institute of Technology, Cambridge, MA, USA, 2009.
31. Khan, Y.U.; Rafiuddin, N.; Farooq, O. Automated seizure detection in scalp EEG using multiple wavelet scales. In Proceedings of the 2012 IEEE International Conference on Signal Processing, Computing and Control (ISPCC 2012), Solan, India, 15–17 March 2012.
32. Yuan, Q.; Zhou, W.; Liu, Y.; Wang, J. Epileptic seizure detection with linear and nonlinear features. *Epilepsy Behav.* **2012**, *24*, 415–421. [CrossRef] [PubMed]
33. Williamson, J.R.; Bliss, D.W.; Browne, D.W.; Narayanan, J.T. Seizure prediction using EEG spatiotemporal correlation structure. *Epilepsy Behav.* **2012**, *25*, 230–238. [CrossRef]
34. Patnaik, L.; Manyam, O.K. Epileptic EEG detection using neural networks and post-classification. *Comput. Methods Programs Biomed.* **2008**, *91*, 100–109. [CrossRef]
35. González Otárula, K.A.; Mikhaeil-Demo, Y.; Bachman, E.M.; Balaguera, P.; Schuele, S. Automated seizure detection accuracy for ambulatory EEG recordings. *Neurology* **2019**, *92*, e1540–e1546. [CrossRef]
36. Marín, M.R.; Martínez, I.V.; Bermúdez, G.R.; Porfiri, M. Integrating old and new complexity measures toward automated seizure detection from long-term video EEG recordings. *iScience* **2021**, *24*, 101997. [CrossRef]
37. Shum, J.; Friedman, D. Commercially available seizure detection devices: A systematic review. *J. Neurol. Sci.* **2021**, *428*, 117611. [CrossRef]
38. Koren, J.; Hafner, S.; Feigl, M.; Baumgartner, C. Systematic analysis and comparison of commercial seizure-detection software. *Epilepsia* **2021**, *62*, 426–438. [CrossRef]
39. Bruno, E.; Viana, P.F.; Sperling, M.R.; Richardson, M.P. Seizure detection at home: Do devices on the market match the needs of people living with epilepsy and their caregivers? *Epilepsia* **2020**, *61*, S11–S24. [CrossRef] [PubMed]
40. Obeid, I.; Picone, J. The Temple University Hospital EEG Data Corpus. *Front. Neurosci.* **2016**, *10*, 196. [CrossRef] [PubMed]
41. Shah, V.; von Weltin, E.; Lopez, S.; McHugh, J.R.; Veloso, L.; Golmohammadi, M.; Obeid, I.; Picone, J. The Temple University Hospital Seizure Detection Corpus. *Front. Neuroinformatics* **2018**, *12*, 83. [CrossRef]
42. Ochal, D.; Rahman, S.; Ferrell, S.; Elseify, T.; Obeid, I.; Picone, J. The Temple University Hospital EEG Corpus: Annotation Guidelines. *Inst. Signal Inf. Process. Rep.* **2020**, *1*, 1–28.
43. Kursa, M.B. Robustness of Random Forest-based gene selection methods. *BMC Bioinform.* **2014**, *15*, 8. [CrossRef]
44. Detti, P.; De Lara, G.Z.M.; Bruni, R.; Pranzo, M.; Sarnari, F.; Vatti, G. A Patient-Specific Approach for Short-Term Epileptic Seizures Prediction through the Analysis of EEG Synchronization. *IEEE Trans. Biomed. Eng.* **2019**, *66*, 1494–1504. [CrossRef]
45. Breiman, L. Random forests. *Mach. Learn.* **2001**, *45*, 5–32. [CrossRef]
46. Friedman, J.H. Greedy function approximation: A gradient boosting machine. *Ann. Stat.* **2001**, *29*, 1189–1232. [CrossRef]
47. Pereira, F.; Mitchell, T.; Botvinick, M. Machine learning classifiers and fMRI: A tutorial overview. *NeuroImage* **2009**, *45*, S199–S209. [CrossRef]
48. Sharma, R.; Pachori, R.B.; Sircar, P. Seizures classification based on higher order statistics and deep neural network. *Biomed. Signal Process. Control* **2020**, *59*, 101921. [CrossRef]
49. Tibdewal, M.N.; Dey, H.R.; Mahadevappa, M.; Ray, A.; Malokar, M. Multiple entropies performance measure for detection and localization of multi-channel epileptic EEG. *Biomed. Signal Process. Control* **2017**, *38*, 158–167. [CrossRef]
50. Biau, G.; Scornet, E. A random forest guided tour. *TEST* **2016**, *25*, 197–227. [CrossRef]
51. Speiser, J.L.; Miller, M.E.; Tooze, J.; Ip, E. A comparison of random forest variable selection methods for classification prediction modeling. *Expert Syst. Appl.* **2019**, *134*, 93–101. [CrossRef]
52. Tsipouras, M.G. Spectral information of EEG signals with respect to epilepsy classification. *EURASIP J. Adv. Signal Process.* **2019**, *2019*, 10. [CrossRef]
53. Pérez-Elvira, R.; Oltra-Cucarella, J.; Carrobles, J.; Teodoru, M.; Bacila, C.; Neamtu, B. Individual Alpha Peak Frequency, an Important Biomarker for Live Z-Score Training Neurofeedback in Adolescents with Learning Disabilities. *Brain Sci.* **2021**, *11*, 167. [CrossRef]

Review

Quantitative Electroencephalogram (qEEG) as a Natural and Non-Invasive Window into Living Brain and Mind in the Functional Continuum of Healthy and Pathological Conditions

Alexander A. Fingelkurts * and Andrew A. Fingelkurts

BM-Science—Brain and Mind Technologies Research Centre, FI-02601 Espoo, Finland
* Correspondence: alexander.fingelkurts@bm-science.com

Featured Application: The presented theoretical–conceptual framework has neurobiological/etiological relevance and focuses on dimensionally conceptualized physiological characteristics, mental functions, and neuropsychopathology and, as such, may provide the better clinical diagnostic and prognostic utility of qEEGs.

Abstract: Many practicing clinicians are time-poor and are unaware of the accumulated neuroscience developments. Additionally, given the conservative nature of their field, key insights and findings trickle through into the mainstream clinical zeitgeist rather slowly. Over many decades, clinical, systemic, and cognitive neuroscience have produced a large and diverse body of evidence for the potential utility of brain activity (measured by electroencephalogram—EEG) for neurology and psychiatry. Unfortunately, these data are enormous and essential information often gets buried, leaving many researchers stuck with outdated paradigms. Additionally, the lack of a conceptual and unifying theoretical framework, which can bind diverse facts and relate them in a meaningful way, makes the whole situation even more complex. To contribute to the systematization of essential data (from the authors' point of view), we present an overview of important findings in the fields of electrophysiology and clinical, systemic, and cognitive neuroscience and provide a general theoretical–conceptual framework that is important for any application of EEG signal analysis in neuropsychopathology. In this context, we intentionally omit detailed descriptions of EEG characteristics associated with neuropsychopathology as irrelevant to this theoretical–conceptual review.

Keywords: quantitative electroencephalogram (qEEG); neurometrics; neurology; psychiatry; disease; health; phenotype; brain; mind

1. Introduction

"The history of EEG studies of mental activity shows that a weak theoretical basis at certain stages can result not only in methodological crises but can also affect empirical data collection and interpretation. An adequate theory can lend strong support to the methodology with "brain-oriented" structuring of psychological tasks and such a theory improves the neurophysiological informative value of the EEG parameters referring to the psychological characteristics of mental processes etc" ([1], p. 384). "It is time to begin the daunting task of relating clinical manifestations of mental disorders to neuroscientific brain dynamics in a comprehensive unifying manner" ([2], p. 942).

In the course of everyday life conditions and within the context of health and disease, people can be evaluated across three different dimensions: *behavioral* (performance), *brain functioning* (neuroimaging), and *introspection* (subjective aspects—psychology). Changes in these different dimensions do not always parallel one another; however, the *common denominator* for all of them is brain functioning, which affects and reflects the other two dimensions—behavior/performance and psychology/subjectivity [1]. Indeed, the physiologic functioning of the brain underlies emotions, cognition, and behavior; hence, in this context, an objective assessment of brain dysfunction is especially critical for neurology and psychiatry [1,3].

Currently, brain dysfunction is assessed according to manuals such as the Diagnostic Statistical Manual in the USA (DSM) or the International Classification of Disorders in Europe (ICD). However, none of the DSM-/ICD-defined syndromes correlate with any neurobiological phenotypic marker or gene that could have etiological relevance or predict the efficacy of medications [2]. It seems that to arrive at a biological basis for disease categories, brain disorders should be classified in association with impairment of brain systems and diagnosed according to deviations from normality in the corresponding brain activity [3].

Several factors need to be considered when choosing an appropriate measure of brain activity. First, we need a *natural* and *non-invasive* 'window' into the *living* brain. This window should give us an 'online' view that *directly* captures the dynamics of *brain activity*, which reflects multiple interacting operational modules hierarchically organized to allow for complex information processing that (a) characterizes the neurophysiological type (combination of traits; *trait* refers to the constitutional characteristics that are *temporally stable* over longer periods of an individual's life-span; an individual with a certain trait characteristic responds similarly over many situations over the period of several months or years; here, trait corresponds to a temporally stable neuro-psycho-physiological system) and multidimensional structure of the functional state of the brain, (b) has high heritability and is, thus, individually specific, (c) reflects individual neurodevelopment ('historicism') and age-related changes, (d) is associated with higher mental and cognitive functions and subjective experience, and (e) reflects or guides neuro- and psychopathology.

Second, from a biophysical perspective, "disease may be regarded not only as a functional or molecular–structural abnormality, as in the classic view, but also (and not by way of contrast) as a disturbance of an entire network of electromagnetic communications. This network is based on long-range interactions between elements [...] which oscillate at frequencies which are coherent and specific and thus capable of resonance. This would be a disturbance of internal oscillators and their communications" ([4], pp. 107–108; see also [5]).

Considering these two aspects, it seems that electroencephalogram (EEG) is the best candidate for measuring brain activity compared to other methods of brain signal acquisition, such as the magnetoencephalogram (MEG), functional magnetic resonance imaging (fMRI), and positron emission tomography (PET).

2. Why the Electroencephalogram (EEG)?

MEG, fMRI, and PET are expensive, non-portable, partially invasive, and are usually associated with high stress due to noise, space confinement, and the need to be motionless. Additionally, fMRI and PET scans provide indirect measures of brain activity, with poor temporal resolution. Further, many types of mental activities, brain disorders, and malfunctions of the brain cannot be registered using fMRI since its effect on the level of oxygenated blood is low [5]. In contrast to these neuroimaging techniques, EEG is, at the *same time*, the cheapest, fastest, and most portable technique that measures neuronal activity directly and non-invasively. EEG does not elicit feelings of claustrophobia, does not require overt cooperative behavior from the person, has a temporal resolution adequate to mental and cognitive processes, and may distinguish between different temporal scales of information processing inherent to mental and cognitive processes.

EEG is a summation of electric voltage fields produced by dendritic and postsynaptic currents of many cortical neurons firing in non-random partial synchrony [6–8]. The aggregate of these electric voltage fields can be detected by electrodes on the scalp. The brainstem and thalamus serve as subcortical generators to synchronize populations of neocortical neurons in both normal and abnormal conditions, thus influencing the EEG. It seems that the activity of subcortical structures can be 'visible' in the EEG either indirectly through their effects on cortical activity or—in contrast to popular belief—more directly via deep sources. See explanations given in [9] (p. 7): "While local field potentials indeed fall off rapidly within the brain, far less attenuation is observed when recording across skull and scalp. The reason is that the lower conductivity of the skull (compared to the brain and scalp) attenuates superficial sources more strongly than deep ones, thus acting like a spatial low-pass filter. This property causes strong blurring and attenuation of the focal superficial fields but has less of an effect on the more diffuse ("low spatial frequency") fields from deeper sources [...] Recent evidence suggests that a considerably larger range of brain structures, layers, and cell types than previously thought can contribute to spontaneous EEG phenomena".

A quantitative electroencephalogram (qEEG) is a mathematically and algorithmically processed digitally recorded EEG that extracts information invisible to 'naked' eye inspections of the signal. For the rest of the paper, we will mostly refer to qEEGs as the majority of studies are performed using qEEGs.

After decades of studies, it is becoming clear that the qEEG is closely related to brain dynamics, with millisecond temporal resolution, functional properties, and global states of brain functioning, information processing, and cognitive activity [10–17]. The interaction of large populations of neurons gives rise to rhythmic electrical events in the brain, which can be observed at several temporal scales—qEEG oscillations. They are the basis of many different behavioral patterns and sensory mechanisms (for a review, see [18]). Indeed, a large body of evidence [19–27] has demonstrated that qEEG oscillations constitute a mechanism by which the brain can actively regulate changes in a state in selected neuronal networks to cause qualitative transitions between modes of information processing [20]. Thus, different qEEG oscillatory patterns are indicative of different information-processing states.

The qEEG has a number of important features that make it especially useful in clinical practice. In the following sections, we present a brief review of these features.

2.1. qEEG Historicism

An adult human qEEG is characterized by 'historicism'—the information about primate phylogeny, pre- and post-natal maturation (individual development), and early life events (utero characteristics and early life stress).

Indeed, phylogenetically (*phylogenesis*—the evolutionary development and diversification of a species or group of organisms), the proportion of power of qEEG oscillations changes as a function of primate phylogeny [5,22,28]. Likewise, ontogenetically (*ontogenesis*—physical and psychological development of an individual organism from inception to maturity), qEEGs undergo significant transformation as a function of *pre*-natal (in utero) development (maternal stress exposure, anxiety, and depression during pregnancy are considered in utero adverse experiences and have been associated with future health problems [29–32]; this is so because intrauterine life events have a much greater impact on epigenetic profiles than stressful exposures during adult life due to heightened brain plasticity that is adversely affected by exposure to environmental insults [33]) [34] as well as a function of *post*-natal maturation (*maturation* refers to the timely appearance or unfolding of brain structures, events, and processes that are the result of the interaction between genes and the environment; brain maturation can be *delayed*, *equal*, or *accelerated* when compared to chronological age) [35–38]. It seems that ontogenetic differences mirror those of phylogenetic differences in the cause of brain development, where there is a gradual increase in qEEG complexity and change in the qEEG oscillations' composition and proportions [17,39–41]. Why is this relevant? The qEEG has been found to have a high prognostic value for identifying the functional level of 'brain maturity' [42,43]. The knowledge of typical qEEG oscillatory patterns for a given phylogenesis/ontogenesis stage gives one the ability to assess the level of qEEG maturation or regression, which often accompanies the development of neuropsychopathology [44]. For example, a person with an immature qEEG is more easily swayed by external influences and has a lower threshold for aggressive and/or antisocial behaviors [45].

Additionally, early life stress (ELS) has been associated with abnormalities in the qEEG of adults and is also paralleled by a range of adverse outcomes in adults, such as personality dimensions, increased vulnerability to substance abuse, depression, anxiety, psychosis, and post-traumatic stress disorder (PTSD) [46–48]. Indeed, ELS such as protein energy malnutrition in the first year of life, extreme social and cognitive deprivation as a result of institutional care, physical or emotional neglect, and low socioeconomic status are all associated with abnormal qEEG characteristics on one hand and with developmental lag or deviation, persistent specific cognitive and behavioral deficits in adulthood, and accelerated cognitive decline [49–54] on another hand. Further, childhood traumas (including childhood sexual abuse) are also associated with adult qEEG deviations in parallel with cognitive dysfunction [55,56]. It seems that changes in catecholamine levels following a traumatic event can impede brain regional development, which, in turn, can compromise later cognitive functioning and emotional regulation and leave a person susceptible to stress later on in life.

Additionally, traumatic brain injury may also be reflected in qEEG deviations that correspond to complaints of cognitive symptoms that can persist anywhere from 11 [57], 22 [58], or even 27 [59] years post-injury, characterizing persistent post-concussive syndrome.

This brief overview suggests that the qEEG contains information that is a historical consequence of individual development, ELS, and significant life events. However, in order to adequately assess qEEG variability associated with pathology, *within-subject stability* over EEG recordings within an EEG session, *test–retest reliability* over time, and *intra-subject specificity* (i.e., the extent to which a qEEG pattern is uniquely associated with a given person) and specificity for different conditions need to be established.

2.2. qEEG Stability, Reliability, and Specificity

Studies have reportedly demonstrated that the majority of qEEG characteristics have high (up to 90%) within-subject stability (internal consistency measured by Cronbach's alpha) within an EEG recording session, high (up to 90%) reproducibility (test–retest reliability) over a period of hours, weeks, months and years, and high (up to 99%) intra-subject specificity, meaning that qEEG can accurately identify subjects from a large group [60–78].

These results suggest that qEEG characteristics possess trait-like qualities (stability over time). In this context, intrinsic properties of brain activity measured by resting qEEG constitute a neural counterpart of personality traits (Section 4.2) and can be regarded as the *statistical neuro-signature* of a person. Such high stability, reliability, and specificity of qEEG characteristics suggest that genetic factors have a strong influence on qEEG variation.

2.3. qEEG Heritability

A large body of studies have suggested that qEEG characteristics and their variability are largely determined by genetics and, thus, are highly heritable (up to 90%) [79–93]. Additionally, it was demonstrated that the correlations for qEEG characteristics between family groups (each consisting of a biologically related father, mother, and two children) were greater than those obtained from the non-family groups (each consisting of biologically unrelated subjects) [94] (see also [89]).

Smit et al. [92] proposed several common genetic sources for EEG: (a) skull and scalp thickness may affect the conductive properties of the tissues surrounding the cortex, (b) genetic influence on cerebral rhythm generators such as the central 'pacemaker' in the septum for hippocampal activity or the thalamocortical and corticocortical generators of cortical rhythmicity, (c) genes directly involved in the bioelectric basis of the EEG signal itself: for example, genes influencing the number of pyramidal cells, the number of dendritic connections, or their orientation with respect to the scalp may directly influence the mass dendritic tree depolarization of pyramidal cells in the cortex that underlies the EEG. Begleiter and Porjesz [95] added another factor: regulatory genes that control the neurochemical processes of the brain and, therefore, influence neural function.

Besides high qEEG heritability, genetic loci underlying the functional organization of human neuroelectric activity and their associated conditions/behavior have also been identified. Below is a short overview of qEEG oscillations, the related genes, and the associated pathological conditions:

(a) qEEG beta oscillations (*beta rhythm* is electromagnetic oscillations in the frequency range of brain activity above 13 Hz)

Winterer et al. [96] reported that three exonic variants of the gene encoding the human gamma-amino butyric acid $(GABA)_B$ receptor on chromosome 6 modify the cortical synchronization measured as scalp-recorded qEEG coherence. Another genetic study indicated the importance of $GABA_A$ receptor genes in the modulation of qEEG beta oscillations in the human brain: Porjesz et al. [97] found a significant genetic linkage between the beta frequency of the human qEEG and a cluster of $GABA_A$ receptor genes on chromosome 4p. Additionally, this same $GABA_A$ receptor gene was found to be associated with a DSM-IV diagnosis of alcohol dependence [98].

(b) qEEG alpha oscillations (*alpha rhythm* is electromagnetic oscillations in the frequency range of 8–13 Hz, arising from the synchronous and coherent electrical activity of neurons in the human brain)

Low voltage qEEG alpha oscillations have also been reported to be linked to (a) the GABAergic system, as an association has been found between the exon 7 variant of the $GABA_B$ receptor gene and alpha voltage [99], (b) a serotonin receptor gene (HTR3B), associated with alcoholism and antisocial behavior [100], and (c) a corticotrophin-releasing binding hormone (CRH-BP) [101,102], associated with depression, anxiety, and alcoholism. Low voltage alpha in females has also been reported to be associated with a genetic variant that leads to low activity of the enzyme that metabolizes dopamine and norepinephrine, catechol-o-methyltransferase (COMT) [103]. Additionally, low voltage alpha has been associated with a subtype of alcohol dependence with anxiety disorders [104,105] and with the brain-derived neurotrophic factor (BDNF) Val66Met polymorphism in depression [106]. High voltage qEEG alpha oscillations are heritable in a simple autosomal dominance manner [79]. The alpha peak frequency (APF) has been associated with the COMT gene, with the Val/Val genotype being marked by a 1.4 Hz slower APF compared to the Met/Met group [107].

(c) qEEG theta oscillations (*theta rhythm* is electromagnetic oscillations in the frequency range of brain activity between 4 and 7.5 Hz)

There is evidence [108] that single nucleotide polymorphisms located in brain-expressed long intergenic non-coding RNAs (lincRNAs) on chromosome 18q23 are associated with posterior interhemispheric theta EEG coherence. These same variants are also associated with alcohol use behavior and posterior corpus callosum volume. Further, the $Val^{158}Met$ polymorphism of the COMT gene is associated with low-frequency oscillation abnormalities in schizophrenia patients [109].

This short overview suggests that there are common genetic links between qEEG oscillation characteristics and specific health conditions. It seems that genetically influenced features of the intrinsic oscillatory activity are related to the structures and functions of the corresponding neural generators and that different features of qEEGs may predict individual differences in brain function and structures.

2.4. qEEG and Structural Integrity of the Brain

Indeed, numerous studies have demonstrated that qEEGs reflect the brain's *structural* characteristics (or '*hardware*'), such as the number of connections between neurons, white matter density, axonal diameter, degree of myelination and white matter integrity, as well as the integrity of corticocortical and thalamocortical circuits, hippocampal volume, the number of active synapses in thalamic nuclei, and the number of potential neural pathways [7,8,110–126]. For example, reduced EEG amplitude is believed to be partially due to a reduced number of synaptic generators and/or reduced integrity of the protein/lipid membranes of neurons [127,128].

2.5. qEEG and Functional Integrity of the Brain

Decades of studies have demonstrated that brain *functional* characteristics (or '*software*'), such as memory performance, attention and processing speed, emotional regulation, individual capacity for information processing, cognitive preparedness, and others, including functional states of the brain, are readily reflected in qEEGs at all ages in both healthy individuals and individuals with neurological or psychiatric conditions [14,21,22,24,115,129–147] (for a review, see [18]). This is so because qEEG oscillatory activity is generated by synchronous neural populations that mirror the firing rate of their constituent neurons [148]: for example, during arousal, task execution, and/or a behavioral act, the underlying neuronal populations will increase spiking with respect to baseline. These increased firing rates will engage non-linear feedback loops, effectively changing the system's response function and the specifics of its emergent oscillations. In contrast, during rest and states of quietness, the spiking activity decreases, which is also reflected by a decrease in oscillatory activity [148]. Further, qEEG oscillations are able to temporally coordinate and control neuronal firing and are proposed to be a basic principle of information processing in the human brain [149,150].

Considering that different qEEG oscillations reflect functionally different components of information processing acting on various temporal scales [140,141], it is possible to map qEEG oscillations onto specific mental and/or behavioral states [151]. qEEG oscillations from the same frequency band may express different functions depending on the conditions they are involved in [152]. This seems biologically plausible: qEEG oscillatory functional diversity creates a rich repertoire of brain activity that can meet the complex computational and communicational demands of the brain during healthy and pathological conditions.

In this context, qEEG measures can provide independent evidence of variations in alertness, attentiveness, memory, emotional regulation, or mental effort. Incorporating them into tests of cognitive function might lead to more sensitive and less ambiguous clinical assessment tools [153,154].

Since information-processing modes depend on the functional integrity of the brain, which, in turn, depends on the orchestrated oscillatory activity of neuronal pools (reflected in the characteristic qEEG rhythms); functional coupling between qEEG oscillations, cognitive functions, and vegetative processes is important.

2.6. qEEG and Vegetative Status/Autonomic Nervous Systems (ANS)

Several studies have demonstrated the association between qEEGs and ANS [155–158]. It seems that the brainstem mediates a functional coupling between the ANS and the central nervous system (CNS) assessed by qEEG [159–161]. A theoretical concept of the integration between the ANS and CNS was presented by Jennings and Coles [162]. The coordination and communication in and between the autonomic vegetative systems and the brain occur with tuned frequencies in the range of qEEG oscillations, suggesting the existence of resonant links in the brain with all organs of the body (for a review and discussion, see [5]; see also [163]). Basar [5] suggested that such mutual resonances form a coordinated dynamic system that maintains survival functions such as blood pressure, respiratory rhythms, cardiac pacemakers, and body temperature (see also [155,156,159–162]).

Since the dynamics of the physiologic variables (autonomic system) and the dynamics of brain activity depend on each other, it is reasonable to hypothesize that reduced variability in the activity of the neural networks should cause a concurrent decrease in the variability of autonomic physiologic functions. Indeed, it was demonstrated that a widespread brain injury that causes a derangement in neural networks leads to a reduced complexity of qEEG (measured by entropy, the dynamic repertoire of the probable qEEG states, and operational architectonics) [164–166] and reduced heart rate variability [167] in unresponsive patients compared to healthy subjects.

It seems that decreased qEEG variability is coupled with a decrease in the variability of other physiologic variables (autonomic system), which results in reduced physiological adaptability. In turn, reduced physiological adaptability can contribute to stress and weakened immunity, which may further impact the qEEG pattern, creating a downward spiral.

2.7. qEEG, Stress, and Immunity

There is a strong link between qEEG oscillatory patterns and stress regulatory systems: the hypothalamic–pituitary–adrenocortical (HPA) axis and the sympathetic–adrenomedullary axis [168]. For example, qEEGs recorded from stressed students before an exam revealed a correlation between greater right hemisphere (RH) activation and higher cortisol levels [169]. This is supported by the following facts: (a) the administration of cortisol to healthy participants has been shown to increase RH frontal activation [170], and (b) greater right-sided activation (measured by resting qEEG) is associated with higher levels of basal cortisol compared to their left-activated counterparts [171]. Cortisol also seems to reduce neural interactions between different areas of the brain. Indeed, an inverse relationship between basal cortisol levels and neural interaction between the frontal and parietal cortex has been demonstrated using qEEG connectivity analysis [172].

Considering the link between qEEG oscillatory patterns and stress regulatory systems, it is not surprising that the association between several factors of the immune system and qEEG activity has also been reported [173–177] (for a recent meta-analysis, see [178]). For example, higher levels of right-prefrontal qEEG activation (a) reliably predicted poorer immune response [176] and (b) are characterized by lower levels of natural killer cell activity [179]. These data support the hypothesis that individuals characterized by a more negative affective style have a weaker immune response and, therefore, may be at greater risk for illness than those with a more positive affective style. Additionally, RH activation is associated with hyprecortisolemia, which contributes to the deterioration of immune system functioning and puts depressed patients at a greater risk of developing other illnesses, accounting for depression's high comorbidity with other diseases [180].

2.8. qEEG and Cerebral Haemodynamics and Metabolism

Studies have suggested that different qEEG characteristics are related to cerebral hemodynamics and metabolism [181–190]. Cerebral cortex metabolism disturbance is associated with and may be responsible for cortical neural synchronization anomalies that may manifest as abnormal qEEG oscillations [191]. Additionally, changes in the characteristics of qEEG oscillations (amplitude, power, frequency) are proportional to cerebrovascular damage (CVD) [119]. The qEEG has been shown to be a reliable marker of the decline in neuronal integrity associated with a decline in blood flow [192–198]. Additionally, studies show a sensitivity greater than 80%, false-positive rates below 5–10%, and correlations of 70% between qEEG and blood flow in ischemic and non-ischemic regions, thus suggesting that the qEEG can reliably detect focal features that can be quite abnormal even if the computer tomography (CT) or MRI scans are still normal (dysfunction without infarction) [199]. Similarly, in patients with subarachnoid hemorrhage, only qEEG could differentiate patients with and without cerebral infarction and not doppler/color-coded duplex sonography [200]. Further, recent meta-analyses have shown that qEEG has prognostic potential in predicting patient independence and stroke severity beyond that afforded by standard clinical assessments [201] (see also [202,203]). Indeed, qEEG changes precede that of multimodal monitoring or confirmation of infarction on CT [204].

Cerebral hemodynamics and metabolism are regulated by a complex interaction between different homeostatic mechanisms where neurotransmitters play a significant role.

2.9. qEEG and Neurotransmitters

Several studies have suggested a relation between different qEEG oscillations and neuromodulator balance [5,205]. This is because peculiarities of qEEG features result from the interaction of numerous resonance loops within the cortex and between the cortex and subcortical structures, and these interactions are significantly influenced by neurotransmitter concentrations in the brain [206]. Indeed, the levels of activity of different neurotransmitter systems (acetylcholinergic (ACh-ergic), noradrenalinergic (NA-ergic), dopaminergic (DA-ergic), serotonergic (ST-ergic), and GABA-ergic), as well as the patterns of their interaction, are important drivers of qEEG oscillations. For example, the activation of the NA-ergic system is associated with the desynchronization of qEEGs during behavioral excitation [207] and an increase in high-frequency qEEG oscillations [208]. It is also believed that the increased activity of the DA-ergic cerebral systems results in shifts of the frequencies of qEEG oscillations toward higher ranges and facilitates the reaction of desynchronization [209]. Additionally, posterior vs. anterior distribution of qEEG theta oscillations is informative on DA levels [210]. Low ST levels result in a higher power of low-frequency qEEG components [206]; conversely, high ST levels result in the decreased power of low-frequency qEEG components and the higher power of high-frequency qEEG components [211]. Higher relative levels of ACh promote qEEG alpha oscillations, whereas an increased tone of inhibitory monoamine receptors is associated with qEEG delta oscillations (*delta rhythm* is electromagnetic oscillations in the frequency range of brain activity between 1.5 and 3.5 Hz) [205]. It seems that for each qEEG oscillatory pattern, there is a correlated neurotransmitter mix [212].

Deficiencies or excesses of any of the neurotransmitters will produce a marked departure from homeostatically regulated normative qEEG oscillatory patterns and may contribute to neuro–psycho pathophysiology [199,205]. Indeed, a large body of data suggests that it is possible to unravel distinctive abnormal qEEG oscillatory profiles in terms of specific neurochemical imbalances in particular brain regions [213].

2.10. qEEG and Neuropsychopathology

The literature indicates that there is a greater proportion of abnormal EEGs in individuals with psychopathology: (a) up to 68% of qEEGs in psychiatric patients display evidence of pathophysiology, and these results have additional utility beyond simply ruling out 'organic brain lesions' [214,215]; (b) up to 73% of nonepileptic adults have qEEG epileptiform discharges (EDs) [216] that are attributable to underlying brain abnormalities (traumatic, vascular, tumor, metabolic), medications, and psychiatric disorders (see, for example, [217]); (c) the mean prevalence of interictal qEEG abnormalities in psychogenic nonepileptic seizures is estimated to be 26% [218–226]; (d) up to 30% of panic attack patients have demonstrable qEEG abnormalities, especially in atypical presentations of panic attacks, and the incidence of abnormal qEEG findings in mood disorders reaches 40% [227]; (e) up to 78% of antisocial and criminal populations have underlying qEEG abnormalities [228] that are more prevalent in subjects with violent crimes, repeated violence, and motiveless crimes; (f) up to 76% of children with reading disabilities but without severe disorders of behavior have EEG abnormalities [229], and (g) 69% of youngsters with behavior disorders with a predominance of aggressiveness have EEG deviations [230]. Additionally, there is evidence that abnormal EEGs are associated with the following clinical conditions: negative histories (13%), severe head injury or neuropsychiatric disorder (46%), psychopathic personality (88%), and family history of seizures (62%) [231].

Basic mechanisms of cerebral rhythmic activities in norm and pathology are described in detail in Steriade et al. [212]. This emphasizes that the presence of qEEG abnormalities should be inferred as '*electrographic markers*' of underlying brain *dysfunction* and is suggestive of the potential usefulness of qEEGs in clinical practice.

Indeed, more recent research shows that certain neuropsychopathologies, such as attention deficit hyperactivity disorder (ADHD), specific learning disabilities, schizophrenia, obsessive–compulsive disorder (OCD), borderline personality disorder (BPD), depression, suicidal ideation, anxiety disorders, traumatic brain injury (TBI), mild cognitive impairment (MCI), Alzheimer's disease (AD), and other disorders are associated with specific qEEG patterns and that these spontaneous electric potentials provide reliable markers of brain function and dysfunction [56,152,232–246] (for reviews, see [199,213,247]).

Given that patients with different disorders display abnormal and distinct qEEG-profiles, it is not surprising that they can be differentially classified utilizing qEEG-variables [248]. For example, qEEG utility in discrimination/differentiation between affective disorders and schizophrenia [249], between Alzheimer and non-Alzheimer dementias [199,250], between sub-types of dementia [251], between depression and dementia [199], between schizophrenia and unipolar and bipolar depression [199], and between panic disorder and depression [199] have been demonstrated. For sensitivity and specificity values of qEEG-based detection/discrimination of patients with specific disorders, see Table 1.

Table 1. qEEG-based detection/discrimination of patients with specific dysfunction/disorder.

Dysfunction/Disorder	Sensitivity	Specificity	References
Cerebrovascular disease	>80%		[199]
ADHD/ADD vs. normal children	90%	94%	[237]
ADD vs. DLD	97%	84.2%	[252]
LD vs. normal children	72%	80%	[253]
Depression	72–93%	75–88%	[254]
Panic disorder	71%	84%	[255]
Dementia	91.9%	92.2%	[256]
AD	71–81%		[257,258]
Declining to MCI	95%	94.1%	[259]
Converting to AD	96.3%	94.1%	[259]
Alcohol and drug abuse predicting relapse	61%	85%	[260]
mTBI vs. sTBI	95.5%	97.4%	[238,261]

ADHD/ADD = attention deficit disorders with or without hyperactivity; DLD = developmental learning disorders; LD = learning disorders; AD = Alzheimer's disease; MCI = minimal cognitive impairment; m(s) TBI = mild(severe) traumatic brain injury.

It is argued that the levels of specificity found in qEEG studies are often higher than those found in routinely used clinical tests, such as mammograms, cervical screenings, and brain scans such as CT or single photon emission computed tomography (SPECT) [199,262,263].

Even within the same disorder, qEEGs may be beneficial in identifying the cause of the abnormal behavior. For example, Kropotov distinguishes five reasons for the neurophysiology of ADHD, stating that "[. . .] mentioned dysfunctions are associated with specific patterns in spontaneous and evoked electrical potentials, recorded from the head by multiple surface electrodes" ([264] (p. 74; see also [3,265]).

Additionally, qEEGs can play a unique role when it comes to dealing with ambiguous or edge cases in clinical practice. It may help to identify/differentiate:

- Electrical changes that precede the clinical onset of a seizure by tens of seconds to minutes—the *early detection of a seizure*. It has been shown that patients go through a preictal transition for approximately 0.5 to 1 h before a seizure occurs [266]. On average, the prediction rate is ~81% and has an average warning time of 63 min [267];
- Whether a given seizure is *epileptic* or *nonepileptic in origin*: For example, there are groups of disorders that produce symptoms similar to an epileptic seizure: (a) cardiac arrhythmias causing syncope, episodes caused by cerebrovascular disease, movement disorders, and unusual manifestations of sleep disorders; (b) events of psychiatric origin (often referred to as psychogenic nonepileptic seizures (PNES)) [268];
- *Subclinical seizures*: Some seizures recorded during prolonged EEG monitoring may be asymptomatic or 'subclinical';
- Whether the cognitive impairments and behavioral problems in question are due to emotional, psychological, or social factors or because of brain dysfunctions or sensory deficits with quantitatively demonstrable abnormalities in brain electrical activity;
- Whether the hyperactive sensation-seeking behavior (typical for ADHD and mania) is due to *hypervigilance* or *vigilance autostabilization behavior*, which is a compensatory behavioral pattern to counter regulate a hypovigilance state, and whether withdrawal

behavior (typical for depression) is due to *hypovigilance* or the result of a compensatory behavioral pattern that counter-regulates hypervigilance [269,270];
- Between a degenerative disorder such as AD and *pseudodementia* due to psychiatric illness [271];
- Between normal and abnormal maturational patterns, such as *brain maturation lag* (characterized by a pattern of qEEG that is typical for younger age) and *brain maturational deviation* (characterized by a pattern of qEEG that is not normal at any age) [253];
- Between *presence* or *absence of consciousness* in minimally and unresponsive patients [166,272–275].

In this context, different spatial–temporal qEEG patterns may reflect different underlying mechanisms/functions/symptoms; this hints at the existence of several clinical sub-types within a given diagnostic group that are not recognized by the current diagnostic systems [276].

Distinct aspects of pathophysiologic mechanisms may be elucidated depending on which qEEG oscillations or their combinations are altered in the qEEG oscillatory pattern of any given neuropsychopathology. It seems that neuropsychopathology manifests through the considerable reorganization of the composition of qEEG oscillations and their ratios over a broad frequency range of 0.5–30 Hz, which constitutes the dynamic repertoires of qEEG states. These qEEG oscillations are 'mixed' or superimposed in proportions that depend on the specific neuromediators and neural circuit disturbance and also depend on the presence of various symptoms and affects. Spatial analysis has revealed that different cortical areas are characterized by varying numbers of qEEG oscillations, with a statistically significant difference in their relative presence and communication within the qEEG oscillatory pattern [152,244].

One aspect that often goes unnoticed by clinicians but is nearly *always* affected by neuropsychopathology is the *experiential Selfhood*.

2.11. qEEG and Experiential Selfhood

Indeed, in various neuropsychopathological conditions, *self-consciousness* alterations dominate the patient's phenomenological experiences and have either a long-term or permanent presence [277–279]. Even though *experiential Selfhood* (also referred to as self-consciousness or self-awareness) is a multi-layered concept that is often conceptualized in different ways by various disciplines [280], the currently emerging consensus is that self-referential processing constitutes the core of Selfhood [281,282]. Empirical evidence from neuroscience [281,283–286] indicates that such self-referential processing is instantiated by a specific *self-referential network* (SRN) within the brain, sometimes also referred to as the default mode network (DMN) [283–287]. Further, it has been documented that specific qEEG oscillations have a significant positive correlation with the SRN [288–292].

Recently, a *three-dimensional* neurophysiological model of the complex experiential Selfhood (which is based on the qEEG analysis) was proposed [286,293,294] (for a detailed description, see [295]). This triad model of Selfhood considers the neurophysiological evidence that *three* major spatially separate yet functionally interacting brain subnets constitute the SRN and account for the phenomenological distinctions between three major aspects of Selfhood, namely, (i) first-person agency (conceptualized as the 'witnessing observer' or simply the '*Self*'), (ii) embodiment (conceptualized as 'representational–emotional agency' or simply '*Me*'), and (iii) reflection/narration (conceptualized as 'reflective agency' or simply '*I*'), all of which commensurate with one another [296] and, together, form a unified sense of Selfhood [286] (see also [297]). Each aspect of the triad can be enhanced or weakened depending on the current physiologic and mental state [286,298], voluntary training [293,294], and neuropsychopathology [299–301]. Since aspects of Selfhood rarely fall under the purview of clinical practice, we present below a few examples of the potential application of qEEGs in the assessment of experiential Selfhood for different neuropsychopathologies.

For example, in its 'pure' unmedicated form, major depressive disorder is associated with functional enhancement (measured by qEEG) of all three aspects ('Self', 'Me', and 'I') [300], thus reflecting the well-documented excessive self-focus, increased rumination, and increased embodiment in patients with depression [302–307]. One could speculate that "these three components of complex Selfhood (indexed by (qEEG)) synergize with one another in a maladaptive loop and, over time, become habitual, leading to a vicious circle that maintains a disordered affective state that clinically manifests as depression" [300] (p. 34). It has been further proposed that the 'Self' plays a chief role here as it organizes, represents, and appraises the salience of interoceptive/emotional/bodily information presented by 'Me' and the narrative and semantic-conceptual information presented by 'I' [300].

PTSD is characterized by rather different Self–Me–I dynamics [301]. Increased activity (measured by qEEG) of the 'Self' aspect was found to be significantly associated with the increased vigilance of PTSD sufferers to their surroundings, with a concurrent shift of their first-person perspective from the current moment in time to the moment of the traumatic event (criterion E, according to DSM-5 [308]). We have speculated [301] that such constant hypervigilance coupled with profound emotional arousal leads to sensory overload and further exacerbates alienation of the Self in such patients [309]. Indeed, the increased activity (measured by qEEG) of 'Me' was found to be significantly linked to enhanced emotional, sensory, and bodily states in PTSD sufferers (criterion D, according to DSM-5), such as fear, stress, frozenness, shivering, shaking, trembling, palpitations, and sweating [310–312]. These feelings and memories are usually reported as intrusive and unwanted (criterion B, according to DSM-5). Additionally, it was observed that the activity (measured by qEEG) of 'I' decreased and that this decrease was associated with a distinct lack of linguistic/contextual information and narrative to accompany the traumatic event (criterion C, according to DSM-5), which is a well-documented phenomenon in PTSD patients [312,313].

A six-year longitudinal analysis of a single patient's recovery of self-consciousness (from a minimally conscious state until full self-consciousness) after a severe traumatic brain injury has revealed that the recovery of first-person agency (or 'Self'), representational–emotional agency (or 'Me'), and reflective agency (or 'I') was paralleled by restoration of functional integrity (measured by qEEG) in the three subnets of the SRN [299]. Of note, the recovery dynamic in the Self–Me–I aspects (and corresponding qEEG metrics) was not linear but followed a unique trajectory for every aspect (some recovered more quickly, while others lagged) and was tightly paralleled by (and significantly correlated with) findings from clinical exams and tests [299,314].

Further, converging evidence for a breakdown of qEEG integrity within the SRN in non- and minimally communicative patients with severe brain injuries was found, and this breakdown was proportional to the degree of expression of clinical self-consciousness [287]. More specifically, it was demonstrated that the strength of qEEG integrity within the SRN was smallest or even absent in patients in a vegetative state (VS), intermediate in patients in a minimally conscious state (MCS), and highest in healthy, fully self-conscious subjects. Curiously the strongest decrease in strength of qEEG integrity as a function of loss of self-consciousness was found in the 'Self' aspect compared to the 'Me' and 'I' SRN modules. The central role of 'Self' was also found for the prediction of self-consciousness recovery: those VS patients who later recovered stable minimal or full self-consciousness in the course of the disease (up to six years post-injury) showed stronger 'Self' functional integrity (measured by qEEG) in the early stage (three months post-injury) compared to those patients who continued to stay in the persistent VS [315].

This brings us to another reason for the clinical and ethical importance of qEEG utility in the assessment of the neurophysiological and neurophenomenological status of *unresponsive* patients.

2.12. qEEG and Disorders of Consciousness

A vegetative state (VS), recently re-termed as 'unresponsive wakefulness syndrome' (UWS) [316], and MCS belong to the so-called disorders of consciousness or DoCs [317]. While, by convention, VS/UWS patients are unresponsive to their external and internal environments and are thus unconscious [318], patients in MCS show some level of overt awareness and fluctuating ability to follow commands non-reflexively [319] (see also [320]).

The factual simplicity of the qEEG assessment, its portability and adaptability for longitudinal protocols, and its relatively low cost have opened up a wide area of qEEG investigations in the recent decade—these assessments aim to study the pathophysiology of DoCs as well as look for prognostication markers for the recovery of consciousness in DoC patients [321,322]. Already, the simple description of standard EEGs (guided by accurate qualitative scales) has shown a robust correlation of such patterns with both the level of consciousness impairment (VS/UWS or MCS) and the degree of short-term consciousness recovery [323]. These studies reveal that the *overall* electrical activity of the brain is *differentially* impaired in patients that fall under different DoCs and that it may be related to the degree of recovery, as follows from the group-analyses [321].

The implementation of more complex numerical computations of the EEG signal—qEEG analyses—has contributed in a much more nuanced way to the evaluation of DoC patients [322], leading to a better understanding of the neural *constituents* of consciousness' impairment [166]. For example, studies on qEEG oscillations have demonstrated that patients in VS/UWS have a considerably reduced repertoire of local qEEG oscillations compared to those in MCS or a fully conscious state [272]. Additionally, unawareness in patients with VS/UWS was associated with an altered composition of qEEG oscillations and their proportions compared with a full consciousness state [272,275]. These results confirmed previous observations that loss of consciousness is associated with altered oscillatory contents of the qEEG [324–326].

In agreement with these findings, it has been proposed that the degree of reduction in the dynamic correlates of neuronal networks' complexity measured by the qEEG may be useful for distinguishing patients with different levels of consciousness impairment (VS/UWS vs. MCS) or even as a prognostic measure [165,275,321,326–328]. Indeed, evaluation of qEEG spatial–temporal patterns (which reflect functionally connected neuronal assemblies and their dynamics over time) [166,327,329,330] in DoC patients demonstrated that neuronal assemblies become considerably smaller, with shortened life-spans, and they became highly unstable and functionally disconnected (desynchronized) in patients in VS/UWS [166]. In contrast, fluctuating (minimal) awareness in patients who are in MCS is paralleled by partial restoration of qEEG functional integrity, whose parameters approach those of the levels found in healthy, fully conscious participants [166]. These studies lead to the conclusion that consciousness is likely to vanish in the presence of many very small, extremely short-lived, and highly unstable neuronal assemblies that perform their operations completely independently of one another (functional disconnection) and, thus, are not capable of supporting any coherent content to be experienced subjectively. Importantly, it has been documented that the observed impairment in the brain's functional integrity in DoC patients is independent of brain damage etiology and, thus, reflects functional (and potentially reversible) damage, as opposed to irreversible structural neuronal loss [273]. As a whole, these findings are in keeping with a recent study [331], where it was shown that, in contrast to MCS, the VS/UWS brain is characterized by small, disconnected networks that do not contribute to higher integrative processes [332].

Another factor that may complicate diagnosis and affect both healthy and diseased individuals is *aging*.

2.13. qEEG and Aging

Since age-related processes affect both the structural and functional integrity of the brain, it is reasonable to suggest that qEEGs possess age-dependent changes that are both pathology-independent (healthy aging) and pathology-dependent (pathological aging).

Indeed, many studies have demonstrated that the aging process is reflected in qEEG changes [63,68,144,333–345] and is associated with age-related conditions such as cognitive decline, Alzheimer's disease, mild cognitive impairment, vascular dementia or other dementias, multiple sclerosis, and cerebral tumors [112,116,117,119,124,185,187,346–348].

Aging, as is well known, eventually results in *death*; and death is no longer understood to be an all-or-nothing state but rather a process, the aspects of which may be captured by qEEG.

2.14. qEEG and Death

Death is often a tragic and somewhat baffling finale of a person's life. Since the person is unresponsive near or during death, we know little (if anything) about it, especially from the neurophenomenological point of view (neurophenomenology is scientific research aimed at combining neuroscience with phenomenology in order to study the human experience [349]). However, recent studies suggest that qEEG may shed some light on this mysterious phenomenon. The data suggest that the mammalian brain has the potential for high levels of internal information processing (consistent with conscious processing) during clinical death [350,351], suggesting that patients near death may generate a replay of memories [352]. This is supported by electrophysiological studies that have demonstrated (a) that the post-mortem human brain may retain latent capacities to respond with potential life-like properties [353], (b) that auditory systems (measured by event-related potentials) respond similarly to those of healthy controls just hours before death [354], and (c) the resting-state default mode—task-positive network anticorrelations were present among unresponsive hospice patients [355], thus suggesting that unresponsive patients may possess functional architecture in the brain that can support internally oriented thought (mind-wandering) at the end of life. Moreover, analysis of qEEG-unresponsive patients just hours before death demonstrated that they might be able to listen to music, despite being unable to overtly indicate their awareness [356].

Furthermore, studies have shown that the prevalence of qEEGs with electrocerebral activity despite a clinical diagnosis of brain death (BD) was 3.5% [357] to 19.6% [358], thus posing a challenge for the diagnostic criteria of BD and stressing the importance of qEEG utility for the confirmation of BD. Further, the association between qEEG patterns and eventual death has been demonstrated [165,328,359], thus suggesting that the qEEG may have potential prognostic value for evaluating near-term patients' survival or death.

2.15. Causality of qEEG Oscillatory Patterns in Neuropsychopathology

The above brief review of qEEG features and properties and their association with neuropsychopathology suggests the existence of *circular causality*, where, on the one hand, different pathological processes affect the qEEG pattern and, on the other hand, changes in the qEEG pattern affect pathological processes. This supposition is supported by converging empirical evidence: (a) central nervous system (CNS)-active drugs that affect known neuromediators change different features of the qEEG oscillatory pattern in a consistent and predictable manner, with a parallel reduction in symptoms [360–363]; (b) specific features of the qEEG oscillatory pattern have better predictive power for medication response compared to a syndrome-based diagnosis [364–371]; for example, the overall predictive accuracy in differentiating treatment responders from non-responders is 84%, with a sensitivity of 77% and a specificity of 92% [372]; (c) different features of the qEEG oscillatory pattern predict future (i) decline within the next 7 years in normal elderly people with subjective cognitive complains (no objective evidence of cognitive deficit) [259], (ii) clinical outcomes in patients in the vegetative state 6 years after brain injury [315,327], and (iii) developments of delinquent (antisocial) behavior [373]; (d) normalization of the distorted structure of the qEEG oscillatory pattern by an exogenous magnetic field stimulation changes the subjective experience of neuropsychopathology, accompanied by a clinical decrease (>50% reduction) of symptom severity [374] (see also [375,376]); (e) normalization of atypical qEEG oscillatory patterns through operant conditioning with neurofeedback results in symptom reduction in neuropsychopathologies such as epilepsy [377,378], depression and anxiety [379], schizophrenia [380], addiction [381], ADHD [382,383], sleep disorders [384], autism [385], chronic pain [386], learning difficulties [387], and dyslexia [388]; last but not least, (f) cognitive enhancement in the elderly by qEEG neurofeedback [389,390].

A substantial corpus of evidence supports the proposition that the successful treatment of psychiatric patients results in the normalization of the previously demonstrated qEEG abnormalities [213].

Such circular causality is possible because the qEEG oscillatory pattern is not just a correlate of information processing, communication, integrated phenomenal experience, and the associated neuropsychopathology but is, indeed, a *constitute (substrate)* of these very things [330,391].

From the above review, it is clear that the qEEG is a natural and non-invasive 'window' into the living brain and mind since only the qEEG permits direct observation of the ongoing dynamics and coordinated processes organized in the patterns of brain activity that reflect the overall architecture of information processing, behavior, and subjective experience during both healthy and pathological conditions [392].

To make sense of the 'view' from this 'window', many different methods have been suggested. However, when processing the qEEG signal, it is essential to remember that it is a neurophysiological phenomenon that has its own peculiarities, regularities, and complex rules of organization which are *functionally relevant* [13,393–396] (for reviews, see [18,329,397–399]). Only when one knows these characteristics is it possible to make proper use of the qEEG as a tool and to give a more neurophysiologically adequate interpretation of the data. In connection to this, a much deeper understanding of the brain dynamics reflected in the qEEG is essential for progress in psychophysiological, cognitive, and clinical sciences.

3. qEEG Functional Structure and Signal Processing

With advances in qEEG signal processing methods, a wide range of statistical and mathematical techniques and analyses has been implemented to analyze complex oscillatory activity in spatial and multi-temporal dimensions. All of these have revealed new insights into the functional neural networks during normal functioning and neuropsychopathology.

Since there are many excellent reviews dedicated to qEEG signal processing methods [6], in this section, we will overview only the most important aspects of the *functional structure* of the qEEG signal that should be considered during processing. "Understanding the [q]EEG "grammar", its internal structural organization would place a "Rozetta stone" in researchers' hands, allowing them to more adequately describe the information processes of the brain in terms of [q]EEG phenomenology" ([400], p. 111), which is functionally relevant for healthy and pathological conditions.

Studies focused on the structural organization of qEEG signals have demonstrated that the qEEG is an extremely non-stationary, highly composite, and very complex signal [18,329,397–399]. This qEEG *multivariability*, in contrast to the popular view, is not noise but a reflection of the underlying integral neurodynamics, thus being functionally significant, information-rich [395,396,401], and individually specific [73]. The qEEG multivariability is characterized by a piecewise stationary structure where stationary processes with different probability characteristics are 'glued' to one another [329,398,399,401–403]. It is proposed that each piecewise stationary qEEG segment reflects the *oscillatory state* of the underlying transient neuronal assembly [13,404–410] that signifies a *functional cortical state* [330,411–413], which can be local (part of the cortex), global (all cortex), micro (ranging from milliseconds to seconds), or macro (ranging from minutes to hours). Here, the qEEG oscillatory state is a steady, transient, and self-organized *operational unit* [414] that has been proposed to present the basic building blocks of cortical activity accompanied by mentation, thinking, and information processing [415]. Activity within each state is stable (or quasi-stable) and is likely to represent a fingerprint of a functionally distinct neuronal network mode. Each qEEG oscillatory state (either local or global) is characterized by multiple qEEG oscillations, where different oscillations are mixed in different proportions depending on the level of vigilance, perceptual, cognitive and mental operations, health, or pathology (for more details, see [152]).

Analysis of the non-stationary behavior of the main qEEG signal characteristics (amplitude, frequency, and phase) has demonstrated that all three change abruptly with the progression of time: qEEG amplitude, frequency, and phase persist for some time around a stable average and then abruptly 'jump' to a new stable average, which, after some time, is replaced by yet another average level (for qEEG amplitude, see: [329,398,399]; for qEEG frequency, see: [395,396,416]; for qEEG phase, see: [417]). These 'jumps' in qEEG characteristics (or *rapid transitional periods (RTPs)*, as we have named them [397,399]) mark the boundaries of segments of relatively stable brain functioning. The abrupt transition from one quasi-stationary qEEG segment to another, in this sense, reflects a 'switching' between brain states (micro, macro, or both) in specific neuronal networks or the whole cortex by the transient formation and disassembling of interconnecting cortical *neuronal assemblies* (*neuronal assembly* is defined as a set of neurons that cooperate (synchronize their activity) to perform a specific computation (operation) required for a specific function or task [418–421]) [394,398,413,422]. During such a transition, there is an abrupt change in the *entropy*, *information*, and *dimensionality* of the neuronal assembly (for details, see [391]). A multitude of different microstates may exist within any one particular macrostate. Consecutive macrostates, in their turn, comprise a new sequence on yet another timescale. Such functional qEEG structures comprise a *nested hierarchical multivariability* that reflects the poly-operational structure of brain activity [329,401,403].

The co-existence of the high multivariability of qEEG characteristics, along with the transient stabilization of these characteristics in time (*metastability*), has been demonstrated (the parameters of temporary stabilization of oscillatory states differ from 'random' EEGs, thus providing evidence for the non-occasional character of stabilization of the main parameters of neuronal activity [395]) [395]. Perhaps the high multivariability of qEEG characteristics is a reflection of the range of the brain states' repertoire and their possible variations. On the other hand, the temporal stabilization of qEEG characteristics reflects the maintenance of some persistent pattern in neurodynamics within a particular time interval on both micro and macro levels. This suggests that the overall brain dynamics is a balancing act between multivariability and metastability [14,401].

Considering that all activities (influences) from multiple primary sources are not just mixed, summed, or averaged in a given cortex area but are integrated within the current state (activity) of this area, the local qEEG is considered to represent a *functional source*, which is defined as the part or parts of the brain that contribute to the activity recorded at a single sensor [423,424]. A functional source is an operational concept that does not have to coincide with a well-defined anatomical part of the brain and is neutral with respect to the problems of localization of primary source and volume conduction [423,424]) In this context, the local EEGs can be described by (a) the *size of the oscillatory state repertoires* (the number of the qEEG quasi-stationary segments types—the neurodynamics diversity); (b) the *life-span* (illustrating the *functional life-span* of a neuronal assembly or the duration of operation produced by this assembly; because a transient neuronal assembly functions during a particular time interval, this period is reflected in the qEEG as a stabilized interval of quasi-stationary activity [425]) *of oscillatory states* of each type (duration of the qEEG quasi-stationary segments of each type or period of the temporal stabilization, which shows the time during which the brain 'maintains' the underlying neurodynamics); (c) the *probability of occurrence of a particular type of oscillatory state* (the number of the most probable types of qEEG quasi-stationary segments, which indicates the most 'preferred oscillations' of the brain); (d) the number of *functionally active oscillatory states* (the types of qEEG quasi-stationary segments that change along the changes in the condition, task, or function); (e) the *relative incidence of change in the type of oscillatory states* (gives an estimation of the rate of relative alteration in the type of qEEG quasi-stationary segment); (f) the *sequence of types of oscillatory states* (consistent groupings or bundling of the types of qEEG quasi-stationary segments, representing more integral blocks of qEEG structural organization); (g) the *size of the neuronal ensemble* (indeed, the more neurons recruited into an assembly through local synchronization of their activity, the higher the resulting amplitude of oscillations in the corresponding qEEG channel [13]) that generates the oscillatory state, measured by the average amplitude within qEEG quasi-stationary segments; (h) *stability of local neuronal synchronization within a neuronal assembly*, estimated by the coefficient of amplitude variability within qEEG quasi-stationary segments; (i) *neuronal assembly growth* (recruitment of new neurons) *or disassembly* (functional elimination of neurons), measured by the average amplitude relation among adjacent qEEG quasi-stationary segments); and (j) the *speed of neuronal assembling or disassembling*, estimated by the average steepness among adjacent qEEG quasi-stationary segments, measured in areas near RTP [329,395,399,401,426].

This is the *first level* of multivariability and metastability [427] (Figure 1), where:

Multivariability is characterized by 'switching' from one local neurodynamic to another, with new oscillatory patterns being continually created, destroyed, and, subsequently, recreated. Therefore, there is an increase in:

- The size of oscillatory states repertoire;
- The number of functionally active oscillatory states;
- The relative incidence of change in oscillatory state types;
- Neuronal assembly disassembling;
- The speed of neuronal assemblies disassembling;

and a decrease in:

- The life-span of oscillatory states;
- The size of the neuronal ensemble;
- The probability of occurrence of a particular type of oscillatory state;
- The stability of local neuronal synchronization within a neuronal assembly;
- The sequence of oscillatory state types.

Metastability is characterized by the temporal stabilization of oscillatory states in sequential combinations. Therefore, there is a decrease in:

- The size of the repertoire of oscillatory states;
- The number of functionally active oscillatory states;
- The relative incidence of change in the type of oscillatory states;

and an increase in:

- The life-span of oscillatory states;
- The size of the neuronal ensemble;
- The probability of occurrence of a particular type of oscillatory state;
- Neuronal assembly growth;
- The speed of neuronal assemblies growing;
- The sequence of oscillatory state types.

Different cortex regions have different dominant qEEG oscillations [150,428] that act as resonant communication networks through large populations of neurons [140,429]. Usually, cortical oscillators communicate only with oscillators that have specific resonance frequencies [430]. They do not communicate with oscillators that have nonresonant frequencies, even though there may be synaptic connections between them. In such a way, various assemblies of oscillators can process information without any cross-interference. By changing the frequency content of bursts and subthreshold oscillations, the brain determines communication at any particular moment [431]. These oscillatory systems may provide a general communication framework that is parallel to and faster than the morphology of sensory networks [140].

It seems that RTPs (see above) also contribute to this communication framework. Studies on the spatial–temporal distribution of RTPs in qEEG amplitude [329,397,399,403], qEEG phase [427,432,433], and qEEG frequency [416] have demonstrated that (a) RTPs observed in different local qEEG signals systematically coincide in time and that (b) this

RTP temporal synchronicity is not random—it occurs at significantly higher or lower levels than is expected by chance alone. This non-random RTP synchrony reflects periods of mutual temporal stabilization of quasi-stationary segments in the multichannel qEEG. At the neurophysiological level, this implies that various neuronal assemblies located in different cortical regions synchronize (temporally coordinate) their operations on a particular timescale [403,427]. Such synchronization is the brain's true *functional connectivity* (as it is defined by Friston et al. [434,435]) and reflects the synchronization of operations; therefore, it is *operational synchrony* [397]. Since operational synchrony has been demonstrated for qEEG amplitude, phase, and frequency, it is reasonable to suggest that operational synchrony is a *universal phenomenon* for different characteristics of the electromagnetic brain field in which complex brain functioning is reflected.

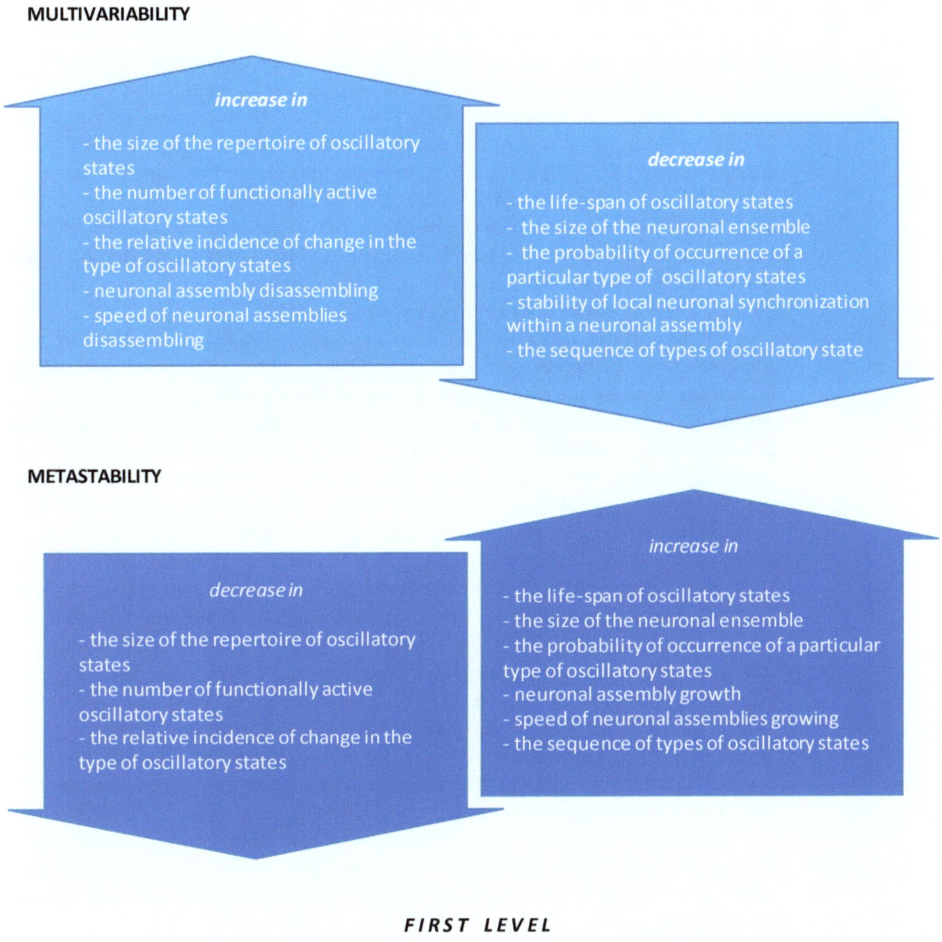

MULTIVARIABILITY

increase in
- the size of the repertoire of oscillatory states
- the number of functionally active oscillatory states
- the relative incidence of change in the type of oscillatory states
- neuronal assembly disassembling
- speed of neuronal assemblies disassembling

decrease in
- the life-span of oscillatory states
- the size of the neuronal ensemble
- the probability of occurrence of a particular type of oscillatory states
- stability of local neuronal synchronization within a neuronal assembly
- the sequence of types of oscillatory state

METASTABILITY

decrease in
- the size of the repertoire of oscillatory states
- the number of functionally active oscillatory states
- the relative incidence of change in the type of oscillatory states

increase in
- the life-span of oscillatory states
- the size of the neuronal ensemble
- the probability of occurrence of a particular type of oscillatory states
- neuronal assembly growth
- speed of neuronal assemblies growing
- the sequence of types of oscillatory states

FIRST LEVEL

Figure 1. The first level of multivariability and metastability measured by qEEG.

It has been demonstrated that the pattern of the functional stabilization of cortical inter-area relations can be expressed as a mosaic of dynamic constellations of different operations executed by remote brain regions—'operational modules' (OMs) [397,399,401,403,436]. The lifetime of such spatial OMs is determined by the duration of the period of joint stabilization of the main dynamic parameters of the activity of neuronal assemblies that are involved in these modules. At the level of the qEEG, this process is reflected in the stabilization of the quasi-stationary qEEG segments in corresponding EEG channels that comprises a metastable state [329,399,401,437]. Here, metastability relates to the phenomenon of a constant interplay/competition between the complementary tendencies of cooperative integration and autonomous fragmentation in the activity of multiple distributed nested neuronal assemblies that defines brain activity dynamics [438] (see also [14,401]). During this metastability, the restriction in the brain's degrees of freedom is what permits the neuronal systems to have the possibility for the interactive information exchange of the essential variables, which are important for reaching a 'consensual decision' that is appropriate for the functional requirements engendered by each successive stage of behavioral performance. It seems that brain areas are able to mutually influence each other in order to reach a common functional state, stabilizing the main parameters of their activities. It is likely that optimal informational processing is possible only under a proper level of the functional stabilization of intercortex relations [244,329].

In this context, each OM is a metastable spatial–temporal pattern of brain activity because the neuronal assemblies that constitute it have different operations/functions and do their own inherent tasks (thus expressing autonomous tendency) while, at the same time, being temporally entangled with each other (and thus expressing coordinated activity) in order to execute a common complex operation or complex cognitive act of a higher hierarchy [401,403,439]. Here, it is important to stress that discrete parts of the cortical networks may gain another functional meaning when they are recruited by other OM and, therefore, take part in the realization of another functional act [397]. This confirms the *dominant principle* of the nervous constellation centers suggested by Ukhtomsky, who discussed the variable functional role of different brain cortical areas depending on their participation in various working constellations [440].

In this context, much like the local oscillatory states described earlier, the *multichannel* EEGs can be described by (a) the *size of the OMs' repertoire* (the number of the types (*OM type* is characterized by the number and topographic locations of the cortex areas that mutually stabilize (temporally synchronize) the main parameters of the neuronal networks involved (temporal synchronization of RTPs)) of spatial configurations of qEEG quasi-stationary segments (temporal synchronization of RTPs)—coordinated neurodynamic diversity); (b) the *life-span of OMs* of every type (duration of spatial configurations of the qEEG quasi-stationary segments of every type—period of RTPs' temporal stabilization—shows the time window during which the brain 'maintains' underlying coordinated neurodynamics); (c) the *probability of occurrence of a particular type of OM* (the number of the most probable types of spatial configurations of qEEG quasi-stationary segments—synchronized RTPs); (d) the number of *functionally active OMs* (the types of spatial configurations of the qEEG quasi-stationary segments that change along with the changes in condition, task, or function); (e) the *relative incidence of change in the type of OMs* (presents an estimation of the rate of relative alteration in the type of spatial configuration of the qEEG quasi-stationary segments); and (f) the *sequence of types of OMs* (consistent grouping or bundling of the types of spatial configurations of the qEEG quasi-stationary segments, reflecting the more integral blocks of qEEG structural coordinated organization) [329,401,403,439].

This is the *second level* of multivariability and metastability (Figure 2) where:

Multivariability is characterized by 'switching' from one coordinated neurodynamic to another, with new OMs being continually created, destroyed, and, subsequently, recreated. Therefore, there is

an increase in:

- The size of the OMs' repertoire;
- The number of functionally active OMs;
- The relative incidence of change in the type of OM;

and a decrease in:

- The life-span of OMs;
- The probability of occurrence of a particular type of OM;
- The stability of the sequence of types of OMs.

Metastability is characterized by the temporary stabilization of RTPs (formation of OMs) in sequential combinations. Therefore, there is

a decrease in:

- The size of the OMs' repertoire;
- The number of functionally active OMs;
- The relative incidence of change in the type of OM;

and an increase in:

- The life-span of OMs;
- The probability of occurrence of a particular type of OM;
- The sequence of types of OMs.

Figure 2. The second level of multivariability and metastability measured by qEEG.

In this context, the participation of cortex areas in the organization of a common functional act is reflected not so much by the presence of a shared qEEG rhythm in different EEG channels (distant neuronal ensembles) but by the systematic synchronization of the moments of switching (RTPs) between qEEG oscillatory modes in the different cortex areas. The fact that operational synchrony is sensitive to the morpho-functional organization of the cortex rather than the volume conduction and/or reference electrode differs from surrogate data (random combination of RTPs) and is functionally sensitive to different cognitive tasks as well as healthy and pathological conditions; this suggests that operational synchrony reflects the remote temporal coordination of brain operations performed by local neuronal assembles (for relevant details, see [244,329,398,399,403]).

It has been suggested that disturbed synchrony in distributed qEEG oscillations may reflect dysfunction within resting-state networks in neuropsychopathology. It seems that a loss of optimal metastable balance between independence and the integrative processes in large-scale cortical activity, which may be due to a dysfunction in large-scale cortical integrative processes and a poverty of regional physiological variation [441], is associated with the 'cognitive disintegration' [442], 'cognitive dysmetria' [443], and 'thalamocortical dysrhythmia' [444] that are typical for a number of neuropsychopathologies [244,276,445]. This view is consistent with the modern concept of brain and mind disorders, where the disease is considered to be a process, with a change in the balance of autonomy (low functional connectivity) and connectedness (high functional connectivity) of different brain systems that sustains health [244,276,438]. Indeed, on one hand, deficits in the ability of cortical areas to coordinate could produce a lack of mutual constraint, leading to excessive expression of local processing unrestrained by large-scale context. On the other hand, an excess of coordination could stifle independent area expression and cause a stereotyped processing rigidity. Thus, the alteration in brain functional connectivity might serve as a contributing factor to the disorganization syndrome [446].

From this overview, it is clear that any qEEG signal processing method should take into account the functional structure of the signal in order to be able to extract neurophysiologically relevant information.

In order to provide accurate and practical methods for separating individuals with cognitive impairments or behavioral disturbances into those *with* and *without* quantitatively demonstrable abnormalities in brain electrical activity (qEEG) and, thus, altered neurophysiology, statistical evaluation of qEEG characteristics relative to *population normative values* is needed. Such possibility was demonstrated by the introduction of so-called '*neurometrics*' [153]. Machine learning algorithms and artificial intelligence techniques are used for this purpose as well. However, the limitation of these methods is their inherent non-explainability: no insight can be obtained from the inferred output. Without the explainability of the learned inference mechanism, not much insight can be gained in terms of the underlying brain activity patterns and mechanisms, which are important to better understand neuropsychopathology. Within this neurometrics, a given qEEG characteristic of the individual is transformed from its original units (voltage, time, latency, coherence, and symmetry) to a common metric reflecting the *relative probability* of that value within a healthy population normative reference. This allows researchers to compare or combine measures that have not initially been dimensionally comparable. From this, '*abnormality*' is derived and defined as statistically improbable values exceeding those expected by chance alone. Thus, a clustering of qEEG deviations provides evidence of underlying functional neurophysiology abnormality that can be associated with the patient's clinical condition or neurological/psychiatric problems. This neurometric analysis of qEEG characteristics was successfully applied to a variety of neuropsychopathologies, with high replicability, specificity, and sensitivity to a wide variety of cognitive, developmental, neurologic, and psychiatric disorders [69,248,413,447–449].

Although much has been learned through systemic, cognitive, and clinical neuroscience about the underlying neural mechanisms and functional correlates of qEEG oscillatory activity, surprisingly little conceptual integration of this knowledge is present in clinical applications of qEEG. Even though the qEEG (a) reflects developmental maturation and aging, anatomical and functional integrities of the brain, including cognitive processes and the functional status of the brain as well as diverse neuropsychopathologies; (b) has high predictive capacity, sensitivity, and specificity for identifying responders and non-responders; and (c) provides qualitative predictions for a patient's state after treatment courses (see Sections 2.1–2.14), all of this knowledge should be put in a wider conceptual context of the health–disease continuum in order to refine the diagnostic capability of qEEGs. In the following section, we will attempt to integrate conceptually recent neuropsychophysiological and electrophysiological findings within a common theoretical framework in order to reduce the divide between state-of-the-art research and current clinical practice.

4. Common/Unifying Theoretical–Conceptual Framework

A general framework needs to be developed in order to allow researchers and clinicians to organize systematically and understand the enormous diversity of observations related to qEEG characteristics in neuropsychopathology. As a first step, a general concept of *dynamic qEEG oscillatory patterns* has been proposed [450]. Within this concept, a dynamic qEEG oscillatory pattern is considered as a spatial–temporally organized superimposition of ongoing multiple qEEG oscillations in many frequency bands [26], where different oscillations are mixed in varying proportions based on the vigilance level; perceptual, cognitive, and mental operations; behavior; and the extent of pathologic processes. It seems that different qEEG descriptors can be combined within the dynamic qEEG oscillatory pattern, which is characterized by (a) *frequency content*, including the composition of delta (0.5–3 Hz), theta (3–7 Hz), alpha (7–13 Hz), beta (13–30 Hz), and gamma (>30 Hz) qEEG oscillations, along with their proportions, dominant frequencies, amplitudes and powers, and (b) *spatial heterogeneity* (expressed in spatially structured extracellular electrical fields), including spatial complexity (amount of brain connectivity), interhemispheric symmetry, and hubs (cortex areas with the highest neuropsychopathology effect or highest functional connectivity). This concept was successfully applied to major depressive disorder [450].

The concept of dynamic qEEG oscillatory patterns is useful since it enables researchers to combine all known and newfound qEEG characteristics within *one entity*. Studies (see above) have suggested that dynamic qEEG oscillatory patterns reflect the structural and functional integrity of the brain, including an information-processing mode that is dynamically regulated by interactions within a homeostatic system that is mediated by many different neurotransmitters on one side and functional activity and various perceptual and cognitive operations associated with a mental or behavioral condition in health and pathology on the other.

The categorization of different *types* of dynamic qEEG oscillatory patterns resulted in the second concept—the *qEEG phenotype*.

4.1. qEEG Phenotype and Neuropsychophysiological Type

qEEG phenotypes are the clusters of commonly occurring qEEG characteristics within qEEG oscillatory patterns found in the general population that are believed to be the net result of genetics, pre- and post-natal individual development, and significant life events such as brain traumas or neuro-diseases (see above). In contrast to the classical view, which sees qEEG phenotypes as classes of qEEG *abnormalities*, where, usually, one qEEG characteristic defines a qEEG phenotype and each phenotype is static [451], we believe that qEEG phenotypes characterize all variabilities of the population, covering the whole *spectrum*, from health to pathology, and, further, that such phenotypes are dynamic. In this view, some of the qEEG phenotypes are characteristic of healthy conditions, while others are typical for different degrees of deviation from the healthy state up to the situation where the qEEG phenotype is pathological. Additionally, every phenotype is defined by the combination of qEEG features (qEEG oscillatory pattern—a *coherent functional whole*) and is dynamic, meaning that every qEEG phenotype may exhibit changes, to some extent, within its own qEEG oscillatory pattern due to dysfunction or development of pathology (Figure 3). This is so because the type of qEEG oscillatory pattern is a phenotypic expression (qEEG phenotype) of (a) cellular and biochemical (dys)function; (b) maturational processes (or delaying factors), partially genetically and epigenetically determined; (c) neurotransmitter (im)balance; (d) regulatory systems (and their disturbances); (e) early subclinical organic brain damage; or (f) morpho-functional disturbances that may be present in neuropsychiatric disorders.

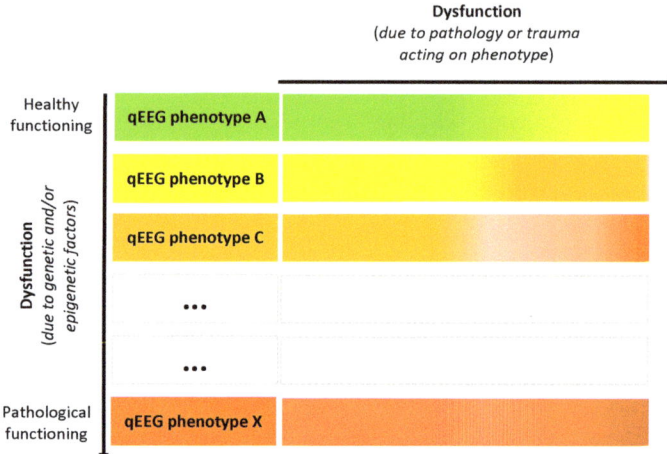

Figure 3. Schematic illustration of qEEG phenotypes' diversity and relative dynamism. Horizontal arrow indicates that every qEEG phenotype may exhibit changes, to some extent, within its own qEEG oscillatory pattern due to pathology, brain development dynamics, or trauma.

Such conceptualization underwent theoretical development and empirical verification in the work of Zhirmunskaya and colleagues [452–456] and was successfully applied to several clinical neuropsychopathologies as well as various modes of pharmacological influence [156,457–462].

In this context, a given qEEG phenotype reflects a *neuropsychophysiological type* (neurodynamic constitution) of the individual. Indeed, numerous studies have demonstrated that qEEG characteristics can encompass functionally different psychophysiological determinants [1,463]. On one hand, a qEEG phenotype reflects the inherent brain functional organization and dynamic structure of brain activity, which are intra-individually stable traits, as evidenced by test–retest reliability (Section 2.2) and genetic studies (Section 2.3). On the other hand, since intrinsic brain activity supports and conditions individual cognitive and information processing, self-regulatory functions, decision-making, behavior, and consciousness, a qEEG phenotype reflects the neurophysiological predispositions of the underlying cognition and personality, temperament, and character factors thus reflecting the individual's psychological and behavioral traits. Indeed, it was found that individual differences in qEEG variability relate strongly to stable indicators of subject identity [464]. Hence, both an individual's neurophysiological and psychological differences can be neuropsychophysiologically interpreted under the same unified notion of the qEEG phenotype [1,465–470].

This brings us to the idea that a qEEG phenotype is not just a concept but a *real phenomenon*. This is supported by the observations that qEEG phenotypes reliably predict the effectiveness of drug interventions, while nosological or behavioral groupings do not [451]: for example, effective treatment of ADHD [366,471], refractory depression [472], and major depressive disorder [369] was achieved when it was based on prospectively identified qEEG phenotypes related to different sub-types within the diagnostic group, thus suggesting that nosological heterogeneity is well-reflected in the multiplicity of spatial–temporal parameters of qEEG oscillatory patterns. Additionally, qEEG phenotypes are a better indicator of pathology vulnerability when compared to classical evaluation: for example, it was demonstrated that qEEG has a better indication of alcoholism susceptibility than the customary dichotomous affection status [473,474].

In summary, a constellation of different qEEG characteristics that are united within the qEEG oscillatory pattern and expressed as the qEEG phenotype should be considered more appropriate for diagnostic and medication/therapy-response purposes. It seems that every qEEG phenotype represents a set of quantitative neuropsychophysiological, cognitive, and behavioral traits that determine an individual's liability or vulnerability to develop or manifest a particular neurodysfunction or disease [3,475,476]. Thus, qEEG phenotypes may aid in revealing disease-specific causal pathways and may aid people in finding a work/lifestyle balance that is more in keeping with their natural predispositions.

4.2. qEEG Phenotype and Neurophysiological, Cognitive, and Behavioral Traits

Studies on intra- and interindividual differences have revealed that qEEG oscillatory patterns reflect individually specific peculiarities of homeostatic and adaptive regulation [468,477–480]. Additionally, it has been demonstrated that the qEEG phenotype reflects the specificity of intracortical and cortico-subcortical interrelations and is, to a significant extent, a neurodynamic substrate of the psychological properties of the personality [481–485]. Together, these suggest that neurophysiological, cognitive, and behavioral traits are reflected in the characteristics of the qEEG phenotype. Indeed, studies have demonstrated the association between personality traits and qEEG characteristics [486–501]. Moreover, normal EEG patterns have been correlated with a well-integrated personality (general personality fitness) [502] (for a review, see [503]).

Personality traits have been proposed to constitute vulnerability factors for psychopathology, including mental diseases and affective disorders [504–511], thus suggesting that personality traits and psychopathology are not distinct entities [512]. They both (a) reflect increased vulnerability for and (b) underly any given psychopathology.

Thus, it seems that, on one hand, the qEEG phenotypes are related to personality types (see above and also [513]) and psychopathology. On the other hand, personality features are related to psychopathology too. This similarity of qEEG relations between psychopathology and personality traits suggests that they are mediated by the same neurophysiological mechanisms and behavioral patterns. Therefore, it is reasonable to suggest that behaviors intrinsic to personality traits are the same as those that are *exaggerated* in psychopathological conditions [514].

There are many different individual traits with varying degrees of significance for neurophysiologic regulation, information and emotions processing, and behavior. For the purpose of this article, we will focus on those traits which are (a) expressed along the entire continuum of functioning, from health to pathology; (b) transdiagnostic; and (c) reflected in the characteristics of qEEG phenotypes, thus being fundamental and primary functions that condition (modify, modulate, or mediate) other dimensions.

The Research Domain Criteria (RDoC) project, launched by the National Institute of Mental Health in 2009 [515], is the first comprehensive attempt to identify fundamental neurobehavioral dimensions that cut across current heterogeneous disorder categories. According to the RDoC, there are five major domains of functioning:

- Negative valence domain;
- Positive valence systems;
- Cognitive systems;
- Systems for social processes;
- Arousal/regulatory systems.

Since some qEEG characteristics can be related to several domains, constructs, or subconstructs of the RDoC matrix and phenomenally different clinical characteristics are, at a more fundamental level, implementations of one and the same process, we suggest the following fundamental and primary domains of functioning that are (a) expressed along the entire continuum of functioning, from health to pathology; (b) transdiagnostic; and (c) reflected in the characteristics of qEEG phenotypes:

- Tonic level of vigilance (corresponds to arousal/regulatory systems in RDoC);
- Speed of information processing (corresponds to cognitive systems in RDoC);
- Directedness of the attention (internal vs. external focus) (corresponds to cognitive systems in RDoC);
- Emotional–motivational tendency (corresponds to positive and negative valence systems in RDoC);

- Sociability (sensory stimulation and excitement tolerability) (corresponds to systems for social processes in RDoC);
- Anxiety tendency (anxious arousal vs. anxious apprehension) (corresponds to negative valence and arousal/regulatory systems in RDoC);
- Stress regulation (resistance and recovery);
- Overall brain resources (resilience).

The fact that individual differences in dispositional moods, stress resilience, behavioral orientation to physical or social objects, temporal processing (primordial sense of flow between events), and brain resources are shaped/present in very early childhood, long before cultural standards and knowledge are internalized by an individual [516–518], supports a *fundamental and primary nature of these domains of functioning*.

One can see that the majority of the suggested domains can be mapped onto the RDoC matrix. In the following sections, we will give a brief overview of these domains.

4.2.1. Tonic Vigilance

Vigilance (or arousal) here refers to the individual's predominant level of supply energy available to the brain's regulatory systems. It is responsible for generating the activation of neural systems appropriate for various contexts and providing appropriate homeostatic regulation of systems such as energy balance and sleep.

There are more or less stable individual differences in the baseline vigilance: some individuals are constantly in a highly activated state (*hypervigilance*), while others have a chronically low level of activity (*hypovigilance*) [519], thus reflecting an individual's *predominant* vigilance state—*tonic vigilance*. The underlying cause is believed to be different levels of activity in the loop connecting the brainstem reticular formation with the cortex. As such, tonic vigilance reflects the predominant baseline cortical and mental arousal that sets the overall level of activity of the entire brain and, as a consequence, the body. In this way, tonic vigilance, as the brain's trait energetic capacity, sets the stage for cognitive and behavioral performance.

It seems that tonic vigilance is one of the most important traits of an individual since it determines the optimal range of activating factors (a sympathovagal balance) on the cerebral cortex that is necessary for the formation of an active, energetic state [520]. Outside this optimal range, work capacity drops and the subjective sense of well-being decreases, while psychopathology vulnerability increases [521]. Indeed, studies have demonstrated that hypervigilance is a common feature of various anxiety and affective disorders (depression), including PTSD [522–525], hyperaroused fatigue (overloading) with reduced sleep propensity, inhibition of drive, and eventual exhaustion [526–528]. Conversely, hypovigilance is associated with hypoaroused fatigue with increased sleepiness, a lack of drive, sickness behavior [526,529–531], burnout [532], impulsivity [533], ADHD [241,534,535], and AD [135]. This is supported by the Arousal Regulation Model of Affective Disorders [269], where hypostable and hyperstable levels of arousal contribute to manic or ADHD and depressive-like behavior, respectively [270,536]. Importantly, tonic vigilance has been suggested to causally contribute to mental illness [269]. Additionally, the relationship between tonic vigilance, personality traits [501,513,537], and psychopathology have been proposed [538,539] (see also Section 4.1).

Besides (a) being expressed along a health–pathology continuum and (b) being transdiagnostic, tonic vigilance is reflected in the characteristics of the qEEG phenotype [157,523,540–547]. Additionally, vigilance-relevant qEEG characteristics are ~80% heritable [548].

To summarize, tonic vigilance affects cognition, emotions, and behavior and is implicated in the etiology of psychiatric disorders and reflected in the characteristics of qEEG phenotypes.

4.2.2. Speed of Information Processing

Even though this domain is dependent on tonic vigilance, it has added significance in relation to normal functioning and neuropsychopathology.

The speed of information processing through periodic cycles of sampling and sensory integration determines the pace of perceptual performance and behavior [148,549,550]. In other words, the speed of information processing determines the resolution at which information is sampled and/or processed by cortical neurons, and, therefore, it limits the capacity for storage, transfer, and retrieval of information in the brain [551–553]. In such a way, the speed of information processing regulates the overall amount of information that reaches the cortex and is related to reaction time, speed, and capacity of cognitive and motor task performance, cognitive preparedness, temporal integration of information across the senses, as well as reasoning ability.

The speed of information processing is an individually stable trait (varies within an individually specific range) and is predictive of the individual's reaction and decision times, attention, and working memory performance across the age span [21,69,78,144,553–561].

The speed of information processing is expressed along the health–pathology continuum, where slowed information flow is observed in sorrow and fear emotions, physiologic and pathologic aging, mild cognitive impairment and AD, vascular dementia, psychosis and schizophrenia, ADHD, chronic fatigue, burnout syndrome, and melancholic depression [117,119,124,530,532,544,562–567], whereas an increased speed of information processing is typical for anger, anxious arousal and panic disorder, hyperaroused fatigue (overloading), anxious depression, and PTSD [523,562]. Additionally, temporality (inner and outer time speed perception) is recognized as an important factor in psychopathology [568–570], where an imbalance between inner and outer time speed perception was demonstrated. For the transdiagnostic character of the speed of information processing, see [518].

Importantly, qEEG characteristics are causally related to the speed of information processing [21,144,148,549,550,559,561] and the temporal integration of information across the senses [552,553,571,572]. Additionally, qEEG characteristics associated with the speed of information processing have a heritability of 81% [89].

To summarize, the speed of information processing is controlled by the characteristics of qEEG phenotypes that are thought to function as a timing or gating mechanism in operation processing [573]: shorter-duration qEEG oscillations provide more gating signals per unit of time and, thus, result in faster information-processing rates and shorter RTs, whereas longer-duration qEEG oscillations provide fewer gating signals and lead to slower processing rates and longer RTs [574]. This is in line with 'physiological lability' (introduced by Wedensky [575]), defined as the capacity of the system to perform a certain amount of complete work cycles per unit of time [576]. Additionally, alterations in the speed of information processing are characteristic of various neuropsychopathologies.

4.2.3. Directedness of the Attention (Internal vs. External Focus)

Attention is one of the basic human cognitive abilities that allow the discrimination of relevant parts of information and the ignoring of others; it usually refers to a more focused activation of the cerebral cortex that enhances information processing [577]. Attention depends on the vigilance level and, therefore, can be predicted through it; however, *directedness* of attention is independent of vigilance and, thus, should be considered separately.

It seems that directedness of attention can be conceptualized as *internal* attention (*self-focus*) and *external* attention (*environment-focus*). Some healthy individuals are generally well tuned into both internal and external events, which helps them to flexibly reallocate attention 'in' and 'out' in order to adapt their behavior to the needs of the current situation. However, others have the propensity to be self-focused or environment-focused. Exaggerated unbalanced attentional focus is typically associated with neuropsychopathology and is known as *attentional bias*. Attentional biases are characterized by an inability to flexibly reallocate attention to relevant internal or external stimuli, for example, an increase in the orienting of attention towards threat-related stimuli and attentional avoidance/difficulty disengaging attention from irrelevant but negatively valanced stimuli [578].

When enhanced focus on the external environment prevails (e.g., family, friends, social and work duties, including the distracting effects of environment), the individual is more alert, anxious, or irritable due to the inability to inhibit irrelevant/inappropriate external information. Therefore, there is a predominance of perceptually guided contents over self- and somatically-guided contents; this is usually associated with anxiety, irritability, stress, and mania [579–582].

When an individual has excessive self-focus, it elevates volitional control (including both execution monitoring and internal focus) but also interferes with automatic actions, which stems from an inability to inhibit irrelevant/inappropriate internal information [583]. Excessive self-focus occurs at the expense of external environmentally oriented contents and their respective social and psychomotor functions. This shift may be manifested in symptoms such as ruminations, various somatic and vegetative symptoms, social withdrawal, lack of motivation, and psychomotor retardation, usually observed in depression and fatigue [300,581,582,584,585]. Moreover, because excessive resources are allocated to processing internal mental contents, other outward-oriented aspects such as attention, working memory, and episodic memory are compromised as well, which can lead to various neuropsychological deficits.

Notice that in both scenarios (excess or deficit of internal attention), perceptually guided contents (environment) are not properly integrated with the somatic- and self-generated contents (self), which results in their disbalance. This disbalance of internal and external contents seems to be abnormally tilted towards either the internal or external content, which, importantly, also goes along with different degrees in the expression of self [300,305].

Converging evidence suggests that directedness of attention is reflected in the characteristics of qEEG phenotypes [21,485,579,586–593].

4.2.4. Emotional–Motivational Tendency

Emotional–motivational tendency reflects the predisposition of an individual to engage in certain types of emotional (positive vs. negative) and motivational (approach vs. withdrawal) responses [594]. Hence, the emotional–motivational tendency may be viewed as a 'diathesis' (trait) that predisposes an individual toward a particular affective and motivational style and establishes risk factors for developing certain psychopathologies [595–600].

Indeed, excessive negative and withdrawal tendencies can result in behavioral inhibition expressed as fatigue, lack of energy, apathy, and slow psychomotor functioning, with a stronger hormonal response to stress (higher cortisol level), where more situations/stimuli are perceived as stressors [169,171,599,601,602]. Likewise, it is accompanied by reduced baseline cellular immune function, negativity bias (alongside reduced reward sensitivity), and sadness, fear, or depression [596,599,603–605]. The tendency toward negativity and withdrawal is associated with a personal pessimism bias—the person believes that negative events are more likely and positive events are less likely to happen to him/her than they are to other people. The tendency toward negativity and withdrawal has also been observed in psychiatrically healthy offspring of individuals with depression [606].

Conversely, excessive positive and approach tendencies can result in the individual having excessive behavioral activation, expressed as increased muscle tension, agitation, and somatic symptoms of arousal, with smaller hormonal responses to stress (fewer situations/stimuli are perceived as stressors), positive emotions, mania, jealousy, anxiety, or anger [599,602,607–610]. The tendency toward positivity and approaching tasks/challenges is associated with a personal optimism bias—the person believes that positive events are more likely and unpleasant events are less likely to happen to him/her than they are to other people [611,612].

Similar to other domains of functioning, converging evidence has demonstrated that emotional–motivational tendency is reflected causally in the characteristics of qEEG phenotypes with high reliability and internal consistency [76,170,613–622] and is associated with a genetic risk for depression [623].

4.2.5. Sociability (Sensory Stimulation and Excitement Tolerability)

Sociability is a trait related to sensation/experience seeking, disinhibition propensity, and boredom susceptibility. It reveals a tendency to interact well with others and the degree to which a person can tolerate/enjoy sensory stimulation from people and situations.

Sociability is a complex facet of introversion–extraversion [499,624], which is related to behavioral inhibition and activation systems (BIS/BAS) [499,625,626] and is associated with cortical arousal via the ascending reticular activating system (ARAS) [627]. In this context, introverts are believed to have a lower threshold for arousal; therefore, they are assumed to be chronically 'over-aroused' and, thus, tend to seek a state of lower arousal. Conversely, extraverts are believed to have a higher threshold for arousal, as they are assumed to be chronically 'under-aroused' and tend to seek a state of higher arousal [537,628].

Both under- and over-sociability are associated with a raft of psychopathologies [629,630]. Thus, lack of sociability can result in the individual preferring low sensory stimulation due to heightened baseline cortical arousal, having a lower threshold for arousal, and, therefore, high behavioral inhibition, increased emotional tension, and depression. Such a person is excited by low-intensity signals and inhibited by high-intensity signals.

In contrast, excess sociability can result in the individual preferring strong sensory stimulation and a propensity for sensations/excitement/novelty-seeking due to baseline cortical under-arousal, having a higher threshold for arousal and, therefore, low behavioral inhibition [631]. Such a person is excited by high-intensity input signals and inhibited by low-intensity ones. In the extreme, over-sociability is associated with psychopathology [632–634], including mania, narcissism, psychopathy, substance abuse, excessive venturesomeness (excitement-seeking), and various forms of externalizing (risk-taking, grandiosity, exhibitionism, manipulativeness) [635], suggesting that excessive sociability potentially represents a vulnerability factor for other conditions.

Converging evidence suggests that sociability is also reflected in the characteristics of qEEG phenotypes [494–496,499,501,636–642].

4.2.6. Anxiety Tendency (Anxious Arousal vs. Anxious Apprehension)

Anxiety is one of the best-known and oldest evolutionary systems evolved in humans. It results from a set of information-gathering reactions that allow the individual to face uncertainty and danger and survive. Despite being adaptive, since it helps us avoid dangerous situations and to achieve our goals, anxiety also causes significant suffering and, in its extreme forms, can be debilitating.

Biologically, there are two subcomponents of anxiety: *somatic* and *cognitive* [643]. Somatic anxiety (or *anxious arousal*) is a physiological component of anxiety, characterized by autonomic arousal and somatic tension (high blood pressure, pounding heart, sweating, dryness of mouth, difficulty breathing) [643–645]. Cognitive anxiety (or *anxious apprehension*) refers to the mental component of anxiety and consists of expectations about and anticipations of a difficult task or threat, success or self-evaluation, worrying, negative self-talk, and disrupted attentional processes [643,645,646]. It is important to stress that each of the two components of anxiety is associated with different psychopathologies.

A person with an anxious arousal tendency is usually alert during the distraction-free resting-state period and experiences rest periods as more aversive and anxiety-inducing. Extreme anxious arousal is typical for neurosis, panic or phobic disorders, and PTSD [647,648].

A person with an anxious apprehension tendency usually has (a) higher expectation for the perceptual events, which reflects higher nonselective readiness for perception and action, even in the absence of any goal-directed task and (b) excessive worry for the future and verbal rumination about those expectations. Highly expressed anxious apprehension is typical for generalized anxiety disorder, obsessive–compulsive disorder, and separation anxiety disorder [647,648].

Again, converging evidence suggests that both components of anxiety are reflected in the characteristics of qEEG phenotypes [497,649–653].

4.2.7. Stress Regulation (Resistance and Recovery)

Considering the link between qEEG oscillatory patterns and stress regulatory systems such as the hypothalamic–pituitary–adrenocortical (HPA) axis and the sympathetic–adrenomedullary axis [168] (see Section 2.7), as well as the association between early life adversity and the subsequent stress regulatory profile in the adult [654,655], it is reasonable to consider stress regulation as a trait.

Stress regulation determines the ability of an individual to withstand, adapt to, and recover from stress [656]. The brain plays a central role here since it perceives and determines what is threatening and executes behavioral and physiological responses to the stressor [657]. When stress regulation is altered (due to chronic stress or individual vulnerability), there is (a) a decreased capacity to resist, adapt, and recover from stress and (b) a tendency to perceive the stressor as a threat (ether to one's physical safety or to one's ego/social sense of self). For example, early life adversity may lead to persistent changes in the neural network balance that increase sensitivity to emotional stimuli [658] and is often associated with blunted HPA-axis reactivity.

Stress dysregulation is associated with a transition from adaptive to maladaptive stress responsivity and stress-related disorders [654,655] and neuropsychopathologies such as inattention, depression, anxiety, and insomnia.

It seems that stress regulation is also reflected in the characteristics of qEEG phenotypes [177,659–663].

4.2.8. Overall Brain Resources (Resilience)

The overall brain resources reflect the brain's morpho-functional integrity, capacity for self-reorganization, self-regulation, and adaptation, and information processing efficiency. The brain resources domain relates to the general capacity of the brain to withstand neuropathological changes before overt behavioral, functional, or cognitive impairment manifests. This domain unites brain and cognitive reserves [664–666]. *Brain reserve* refers to quantitative aspects of the brain (structure or '*hardware*'), including but not limited to the number and size of neurons, the number of connections between neurons, fiber density, axonal diameter, the degree of myelination, the integrity of corticocortical and thalamocortical circuits, hippocampal volume, the number of active synapses in the thalamic nuclei, and the number of potential neural pathways [664,667–669]. *Cognitive reserve* refers to the 'neuropsychological competence' aspects of the brain, that is, how well the underlying 'hardware' is used (functions or '*software*'). It reflects a process where the brain actively attempts to cope with brain challenges or damage by using pre-existing neural networks and/or by recruiting additional or alternative brain regions to support the task network. Cognitive reserve also includes the processes of network efficiency and neural compensation [664,666,669,670].

In this view, the dualism of brain and cognitive reserves is considered within a single framework—*brain resources*—where the term 'resource' refers to the joint structural and functional characteristics of brain networks that offer cognitive protection in disease.

A person with high overall brain resources has increased compensatory and neuroprotective mechanisms that give the person increased capacity for effective cognitive functioning *in spite of* neuropathophysiological challenges or aging. A person with high brain resources has a younger brain phenotype and is more likely to remain within normal limits for a longer period of time despite the possible parallel progression of underlying disease [671]. Therefore, a person with high brain resources and a high disease burden may remain asymptomatic due to compensatory and resistant adaptations of the functional brain network [667,672,673] (Figure 4).

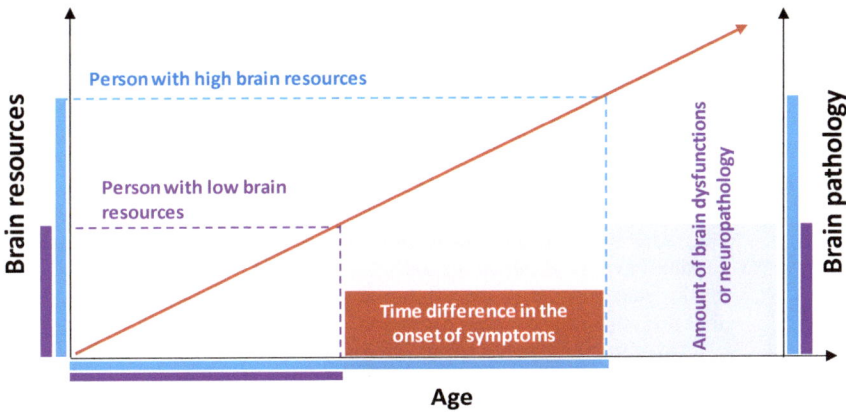

Figure 4. Diagrammatic representation of the interrelation between brain resources, age, and amount of brain dysfunctions or neuropathology. Red arrow = brain dysfunctions or neuropathology. Individuals with high brain resources are able to compensate for dysfunctions or structural damage for a longer period of time, and thus, symptoms do not manifest at low/moderate degrees of neural dysfunction/damage. However, after reaching a critical threshold at which the compensatory mechanisms are exhausted, clinical impairment progresses quickly. In contrast, slight brain damage or dysfunction is sufficient to produce clinical symptoms in individuals with low brain resources due to a weaker deployment of functional compensatory mechanisms. Consequently, they exhibit clinical symptoms at an earlier age and stage of the disease; however, these clinical symptoms progress slowly as pathologic neural changes slowly accrue. The period of time that elapses between the onset of clinical symptoms and the advanced stages of cognitive impairment is shorter in individuals with high brain resources compared to individuals with low brain resources because individuals with high brain resources remain asymptomatic during the early stages, before compensatory mechanisms are exhausted, but individuals with low brain resources already show clinical symptoms during those stages (description is modified from [674]).

A person with low overall brain resources has decreased compensatory and neuroprotective mechanisms that decrease the capacity for effective cognitive functioning even without neuropathophysiological challenges. A person with fewer brain resources has an older brain phenotype, so he or she may express symptoms after a trivial brain dysfunction due to redundant neural pathways and an inflexible or intolerant functional network. Likewise, the brain has a lower threshold for the expression of functional impairments following the onset of brain pathology [667,672,673] (Figure 4).

Since the qEEG oscillatory pattern reflects structural and functional integrity of the brain (see Sections 2.4 and 2.5) and causally relates to neuropsychopathology (see Section 2.15), it is not surprising that overall brain resources are reflected in the characteristics of the qEEG phenotype [8,112,113,116,119,121,122,136,143,144,185,187,671,675–678].

From this brief overview of the fundamental and primary domains of functioning, it is clear that each (a) exhibits trait properties; (b) is expressed along the entire continuum of functioning, from health to pathology; (c) is transdiagnostic; and (d) is reflected in the characteristics of qEEG phenotypes.

It has been proposed that trait characteristics and qEEG phenotypes should only be evaluated during a resting state [453,519].

4.3. Why the Resting State?

Studies of the *closed-eyes resting condition* provide an important opportunity to examine baseline qEEG patterns unbiased by any task. The resting-state condition avoids the confounding effects of visual scenes, instructions, and task execution (i.e., expectation matching, strategies employed, motivation or lack of it, fatigue, and anxiety associated with task performance). Additionally, the resting state seems more *self-relevant* than standard cognitive tasks, which typically drive subjects to direct their attention away from their personal concerns [50]. The resting-state condition permits the assessment of the *'pure'* self-relevant *baseline* brain and mind activity [51]. This activity reflects the individual type of spontaneous processing of an internal mental context (top-down processing) [52], such as random episodic memory [53] and related imagery [54], conceptual processing [55], stimulus-independent thought [56], self-reflection, internal narrative, and autobiographical self [57–59]. The frequently expressed concern that unconstrained brain activity varies unpredictably does not apply to the passive resting-state condition of the human brain. Studies have shown that "it is rather *intrinsically constrained by the default functionality* of the resting-state condition [60]" (references in this citation can be found in [450], p. 1051) and that this constrained default functionality comprises the *individual neurophysiological type*.

Indeed, numerous studies have demonstrated that the resting-state qEEG represents the *default functional infrastructure* of the brain that is involved in information processing related to inherent and relatively stable capacities of the individual, such as emotional regulation, cognition, and behavior as well as individually specific neurocognitive mechanisms underlying adaptation to motivationally and intellectually challenging tasks or conditions and the systemic self-regulation of brain functions [1,11,679–681]. Additionally, relations between personality traits and resting-state activity have been found for each of the Big-5 personality traits (for a review, see [682]). The resting qEEG has a non-random complex spatial–temporal structure (see Section 3), is highly predictable (Section 2.2), and is regulated by a complex neuroanatomical and neurochemical homeostatic system (see Section 2.9). This system is genetically based (Section 2.3) but also demonstrates some flexibility, e.g., epigenetic changes as a result of lifestyle changes (Section 2.1). The resting-state qEEG may thus serve as an intrinsic functional *'fingerprint'* that captures trait-like features of brain organization that are relevant for neurocognitive functioning in healthy and pathological conditions. This intrinsic functional 'fingerprint' reflects an individual's *neurophysiological type*, which can be captured by the *qEEG phenotype* [1,392,421].

The importance of the resting-state qEEG is supported by evidence that functional brain organization, as measured in the resting condition, is *predictive* of task execution, performance, and behavioral reactions during the actual activity or task [329,683–686]. Additionally, most of the energy used by the brain goes into supporting resting and ongoing neuronal activity [687–689]. A task-related increase in neuronal metabolism is usually small when compared with this large resting energy consumption [689]. These facts also support the importance of the resting-state neuronal activity that consumes most of the brain's energy.

In this context, the resting-state qEEG manifests baseline trait mechanics of self-organization that regulate multiple brain systems, thus adapting the brain and body to an ever-changing environment [690,691]. Thus, the resting-state qEEG reflects *intrinsic baseline/default activity* that instantiates the maintenance of information for interpreting, responding to, and even predicting environmental demands. Here, the resting-state constitutes a *reference baseline*, relative to which all cognitive and physiological states in healthy and pathological conditions can be considered [288,289,680].

Thus, following Fox and Raichle [692], it can be suggested that the pattern of the resting-state qEEG (qEEG phenotype) serves as a *functional localizer* ('content'), providing a priori information about the way in which the brain will respond across a wide variety of tasks during healthy and pathological conditions ('context'). Indeed, in all of the neuropsychopathologic conditions we studied, none of the participants could reach a proper resting state that is typical for a healthy brain [152,244]. The corollary is that such a system is less able to cope with the demands of a constantly changing environment.

Based on this logic, we consider abnormality in a closed-eyes resting qEEG to be a core feature of any neuropsychodysfunction or -pathology [152,244]. In this context, alteration in the closed-eyes resting qEEG oscillatory pattern (qEEG phenotype) may constitute a tonic component of qEEG microstructural organization that can serve as the *field of action* for abnormalities governed by multiple causes [450].

4.4. qEEG Phenotype, Personality Traits, and Norm-Pathology Continuum

The above review (Section 4.2) demonstrates that variability in personality traits and their associated characteristics of qEEG phenotypes are expressed in both healthy and pathological conditions. This suggests that personality traits and neuropsychopathology are not distinct entities, but rather, they manifest along a unified continuum of functioning, where mental diseases can be considered to fall along multiple continuous trait dimensions, with traits (and corresponding characteristics of qEEG phenotype) ranging from normal to extreme (pathological). Here, neuropsychopathology is considered in terms of dysfunctions of various kinds and degrees. In other words, neuropsychopathology is the *over*-expression or *under*-expression of personality traits and the associated characteristics of qEEG phenotypes that are otherwise moderately expressed in healthy conditions.

Such a view assumes a *dimensional* approach to neuropsychopathology, which considers the full range of variation, from normal to abnormal functioning, where both extremes of a dimension may be considered 'abnormal' [515]. For example, when anxiety increases beyond an optimal level, (a) one's perceptual field narrows and attention to task-relevant cues fails [693], (b) the ability to store and retrieve task-relevant information deteriorates [694,695], and (c) distracting and task-irrelevant thoughts increase [696]. In its extreme form, it is associated with anxiety disorders. On the opposite end of the dimension, a complete lack of anxiety may be associated with aggressive or psychopathic behavior. Consider another example from [697] (p. 116): "[. . .], persons that fall in the high-end of sociability dimension, have positive emotional and high approach-motivation tendencies are also high in the novelty-seeking trait and low in the harm-avoidance trait (Davies, 2012; Eysenck, 1990). Such people are likely to find extravagance, novelty, and excitement motivating but will be relatively insensitive to the feelings of others, punishment for breaching rules, or the possibility of failure (Cavanagh, 2005)" (references in this citation can be found in [697]). In the extreme, it may be associated with mania, narcissism, psychopathy, substance abuse, excessive venturesomeness (excitement-seeking), risk-taking, grandiosity, exhibitionism, and manipulativeness [635]. On the opposite end of the dimension, lack of sociability may be associated with emotional tension or depression.

Considering a dimensional approach to neuropsychopathology, the following *functional continuum*, through normality (health) and mental disorders (psychopathology), can be proposed (Figure 5).

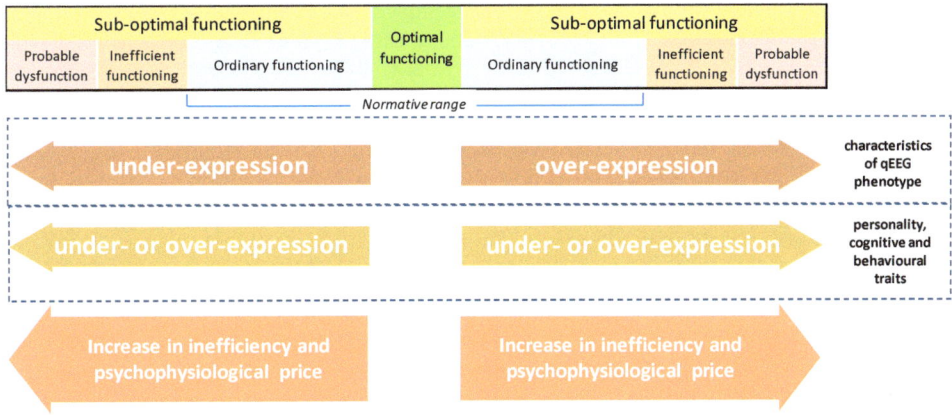

Figure 5. The norm-pathology continuum.

Here, the focus is on the expression of the characteristics of qEEG phenotypes (neural mechanism) over a continuous range. A corresponding mental characteristic that is linked to a particular qEEG phenotype (or its features) would also show a range of expression. Importantly, 'functionality' is the degree to which the individual is able to function, given various degrees of expression of the mental and associated qEEG phenotype's characteristics [519,582]. Hence, the degree of functionality associated with the expression of the mental characteristic corresponds to the degree of the expression of a given characteristic(s) of the qEEG phenotype. The expression of the characteristics of qEEG phenotypes in the middle range (green, Figure 5) is expected to correspond to optimal neurophysiological and mental functioning. Indeed, it has been shown that optimal characteristics of the qEEG phenotype are associated with optimal cognition, good personality structure, and overall well-being [506,582,612,677,698,699]. Similarly, it has been shown that (a) moderate values of self-esteem are optimal for psychological functioning, where both high and low values lead to dysfunction [700], (b) moderate anxiety is associated with the best performance [701,702], and (c) moderate vigilance/arousal is likewise associated with optimal performance [703–705]. For more examples, see [582].

Expression of qEEG phenotype characteristics outside the optimal range is expected to be associated with changes in the functionality of mental characteristics that are 'normal' but fall in a sub-optimal zone—*ordinary functioning* (in blue, Figure 5), representing a state of increased risk for mental dysfunction. The *inefficient functioning* range (orange area, Figure 5) on either end of the spectrum of the 'normal' range can be seen as a transition zone towards probable psychopathology. It can be defined as a range where personal features and behaviors cause difficulties but are nonetheless adaptable enough to avoid being classified as disordered [706]. Further extremes in the expression of the qEEG phenotype characteristics (purple area, Figure 5) are associated with states that are *dysfunctional* and may occur in a *pathological condition* as it leads to impairments in behavior. Extremes are always maladaptive, with no exception. Indeed, studies on healthy people showed that extreme neural values, even in otherwise normal individuals, impair their functionality and that the amelioration of extreme readings leads to improved functioning, whereas the middle values favor optimal functioning [582].

From an information-processing point of view, neurocognitive capacities follow a normal distribution in the human population, varying from being extremely efficient to extremely inefficient, depending on the underlying genetic makeup, trait predispositions, learning history, life events and style, and state variables. Since neurocognitive processes form and shape individual behaviors, sub-optimal neurocognitive capacity can translate into behavioral symptoms. The frequency and intensity of these symptoms will be continuously distributed in the general population—with most below (or above) the threshold for clinical significance. Here, normal function becomes impaired when symptoms escalate, making it difficult to maintain normal relationships and occupational productivity.

In the proposed continuum, the *optimal range* has special importance and, thus, demands more explanation. The *optimal range* of the qEEG phenotype characteristics represents certain '*idealized*' characteristics displayed by the majority of healthy subjects within the same age group without current or past neurologic or mental health complaints or other illnesses or traumas that might be associated with brain dysfunction and without a family history of neurologic and psychiatric diseases [455,707,708]. It is assumed that the optimal range of the qEEG phenotype characteristics is a zone in which an individual has a higher probability of achieving an optimal performance compared to a qEEG phenotype with characteristics that fall outside the optimal zone (see above). Optimal functioning ensures the efficient recruitment of resources, the adequate mobilization of energy, and the utilization of skills in proportion to the task at hand. In this context, *optimal functioning* includes:

- Successful performance;
- Maximum efficiency;
- Minimal cost;

- Temporal adequacy.

Here, *optimality* is understood in terms of a *trade-off* that balances the accuracy or benefit (result) of performance against an appropriate cost (the time, energy, memory, and computational resources). It is consistent with (a) an evolutionary trade-off approach, according to which a biological system (here, the brain) maximizes a specific fitness function that results in an optimal phenotype [709] that reflects trade-offs among traits to *optimize fitness* and (b) the *principle of optimality* in biology (formulated by Liberman et al. [710]) that relates to the establishment of *spatiotemporal patterns* that are maximally predictable and can hold the living state for a prolonged time [711]. In this context, optimal functioning can be considered a complex trait-like construct—an *optimal phenotype*.

Since there are *neurophysiological limits* as to how high or low qEEG characteristics are able to go, only the optimal range offers enough 'operating room' to increase or decrease a given qEEG characteristic (deviation to both sides away from the optimal range) when faced with a task and in dependence on the internal or external environment; thus, the optimal (mid) range is by definition maximally adaptive.

In between optimality and dysfunction, there lies a whole raft of conditions associated with varying kinds and degrees of functionality. Therefore, health and diseases can be conceptualized *non-categorically* as the heterogeneity of phenotypes that exist along a *continuum* between optimality and dysfunction. In this context, *health* can be defined as the *optimal*, flexible, and successful interaction of an organism with a complex and changing environment. Thus, *brain health* can be defined as the development and preservation of *optimal* brain integrity and neural network functioning for a given age [712] and *not* merely the absence of disease. *Disease* can be defined as the rigid persistence of context-insensitive operational set-points. Here, *healthiness* is not a difference in *kind* from the clinical population but merely a matter of *dimensional divergence distance*.

Deviation of qEEG characteristics from the optimal range does not necessarily reflect gross abnormality or a pathologic process. Deviation means that the brain functions *less efficiently*, thus spending more energy and resources to achieve a needed function, operation, or task (Figure 5) [153,449,707] or having increasing difficulties in finding resources available for compensation in order to preserve core cognitive functions [713]. The latter places more stress on available resources. Inefficiency in handling the normal challenges of daily life, as well as the adverse physiological consequences, can be genetically or developmentally programmed or mediated by lifestyle choices. Additionally, depending on the degree, deviations may limit the range of the cognitive, emotional, and behavioral repertoire accessible to the individual.

If the *compensatory* mechanisms of the brain are *intact*, then a small deviation is unlikely to be pathologically significant and lies within the bounds of normal individual variability (despite being *sub-optimal*). Indeed, the brain may have many built-in alternative solutions, which are revealed and manifested via neurological cases of resilience. Consider, for example, the mechanisms of dynamic rerouting of information streams in the brain that are necessary to maintain functional integrity in the face of structural network failure. However, the pathogenic significance of qEEG deviations from the optimal range increases when compensatory mechanisms of the brain are decreased or *exhausted*. This is usually associated with strong and very strong deviations from the optimal range (Figure 5). Indeed, since neuro-cognitive processes form and shape individual behaviors, sub-optimal neurocognitive capacity can translate into behavioral symptoms and particular dysfunctions or abnormalities that may already be associated with neurological, developmental, and psychiatric disorders when the clinical significance threshold is crossed [455,707,714].

An important consideration regarding dimensionality is that (a) the relationship between increasing disruptions in functional mechanisms reflected in the characteristics of qEEG phenotypes and the severity of symptoms may be *non-linear*, with set-points that mark a transition to more severe pathology [515], and (b) a particular characteristic of the qEEG phenotype may have a different meaning and reflect a different underlying pathogenetic process as a function of the overall *context* within which it emerges ('contextual functionality') [715].

Since there is abundant scientific evidence (see above) that normative qEEG values are the result of brain electrical rhythm autoregulation by a complex homeostatic system [248,449] that displays a characteristic metastability around certain homeostatic levels, then deviations in characteristics of qEEG phenotypes may reflect a departure from homeostatic regulation that is a new stable state of altered brain activity [716] and can be a pathological *'set point'* if this deviation is chronic and outside the normal range. This adaptation is referred to as *'allostasis'* and is defined as *'homeostasis through change'* [716,717], where normal mechanisms for homeostatic regulation have spun out of the physiological range, which can lead to a chronic condition of heightened vulnerability to pathology. The *price* the brain pays to adapt to adverse psychosocial or physical situations (environments), which is related to how inefficient the response is and how many challenges the brain experiences for a sustained period of time, is defined as the *allostatic load* [718,719]. Allostatic load is mediated by factors such as genes and early development, as well as learned behaviors, reflecting lifestyle choices of diet, exercise, drug use, and so on [716], all of which are reflected in the qEEG phenotype (see above), which is a long lasting 'brain signature'.

Considering (a) the dimensional approach to neuropsychopathology (see above), (b) the functional continuum of normality–pathology (see above), and (c) the fact that the same symptoms (though with different expressions) are present in the majority of psychiatric disorders [720] due to the transdiagnostic nature of primary domains of functioning (traits) and their associated qEEG phenotypes' characteristics (see above), *neuropsychopathology* can be conceptualized as the *degree of deviations* across every domain (assessed by qEEG phenotypes' characteristics) and their *disposition* (stabilized relative to each other's positions on the normality–pathology continuum), where the domain with the largest deviation (or combination of several deviant domains) may represent the leading pathogenic marker or risk factor that manifests as either in the permanent mental constitution or a heightened vulnerability to mental disorder. This is in line with animal models of neuropsychiatric disorders, where domain interplay was demonstrated [721].

In this context, developing a new diagnostic model should involve building a brain-based conceptualization of psychiatric illnesses [722], where *neuropsychopathology* is quantified through qEEG phenotypes that are defined by the *degree of deviation* from the normative (age- and sex-matched) data and the *disposition* of qEEG characteristics and the associated transdiagnostic primary domains of functioning (traits) that are placed in the *functional continuum* of normality–pathology. Here, the deviation from the normative values provides a probabilistic measure of the likelihood that the individual's electrical activity reflects abnormal brain functioning. Thus, concepts of *'normal'* or *'abnormal'* can be redefined based on the probability of qEEG characteristics associated with transdiagnostic primary domains of functioning relative to normative data, where *'abnormality'* is defined statistically as improbable values exceeding those expected randomly [153]. It is assumed that the more statistically unusual the observation, the more likely it is that the underlying brain system is clinically abnormal. This provides a quantitative and objective criterion for the severity of brain dysfunction in an individual. From this viewpoint, the distinction between 'normality' and 'abnormality' depends upon the threshold value established in any particular set of transdiagnostic primary domains of functioning [153] along the normality–pathology continuum. The proposed framework of brain profiling diagnosis provides valuable insights into the etiology of psychiatric diseases and allows researchers to understand how usually adaptive processes may become part of vicious circles that result in pathology; it also has important implications for identifying at-risk individuals, initiating early prevention, and tailoring treatments, thereby providing more cost-effective and efficient diagnostic tools.

5. Summary and Concluding Remarks

There is mounting evidence that the problems experienced by the current paradigm of psychiatric diagnoses are due to a lack of brain-related etiological knowledge about neuropsychopathology [2,276,722–729]. An important step towards solving these problems is through formulating a theoretical–conceptual framework that relates clinical manifestations of mental disorders to individual history, lifestyles, and brain dynamics in a comprehensive unifying manner.

Here, such a framework is proposed, and it is based on knowledge accumulated over many decades (including recent developments) from systemic, cognitive, and clinical neuroscience, where mathematically and statistically derived analyses (neurometrics) of qEEGs provide quantitative information about brain activity that is related to anatomical and functional integrity, developmental maturation, and the mediation of sensory, perceptual, and cognitive processes as well as clinical manifestations of mental disorders. The resulting patterns of qEEG characteristics—qEEG phenotypes—are placed on the functional continuum of health and pathology of primary domains of functioning (traits), which have a dimensional nature and are transdiagnostic. In this context, the typicality or atypicality of each qEEG phenotype is quantified by the disposition of qEEG characteristics on the distribution of phenotypic parameters along the health–pathology continuum, where more atypical phenotypes have more extreme positions. This theoretical–conceptual framework has neurobiological/etiological relevance and uses dimensionally parameterized physiological characteristics, mental functions, and neuropsychopathology; we believe it provides better clinical diagnostic and prognostic utility of qEEGs.

However, more work is needed to arrive at (a) the optimal number and types of primary domains of functioning (traits), (b) the optimal repertoire of qEEG phenotypes, (c) the 'library' of qEEG characteristics within the qEEG phenotypes associated with the primary domains of functioning, and (d) adequate thresholds in the values of qEEG characteristics (age- and sex-stratified) that signal the transition from one functioning condition to another along the health–pathology continuum.

Author Contributions: Conceptualization, A.A.F. (Alexander A. Fingelkurts) and A.A.F. (Andrew A. Fingelkurts); methodology, A.A.F. (Alexander A. Fingelkurts) and A.A.F. (Andrew A. Fingelkurts); investigation, A.A.F. (Alexander A. Fingelkurts) and A.A.F. (Andrew A. Fingelkurts); resources, A.A.F. (Alexander A. Fingelkurts); writing—original draft preparation, A.A.F. (Alexander A. Fingelkurts); writing—review and editing, A.A.F. (Andrew A. Fingelkurts); visualization, A.A.F. (Alexander A. Fingelkurts) and A.A.F. (Andrew A. Fingelkurts) All authors have read and agreed to the published version of the manuscript.

Funding: This research received no external funding.

Institutional Review Board Statement: Not applicable.

Informed Consent Statement: Not applicable.

Data Availability Statement: Not applicable.

Acknowledgments: The authors thank D. Skarin for English editing.

Conflicts of Interest: A.A.F. (Alexander A. Fingelkurts) and A.A.F. (Andrew A. Fingelkurts) hold senior researcher positions at BM-Science and are involved in fundamental and applied neuroscience research and the development of qEEG-based brain analyses and well-being applications.

References

1. Lazarev, V.V. The relationship of theory and methodology in EEG studies of mental activity. *Int. J. Psychophysiol.* **2006**, *62*, 384–393. [CrossRef] [PubMed]
2. Peled, A. Brain profiling and clinical-neuroscience. *Med. Hypotheses* **2006**, *67*, 941–946. [CrossRef] [PubMed]
3. Kropotov, J.D. *Quantitative EEG, Event-Related Potentials and Neurotherapy*; Elsevier: Oxford, UK, 2009; p. 531.
4. Bellavite, P.; Signorini, A. Pathology, complex systems, and resonance. In *Fundamental Research in Ultra-High Dilution and Homoeopathy*; Schulte, J., Endler, P.C., Eds.; Kluwer Academic Publishers: Dordrecht, The Netherlands, 1998; pp. 105–116.
5. Basar, E. *Brain-Body-Mind in the Nebulous Cartesian System: A Holistic Approach by Oscillations*; Springer: New York, NY, USA, 2011; p. 523.
6. Motamedi-Fakhr, S.; Moshrefi-Torbati, M.; Hill, M.; Hill, C.M.; White, P.R. Signal processing techniques applied to human sleep EEG signals—A review. *Biomed. Signal Process. Control.* **2014**, *10*, 21–33. [CrossRef]

7. Da Silva, F.L. Neural mechanisms underlying brain waves: From neural membranes to networks. *Electroencephalogr. Clin. Neurophysiol.* **1991**, *79*, 81–93. [CrossRef]
8. Nunez, P.L. *Neocortical Dynamics and Human EEG Rhythms*; Oxford University Press: New York, NY, USA, 1995; p. 730.
9. Michel, C.M.; Brandeis, D. The sources and temporal dynamics of scalp electric fields. In *Simultaneous EEG and fMRI. Recording, Analysis, and Application*; Ullsperger, M., Debener, S.S., Eds.; Oxford University Press: New York, NY, USA, 2010; pp. 1–19.
10. Freeman, W.J. *Mass Action in the Nervous System. Examination of the Neurophysiological Basis of Adaptive Behavior through the EEG*; Academic Press: New York, NY, USA, 1975; p. 489.
11. Corsi-Cabrera, M.; Herrera, P.; Malvido, M. Correlation between EEG and cognitive abilities: Sex differences. *Int. J. Neurosci.* **1989**, *45*, 133–141. [CrossRef]
12. Tsodyks, M.; Kenet, T.; Grinvald, A.; Arieli, A. Linking spontaneous activity of single cortical neurons and the underlying functional architecture. *Science* **1999**, *286*, 1943–1946. [CrossRef]
13. Nunez, P.L. Toward a quantitative description of large-scale neocortical dynamic function and EEG. *Behav. Brain Sci.* **2000**, *23*, 371–437. [CrossRef]
14. Bressler, S.L.; Kelso, J.A.S. Cortical coordination dynamics and cognition. *Trends Cogn. Sci.* **2001**, *5*, 26–36. [CrossRef]
15. Moran, R.J.; Stephan, K.E.; Kiebel, S.J.; Rombach, N.; O'Connor, W.T.; Murphy, K.J.; Reilly, R.B.; Friston, K.J. Bayesian estimation of synaptic physiology from the spectral responses of neural masses. *Neuroimage* **2008**, *42*, 272–284. [CrossRef]
16. Hadjipapas, A.; Casagrande, E.; Nevado, A.; Barnes, G.R.; Green, G.; Holliday, I.E. Can we observe collective neuronal activity from macroscopic aggregate signals? *Neuroimage* **2009**, *44*, 1290–1303. [CrossRef]
17. van Albada, S.J.; Kerr, C.C.; Chiang, A.K.I.; Rennie, C.J.; Robinson, P.A. Neurophysiological changes with age probed by inverse modelling of EEG spectra. *Clin. Neurophysiol.* **2010**, *121*, 21–38. [CrossRef] [PubMed]
18. Fingelkurts, A.A.; Fingelkurts, A.A. Short-term EEG spectral pattern as a single event in EEG phenomenology. *Open Neuroimag. J.* **2010**, *4*, 130–156. [CrossRef] [PubMed]
19. Klimesch, W. Memory processes, brain oscillations and EEG synchronization. *Int. J. Psychophysiol.* **1996**, *24*, 61–100. [CrossRef]
20. Da Silva, F.H.L. The generation of electric and magnetic signals of the brain by local networks. In *Comprehensive Human Physiology*; Greger, R., Windhorst, U., Eds.; Springer: Berlin/Heidelberg, Germany, 1996; Volume 1, pp. 509–528.
21. Klimesch, W. EEG alpha and theta oscillations reflect cognitive and memory performance: A review and analysis. *Brain Res. Rev.* **1999**, *29*, 169–195. [CrossRef]
22. Basar, E. *Brain Function and Oscillations. I Vol. Brain Oscillations: Principles and Approaches*; Springer: Berlin/Heidelberg, Germany, 1998; p. 396.
23. Basar, E.; Basar-Eroglu, C.; Karakas, S.; Schurmann, M. Are cognitive processes manifested in event-related gamma, alpha, theta and delta oscillations in the EEG? *Neurosci. Lett.* **1999**, *259*, 165–168. [CrossRef]
24. Basar, E.; Basar-Eroglu, C.; Karakas, S.; Schurmann, M. Brain oscillations in perception and memory. *Int. J. Psychophysiol.* **2000**, *35*, 95–124. [CrossRef]
25. Klimesch, W. Interindividual differences in oscillatory EEG activity and cognitive performance. In *The Cognitive Neuroscience of Individual Differences*; Reinvang, I., Greenlee, M., Herrmann, M., Eds.; BIS: Oldenburg, Germany, 2003; pp. 87–99.
26. Basar, E.; Özgören, M.; Karakas, S.; Basar-Eroglu, C. Super-synergy in the brain: The grandmother percept is manifested by multiple oscillations. *Int. J. Bifurcat. Chaos* **2004**, *14*, 453–491. [CrossRef]
27. Klimesch, W.; Schack, B.; Sauseng, P. The functional significance of theta and upper alpha oscillations. *Exp. Psychol.* **2005**, *52*, 99–108. [CrossRef]
28. Jurko, M.F.; Giurintano, L.P.; Giurintano, S.L.; Andy, O.J. Spontaneous awake EEG patterns in three lines of primate evolution. *Behav. Biol.* **1974**, *10*, 377–384. [CrossRef]
29. Van den Bergh, B.R.; Mulder, E.J.; Mennes, M.; Glover, V. Antenatal maternal anxiety and stress and the neurobehavioural development of the fetus and child: Links and possible mechanisms. A review. *Neurosci. Biobehav. Rev.* **2005**, *29*, 237–258. [CrossRef]
30. Class, Q.A.; Lichtenstein, P.; Langstrom, N.; D'Onofrio, B.M. Timing of prenatal maternal exposure to severe life events and adverse pregnancy outcomes: A population study of 2.6 million pregnancies. *Psychosom. Med.* **2011**, *73*, 234–241. [CrossRef] [PubMed]
31. Grigoriadis, S.; VonderPorten, E.H.; Mamisashvili, L.; Tomlinson, G.; Dennis, C.L.; Koren, G.; Steiner, M.; Mousmanis, P.; Cheung, A.; Radford, K.; et al. The impact of maternal depression during pregnancy on perinatal outcomes: A systematic review and meta-analysis. *J. Clin. Psychiatr.* **2013**, *74*, e321–e341. [CrossRef] [PubMed]
32. Slykerman, R.F.; Thompson, J.; Waldie, K.; Murphy, R.; Wall, C.; Mitchell, E.A. Maternal stress during pregnancy is associated with moderate to severe depression in 11-year-old children. *Acta Paediatr.* **2015**, *104*, 68–74. [CrossRef] [PubMed]
33. Papadopoulou, Z.; Vlaikou, A.M.; Theodoridou, D.; Markopoulos, G.S.; Tsoni, K.; Agakidou, E.; Drosou-Agakidou, V.; Turck, C.W.; Filiou, M.D.; Syrrou, M. Stressful newborn memories: Pre-conceptual, in utero, and postnatal events. *Front. Psychiatr.* **2019**, *10*, 220. [CrossRef] [PubMed]
34. Prichep, L.S.; Kowalik, S.C.; Alper, K.; de Jesus, C. Quantitative EEG characteristics of children exposed in utero to cocaine. *Clin. Electroencephalogr.* **1995**, *26*, 166–172. [CrossRef]
35. Matousek, M.; Petersen, I. Frequency analysis of the EEG in normal children and adolescents. In *Automation of Clinical Electroencephalography*; Kellaway, P., Petersen, I., Eds.; Raven Press: New York, NY, USA, 1973; pp. 75–102.

36. Cragg, L.; Kovacevic, N.; McIntosh, A.R.; Poulsen, C.; Martinu, K.; Leonard, G.; Paus, T. Maturation of EEG power spectra in early adolescence: A longitudinal study. *Dev. Sci.* **2011**, *14*, 935–943. [CrossRef]
37. Scraggs, T.L. EEG maturation: Viability through adolescence. *Neurodiagn. J.* **2012**, *52*, 176–203.
38. Kaminska, A.; Eisermann, M.; Plouin, P. Child EEG (and maturation). In *Handbook of Clinical Neurology, Clinical Neurophysiology: Basis and Technical Aspects*, 3rd ed.; Levin, K.H., Chauvel, P., Eds.; Elsevier B.V.: Amsterdam, The Netherlands, 2019; Volume 160, pp. 125–142.
39. Zhadin, M.N. Rhythmic processes in the cerebral cortex. *J. Theor. Biol.* **1984**, *108*, 565–595. [CrossRef]
40. Anokhin, A.P.; Birbaumer, N.; Lutzenberger, W.; Nikolaev, A.; Vogel, F. Age increases brain complexity. *Electroencephalogr. Clin. Neurophysiol.* **1996**, *99*, 63–68. [CrossRef]
41. Schutter, D.J.L.G.; Leitner, C.; Kenemans, J.L.; van Honk, J. Electrophysiological correlates of cortico-subcortical interaction: A cross-frequency spectral EEG analysis. *Clin. Neurophysiol.* **2006**, *117*, 381–387. [CrossRef]
42. Marosi, E.; Harmony, T.; Sánchez, L.; Becker, J.; Bernal, J.; Reyes, A.; de León, A.E.D.; Rodríguez, M.; Fernández, T. Maturation of the coherence of EEG activity in normal and learning-disabled children. *EEG Clin. Neurophysiol.* **1992**, *83*, 350–357. [CrossRef]
43. Lukashevich, I.P.; Machinskaya, R.I.; Fishman, M.N. Diagnosis of the functional state of the brain in young school-age children with learning difficulties. *Hum. Physiol. (Fiziol Cheloveka)* **1994**, *20*, 34–46.
44. Ulrich, G. Psychiatric Electroencephalography. In *Updated and Revised Edition (2002) of the Original Textbook Psychiatrische Elektroenzephalographie (in German)*; Gustav Fischer Verlag: New York, NY, USA, 1994; p. 343.
45. Zentner, M. Antisocial personalities. In *Adult Psychopathology. A Social Work Perspective*; Turner, F.J., Ed.; The Free Press: New York, NY, USA, 1984; pp. 345–363.
46. Davies, R.K. Incest: Some neuropsychiatric findings. *Int. J. Psychiatry Med.* **1979**, *9*, 117–121. [CrossRef] [PubMed]
47. McFarlane, A.; Clark, C.R.; Bryant, R.A.; Williams, L.M.; Niaura, R.; Paul, R.H.; Hitsman, B.L.; Stroud, L.; Alexander, D.M.; Gordon, E. The impact of early life stress on psychophysiological, personality and behavioural measures in 740 non-clinical subjects. *J. Integr. Neurosci.* **2005**, *4*, 27–40. [CrossRef] [PubMed]
48. Taylor, S.E. Mechanisms linking early life stress to adult health outcomes. *Proc. Natl. Acad. Sci. USA* **2010**, *107*, 8507–8512. [CrossRef]
49. Harmony, T.; Alvarez, A.; Pascual, R.; Ramos, A.; Marosi, E.; De León, A.E.D.; Valdés, P.; Becker, J. EEG maturation on children with different economic and psychosocial characteristics. *Int. J. Neurosci.* **1988**, *41*, 103–113. [CrossRef]
50. Otero, G.A.; Pliego-Rivero, F.B.; Fernández, T.; Ricardo, J. EEG development in children with sociocultural disadvantages: A follow-up study. *Clin. Neurophysiol.* **2003**, *114*, 1918–1925. [CrossRef]
51. Marshall, P.J.; Fox, N.A. Bucharest early intervention project core group. A comparison of the electroencephalogram between institutionalized and community children in Romania. *J. Cogn. Neurosci.* **2004**, *16*, 1327–1338. [CrossRef]
52. Howells, F.M.; Stein, D.J.; Russell, V.A. Childhood trauma is associated with altered cortical arousal: Insights from an EEG study. *Front. Integr. Neurosci.* **2012**, *6*, 120. [CrossRef]
53. Vanderwert, R.E.; Zeanah, C.H.; Fox, N.A.; Nelson, C.A. Normalization of EEG activity among previously institutionalized children placed into foster care: A 12-year follow-up of the Bucharest Early Intervention Project. *Dev. Cogn. Neurosci.* **2016**, *17*, 68–75. [CrossRef]
54. Bosch-Bayard, J.; Razzaq, F.A.; Lopez-Naranjo, C.; Wang, Y.; Li, M.; Galan-Garcia, L.; Calzada-Reyes, A.; Virues-Alba, T.; Rabinowitz, A.G.; Suarez-Murias, C.; et al. Early protein energy malnutrition impacts life-long developmental trajectories of the sources of EEG rhythmic activity. *NeuroImage* **2022**, *254*, 119144. [CrossRef] [PubMed]
55. Black, L.M.; Hudspeth, W.J.; Townsend, A.; Bodenhamer-Davis, E. EEG Connectivity Patterns in Childhood Sexual Abuse: A Multivariate Application Considering Curvature of Brain Space. *J. Neurother.* **2008**, *12*, 141–160. [CrossRef]
56. Lee, S.-H.; Park, Y.; Jin, M.J.; Lee, Y.J.; Hahn, S.W. Childhood trauma associated with enhanced high frequency band powers and induced subjective inattention of adults. *Front. Behav. Neurosci.* **2017**, *11*, 148. [CrossRef]
57. Thatcher, R.W.; Walker, R.A.; Gerson, I.; Geisler, F.H. EEG discriminant analysis of mild head trauma. *Electroencephalogr. Clin. Neurophysiol.* **1989**, *73*, 10–94. [CrossRef]
58. Hooshmand, H.; Beckner, E.; Radfar, R. Technical and clinical aspects of topographic brain mapping. *Clin. Electroencephalogr.* **1989**, *20*, 235–247. [CrossRef] [PubMed]
59. Thornton, K.E. The electrophysiological effects of a brain injury on auditory memory functioning: The QEEG correlates of impaired memory. *Arch. Clin. Neuropsychol.* **2003**, *18*, 363–378. [CrossRef] [PubMed]
60. Stassen, H.H. Computerized recognition of persons by EEG spectral patterns. *Electroencephalogr. Clin. Neurophysiol.* **1980**, *49*, 190–194. [CrossRef]
61. Gasser, T.; Bacher, P.; Steinberg, H. Test–retest reliability of spectral parameters of the EEG. *Electroencephalogr. Clin. Neurophysiol.* **1985**, *60*, 312–319. [CrossRef]
62. Salinsky, M.C.; Oken, B.S.; Morehead, L. Test–retest reliability in EEG frequency analysis. *Electroencephalogr. Clin. Neurophysiol.* **1991**, *79*, 382–392. [CrossRef]
63. Pollock, V.E.; Schneider, L.S.; Lyness, S.A. Reliability of topographic quantitative EEG amplitude in healthy late-middle-aged and elderly subjects. *Electroencephalogr. Clin. Neurophysiol.* **1991**, *79*, 20–26. [CrossRef]
64. Burgess, A.; Gruzelier, J. Individual reliability of amplitude distribution in topographical mapping of EEG. *Electroencephalogr. Clin. Neurophysiol.* **1993**, *86*, 219–223. [CrossRef]

65. Harmony, T.; Fernandez, T.; Rodriguez, M.; Reyes, A.; Marosi, E.; Bernal, J. Test–retest reliability of EEG spectral parameters during cognitive tasks: II. Coherence. *Int. J. Neurosci.* **1993**, *68*, 263–271. [CrossRef] [PubMed]
66. Lund, T.R.; Sponheim, S.R.; Iacono, W.G.; Clementz, B.A. Internal consistency reliability of resting EEG power spectra in schizophrenic and normal subjects. *Psychophysiology* **1995**, *32*, 66–71. [CrossRef] [PubMed]
67. Stassen, H.H.; Bomben, G.; Hell, D. Familial brain wave patterns: Study of a 12-sib family. *Psychiatr. Genet.* **1998**, *8*, 141–153. [CrossRef]
68. Dustman, R.E.; Shearer, D.E.; Emmerson, R.Y. Life-span changes in EEG spectral amplitude, amplitude variability and mean frequency. *Clin. Neurophysiol.* **1999**, *110*, 1399–1409. [CrossRef]
69. Kondacs, A.; Szabo, M. Long-term intra-individual variability of the background EEG in normals. *Clin. Neurophysiol.* **1999**, *110*, 1708–1716. [CrossRef]
70. Dünki, R.M.; Schmid, B.; Stassen, H.H. Intraindividual specificity and stability of human EEG: Comparing a linear vs. a onlinear approach. *Methods Inf. Med.* **2000**, *39*, 78–82.
71. Poulos, M.; Rangoussi, M.; Alexandris, N.; Evangelou, A. Person identification from the EEG using nonlinear signal classification. *Methods Inf. Med.* **2002**, *41*, 64–75.
72. Maltez, J.; Hyllienmark, L.; Nikulin, V.V.; Brismar, T. Time course and variability of power in different frequency bands of EEG during resting conditions. *Neurophysiol. Clin.* **2004**, *34*, 195–202. [CrossRef]
73. Fingelkurts, A.A.; Fingelkurts, A.A.; Ermolaev, V.A.; Kaplan, A.Y. Stability, reliability and consistency of the compositions of brain oscillations. *Int. J. Psychophysiol.* **2006**, *59*, 116–126. [CrossRef]
74. Vuga, M.; Fox, N.A.; Cohn, J.F.; George, C.J.; Levenstein, R.M.; Kovacs, M. Long-term stability of frontal electroencephalographic asymmetry in adults with a history of depression and controls. *Int. J. Psychophysiol.* **2006**, *59*, 107–115. [CrossRef] [PubMed]
75. Näpflin, M.; Wildi, M.; Sarnthein, J. Test-retest reliability of resting EEG spectra validates a statistical signature of persons. *Clin. Neurophysiol.* **2007**, *118*, 2519–2524. [CrossRef] [PubMed]
76. Towers, D.N.; Allen, J.J. A better estimate of the internal consistency reliability of frontal EEG asymmetry scores. *Psychophysiology* **2009**, *46*, 132–142. [CrossRef] [PubMed]
77. Cannon, R.L.; Baldwin, D.R.; Shaw, T.L.; Diloreto, D.J.; Phillips, S.M.; Scruggs, A.M.; Riehl, T.C. Reliability of quantitative EEG (qEEG) measures and LORETA current source density at 30 days. *Neurosci. Lett.* **2012**, *518*, 27–31. [CrossRef]
78. Grandy, T.H.; Werkle-Bergner, M.; Chicherio, C.; Schmiedek, F.; Lövdén, M.; Lindenberger, U. Peak individual alpha frequency qualifies as a stable neurophysiological trait marker in healthy younger and older adults. *Psychophysiology* **2013**, *50*, 570–582. [CrossRef]
79. Vogel, F. The genetic basis of the normal human electroencephalogram (EEG). *Humangenetik* **1970**, *10*, 91–114. [CrossRef]
80. Lykken, D.T.; Tellegen, A.; Thorkelson, K. Genetic determination of EEG frequency spectra. *Biol. Psychol.* **1974**, *1*, 245–259. [CrossRef]
81. Lykken, D.T.; Tellegen, A.; Iacono, W.G. EEG spectra in twins: Evidence for a neglected mechanism of genetic determination. *Physiol. Psychol.* **1982**, *10*, 60–65. [CrossRef]
82. Stassen, H.H.; Bomben, G.; Propping, P. Genetic aspects of the EEG: An investigation into the within-pair similarity of monozygotic and dyzigotic twins with a new method of analysis. *Electroencephalogr. Clin. Neurophysiol.* **1987**, *66*, 489–501. [CrossRef]
83. van Beijsterveldt, C.E.; Boomsma, D.I. Genetics of the human electroencephalogram (EEG) and event-related brain potentials (ERPs): A review. *Hum. Genet.* **1994**, *94*, 319–330. [CrossRef]
84. Christian, J.C.; Morzorati, S.; Norton, J.A., Jr.; Williams, C.J.; O'Connor, S.; Li, T.K. Genetic analysis of the resting electroencephalographic power spectrum in human twins. *Psychophysiology* **1996**, *33*, 584–591. [CrossRef] [PubMed]
85. van Baal, G.C.; De Geus, E.J.; Boomsma, D.I. Genetic architecture of EEG power spectra in early life. *Electroencephalogr. Clin. Neurophysiol.* **1996**, *98*, 502–514. [CrossRef]
86. van Beijsterveldt, C.E.; Molenaar, P.C.; De Geus, E.J.; Boomsma, D.I. Heritability of human brain functioning as assessed by electroencephalography. *Am. J. Hum. Genet.* **1996**, *58*, 562–573. [PubMed]
87. Posthuma, D.; Neale, M.C.; Boomsma, D.I.; De Geus, E.J.C. Are smarter brains running faster? Heritability of alpha peak frequency, IQ, and their interrelation. *Behav. Genet.* **2001**, *31*, 567–579. [CrossRef]
88. van Baal, G.; van Beijsterveldt, C.; Molenaar, P.; Boomsma, D.; De Geus, E. A genetic perspective on the developing brain: Electrophysiological indices of neural functioning in young and adolescent twins. *Eur. Psychol.* **2001**, *6*, 254–263. [CrossRef]
89. van Beijsterveldt, C.E.M.; van Baal, G. Twin and family studies of the human electroencephalogram: A review and a meta-analysis. *Biol. Psychol.* **2002**, *61*, 111–138. [CrossRef]
90. Smit, D.J.A.; Posthuma, D.; Boomsma, D.I.; De Geus, E.J.C. Heritability of background EEG across the power spectrum. *Psychophysiology* **2005**, *42*, 691–697. [CrossRef]
91. Anokhin, A.P.; Müller, V.; Lindenberger, U.; Heath, A.C.; Myers, E. Genetic influences on dynamic complexity of brain oscillations. *Neurosci. Lett.* **2006**, *397*, 93–98. [CrossRef]
92. Smit, C.M.; Wright, M.J.; Hansell, N.K.; Geffen, G.M.; Martin, N.G. Genetic variation of individual alpha frequency (IAF) and alpha power in a large adolescent twin sample. *Int. J. Psychophysiol.* **2006**, *61*, 235–243. [CrossRef]
93. Tang, Y.; Chorlian, D.B.; Rangaswamy, M.; O'Connor, S.; Taylor, R.; Rohrbaugh, J.; Porjesz, B.; Begleiter, H. Heritability of bipolar EEG spectra in a large sib-pair population. *Behav. Genet.* **2007**, *37*, 302–313. [CrossRef]
94. Eischen, S.E.; Luckritz, J.Y.; Polich, J. Spectral analysis of EEG from families. *Biol. Psychol.* **1995**, *41*, 61–68. [CrossRef]

95. Begleiter, H.; Porjesz, B. Genetics of human brain oscillations. *Int. J. Psychophysiol.* **2006**, *60*, 162–171. [CrossRef] [PubMed]
96. Winterer, G.; Smolka, M.; Samochowiec, J.; Ziller, M.; Mahlberg, R.; Gallinat, J.; Rommelspacher, H.P.; Herrmann, W.M.; Sander, T. Association of EEG coherence and an exonic GABA(B)R1 gene polymorphism. *Am. J. Med. Genet.* **2003**, *117B*, 51–56. [CrossRef] [PubMed]
97. Porjesz, B.; Almasy, L.; Edenberg, H.; Wang, K.; Chorlian, D.B.; Foroud, T.; Goate, A.; Rice, J.P.; O'Connor, S.; Rohrbaugh, J.; et al. Linkage disequilibrium between the beta frequency of the human EEG and a GABAa receptor gene locus. *Proc. Natl. Acad. Sci. USA* **2002**, *99*, 3729–3733. [CrossRef] [PubMed]
98. Edenberg, H.J.; Dick, D.M.; Xuei, X.; Tian, H.; Almasy, L.; Bauer, L.O.; Crowe, R.R.; Goate, A.; Hesselbrock, V.; Jones, K.; et al. Variations in GABRA2, encoding the alpha 2 subunit of the GABA(A) receptor, are associated with alcohol dependence and with brain oscillations. *Am. J. Hum. Genet.* **2004**, *74*, 705–714. [CrossRef] [PubMed]
99. Winterer, G.; Mahlberg, R.; Smolka, M.N.; Samochowiec, J.; Ziller, M.; Rommelspacher, H.P.; Herrmann, W.M.; Schmidt, L.G.; Sander, T. Association analysis of exonic variants of the GABA(B)-receptor gene and alpha electroencephalogram voltage in normal subjects and alcohol-dependent patients. *Behav. Genet.* **2003**, *33*, 7–15. [CrossRef]
100. Ducci, F.; Enoch, M.A.; Yuan, Q.; Shen, P.H.; White, K.V.; Hodgkinson, C.; Goldman, D. HTR3B is associated with alcoholism with antisocial behavior and alpha EEG power–an intermediate phenotype for alcoholism and co-morbid behaviors. *Alcohol* **2009**, *43*, 73–84. [CrossRef]
101. Enoch, M.A.; White, K.V.; Waheed, J.; Goldman, D. Neurophysiological and genetic distinctions between pure and comorbid anxiety disorders. *Depress. Anxiety* **2008**, *25*, 383–392. [CrossRef]
102. Enoch, M.A.; Shen, P.H.; Ducci, F.; Yuan, Q.; Liu, J.; White, K.V.; Albaugh, B.; Hodgkinson, C.A.; Goldman, D. Common genetic origins for EEG, alcoholism and anxiety: The role of CRH-BP. *PLoS ONE* **2008**, *3*, e3620. [CrossRef]
103. Enoch, M.A.; Xu, K.; Ferro, E.; Harris, C.R.; Goldman, D. Genetic origins of anxiety in women: A role for a functional catechol-O-methyltransferase polymorphism. *Psychiatr. Genet.* **2003**, *13*, 33–41. [CrossRef]
104. Enoch, M.A.; Rohrbaugh, J.W.; Davis, E.Z.; Harris, C.R.; Ellingson, R.J.; Andreason, P.; Moore, V.; Varner, J.L.; Brown, G.L.; Eckardt, M.J. Relationship of genetically transmitted alpha EEG traits to anxiety disorders and alcoholism. *Am. J. Med. Genet.* **1995**, *60*, 400–408. [CrossRef] [PubMed]
105. Enoch, M.A.; White, K.V.; Harris, C.R.; Robin, R.W.; Ross, J.; Rohrbaugh, J.W.; Goldman, D. Association of low-voltage alpha EEG with a subtype of alcohol use disorders. *Alcohol. Clin. Exp. Res.* **1999**, *23*, 1312–1319. [CrossRef] [PubMed]
106. Zoon, H.F.; Veth, C.P.; Arns, M.; Drinkenburg, W.H.; Talloen, W.; Peeters, P.J.; Kenemans, J.L. EEG alpha power as an intermediate measure between brain-derived neurotrophic factor Val66Met and depression severity in patients with major depressive disorder. *J. Clin. Neurophysiol.* **2013**, *30*, 261–267. [CrossRef] [PubMed]
107. Bodenmann, S.; Rusterholz, T.; Dürr, R.; Stoll, C.; Bachmann, V.; Geissler, E.; JaggiSchwarz, K.; Landolt, H.P. The functional val158met polymorphism of COMT predicts interindividual differences in brain alpha oscillations in young men. *J. Neurosci.* **2009**, *29*, 10855–10862. [CrossRef]
108. Meyers, J.L.; Zhang, J.; Chorlian, D.B.; Pandey, A.K.; Kamarajan, C.; Wang, J.-C.; Wetherill, L.; Lai, D.; Chao, M.; Chan, G.; et al. A genome-wide association study of interhemispheric theta EEG coherence: Implications for neural connectivity and alcohol use behavior. *Mol. Psychiatr.* **2021**, *26*, 5040–5052. [CrossRef]
109. Venables, N.C.; Bernat, E.M.; Sponheim, S.R. Genetic and disorder-specific aspects of resting state EEG abnormalities in schizophrenia. *Schizophr. Bull.* **2009**, *35*, 826–839. [CrossRef]
110. da Silva, F.H.L.; van Rotterdam, A.; Barts, P.; van Heusden, E.; Burr, W. Models of neuronal populations: The basic mechanism of rhythmicity. In *Perspectives of Brain Research. Progress in Brain Research*; Corner, M.A., Swaab, D.F., Eds.; Elsevier: Amsterdam, The Netherlands, 1976; Volume 45, pp. 281–308.
111. Hughes, J.R.; Cayaffa, J.J. The EEG in patients at different ages without organic cerebral disease. *Electroencephalogr. Clin. Neurophysiol.* **1977**, *42*, 776–784. [CrossRef]
112. Goldensohn, E.S. Use of EEG for evaluation of focal intracranial lesions. In *Current Practice of Clinical Electroencephalography*; Klass, D.W., Daly, D.D., Eds.; Raven: New York, NY, USA, 1979; pp. 307–341.
113. da Silva, F.H.L.; Vos, J.E.; Mooibroek, J.; van Rotterdam, A. Relative contributions of intracortical and thalamo-cortical processes in the generation of alpha rhythms, revealed by partial coherence analysis. *Electroencephalogr. Clin. Neurophysiol.* **1980**, *50*, 449–456. [CrossRef]
114. Steriade, M.; Llinas, R.R. The functional states of the thalamus and the associated neuronal interplay. *Physiol. Rev.* **1988**, *68*, 649–742. [CrossRef]
115. Anokhin, A.; Vogel, F. EEG alpha rhythm frequency and intelligence in normal adults. *Intelligence* **1996**, *23*, 1–14. [CrossRef]
116. Leocani, L.; Locatelli, T.; Martinelli, V.; Rovaris, M.; Falautano, M.; Filippi, M.; Magnani, G.; Comi, G. Electroencephalographic coherence analysis in multiple sclerosis: Correlation with clinical, neuropsychological, and MRI findings. *J. Neurol. Neurosurg. Psychiatry* **2000**, *69*, 192–198. [CrossRef] [PubMed]
117. Moretti, D.V.; Babiloni, C.; Binetti, G.; Cassetta, E.; Dal Forno, G.; Ferreric, F.; Ferri, R.; Lanuzza, B.; Miniussi, C.; Nobili, F.; et al. Individual analysis of EEG frequency and band power in mild Alzheimer's disease. *Clin. Neurophysiol.* **2004**, *115*, 299–308. [CrossRef]
118. Nunez, P.L.; Srinivasan, R. A theoretical basis for standing and traveling brain waves. *Clin. Neurophysiol.* **2006**, *117*, 2425–2435. [CrossRef] [PubMed]

119. Moretti, D.V.; Miniussi, C.; Frisoni, G.; Zanetti, O.; Binetti, G.; Geroldi, C.; Galluzzi, S.; Rossini, P.M. Vascular damage and EEG markers in subjects with mild cognitive impairment. *Neurophysiol. Clin.* **2007**, *118*, 1866–1876. [CrossRef]
120. Babiloni, C.; Frisoni, G.B.; Pievani, M.; Vecchio, F.; Lizio, R.; Buttiglione, M.; Geroldi, C.; Fracassi, C.; Eusebi, F.; Ferri, R.; et al. Hippocampal volume and cortical sources of EEG alpha rhythms in mild cognitive impairment and Alzheimer disease. *Neuroimage* **2009**, *44*, 123–135. [CrossRef]
121. Valdés-Hernández, P.A.; Ojeda-González, A.; Martínez-Montes, E.; Lage-Castellanos, A.; Virués-Alba, T.; Valdés-Urrutia, L.; Valdes-Sosa, P.A. White matter architecture rather than cortical surface area correlates with the EEG alpha rhythm. *NeuroImage* **2010**, *49*, 2328–2339. [CrossRef]
122. Bhattacharya, B.S.; Coyle, D.; Maguire, L.P. A thalamo-cortico-thalamic neural mass model to study alpha rhythms in Alzheimer's disease. *Neural Netw.* **2011**, *24*, 631–645. [CrossRef]
123. Jann, K.; Federspiel, A.; Giezendanner, S.; Andreotti, J.; Kottlow, M.; Dierks, T.; Koenig, T. Linking brain connectivity across different time scales with electroencephalogram, functional magnetic resonance imaging, and diffusion tensor imaging. *Brain Connect* **2012**, *2*, 11–20. [CrossRef]
124. Garcés, P.; Vicente, R.; Wibral, M.; Pineda-Pardo, J.Á.; López, M.E.; Aurtenetxe, S.; Marcos, A.; de Andrés, M.E.; Yus, M.; Sancho, M.; et al. Brain-wide slowing of spontaneous alpha rhythms in mild cognitive impairment. *Front. Ageing Neurosci.* **2013**, *5*, 100. [CrossRef]
125. Babiloni, C.; Carducci, F.; Lizio, R.; Vecchio, F.; Baglieri, A.; Bernardini, S.; Cavedo, E.; Bozzao, A.; Buttinelli, C.; Esposito, F.; et al. Resting state cortical electroencephalographic rhythms are related to gray matter volume in subjects with mild cognitive impairment and Alzheimer's disease. *Hum. Brain Mapp.* **2013**, *34*, 1427–1446. [CrossRef]
126. Hindriks, R.; van Putten, M.J.A.M. Thalamo-cortical mechanisms underlying changes in amplitude and frequency of human alpha oscillations. *NeuroImage* **2013**, *70*, 150–163. [CrossRef] [PubMed]
127. Thatcher, R.W.; Biver, C.; McAlaster, R.; Salazar, A.M. Biophysical linkage between MRI and EEG coherence in traumatic brain injury. *NeuroImage* **1998**, *8*, 307–326. [CrossRef] [PubMed]
128. Thatcher, R.W.; Biver, C.L.; Gomez-Molina, J.F.; North, D.; Curtin, R.; Walker, R.W.; Salazar, A. Estimation of the EEG power spectrum by MRI T2 relaxation time in traumatic brain injury. *Clin. Neurophysiol.* **2001**, *112*, 1729–1745. [CrossRef]
129. Ray, W.; Cole, H. EEG alpha activity reflects attentional demands and beta activity reflects emotional and cognitive processes. *Science* **1985**, *228*, 750–752. [CrossRef]
130. Lazarev, V.V. Factorial structure of the principal EEG parameters during intellectual activity. I. Local characteristics of nonhomogeneity of functional states. *Hum. Physiol.* **1986**, *12*, 375–382, (A translation of Fiziol. Cheloveka).
131. Lazarev, V.V. Factorial structure of the principal EEG parameters during intellectual activity. II. Topography of functional states. *Hum. Physiol.* **1987**, *13*, 9–12, (A translation of Fiziol. Cheloveka).
132. Lazarev, V.V. On the intercorrelation of some frequency and amplitude parameters of the human EEG and its functional significance. Com. I. Multidimensional neurodynamic organization of functional states of the brain during intellectual, perceptive and motor activity in normal subjects. *Int. J. Psychophysiol.* **1998**, *28*, 77–98.
133. Lazarev, V.V. On the intercorrelation of some frequency and amplitude parameters of the human EEG and its functional significance. Com. II. Neurodynamic imbalance in endogenous asthenic-like disorders. *Int. J. Psychophysiol.* **1998**, *29*, 277–289. [CrossRef]
134. Mizuki, Y. Frontal lobe: Mental function and EEG. *Am. J. EEG Technol.* **1987**, *27*, 91–101. [CrossRef]
135. Klimesch, W.; Schimke, H.; Ladurner, G.; Pfurtscheller, G. Alpha frequency and memory performance. *J. Psychophysiol.* **1990**, *4*, 381–390.
136. Klimesch, W.; Schimke, H.; Pfurtscheller, G. Alpha frequency, cognitive load and memory performance. *Brain Topogr.* **1993**, *5*, 241–251. [CrossRef] [PubMed]
137. Harmony, T.; Fernandez, T.; Silva, J.; Bernal, J.; Díaz-Comas, L.; Reyes, A.; Marosi, E.; Rodríguez, M.; Rodríguez, M. EEG delta activity: An indicator of attention to internal processing during performance of mental tasks. *Int. J. Psychophysiol.* **1996**, *24*, 161–171. [CrossRef]
138. Doppelmayr, M.; Klimesch, W.; Schwaiger, J.; Auinger, P.; Winkler, T. Theta synchronization in the human EEG and episodic retrieval. *Neurosci. Lett.* **1998**, *257*, 41–44. [CrossRef]
139. Basar, E. *Brain Function and Oscillations. II Vol. Integrative Brain Function. Neurophysiology and Cognitive Processes*; Springer: Berlin/Heidelberg, Germany, 1999; p. 515.
140. Basar, E.; Basar-Eroglu, C.; Karakas, S.; Schurmann, M. Gamma, alpha, delta, and theta oscillations govern cognitive processes. *Int. J. Psychophysiol.* **2001**, *39*, 241–248. [CrossRef]
141. Basar, E.; Schurmann, M.; Sakowitz, O. The selectively distributed theta system: Functions. *Int. J. Psychophysiol.* **2001**, *39*, 197–212. [CrossRef]
142. Angelakis, E.; Lubar, J.F.; Stathopoulou, S. Electroencephalographic peak alpha frequency correlates of cognitive traits. *Neurosci. Lett.* **2004**, *371*, 60–63. [CrossRef] [PubMed]
143. Angelakis, E.; Lubar, J.F.; Stathopoulou, S.; Kounios, J. Peak alpha frequency: An electroencephalographic measure of cognitive preparedness. *Clin. Neurophysiol.* **2004**, *115*, 887–897. [CrossRef]
144. Clark, C.R.; Veltmeyer, M.D.; Hamilton, R.J.; Simms, E.; Paul, R.; Hermens, D.; Gordon, E. Spontaneous alpha peak frequency predicts working memory performance across the age span. *Int. J. Psychophysiol.* **2004**, *53*, 1–9. [CrossRef]

145. Knyazev, G.G. Motivation, emotion, and their inhibitory control mirrored in brain oscillations. *Neurosci. Biobehav. Rev.* **2007**, *31*, 377–395. [CrossRef]
146. Basar, E. A review of alpha activity in integrative brain function: Fundamental physiology, sensory coding, cognition and pathology. *Int. J. Psychophysiol.* **2012**, *86*, 1–24. [CrossRef]
147. Grandy, T.H.; Werkle-Bergner, M.; Chicherio, C.; Lövdén, M.; Schmiedek, F.; Lindenberger, U. Individual alpha peak frequency is related to latent factors of general cognitive abilities. *NeuroImage* **2013**, *79*, 10–18. [CrossRef] [PubMed]
148. Mierau, A.; Klimesch, W.; Lefebvre, J. State-dependent alpha peak frequency shifts: Experimental evidence, potential mechanisms and functional implications. *Neuroscience* **2017**, *360*, 146–154. [CrossRef] [PubMed]
149. Engel, A.K.; Fries, P.; Singer, W. Dynamic predictions: Oscillations and synchrony in top-down processing. *Nat. Rev. Neurosci.* **2001**, *2*, 704–716. [CrossRef]
150. Buzsáki, G.; Draguhn, A. Neuronal oscillations in cortical networks. *Science* **2004**, *304*, 1926–1929. [CrossRef] [PubMed]
151. Gazzaniga, M.S.; Ivry, R.B.; Mangun, G.R. *Cognitive Neuroscience: The Biology of The Mind*, 2nd ed.; W.W. Norton &Company: New York, NY, USA, 2002; p. 681.
152. Fingelkurts, A.A.; Fingelkurts, A.A. EEG oscillatory states: Universality, uniqueness and specificity across healthy-normal, altered and pathological brain conditions. *PLoS ONE* **2014**, *9*, e87507. [CrossRef]
153. John, E.R.; Karmel, B.Z.; Corning, W.C.; Easton, P.; Brown, D.; Ahn, H.; John, M.; Harmony, T.; Prichep, L.; Toro, A.; et al. Neurometrics: Numerical taxonomy identifies different profiles of brain functions within groups of behaviourally similar people. *Science* **1977**, *196*, 1393–1410. [CrossRef]
154. Gevins, A. Electrophysiological imaging of brain function. In *Brain Mapping. The Methods*, 2nd ed.; Toga, A.W., Mazzoitta, J.C., Eds.; Elsevier Science: New York, NY, USA, 2002; pp. 175–188.
155. Gebber, G.L.; Zhong, S.; Barman, S.M. The functional significance of the 10-Hz sympathetic rhythm: A hypothesis. *Clin. Exp. Hypertens.* **1995**, *17*, 181–195. [CrossRef]
156. Osintseva, Y.V.; Nadezhdina, M.V.; Zhezher, M.N.; Kurus, O.S.; Skulskaya, N.I. The vegetative status and bioelectric activity of the brain in different terms of the remote period of a fighting craniocereberal trauma. *Bull. Sib. Med.* **2010**, *4*, 84–88. [CrossRef]
157. Olbrich, S.; Sander, C.; Matschinger, H.; Mergl, R.; Trenner, M.; Schönknecht, P.; Hegerl, U. Brain and body. Associations between EEG-vigilance and the autonomic nervous system activity during rest. *J. Psychophysiol.* **2011**, *25*, 190–200. [CrossRef]
158. Duschek, S.; Wörsching, J.; del Paso, G.A.R. Autonomic cardiovascular regulation and cortical tone. *Clin. Physiol. Funct. Imaging* **2014**, *35*, 383–392. [CrossRef]
159. Langhorst, P.; Stroh-Werz, M.; Dittmar, K.; Camerer, H. Facultative coupling of reticular neuronal activity with peripheral cardiovascular and central cortical rhythms. *Brain Res.* **1975**, *87*, 407–418. [CrossRef]
160. Langhorst, P.; Schulz, P.; Lambertz, M.; Schulz, G.; Camerer, H. Dynamic characteristics of the "unspecific brain stem system". In *Central Interaction between Respiratory and Cardiovascular Control System*; Koepchen, H.P., Hilton, S.M., Trzebski, A., Eds.; Springer: New York, NY, USA, 1980; pp. 30–41.
161. Achimowicz, J.Z. Evaluation of pilot psychophysiological state in real time by analysis of spectral dynamics in EEG and ERP correlates of sensory and cognitive brain functions and its possible coupling with autonomic nervous system. In *Human System Division, Research Proposal Draft Version 10.5.*; H.G. Armstrong Aero-Space Medical Research Laboratory, Wright–Petterson Air Force Base: Dayton, OH, USA, 1992.
162. Jennings, J.R.; Coles, M.G.H. *Handbook of Cognitive Psychophysiology, Central and Autonomic Nervous System Approaches*; Wiley Psychophysiology Handbooks; Wiley: Chichester, UK, 1991; p. 762.
163. Walker, B.B.; Walker, J.M. Phase relationship between cariotid pressure and ongoing electrocortical activity. *Int. J. Psychophysiol.* **1983**, *1*, 65–73. [CrossRef]
164. Sarà, M.; Pistoia, F. Complexity loss in physiological time series of patients in a vegetative state. *Nonlinear Dyn. Psychol. Life Sci.* **2010**, *14*, 1–13.
165. Fingelkurts, A.A.; Fingelkurts, A.A.; Bagnato, S.; Boccagni, C.; Galardi, G. Life or death: Prognostic value of a resting EEG with regards to survival in patients in vegetative and minimally conscious states. *PLoS ONE* **2011**, *6*, e25967. [CrossRef] [PubMed]
166. Fingelkurts, A.A.; Fingelkurts, A.A.; Bagnato, S.; Boccagni, C.; Galardi, G. Toward operational architectonics of consciousness: Basic evidence from patients with severe cerebral injuries. *Cogn. Process.* **2012**, *13*, 111–131. [CrossRef] [PubMed]
167. Sarà, M.; Sebastiano, F.; Sacco, S.; Pistoia, F.; Onorati, P.; Albertini, G.; Carolei, A. Heart rate nonlinear dynamics in patients with persistent vegetative state: A preliminary report. *Brain Inj.* **2008**, *22*, 33–37. [CrossRef]
168. Wittling, W. The right hemisphere and the human stress response. *Acta Physiol. Scand. Suppl.* **1997**, *640*, 55–59.
169. Hewig, J.; Schlotz, W.; Gerhards, F.; Breitenstein, C.; Lürken, A.; Naumann, E. Associations of the cortisol awakening response (CAR) with cortical activation asymmetry during the course of an exam stress period. *Psychoneuroendocrinology* **2008**, *33*, 83–91. [CrossRef]
170. Tops, M.; Wijers, A.A.; van Staveren, A.S.; Bruin, K.J.; Den Boer, J.A.; Meijman, T.F.; Korf, J. Acute cortisol administration modulates EEG alpha asymmetry in volunteers: Relevance to depression. *Biol. Psychol.* **2005**, *69*, 181–193. [CrossRef]
171. Buss, K.A.; Malmstadt, J.R.; Dolski, I.; Kalin, N.H.; Goldsmith, H.H.; Davidson, R.J. Right frontal brain activity, cortisol, and withdrawal behavior in 6-month-old infants. *Behav. Neurosci.* **2003**, *117*, 11–20. [CrossRef]
172. Schutter, D.J.L.G.; Van Honk, J.; Koppeschaar, H.P.F.; Kahn, R.S. Cortisol and reduced interhemispheric coupling between the left prefrontal and the right parietal cortex. *J. Neuropsychiatr. Clin. Neurosci.* **2002**, *14*, 89–90. [CrossRef] [PubMed]

173. Birmanns, B.; Saphier, D.; Abramsky, O. a-Interferon modifies cortical EEG activity: Dose-dependence and antagonism by naloxone. *J. Neurol. Sci.* **1990**, *100*, 22–26. [CrossRef]
174. Saphier, D.; Ovadia, H.; Abramsky, O. Neural responses to antigenic challenges and immunomodulatory factors. *Yale J. Biol. Med.* **1990**, *63*, 109–119.
175. Kang, D.H.; Davidson, R.J.; Coe, C.L.; Wheeler, R.F.; Tomarken, A.J.; Ershler, W. Frontal brain asymmetry and immune function. *Behav. Neurosci.* **1991**, *105*, 860–869. [CrossRef] [PubMed]
176. Rosenkranz, M.A.; Jackson, D.C.; Dalton, K.M.; Dolski, I.; Ryff, C.D.; Singer, B.H.; Muller, D.; Kalin, N.H.; Davidson, R.J. Affective style and in vivo immune response: Neurobehavioral mechanisms. *Proc. Natl. Acad. Sci. USA* **2003**, *100*, 11148–11152. [CrossRef] [PubMed]
177. Seo, S.-H.; Lee, J.-T. Stress and EEG. In *Convergence and Hybrid Information Technologies*; Crisan, M., Ed.; INTECH: Rijeka, Croatia, 2010; pp. 413–426.
178. Vanhollebeke, G.; De Smet, S.; De Raedt, R.; Baeken, C.; van Mierlo, P.; Vanderhasselt, M.A. The neural correlates of psychosocial stress: A systematic review and meta-analysis of spectral analysis EEG studies. *Neurobiol. Stress* **2022**, *18*, 100452. [CrossRef] [PubMed]
179. Davidson, R.J.; Coe, C.C.; Dolski, I.; Donzella, B. Individual differences in prefrontal activation asymmetry predict natural killer cell activity at rest and in response to challenge. *Brain Behav. Immun.* **1999**, *13*, 93–108. [CrossRef]
180. Hecht, D. Depression and the hyperactive right-hemisphere. *Neurosci. Res.* **2010**, *68*, 77–87. [CrossRef]
181. Davis, P.A. Effect on the EEG of changing the blood sugar level. *Arch. Neurol. Psychiatr.* **1943**, *49*, 186–194. [CrossRef]
182. Sulg, I.A.; Sotaniemi, K.A.; Tolonen, U.; Hokkanen, E. Dependence between cerebral metabolism and blood flow as reflected in the quantitative EEG. *Adv. Biol. Psychiatr.* **1981**, *6*, 102–108.
183. Köpruner, V.; Pfurtscheller, G.; Auer, L.M. Quantitative EEG in normals and in patients with cerebral ischemia. *Prog. Brain Res.* **1984**, *62*, 29–50. [PubMed]
184. Knyazeva, M.G.; Vil'davskii, V.U. Correspondence of spectral characteristics of EEG and regional blood circulation in 9-14 years old children. *Hum. Physiol. (Physiol. Cheloveka)* **1986**, *12*, 387–394.
185. Passero, S.; Rocchi, R.; Vatti, G.; Burgalassi, L.; Battistini, N. Quantitative EEG mapping, regional cerebral blood flow and neuropsychological function in Alzheimer's disease. *Dementia* **1995**, *6*, 148–156. [CrossRef] [PubMed]
186. Kraaier, V.; van Huffelen, A.C.; Wieneke, G.H. Changes in quantitative EEG and blood flow velocity due to standardized hyperventilation; a model of transient ischaemia in young human subjects. *Electroencephalogr. Clin. Neurophysiol.* **1988**, *70*, 377–387. [CrossRef]
187. Szelies, B.; Mielke, R.; Kessler, J.; Heiss, W.D. EEG power changes are related with regional cerebral glucose metabolism in vascular dementia. *Clin. Neurophysiol.* **1999**, *110*, 615–620. [CrossRef]
188. Alper, K.R.; John, E.R.; Brodie, J.; Günther, W.; Daruwala, R.; Prichep, L.S. Correlation of PET and qEEG in normal subjects. *Psychiatr. Res.* **2006**, *146*, 271–282. [CrossRef]
189. Jann, K.; Koenig, T.; Dierks, T.; Boesch, C.; Federspiel, A. Association of individual resting state EEG alpha frequency and cerebral blood flow. *NeuroImage* **2010**, *51*, 365–372. [CrossRef]
190. O'Gorman, R.L.; Poil, S.S.; Brandeis, D.; Klaver, P.; Bollmann, S.; Ghisleni, C.; Lüchinger, R.; Martin, E.; Shankaranarayanan, A.; Alsop, D.C.; et al. Coupling between resting cerebral perfusion and EEG. *Brain Topogr.* **2013**, *26*, 442–457. [CrossRef]
191. Babiloni, C.; Del Percio, C.; Caroli, A.; Salvatore, E.; Nicolai, E.; Marzano, N.; Lizio, R.; Cavedo, E.; Landau, S.; Chen, K.; et al. Cortical sources of resting state EEG rhythms are related to brain hypometabolism in subjects with alzheimer's disease: An EEG-Pet study. *Neurobiol. Aging* **2016**, *48*, 122–134. [CrossRef]
192. Cohn, R.; Raines, G. Cerebral vascular lesions: Electroencephalographic and neuropathologic correlations. *Arch. Neurol. Psychiatr.* **1948**, *60*, 165–181. [CrossRef]
193. Ingvar, D.H.; Sjolund, B.; Ardo, A. Correlation between dominant EEG frequency, cerebral oxygen uptake and blood flow. *Electroencephalogr. Clin. Neurophysiol.* **1976**, *41*, 268–276. [CrossRef]
194. Blume, W.T.; Ferguson, G.G.; McNeill, D.K. Significance of EEG changes at carotid endarterectomy. *Stroke* **1985**, *17*, 891–897. [CrossRef]
195. Jonkman, E.J.; Poortvliet, D.C.J.; Veering, M.M.; De Weerd, A.W.; John, E.R. The use of neurometrics in the study of patients with cerebral ischemia. *Electroencephalogr. Clin. Neurophysiol.* **1985**, *61*, 333–341. [CrossRef]
196. Nagata, K. Topographic EEG in brain ischemia: Correlation with blood flow and metabolism. *Brain Topogr.* **1988**, *1*, 97–106. [CrossRef]
197. Nagata, K.; Tagwa, K.; Hiroi, S.; Shishido, F.; Uemura, K. Electroencephalographic correlates of blood flow and oxygen metabolism provided by positron emission tomography in patients with cerebral infarction. *Electroencephalogr. Clin. Neurophysiol.* **1989**, *72*, 16–30. [CrossRef]
198. Claassen, J.; Hirsch, L.J.; Kreiter, K.T.; Du, E.Y.; Connolly, S.E.; Emerson, R.G.; Mayer, S.A. Quantitative continuous EEG for detecting delayed cerebral ischemia in patients with poor-grade subarachnoid hemorrhage. *Clin. Neurophysiol.* **2004**, *115*, 2699–2710. [CrossRef]
199. Hughes, J.R.; John, E.R. Conventional and quantitative electroencephalography in psychiatry. *J. Neuropsychiatr. Clin. Neurosci.* **1999**, *11*, 190–208. [CrossRef]

200. Mueller, T.M.; Gollwitzer, S.; Hopfengärtner, R.; Rampp, S.; Lang, J.D.; Stritzelberger, J.; Madžar, D.; Reindl, C.; Sprügel, M.I.; Onugoren, M.D.; et al. Alpha power decrease in quantitative EEG detects development of cerebral infarction after subarachnoid hemorrhage early. *Clin. Neurophysiol.* **2021**, *132*, 1283–1289. [CrossRef]
201. Vatinno, A.A.; Simpson, A.; Ramakrishnan, V.; Bonilha, H.S.; Bonilha, L.; Seo, N.J. The prognostic utility of electroencephalography in stroke recovery: A systematic review and meta-analysis. *Neurorehabil. Neural Repair* **2022**, *36*, 255–268. [CrossRef]
202. Gollwitzer, S.; Groemer, T.; Rampp, S.; Hagge, M.; Olmes, D.; Huttner, H.B.; Schwab, S.; Madžar, D.; Hopfengaertner, R.; Hamer, H.M. Early prediction of delayed cerebral ischemia in subarachnoid hemorrhage based on quantitative EEG: A prospective study in adults. *Clin. Neurophysiol.* **2015**, *126*, 1514–1523. [CrossRef]
203. Rots, M.L.; van Putten, M.J.; Hoedemaekers, C.W.; Horn, J. Continuous EEG monitoring for early detection of delayed cerebral ischemia in subarachnoid hemorrhage: A pilot study. *Neurocrit. Care* **2016**, *24*, 207–216. [CrossRef]
204. Balança, B.; Dailler, F.; Boulogne, S.; Ritzenthaler, T.; Gobert, F.; Rheims, S.; Andre-Obadia, N. Diagnostic accuracy of quantitative EEG to detect delayed cerebral ischemia after subarachnoid hemorrhage: A preliminary study. *Clin. Neurophysiol.* **2018**, *129*, 1926–1936. [CrossRef]
205. Vakalopoulos, C. The EEG as an index of neuromodulator balance in memory and mental illness. *Front. Neurosci.* **2014**, *8*, 63. [CrossRef]
206. Lubar, J.F. Neocortical dynamics: Implication for understanding the role of neurofeedback and related techniques for the enhancement of attention. *Appl. Psychophysiol. Biofeedback* **1997**, *22*, 111–126. [CrossRef]
207. Nistico, G.; Nappy, G. Locus coeruleus, an integrative station involved in the control of several vital functions. *Funct. Neurol.* **1993**, *8*, 5–25.
208. Panyushkina, S.V.; Kurova, N.S.; Egorov, S.F.; Koshelev, V.V. Individual EEG reactions of healthy humans to mutually antagonistic noradrenotropic influences. *Zh Vyss. Nerv Deyat* **1994**, *44*, 457–469.
209. Sadato, N.; Nakamura, S.; Oohashi, T. Neural networks for generation and suppression of alpha rhythm: A PET study. *NeuroReport* **1998**, *9*, 893–897. [CrossRef]
210. Chavanon, M.-L.; Wacker, J.; Stemmler, G. Paradoxical dopaminergic drug effects in extraversion: Dose- and time-dependent effects of sulpiride on EEG theta activity. *Front. Hum. Neurosci.* **2013**, *7*, 117. [CrossRef]
211. Knott, V.J.; Hovson, A.L.; Perugimi, M. The effect of acute tryptophan depletion and fenfluramine on quantitative EEG and mood in healthy male subjects. *Biol. Psychiatr.* **1999**, *46*, 229–238. [CrossRef]
212. Steriade, M.; Gloor, P.; Llinas, R.R.; da Silva, F.H.L.; Mesulam, M.-M. Basic mechanisms of cerebral rhythmic activities. Report of IFCN Committee on Basic Mechanisms. *Electroencepahlogr. Clin. Neurophysiol.* **1990**, *76*, 481–508. [CrossRef]
213. John, E.R.; Prichep, L.S.; Winterer, G.; Herrmann, W.M.; diMichele, F.; Halper, J.; Bolwig, T.G.; Cancro, R. Electrophysiological subtypes of psychotic states. *Acta Psychiatr. Scand.* **2007**, *116*, 17–35. [CrossRef]
214. Small, J.G. Psychiatric disorders and EEG. In *Electroencephalography: Basic Principles, Clinical Applications, and Related Fields*; Niedermeyer, E., da Silva, F.L., Eds.; Williams and Wilkins: Baltimore, MD, USA, 1993; pp. 581–596.
215. Hughes, J.R. The EEG in psychiatry: An outline with summarized points and references. *Clin. Electroencephalogr.* **1995**, *26*, 92–101. [CrossRef]
216. Sam, M.C.; So, E.L. Significance of epileptiform discharges in patients without epilepsy in the community. *Epilepsia* **2001**, *42*, 1273–1278. [CrossRef]
217. Zivin, L.; Marsan, C.A. Incidence and prognostic significance of "epileptiform" activity in the EEG of nonepileptic subjects. *Brain* **1968**, *91*, 751–777. [CrossRef]
218. Standage, K.F. The etiology of hysterical seizures. *Can. Psychiatr. Assoc. J.* **1975**, *20*, 67–73. [CrossRef]
219. Cohen, R.J.; Suter, C. Hysterical seizures: Suggestion as a provocative EEG test. *Ann. Neurol.* **1982**, *11*, 391–395. [CrossRef]
220. King, D.W.; Gallagher, B.B.; Murvin, A.J.; Smith, D.B.; Marcus, D.J.; Hartlage, L.C.; Ward, L.C., 3rd. Pseudoseizures: Diagnostic evaluation. *Neurology* **1982**, *32*, 18–23. [CrossRef]
221. Luther, J.S.; McNamara, J.O.; Carwile, S.; Miller, P.; Hope, V. Pseudoepileptic seizures: Methods and video analysis to aid diagnosis. *Ann. Neurol.* **1982**, *12*, 458–462. [CrossRef]
222. Wilkus, R.J.; Dodrill, C.B.; Thompson, P.M. Intensive EEG monitoring and psychological studies of patients with pseudoepileptic seizures. *Epilepsia* **1984**, *25*, 100–107. [CrossRef]
223. Wilkes, R.J.; Thompson, P.M.; Vossler, D.G. Bizarre ictal automatisms: Frontal lobe epileptic or psychogenic seizures? *J. Epilepsy* **1990**, *3*, 297–313. [CrossRef]
224. Lelliott, P.T.; Fenwick, P. Cerebral pathology in pseudoseizures. *Acta Neurol. Scand.* **1991**, *83*, 29–132. [CrossRef]
225. Bowman, E.S. Etiology and clinical course of pseudoseizures: Relationship to trauma, depression, and dissociation. *Psychosomatics* **1993**, *34*, 333–342. [CrossRef]
226. Devinsky, O.; Sanchez-Villasenor, F.; Vazquez, B.; Kothari, M.; Alper, K.; Luciano, D. Clinical profile of patients with epileptic and nonepileptic seizures. *Neurology* **1996**, *46*, 1530–1533. [CrossRef] [PubMed]
227. Shelley, B.P.; Trimble, M.R.; Boutros, N.N. Electroencephalographic cerebral dysrhythmic abnormalities in the trinity of nonepileptic general population, neuropsychiatric, and neurobehavioral disorders. *J. Neuropsychiatr. Clin. Neurosci.* **2008**, *20*, 7–22. [CrossRef]
228. Pillmann, F.; Rohde, A.; Ullrich, S.; Draba, S.; Sannemüller, U.; Marneros, A. Violence, criminal behavior, and the EEG: Significance of left hemispheric focal abnormalities. *J. Neuropsychiatr. Clin. Neurosci.* **1999**, *11*, 454–457. [CrossRef]

229. Hughes, J.; Leander, R.; Ketchum, G. Electroencephalographic study of specific reading disabilities. EEG. Clin. Neurophysiol. **1949**, *1*, 377.
230. Ribas, J.C.; Baptistete, E.; Fonseca, C.A.; Tiba, I.; Filho, H.S.C. Behavior disorders with predominance of aggressiveness, irritability, impulsiveness, and instability: Clinical electroencephalographic study of 100 cases. *Arq. De Neuro-Psiquiatr.* **1974**, *32*, 187–194. [CrossRef]
231. Harty, J.E.; Gibbs, E.L.; Gibbs, F.A. Electroencephalographic study of two hundred and seventy-five candidates for military service. *War Med.* **1942**, *2*, 923–930.
232. Socanski, D.; Herigstad, A.; Thomsen, P.H.; Dag, A.; Larsen, T.K. Epileptiform abnormalities in children diagnosed with attention deficit/hyperactivity disorder. *Epilepsy Behav.* **2010**, *19*, 483–486. [CrossRef]
233. Dierks, T.; Ihl, R.; Frolich, L.; Maurer, K. Dementia of the Alzheimer type: Effects on the spontaneous EEG described by dipole sources. *Psychiatr. Res.* **1993**, *50*, 51–162. [CrossRef]
234. Prichep, L.S.; Mas, F.; Hollander, E.; Liebowitz, M.; John, E.R.; Almas, M.; DeCaria, C.M.; Levine, R.H. Quantitative electroencephalographic (QEEG) subtyping of obsessive-compulsive disorder. *Psychiatr. Res.* **1993**, *50*, 25–32. [CrossRef]
235. Inui, K.; Motomura, E.; Okushima, R.; Kaige, H.; Inoue, K.; Nomura, J. Electroencephalographic findings in patients with DSM-IV mood disorder, schizophrenia, and other psychotic disorders. *Biol. Psychiatr.* **1998**, *43*, 69–75. [CrossRef]
236. Huang, C.; Wahlund, L.O.; Dierks, T.; Julin, P.; Winblad, B.; Jelic, V. Discrimination of Alzheimer's disease and mild cognitive impairment by equivalent EEG sources: A cross-sectional and longitudinal study. *Clin. Neurophysiol.* **2000**, *11*, 1961–1967. [CrossRef]
237. Monastra, V.J.; Lubar, J.F.; Linden, M. The development of a quantitative electroencephalographic scanning process for attention deficit-hyperactivity disorder: Reliability and validity studies. *Neuropsychology* **2001**, *15*, 136–144. [CrossRef] [PubMed]
238. Thatcher, R.W.; North, D.M.; Curtin, R.T.; Walker, R.A.; Biver, C.J.; Gomez, J.F.; Salazar, A.M. An EEG severity index of traumatic brain injury. *J. Neuropsychiatr. Clin. Neurosci.* **2001**, *13*, 77–87. [CrossRef] [PubMed]
239. Karadag, F.; Oguzhanoglu, N.K.; Kurt, T.; Oguzhanoglu, A.; Atesci, F.; Ozdel, O. Quantitative EEG analysis in obsessive compulsive disorder. *Int. J. Neurosci.* **2003**, *113*, 833–847. [CrossRef]
240. Boutros, N.N.; Torello, M.; McGlashan, T.H. Electrophysiological aberrations in borderline personality disorder: State of the evidence. *J. Neuropsychiatr. Clin. Neurosci.* **2003**, *15*, 145–154. [CrossRef]
241. Rowe, D.L. Biophysical modeling of tonic cortical electrical activity in attention deficit hyperactivity disorder. *Int. J. Neurosci.* **2005**, *115*, 1273–1305. [CrossRef]
242. Babiloni, C.; Benussi, L.; Binetti, G.; Cassetta, E.; Dal Forno, G.; Del Percio, C.; Ferreri, F.; Ferri, R.; Frisoni, G.; Ghidoni, R.; et al. Apolipoprotein E and alpha brain rhythms in mild cognitive impairment: A multicentric electroencephalogram study. *Ann. Neurol.* **2006**, *59*, 323–334. [CrossRef]
243. Başar, E.; Güntekin, B. A review of brain oscillations in cognitive disorders and the role of neurotransmitters. *Brain Res.* **2008**, *1235*, 172–193. [CrossRef]
244. Fingelkurts, A.A.; Fingelkurts, A.A. Alpha rhythm operational architectonics in the continuum of normal and pathological brain states: Current state of research. *Int. J. Psychophysiol.* **2010**, *76*, 93–106. [CrossRef] [PubMed]
245. Schultz, E.V.; Baburin, I.N.; Karavaeva, T.A.; Karvasarsky, B.D.; Slezin, V.B. Bioelectric brain activity in patients with neurotic and neurosis-like disorders (according to a spectral analysis). *Bekhterev. Rev. Psychiatr. Med. Psychol.* **2010**, *3*, 26–31.
246. Lee, S.M.; Jang, K.-I.; Chae, G.-H. Electroencephalographic correlates of suicidal ideation in the theta band. *Clin. EEG Neurosci.* **2017**, *48*, 316–321. [CrossRef]
247. Kanda, P.A.M.; Anghinah, R.; Smidth, M.T.; Silva, J.M. The clinical use of quantitative EEG in cognitive disorders. *Dement. Neuropsychol.* **2009**, *3*, 195–203. [CrossRef] [PubMed]
248. John, E.R. The role of quantitative EEG topographic mapping or 'neurometrics' in the diagnosis of psychiatric and neurological disorders: The pros. *Electroencephalogr. Clin. Neurophysiol.* **1989**, *73*, 2–4. [CrossRef]
249. Abrams, R.; Taylor, M.A. Differential EEG patterns in affective disorder and schizophrenia. *Arch. Gen. Psychiatr.* **1979**, *36*, 1355–1358. [CrossRef]
250. Giannitrapani, D.; Collins, J. EEG differentiation between Alzheimer's and non-Alzheimer's dementias. In *The EEG of Mental Activities*; Giannitrapani, D., Murri, L., Eds.; Karger: New York, NY, USA, 1988; pp. 26–41.
251. Goodin, D.S.; Aminoff, M.J. Electrophysiological differences between subtypes of dementia. *Brain* **1986**, *109*, 1102–1113. [CrossRef]
252. Chabot, R.J.; Serfontein, G. Quantitative EEG profiles of children with attention deficit disorder. *Biol. Psychiatr.* **1996**, *40*, 951–963. [CrossRef]
253. John, E.R.; Prichep, L.; Ahn, H.; Easton, P.; Fridman, J.; Kaye, H. Neurometric evaluation of cognitive dysfunctions and neurological disorders in children. *Prog. Neurobiol.* **1983**, *21*, 239–290. [CrossRef]
254. Coburn, K.L.; Lauterbach, E.C.; Boutros, N.N.; Black, K.J.; Arciniegas, D.B.; Coffey, C.E. The value of quantitative electroencephalography in clinical psychiatry: A report by the Committee on Research of the American Neuropsychiatric Association. *J. Neuropsychiatry Clin. Neurosci.* **2006**, *18*, 460–500. [CrossRef]
255. Knott, V.; Bakish, D.; Lusk, S.; Barkely, J.; Perugini, M. Quantitative EEG correlates of panic disorder. *Psychiatr. Res.* **1996**, *68*, 31–39. [CrossRef]

256. Deslandes, A.; Veiga, H.; Cagy, M.; Fiszman, A.; Piedade, R.; Ribeiro, P. Quantitative electroencephalography (qEEG) to discriminate primary degenerative dementia from major depression disorder (depression). *Arq. Neuropsiquiatr.* **2004**, *62*, 44–50. [CrossRef] [PubMed]
257. Pucci, E.; Belardinelli, N.; Cacchiò, G.; Signorino, M.; Angeleri, F. EEG power spectrum differences in early and late onset forms of Alzheimer's disease. *Clin. Neurophysiol.* **1999**, *110*, 621–631. [CrossRef]
258. Dierks, T.; Perisic, I.; Frölich, L.; Ihl, R.; Maurer, K. Topography of the qEEG in dementia of Alzheimer type: Relation to severity of dementia. *Psychol. Res.* **1991**, *40*, 181–194.
259. Prichep, L.S.; John, E.R.; Ferris, S.H.; Rausch, L.; Fang, Z.; Cancro, R.; Torossian, C.; Reisberg, B. Prediction of longitudinal cognitive decline in normal elderly with subjective complaints using electrophysiological imaging. *Neurobiol. Aging* **2006**, *27*, 471–481. [CrossRef]
260. Bauer, L.O. Predicting relapse to alcohol and drug abuse via quantitative electroencephalography. *Neuropsychopharmacology* **2001**, *25*, 332–340. [CrossRef]
261. Naunheim, R.S.; Treaster, M.; English, J.; Casner, T.; Chabot, R. Use of brain electrical activity to quantify traumatic brain injury in the emergency department. *Brain Inj.* **2010**, *24*, 1324–1329. [CrossRef]
262. Ritchlin, C.T.; Chabot, R.J.; Alper, K.; Buyon, J.; Belmont, H.M.; Roubey, R.; Abramson, S.B. Quantitative electroencephalography: A new approach to the diagnosis of cerebral dysfunction in systemic lupus erythematosus. *Arth. Rheumat.* **1992**, *35*, 1330–1342. [CrossRef]
263. Sloan, E.P.; Fenton, G.W.; Kennedy, J.S.J.; MacLennan, J.M. Electroencephalography and single photon emission computed tomography in dementia: A comparative study. *Psychol. Med.* **1995**, *25*, 631–638. [CrossRef]
264. Kropotov, J.D.; Pąchalska, M.; Mueller, A. New neurotechnologies for the diagnosis and modulation of brain dysfunctions. *Health Psychol. Rep.* **2014**, *2*, 73–82. [CrossRef]
265. Kropotov, J.D.; Müller, A.; Candrian, G.; Valery, P. *Neurobiology of ADHD: A New Diagnostic Approach Based on Electrophysiological Endophenotypes*; Springer: London, UK, 2013; p. 300.
266. Fisher, N.K.; Talathi, S.S.; Cadotte, A.; Carney, P.R. Epilepsy detection and monitoring. In *Quantitative EEG Analysis Methods and Clinical Applications*; Tong, S., Thakor, N.V., Eds.; Artech House: Norwood, MA, USA, 2009; pp. 141–167.
267. Pardalos, P.M. Seizure warning algorithm based on optimization and nonlinear dynamics. *Math. Program* **2004**, *101*, 365–385. [CrossRef]
268. Drislane, F.W. The clinical use of ambulatory EEG. In *Atlas of Ambulatory EEG*; Chang, B.S., Schachter, S.C., Schomer, D.L., Eds.; Elsevier: Amsterdam, The Netherlands, 2005; pp. 17–25.
269. Hegerl, U.; Hensch, T. The vigilance regulation model of affective disorders and ADHD. *Neurosci. Biobehav. Rev.* **2014**, *44*, 45–57. [CrossRef] [PubMed]
270. Wittekind, D.A.; Spada, J.; Gross, A.; Hensch, T.; Jawinski, P.; Ulke, C.; Sander, C.; Hegerl, U. Early report on brain arousal regulation in manic vs. depressive episodes in bipolar disorder. *Bipolar Disord.* **2016**, *18*, 502–510. [CrossRef]
271. Brenner, R.P. EEG and dementia, Chapter 19. In *Electroencephalography, Basic Principles, Clinical Applications, and Related Fields*, 4th ed.; Niedermeyer, E., da Silva, F.L., Eds.; Williams and Wilkins: Baltimore, MD, USA, 1999; pp. 349–359.
272. Fingelkurts, A.A.; Fingelkurts, A.A.; Bagnato, S.; Boccagni, C.; Galardi, G. EEG oscillatory states as neuro-phenomenology of consciousness as revealed from patients in vegetative and minimally conscious states. *Conscious Cogn.* **2012**, *21*, 149–169. [CrossRef]
273. Fingelkurts, A.A.; Fingelkurts, A.A.; Bagnato, S.; Boccagni, C.; Galardi, G. Dissociation of vegetative and minimally conscious patients based on brain operational architectonics: Factor of etiology. *Clin. EEG Neurosci.* **2013**, *44*, 209–220. [CrossRef]
274. Fingelkurts, A.A.; Fingelkurts, A.A.; Bagnato, S.; Boccagni, C.; Galardi, G. Prognostic value of resting-state electroencephalography structure in disentangling vegetative and minimally conscious states: A preliminary study. *Neurorehabil. Neural Repair* **2013**, *27*, 345–354. [CrossRef]
275. Fingelkurts, A.A.; Fingelkurts, A.A.; Bagnato, S.; Boccagni, C.; Galardi, G. The value of spontaneous EEG oscillations in distinguishing patients in vegetative and minimally conscious states, chapter 5. In *Application of Brain Oscillations in Neuropsychiatric Diseases (Supplements to Clinical Neurophysiology)*; Basar, E., Basar-Eroglu, C., Ozerdem, A., Rossini, P.M., Yener, G.G., Eds.; Elsevier B.V.: Amsterdam, The Netherlands, 2013; Volume 62, pp. 81–99.
276. Fingelkurts, A.A.; Fingelkurts, A.A. Brain space and time in mental disorders: Paradigm shift in biological psychiatry. *Int. J. Psychiatr. Med.* **2019**, *54*, 53–63. [CrossRef]
277. Dittrich, A. The standardized psychometric assessment of altered states of consciousness (ASCs) in humans. *Pharmacopsychiatry* **1998**, *31*, 80–84. [CrossRef]
278. Parnas, J.; Møller, P.; Kircher, T.; Thalbitzer, J.; Jansson, L.; Handest, P.; Zahavi, D. EASE: Examination of anomalous self-experience. *Psychopathology* **2005**, *38*, 236–258. [CrossRef]
279. Beck, A.T. The evolution of the cognitive model of depression and its neurobiological correlates. *Am. J. Psychiatr.* **2008**, *165*, 969–977. [CrossRef]
280. Musholt, K. *Thinking about Oneself: From Nonconceptual Content to the Concept of a Self*; MIT Press: Cambridge, UK, 2015; p. 232.
281. Northoff, G.; Heinzel, A.; de Greck, M.; Bermpohl, F.; Dobrowolny, H.; Panksepp, J. Self-referential processing in our brain. A meta-analysis of imaging studies on the self. *NeuroImage* **2006**, *31*, 440–457. [CrossRef] [PubMed]

282. Northoff, G. Is the self a higher-order or fundamental function of the brain? The 'basis model of self-specificity' and its encoding by the brain's spontaneous activity. *Cogn. Neurosci.* **2016**, *7*, 203–222. [CrossRef] [PubMed]
283. Raichle, M.E.; MacLeod, A.M.; Snyder, A.Z.; Powers, W.J.; Gusnard, D.A.; Shulman, G.L. A default mode of brain function. *Proc. Natl. Acad. Sci. USA* **2001**, *98*, 676–682. [CrossRef] [PubMed]
284. Gusnard, D.A. Being a self: Considerations from functional imaging. *Conscious Cogn.* **2005**, *14*, 679–697. [CrossRef]
285. Schilbach, L.; Eickhoff, S.B.; Rotarska-Jagiela, A.; Fink, G.R.; Vogeley, K. Minds at rest? Social cognition as the default mode of cognizing and its putative relationship to the "default system" of the brain. *Conscious Cogn.* **2008**, *17*, 457–467. [CrossRef]
286. Fingelkurts, A.A.; Fingelkurts, A.A. Persistent operational synchrony within brain default-mode network and self-processing operations in healthy subjects. *Brain Cogn.* **2011**, *75*, 79–90. [CrossRef]
287. Fingelkurts, A.A.; Fingelkurts, A.A.; Bagnato, S.; Boccagni, C.; Galardi, G. DMN Operational Synchrony Relates to Self-Consciousness: Evidence from Patients in Vegetative and Minimally Conscious States. *Open Neuroimag. J.* **2012**, *6*, 55–68. [CrossRef]
288. Laufs, H.; Kleinschmidt, A.; Beyerle, A.; Eger, E.; Salek-Haddadi, A.; Preibisch, C.; Krakow, K. EEG-correlated fMRI of human alpha activity. *Neuroimage* **2003**, *19*, 1463–1476. [CrossRef]
289. Mantini, D.; Perrucci, M.G.; Del Gratta, C.; Romani, G.L.; Corbetta, M. Electrophysiological signatures of resting state networks in the human brain. *Proc. Natl. Acad. Sci. USA* **2007**, *104*, 13170–13175. [CrossRef]
290. Jann, K.; Dierks, T.; Boesch, C.; Kottlow, M.; Strik, W.; Koenig, T. BOLD correlates of EEG alpha phase-locking and the fMRI default mode network. *Neuroimage* **2009**, *45*, 903–916. [CrossRef]
291. Knyazev, G.G.; Slobodskoj-Plusnin, J.Y.; Bocharov, A.V.; Pylkova, L.V. The default mode network and EEG α oscillations: An independent component analysis. *Brain Res.* **2011**, *1402*, 67–79. [CrossRef] [PubMed]
292. Knyazev, G.G.; Savostyanov, A.N.; Volf, N.V.; Liou, M.; Bocharov, A.V. EEG correlates of spontaneous self-referential thoughts: A cross-cultural study. *Int. J. Psychophysiol.* **2012**, *86*, 173–181. [CrossRef] [PubMed]
293. Fingelkurts, A.A.; Fingelkurts, A.A.; Kallio-Tamminen, T. Long-term meditation training induced changes in the operational synchrony of default mode network modules during a resting state. *Cogn. Process.* **2016**, *17*, 27–37. [CrossRef] [PubMed]
294. Fingelkurts, A.A.; Fingelkurts, A.A.; Kallio-Tamminen, T. Trait lasting alteration of the brain default mode network in experienced meditators and the experiential selfhood. *Self Identity* **2016**, *15*, 381–393. [CrossRef]
295. Fingelkurts, A.A.; Fingelkurts, A.A.; Kallio-Tamminen, T. Selfhood triumvirate: From phenomenology to brain activity and back again. *Conscious Cogn.* **2020**, *86*, 103031. [CrossRef]
296. Gallagher, S. A pattern theory of self. *Front. Hum. Neurosci.* **2013**, *7*, 443. [CrossRef]
297. Gallagher, S.; Daly, A. Dynamical relations in the self-pattern. *Front. Psychol.* **2018**, *9*, 664. [CrossRef]
298. Fingelkurts, A.A.; Fingelkurts, A.A.; Kallio-Tamminen, T. Self, Me and I in the repertoire of spontaneously occurring altered states of Selfhood: Eight neurophenomenological case study reports. *Cogn. Neurodyn.* **2022**, *16*, 255–282. [CrossRef]
299. Fingelkurts, A.A.; Fingelkurts, A.A. Longitudinal dynamics of 3-dimensional components of selfhood after severe traumatic brain injury: A qEEG case study. *Clin. EEG Neurosci.* **2017**, *48*, 327–337. [CrossRef]
300. Fingelkurts, A.A.; Fingelkurts, A.A. Three-dimensional components of selfhood in treatment-naive patients with major depressive disorder: A resting-state qEEG imaging study. *Neuropsychologia* **2017**, *99*, 30–36. [CrossRef]
301. Fingelkurts, A.A.; Fingelkurts, A.A. Alterations in the three components of selfhood in persons with post-traumatic stress disorder symptoms: A pilot qEEG neuroimaging study. *Open Neuroimag. J.* **2018**, *12*, 42–54. [CrossRef] [PubMed]
302. Beck, A.T. Cognitive models of depression. *J. Cogn. Psychother.* **1987**, *1*, 5–37.
303. Damasio, A.R. *The Feeling of What Happens: Body and Emotion in the Making of Consciousness*; Harcourt Brace: San Diego, CA, USA, 1999; p. 400.
304. Rimes, K.A.; Watkins, E. The effects of self-focused rumination on global negative self-judgements in depression. *Behav. Res. Ther.* **2005**, *43*, 1673–1681. [CrossRef]
305. Northoff, G. Psychopathology and pathophysiology of the self in depression-neuropsychiatric hypothesis. *J. Affect. Disord.* **2007**, *104*, 1–14. [CrossRef] [PubMed]
306. Nolen-Hoeksema, S.; Wisco, B.E.; Lyubomirsky, S. Rethinking rumination. *Perspect. Psychol. Sci.* **2008**, *3*, 400–424. [CrossRef] [PubMed]
307. Paulus, M.P.; Stein, M.B. Interoception in anxiety and depression. *Brain Struct. Funct.* **2010**, *214*, 451–463.
308. *Diagnostic and Statistical Manual of Mental Disorders*, 5th ed.; The American Psychiatric Association: Arlington, VA, USA, 2013.
309. Zepinic, V. *Understanding and Treating Complex Trauma*; Xlibris: London, UK, 2011.
310. van der Kolk, B.A. The body keeps the score: Memory and the evolving psychobiology of posttraumatic stress. *Harv. Rev. Psychiatry* **1994**, *1*, 253–265. [CrossRef]
311. McNally, R.J. *Remembering Trauma*; Belknap Press/Harvard University Press: Cambridge, MA, USA, 2003; p. 448.
312. Ataria, Y. Traumatic memories as black holes: A qualitative-phenomenological approach. *Qual. Psychol.* **2014**, *1*, 123–140. [CrossRef]
313. van der Kolk, B.A.; Fisler, R. Dissociation and the Fragmentary Nature of Traumatic Memories: Overview and Exploratory Study. 1995. Available online: http://www.trauma-pages.com/a/vanderk2.php (accessed on 1 May 2022).
314. Kullberg-Turtiainen, M.; Vuorela, K.; Huttula, L.; Turtiainen, P.; Koskinen, S. Individualized goal directed dance rehabilitation in chronic state of severe traumatic brain injury: A case study. *Heliyon* **2019**, *5*, e01184. [CrossRef]

315. Fingelkurts, A.A.; Fingelkurts, A.A.; Bagnato, S.; Boccagni, C.; Galardi, G. The chief role of frontal operational module of the brain default mode network in the potential recovery of consciousness from the vegetative state: A preliminary comparison of three case reports. *Open Neuroimag. J.* **2016**, *10* (Suppl. S1, M4), 41–51. [CrossRef]
316. Laureys, S.; Celesia, G.G.; Cohadon, F.; Lavrijsen, J.; León-Carrión, J.; Sannita, W.G.; Sazbon, L.; Schmutzhard, E.; von Wild, K.R.; Zeman, A.; et al. European Task Force on disorders of consciousness, unresponsive wakefulness syndrome: A new name for the vegetative state or apallic syndrome. *BMC Med.* **2010**, *8*, 68. [CrossRef] [PubMed]
317. Peterson, A.; Bayne, T. Post-comatose disorders of consciousness. In *The Routledge Handbook of Consciousness*; Gennaro, R., Ed.; Routledge: Abingdon, UK, 2018; pp. 351–365.
318. Jennett, B.; Plum, F. Persistent vegetative state after brain damage. A syndrome in search of a name. *Lancet* **1972**, *1*, 734–737. [CrossRef]
319. Giacino, J.T.; Ashwal, S.; Childs, N.; Cranford, R.; Jennett, B.; Katz, D.I.; Kelly, J.P.; Rosenberg, J.H.; Whyte, J.; Zafonte, R.D.; et al. The minimally conscious state: Definition and diagnostic criteria. *Neurology* **2002**, *58*, 349–353. [CrossRef] [PubMed]
320. Naccache, L. Minimally conscious state or cortically mediated state? *Brain* **2018**, *141*, 949–960. [CrossRef]
321. Bagnato, S.; Boccagni, C.; Sant'Angelo, A.; Fingelkurts, A.A.; Fingelkurts, A.A.; Galardi, G. Emerging from an unresponsive wakefulness syndrome: Brain plasticity has to cross a threshold level. *Neurosci. Biobehav. Rev.* **2013**, *37*, 2721–2736. [CrossRef]
322. Porcaro, C.; Nemirovsky, I.E.; Riganello, F.; Mansour, Z.; Cerasa, A.; Tonin, P.; Stojanoski, B.; Soddu, A. Diagnostic developments in differentiating unresponsive wakefulness syndrome and the minimally conscious state. *Front. Neurol.* **2022**, *12*, 778951. [CrossRef]
323. Bagnato, S.; Boccagni, C.; Prestandrea, C.; Sant'Angelo, A.; Castiglione, A.; Galardi, G. Prognostic value of standard EEG in traumatic and non-traumatic disorders of consciousness following coma. *Clin. Neurophysiol.* **2010**, *121*, 274–280. [CrossRef]
324. Brenner, R.P. The interpretation of the EEG of stupor and coma. *Neurologist* **2005**, *11*, 271–284. [CrossRef]
325. Gosseries, O.; Schnakers, C.; Ledoux, D.; Vanhaudenhuyse, A.; Bruno, M.A.; Demertzi, A.; Noirhomme, Q.; Lehembre, R.; Damas, P.; Goldman, S.; et al. Automated EEG entropy measurements in coma, vegetative state/unresponsive wakefulness syndrome and minimally conscious state. *Funct. Neurol.* **2011**, *26*, 25–30.
326. Lehembre, R.; Gosseries, O.; Lugo, Z.; Jedidi, Z.; Chatelle, C.; Sadzot, B.; Laureys, S.; Noirhomme, Q. Electrophysiological investigations of brain function in coma, vegetative and minimally conscious patients. *Arch. Ital. Biol.* **2012**, *150*, 122–139. [CrossRef]
327. Fingelkurts, A.A.; Fingelkurts, A.A.; Bagnato, S.; Boccagni, C.; Galardi, G. Long-term (six years) clinical outcome discrimination of patients in the vegetative state could be achieved based on the operational architectonics EEG analysis: A pilot feasibility study. *Open Neuroimag. J.* **2016**, *10* (Suppl. S1, M6), 69–79. [CrossRef] [PubMed]
328. Sarà, M.; Pistoia, F.; Pasqualetti, P.; Sebastiano, F.; Onorati, P.; Rossini, P.M. Functional isolation within the cerebral cortex in the vegetative state: A nonlinear method to predict clinical outcomes. *Neurorehabil. Neural Repair* **2011**, *25*, 35–42. [CrossRef] [PubMed]
329. Fingelkurts, A.A.; Fingelkurts, A.A. Operational architectonics methodology for EEG analysis: Theory and results. *Neuromethods* **2015**, *91*, 1–59. [CrossRef]
330. Fingelkurts, A.A.; Fingelkurts, A.A.; Neves, C.F.H. Natural world physical, brain operational, and mind phenomenal space–time. *Phys. Life Rev.* **2010**, *7*, 195–249. [CrossRef]
331. Cacciola, A.; Naro, A.; Milardi, D.; Bramanti, A.; Malatacca, L.; Spitaleri, M.; Leo, A.; Muscoloni, A.; Cannistraci, C.V.; Bramanti, P.; et al. Functional brain network topology discriminates between patients with minimally conscious state and unresponsive wakefulness syndrome. *J. Clin. Med.* **2019**, *8*, 306. [CrossRef] [PubMed]
332. Tononi, G.; Boly, M.; Massimini, M.; Koch, C. Integrated information theory: From consciousness to its physical substrate. *Nat. Rev. Neurosci.* **2016**, *17*, 450–461. [CrossRef]
333. Hubbard, O.; Sunde, D.; Goldensohn, E.S. The EEG in centenarians. *Electroencephalogr. Clin. Neurophysiol.* **1976**, *40*, 407–417. [CrossRef]
334. Matejcek, M. Some relationships between occipital EEG activity and age. A spectral analytic study. *Rev. Electroencephalogr. Neurophysiol. Clin.* **1980**, *10*, 122–130. [CrossRef]
335. Matthis, P.; Scheffner, D.; Benninger, C.; Lipinski, C.; Stolzis, L. Changes in the background activity of the electroencephalogram according to age. *Electroencephalogr. Clin. Neurophysiol.* **1980**, *49*, 626–635. [CrossRef]
336. Marciani, M.G.; Maschio, M.; Spanedda, F.; Caltagirone, C.; Gigli, G.; Bernardi, G. Quantitative EEG evaluation in normal elderly subjects during mental processes: Age-related changes. *Int. J. Neurosci.* **1994**, *76*, 131–140. [CrossRef]
337. Shigeta, M.; Julin, P.; Almkvist, O.; Basun, H.; Rudberg, U.; Wahlund, L.-O. EEG in successful ageing; a 5-year follow-up study from the eighth to ninth decade of life. *Electroencephalogr. Clin. Neurophysiol.* **1995**, *95*, 77–83. [CrossRef]
338. Li, D.; Sun, F.; Jiao, Y. Frontal EEG characters in ageing and the correlativity with some cognitive abilities. *Acta Psychol. Sin.* **1996**, *28*, 76–81.
339. Widagdo, M.; Pierson, J.; Helme, R. Age-related changes in qEEG during cognitive tasks. *Int. J. Neurosci.* **1998**, *95*, 63–75. [CrossRef]
340. Van Sweden, B.; Wauquier, A.; Niedermeyer, E. Normal ageing and transient cognitive disorders in the elderly. In *Electroencephalography: Basic Principles, Clinical Applications, and Related Fields*; Niedermeyer, E., da Silva, F.H.L., Eds.; Williams and Wilkins: Baltimore, MD, USA, 1999; pp. 340–348.

341. Kikuchi, M.; Wada, Y.; Koshino, Y.; Nanbu, Y.; Hashimoto, T. Effect of normal ageing upon interhemispheric EEG coherence: Analysis during rest and photic stimulation. *Clin. Electroencephalogr.* **2000**, *31*, 170–174. [CrossRef] [PubMed]
342. Babiloni, C.; Binetti, G.; Cassarino, A.; Dal Forno, G.; Del Percio, C.; Ferreri, F.; Ferri, R.; Frisoni, G.; Galderisi, S.; Hirata, K.; et al. Sources of cortical rhythms in adults during physiological ageing: A multicentric EEG study. *Hum. Brain Mapp.* **2006**, *27*, 162–172. [CrossRef]
343. Boha, R.; Stam, C.J.; Molnár, M. Age-dependent features of EEG-reactivity-spectral, complexity, and network characteristics. *Neurosci. Lett.* **2010**, *479*, 79–84.
344. Peltz, C.B.; Kim, H.L.; Kawas, C.H. Abnormal EEGs in cognitively and physically healthy oldest-old: Findings from the 90þ study. *J. Clin. Neurophysiol.* **2010**, *27*, 292–295. [CrossRef]
345. Knyazeva, M.G.; Barzegaran, E.; Vildavski, V.Y.; Demonet, J.-F. Ageing of human alpha rhythm. Neurobiol. *Ageing* **2018**, *69*, 261–273.
346. Markand, O.N. Electroencephalogram in dementia. *Am. J. EEG Technol.* **1986**, *26*, 3–17. [CrossRef]
347. Samson-Dollfus, D.; Delapierre, G.; Do Marcolino, C.; Blondeau, C. Normal and pathological changes in alpha rhythms. *Int. J. Psychophysiol.* **1997**, *26*, 395–409. [CrossRef]
348. Fernández, A.; Hornero, R.; Mayo, A.; Poza, J.; Gil-Gregorio, P.; Ortiz, T. EEG spectral profile in Alzheimer's disease and mild cognitive impairment. *Clin. Neurophysiol.* **2006**, *117*, 306–314. [CrossRef] [PubMed]
349. Varela, F. Neurophenomenology: A methodological remedy for the hard problem. *J. Conscious Stud.* **1996**, *3*, 330–349.
350. Borjigin, J.; Lee, U.; Liu, T.; Pal, D.; Huff, S.; Klarr, D.; Sloboda, J.; Hernandez, J.; Wang, M.M.; Mashour, G.A. Surge of neurophysiological coherence and connectivity in the dying brain. *Proc. Natl. Acad. Sci. USA* **2013**, *110*, 14432–14437. [CrossRef] [PubMed]
351. Vicente, R.; Rizzuto, M.; Sarica, C.; Yamamoto, K.; Sadr, M.; Khajuria, T.; Fatehi, M.; Moien-Afshari, F.; Haw, C.S.; Llinas, R.R.; et al. Enhanced interplay of neuronal coherence and coupling in the dying human brain. *Front. Aging Neurosci.* **2022**, *14*, 813531. [CrossRef]
352. Chawla, L.S.; Akst, S.; Junker, C.; Jacobs, B.; Seneff, M.G. Surges of electroencephalogram activity at the time of death: A case series. *J. Palliat. Med.* **2009**, *12*, 1095–1100. [CrossRef] [PubMed]
353. Persinger, M.A.; Rouleau, N.; Murugan, N.J.; Tessaro, L.W.E.; Costa, J.N. When is the brain dead? Living-like electrophysiological responses and photon emissions from applications of neurotransmitters in fixed post-mortem human brains. *PLoS ONE* **2016**, *11*, e0167231. [CrossRef]
354. Blundon, E.G.; Gallagher, R.E.; Ward, L.M. Electrophysiological evidence of preserved hearing at the end of life. *Sci. Rep.* **2020**, *10*, 10336. [CrossRef]
355. Blundon, E.G.; Gallagher, R.E.; Ward, L.M. Resting state network activation and functional connectivity in the dying brain. *Clin. Neurophysiol.* **2022**, *135*, 166–178. [CrossRef]
356. Blundon, E.G.; Gallagher, R.E.; Ward, L.M. Electrophysiological evidence of sustained attention to music among conscious participants and unresponsive hospice patients at the end of life. *Clin. Neurophysiol.* **2022**, *139*, 9–22. [CrossRef]
357. Fernández-Torre, J.L.; Hernández-Hernández, M.A.; Muñoz-Esteban, C. Non confirmatory electroencephalography in patients meeting clinical criteria for brain death: Scenario and impact on organ donation. *Clin. Neurophysiol.* **2013**, *124*, 2362–2367. [CrossRef]
358. Grigg, M.M.; Kelly, M.A.; Celesia, G.G.; Ghobrial, M.W.; Ross, E.R. Electroencephalographic activity after brain death. *Arch. Neurol.* **1987**, *44*, 948–954. [CrossRef] [PubMed]
359. Sutter, R.; Stevens, R.D.; Kaplan, P.W. Significance of triphasic waves in patients with acute encephalopathy: A nine-year cohort study. *Clin. Neurophysiol.* **2013**, *124*, 1952–1958. [CrossRef] [PubMed]
360. Itil, T.M. Quantitative pharmacoelectroencephalography. In *Psychotropic Drugs and the Human EEG: Modern Problems in Pharmachopsychiatry*; Itil, T.M., Ed.; Karger: New York, NY, USA, 1974; Volume 8, pp. 43–75.
361. Herrmann, W.M.; Schaerer, E. Pharmaco-EEG: Computer EEG analysis to describe the projection of drug effects on a functional cerebral level in humans. In *Handbook of Electroencephalography and Clinical Neurophysiology*; Silva, F.H.L., Leeuwen, W.S., Rémond, A., Eds.; Elsevier: Amsterdam, The Netherlands, 1986; Volume 2, pp. 386–445.
362. Saletu, B. The use of pharmaco-EEG in drug profiling. In *Human Psychopharmacology Measures and Methods*; Hindmarch, I., Stonier, P.D., Eds.; John Wiley: New York, NY, USA, 1987; Volume 1, pp. 173–200.
363. Mandema, J.W.; Danhof, M. Electroencephalogram effect measures and relationships between pharmacokinetics and pharmacodynamics of centrally acting drugs. *Clin. Pharmacokinet.* **1992**, *23*, 191–215. [CrossRef]
364. Bruder, G.E.; Stewart, J.W.; Tenke, C.E.; McGrath, P.J.; Leite, P.; Bhattacharya, N.; Quitkin, F.M. Electroencephalographic and perceptual asymmetry differences between responders and nonresponders to an SSRI antidepressant. *Biol. Psychiatr.* **2001**, *49*, 416–425. [CrossRef]
365. Hermens, D.F.; Cooper, N.J.; Kohn, M.; Clarke, S.; Gordon, E. Predicting stimulant medication response in ADHD: Evidence from an integrated profile of neuropsychological, psychophysiological and clinical factors. *J. Integr. Neurosci.* **2005**, *4*, 107–121. [CrossRef] [PubMed]
366. Arns, M.; Gunkelman, J.; Breteler, M.; Spronk, D. EEG phenotypes predict treatment outcome to stimulants in children with ADHD. *J. Integr. Neurosci.* **2008**, *7*, 421–438. [CrossRef]

367. Bruder, G.E.; Sedoruk, J.P.; Stewart, J.W.; McGrath, P.J.; Quitkin, F.M.; Tenke, C.E. Electroencephalographic alpha measures predict therapeutic response to a selective serotonin reuptake inhibitor antidepressant: Pre- and post-treatment findings. *Biol. Psychiatr.* **2008**, *63*, 1171–1177. [CrossRef]
368. Iosifescu, D.V.; Greenwald, S.; Devlin, P.; Perlis, R.H.; Denninger, J.W.; Alpert, J.E.; Fava, M. Pretreatment frontal EEG and changes in suicidal ideation during SSRI treatment in major depressive disorder. *Acta Psychiatr. Scand.* **2008**, *117*, 271–276. [CrossRef]
369. Leuchter, A.F.; Cook, I.A.; Gilmer, W.S.; Marangell, L.B.; Burgoyne, K.S.; Howland, R.H.; Trivedi, M.H.; Zisook, S.; Jain, R.; Fava, M.; et al. Effectiveness of a quantitative electroencephalographic biomarker for predicting differential response or remission with escitalopram and bupropion in major depressive disorder. *Psychiatr. Res.* **2009**, *169*, 132–138. [CrossRef]
370. Leuchter, A.F.; Cook, I.A.; Marangell, L.B.; Gilmer, W.S.; Burgoyne, K.S.; Howland, R.H.; Trivedi, M.H.; Zisook, S.; Jain, R.; McCracken, J.T.; et al. Comparative effectiveness of biomarkers and clinical indicators for predicting outcomes of SSRI treatment in Major Depressive Disorder: Results of the BRITE-MD study. *Psychiatr. Res.* **2009**, *169*, 124–131. [CrossRef]
371. Iznak, A.F.; Iznak, E.V. EEG predictors of therapeutic responses in psychiatry. *Neurosci. Behav. Physiol.* **2022**, *52*, 207–212. [CrossRef]
372. Cook, I.A.; Hunter, A.M.; Korb, A.; Farahbod, H.; Leuchter, A.F. EEG signals in psychiatry: Biomarkers for depression management. In *Quantitative EEG Analysis Methods and Clinical Applications*; Tong, S., Thakor, N.V., Eds.; Artech House: Norwood, MA, USA, 2009; pp. 289–315.
373. Mednick, S.A.; Vka, J.V.; Gabrielli, J.W.F.; Itil, T.M. EEG as a predictor of antisocial behavior. *Criminology* **1981**, *19*, 219–230. [CrossRef]
374. Fingelkurts, A.A.; Fingelkurts, A.A.; Neves, C.F.H. The structure of brain electromagnetic field relates to subjective experience: Exogenous magnetic field stimulation study. In Proceedings of the Neuroscience Finland 2013 Meeting: Optogenetics and Brain Stimulation, Helsinki, Finland, 22 March 2013.
375. Daskalakis, Z.J.; Levinson, A.J.; Fitzgerald, P.B. Repetitive transcranial magnetic stimulation for major depressive disorder: A review. *Can. J. Psychiatr.* **2008**, *53*, 555–566. [CrossRef]
376. Lam, R.W.; Chan, P.; Wilkins-Ho, M.; Yatham, L.N. Repetitive transcranial magnetic stimulation for treatment-resistant depression: A systematic review and meta-analysis. *Can. J. Psychiatr.* **2008**, *53*, 621–631. [CrossRef]
377. Walker, J.E.; Kozlowski, G.P. Neurofeedback treatment of epilepsy. *Child Adolesc. Psychiatr. Clin. N. Am.* **2005**, *14*, 163–176. [CrossRef]
378. Tan, G.; Thornby, J.; Hammond, D.C.; Strehl, U.; Canady, B.; Arnemann, K.; Kaiser, D.A. Meta-analysis of EEG biofeedback in treating epilepsy. *Clin. EEG Neurosci.* **2009**, *40*, 173–179. [CrossRef]
379. Hammond, D.C. Neurofeedback with anxiety and affective disorders. *Child Adolesc. Psychiatr. Clin. N. Am.* **2005**, *14*, 105–123. [CrossRef]
380. Surmeli, T.; Ertem, A.; Eralp, E.; Kos, I.H. Schizophrenia and the efficacy of qEEG-guided neurofeedback treatment: A clinical case series. *Neurosci. Lett.* **2011**, *500S*, e16. [CrossRef]
381. Dehghani-Arani, F.; Rostami, R.; Nadali, H. Neurofeedback training for opiate addiction: Improvement of mental health and craving. *Appl. Psychophysiol. Biofeedback* **2013**, *38*, 133–141. [CrossRef]
382. Arns, M.; De Ridder, S.; Strehl, U.; Breteler, M.; Coenen, A. Efficacy of neurofeedback treatment in ADHD: The effects on inattention, impulsivity and hyperactivity: A meta-analysis. *Clin. EEG Neurosci.* **2009**, *40*, 180–189. [CrossRef]
383. Gevensleben, H.; Holl, B.; Albrecht, B.; Schlamp, D.; Kratz, O.; Studer, P.; Wangler, S.; Rothenberger, A.; Moll, G.H.; Heinrich, H. Distinct EEG effects related to neurofeedback training in children with ADHD: A randomized controlled trial. *Int. J. Psychophysiol.* **2009**, *74*, 149–157. [CrossRef]
384. Hammer, B.U.; Colbert, A.P.; Brown, K.A.; Ilioi, E.C. Neurofeedback for insomnia: A pilot study of Z-score SMR and individualized protocols. *Appl. Psychophysiol. Biofeedback* **2011**, *36*, 251–264. [CrossRef]
385. Kouijzer, M.E.J.; van Schie, H.T.; de Moor, J.M.H.; Gerrits, B.J.L.; Buitelaar, J.K. Neurofeedback treatment in autism. Preliminary findings in behavioral, cognitive, and neurophysiological functioning. *Res. Autism Spectr. Disord.* **2010**, *4*, 386–399. [CrossRef]
386. Ibric, V.L.; Dragomirescu, L.G. Neurofeedback in pain management. In *Introduction to Quantitative EEG Neurofeedback Advanced Theory and Application*, 2nd ed.; Budzynski, T.H., Budzynski, H.K., Evans, J.R., Abarbanel, A., Eds.; Elsevier: New York, NY, USA, 2009; pp. 417–451. [CrossRef]
387. Orlando, P.C.; Rivera, R.O. Neurofeedback for elementary students with identified learning problems. *J. Neurother.* **2004**, *8*, 5–19. [CrossRef]
388. Breteler, M.H.; Arns, M.; Peters, S.; Giepmans, I.; Verhoeven, L. Improvements in spelling after QEEG-based neurofeedback in dyslexia: A randomized controlled treatment study. *Appl. Psychophysiol. Biofeedback* **2010**, *35*, 5–11. [CrossRef]
389. Hanslmayr, S.; Sauseng, P.; Doppelmayr, M.; Schabus, M.; Klimesch, W. Increasing individual upper alpha power by neurofeedback improves cognitive performance in human subjects. *Appl. Psychophysiol. Biofeedback* **2005**, *30*, 1–10. [CrossRef]
390. Angelakis, E.; Stathopoulou, S.; Frymiare, J.L. EEG neurofeedback: A brief overview and an example of peak alpha frequency training for cognitive enhancement in the elderly. *Clin. Neuropsychol.* **2007**, *21*, 110–129. [CrossRef]
391. Fingelkurts, A.A.; Fingelkurts, A.A.; Neves, C.F.H. Consciousness as a phenomenon in the operational architectonics of brain organization: Criticality and self-organization considerations. *Chaos Solitons Fractals* **2013**, *55*, 13–31. [CrossRef]
392. Thatcher, R.W.; John, E.R. *Functional Neuroscience: I. Foundations of Cognitive Processes*; Lawrence Erlbaum: Boca Raton, NJ, USA, 1977; p. 370.

393. Bodunov, M.V. The EEG "alphabet": The typology of human EEG stationary segments. In *Individual and Psychological Differences and Bioelectrical Activity of Human Brain*; Rusalov, V.M., Ed.; Nauka: Moscow, Russia, 1988; pp. 56–70. (In Russian)
394. Jansen, B.H.; Cheng, W.-K. Structural EEG analysis: An explorative study. *Int. J. Biomed. Comput.* **1988**, *23*, 221–237. [CrossRef]
395. Fingelkurts, A.A.; Fingelkurts, A.A.; Kaplan, A.Y. The regularities of the discrete nature of multi-variability of EEG spectral patterns. *Int. J. Psychophysiol.* **2003**, *47*, 23–41. [CrossRef]
396. Fingelkurts, A.A.; Fingelkurts, A.A.; Krause, C.M.; Kaplan, A.Y. Systematic rules underlying spectral pattern variability: Experimental results and a review of the evidences. *Int. J. Neurosci.* **2003**, *113*, 1447–1473. [CrossRef] [PubMed]
397. Fingelkurts, A.A.; Fingelkurts, A.A. Operational Architectonics of the human brain biopotential field: Towards solving the mind-brain problem. *BrainMind* **2001**, *2*, 261–296. Available online: http://www.bm-science.com/team/art18.pdf (accessed on 1 May 2022).
398. Kaplan, A.Y.; Fingelkurts, A.A.; Fingelkurts, A.A.; Borisov, S.V.; Darkhovsky, B.S. Nonstationary nature of the brain activity as revealed by EEG/MEG: Methodological, practical and conceptual challenges. *Signal Process.* **2005**, *85*, 2190–2212. [CrossRef]
399. Fingelkurts, A.A.; Fingelkurts, A.A. Brain-mind Operational Architectonics imaging: Technical and methodological aspects. *Open Neuroimag. J.* **2008**, *2*, 73–93. [CrossRef]
400. Fingelkurts, A.A.; Fingelkurts, A.A. Editorial: EEG Phenomenology and Multiple Faces of Short-term EEG Spectral Pattern. *Open Neuroimag. J.* **2010**, *4*, 111–113. [PubMed]
401. Fingelkurts, A.A.; Fingelkurts, A.A. Making complexity simpler: Multivariability and metastability in the brain. *Int. J. Neurosci.* **2004**, *114*, 843–862. [CrossRef]
402. Barlow, J.S. Methods of analysis of nonstationary EEGs, with emphasis on segmentation techniques: A comparative review. *J. Clin. Neurophysiol.* **1985**, *2*, 267–304. [CrossRef]
403. Fingelkurts, A.A.; Fingelkurts, A.A. Mapping of the brain operational architectonics. In *Focus on Brain Mapping*; Chen, F.J., Ed.; Research Nova Science Publishers, Inc.: Hauppauge, NY, USA, 2005; pp. 59–98. Available online: http://www.bm-science.com/team/chapt3.pdf (accessed on 1 May 2022).
404. Friston, K. The labile brain. I. Neuronal transients and nonlinear coupling. *Philos. Trans. R. Soc. Lond. B Biol. Sci.* **2000**, *355*, 215–236. [CrossRef]
405. Triesch, J.; von der Malsburg, C. Democratic integration: Self-organized integration of adaptive cues. *Neural Comput.* **2001**, *13*, 2049–2074. [CrossRef]
406. Buzsaki, G. Large-scale recording of neuronal ensembles. *Nat. Neurosci.* **2004**, *7*, 446–451. [CrossRef]
407. Averbeck, B.B.; Lee, D. Coding and transmission of information by neural ensembles. *Trends Neurosci.* **2004**, *27*, 225–230. [CrossRef]
408. Freeman, W.J.; Vitiello, G. Nonlinear brain dynamics and many-body field dynamics. *Electromagn. Biol. Med.* **2005**, *24*, 233–241. [CrossRef]
409. Stam, C.J. *Nonlinear Brain Dynamics*, 1st ed.; Nova Science Publishers, Inc.: New York, NY, USA, 2006; p. 148.
410. Freeman, W.J. Indirect biological measures of consciousness from field studies of brains as dynamical systems. *Neural Netw.* **2007**, *20*, 1021–1031. [CrossRef]
411. Lehmann, D. Brain electric microstates and cognition: The atoms of thought. In *Machinery of the Mind*; John, E.R., Ed.; Birkhauser: Boston, MA, USA, 1990; pp. 209–224.
412. Freeman, W.J. Tutorial on neurobiology: From single neurons to brain chaos. *Int. J. Bifurcat. Chaos* **1992**, *2*, 451–482. [CrossRef]
413. John, R.; Easton, P.; Isenhart, R. Consciousness and cognition may be mediated by multiple independent coherent ensembles. *Con Cogn.* **1997**, *6*, 3–39. [CrossRef]
414. Wright, J.J.; Kydd, R.R.; Lees, G.J. State-changes in the brain viewed as linear steady-states and non-linear transitions between steady-states. *Biol. Cybern.* **1985**, *53*, 11–17. [CrossRef]
415. Lehmann, D.; Strik, W.K.; Henggeler, B.; Koenig, T.; Koukkou, M. Brain electric microstates and momentary conscious mind states as building blocks of spontaneous thinking: I. Visual imagery and abstract thoughts. *Int. J. Psychophysiol.* **1998**, *29*, 1–11. [CrossRef]
416. Fingelkurts, A.A.; Fingelkurts, A.A. Topographic mapping of rapid transitions in EEG multiple frequencies: EEG frequency domain of operational synchrony. *Neurosci. Res.* **2010**, *68*, 207–224. [CrossRef]
417. Freeman, W.J. On the problem of anomalous dispersion in chaoto-chaotic phase transitions of neural masses, and its significance for the management of perceptual information in brains. In *Synergetics of Cognition*; Haken, H., Stadler, M., Eds.; Springer: Berlin/Heidelberg, Germany, 1990; Volume 45, pp. 126–143.
418. Palm, G. Cell assemblies as a guideline for brain research. *Concepts Neurosci.* **1990**, *1*, 133–147.
419. Eichenbaum, H. Thinking about brain cell assemblies. *Science* **1993**, *261*, 993–994. [CrossRef]
420. von der Malsburg, C. The what and why of binding: The modeler's perspective. *Neuron* **1999**, *24*, 95–104. [CrossRef]
421. Buzsáki, G. *Rhythms of the Brain*; Oxford University Press: Oxford, UK, 2006; p. 448.
422. Lehmann, D. Multichannel topography of human alpha EEG fields. *Electroencephalogr. Clin. Neurophysiol.* **1971**, *31*, 439–449. [CrossRef]
423. Stam, C.J. Nonlinear dynamical analysis of EEG and MEG: Review of an emerging field. *Clin. Neurophysiol.* **2005**, *116*, 2266–2301. [CrossRef]

424. Wackermann, J.; Allefeld, C. On the meaning and interpretation of global descriptors of brain electrical activity. Including a reply to X. Pei et al. *Int. J. Psychophysiol.* **2007**, *64*, 199–210. [CrossRef]
425. Freeman, W.J. The wave packet: An action potential for the 21st Century. *J. Integr. Neurosci.* **2003**, *2*, 3–30. [CrossRef]
426. Fingelkurts, A.A.; Fingelkurts, A.A. Morphology and dynamic repertoire of EEG short-term spectral patterns in rest: Explorative study. *Neurosci. Res.* **2010**, *66*, 299–312. [CrossRef]
427. Freeman, W.J.; Holmes, M.D. Metastability, instability, and state transition in neocortex. *Neural Netw.* **2005**, *18*, 497–504. [CrossRef]
428. Penttonen, M.; Buzsaki, G. Natural logarithmic relationship between brain oscillators. *Thalamus Relat. Syst.* **2003**, *2*, 145–152. [CrossRef]
429. Hoppensteadt, F.C.; Izhikevich, E.M. *Weakly Connected Neural Networks*; Springer: New York, NY, USA, 1997; p. 402.
430. Izhikevich, E.M. Weakly connected quasi-periodic oscillators, FM interactions, and multiplexing in the brain. *SIAM J. Appl. Math.* **1999**, *59*, 2193–2223. [CrossRef]
431. Izhikevich, E.M.; Desai, N.S.; Walcott, E.C.; Hoppensteadt, F.C. Bursts as a unit of neural information: Selective communication via resonance. *Trends Neurosci.* **2003**, *26*, 161–167. [CrossRef]
432. Freeman, W.J.; Rogers, L.J. Fine temporal resolution of analytic phase reveals episodic synchronization by state transitions in gamma EEGs. *J. Neurophysiol.* **2002**, *87*, 937–945. [CrossRef]
433. Kozma, R.; Puljic, M.; Balister, P.; Bollobas, B.; Freeman, W.J. Phase transitions in the neuropercolation model of neural populations with mixed local and nonlocal interactions. *Biol. Cyber.* **2005**, *92*, 367–379. [CrossRef]
434. Friston, K.J.; Frith, C.D.; Liddle, P.F.; Frackowiak, R.S.J. Functional connectivity: The principal component analysis of large (PET) data sets. *J. Cereb. Blood Flow Metab.* **1993**, *13*, 5–14. [CrossRef]
435. Friston, K.J.; Frith, C.D.; Fletcher, P.; Liddle, P.F.; Frackowiak, R.S.J. Functional topography: Multidimensional scaling and functional connectivity in the brain. *Cereb. Cortex* **1996**, *6*, 156–164. [CrossRef]
436. Nunez, P.L. Generation of human EEG by a combination of long- and short-range neocortical interactions. *Brain Topogr.* **1989**, *1*, 199–215. [CrossRef]
437. Werner, G. Metastability, criticality and phase transitions in brain and its models. *Biosystems* **2007**, *90*, 496–508. [CrossRef]
438. Kelso, J.A.S. *Dynamic Patterns: The Self-Organization of Brain and Behavior*; MIT Press: Cambridge, MA, USA, 1995; p. 334.
439. Fingelkurts, A.A.; Fingelkurts, A.A.; Neves, C.F.H. Phenomenological architecture of mind and operational architectonics of the brain: The unified metastable continuum. *New Math. Nat. Comput.* **2009**, *5*, 221–244. [CrossRef]
440. Ukhtomsky, A.A. *Selected Works*; Nauka: Lenengrad, Russia, 1978; p. 358. (In Russian)
441. Breakspear, M.; Terry, J.R.; Friston, K.J.; Harris, A.W.F.; Williams, L.M.; Brown, K.; Brennan, J.; Gordona, E. A disturbance of nonlinear interdependence in scalp EEG of subjects with first episode schizophrenia. *NeuroImage* **2003**, *20*, 466–478. [CrossRef]
442. Tononi, G.; Edelman, G.M. Schizophrenia and the mechanisms of conscious integration. *Brain Res. Rev.* **2000**, *31*, 391–400. [CrossRef]
443. Andreasen, N.C. A unitary model of schizophrenia: Bleuler's "fragmented phrene" as schizencephaly. *Arch. Gen. Psychol.* **1999**, *56*, 781–787. [CrossRef]
444. Schulman, J.J.; Cancro, R.; Lowe, S.; Lu, F.; Walton, K.D.; Llinás, R.R. Imaging of thalamocortical dysrhythmia in neuropsychiatry. *Front. Hum. Neurosci.* **2011**, *5*, 69. [CrossRef]
445. Buckholtz, J.W.; Meyer-Lindenberg, A. Psychopathology and the human connectome: Toward a transdiagnostic model of risk for mental illness. *Neuron* **2012**, *74*, 990–1004. [CrossRef]
446. Dawson, K.A. Temporal organization of the brain: Neurocognitive mechanisms and clinical implications. *Brain Cogn.* **2004**, *54*, 75–94. [CrossRef]
447. Gasser, T.; Bacher, P.; Mochs, J. Transformation towards the normal distribution of broad band spectral parameters of the EEG. *Electroencephalogr. Clin. Neurophysiol.* **1982**, *53*, 119–124. [CrossRef]
448. John, E.R.; Prichep, L.S.; Easton, P. Normative data banks and neurometrics: Basic concepts, methods and results of norm construction. In *Handbook of Electroencephalography and Clinical Neurophysiology*; Gevins, A.S., Remond, A., Eds.; Elsevier: Amsterdam, The Netherlands, 1987; Volume I, pp. 449–495.
449. John, E.R.; Prichep, L.S. Principles of neurometrics and neurometric analysis of EEG and evoked potentials. In *EEG: Basic Principles, Clinical Applications and Related Fields*; Niedermeyer, E., Da Silva, F.L., Eds.; Williams & Wilkins: Baltimore, MD, USA, 1993; pp. 989–1003.
450. Fingelkurts, A.A.; Fingelkurts, A.A. Altered structure of dynamic electroencephalogram oscillatory pattern in major depression. *Biol. Psychiatr.* **2015**, *77*, 1050–1060. [CrossRef]
451. Johnstone, J.; Gunkelman, J.; Lunt, J. Clinical database development: Characterization of EEG phenotypes. *Clin. EEG Neurosci.* **2005**, *36*, 99–107. [CrossRef]
452. Zhirmunskaya, E.K.; Losev, B.C.; Maslov, V.K. Mathematical analysis of EEG type and interhemispheric EEG asymmetry. *Hum. Physiol.* **1978**, *5*, 791–799.
453. Zhirmunskaya, E.A.; Losev, V.S. The concept of type in the classification of electroencephalograms. *Hum. Physiol.* **1980**, *6*, 1039–1045.
454. Zhirmunskaya, E.A.; Losev, B.C. *Description Systems and Classification of Human Electroencephalograms*; Nauka: Moscow, Russia, 1984; p. 81.
455. Zhirmunskaya, E.A. *In Search of an Explanation of EEG Phenomena*; Biola: Moscow, Russia, 1996; p. 117.

456. Misyuk, N.N. Diagnostic efficiency and basic principles of classification of electroencephalograms. *Med. News* **2006**, *1*, 24–33.
457. Bochkarev, V.C.; Panyushkin, S.V. Electroencephalographic studies in borderline conditions. In *Borderline Mental Disorders*, 3rd ed.; Aleksandrovsky, Y.A., Ed.; Moscow, Russia, 2000; pp. 120–133.
458. Chemiy, V.I.; Ostrovaya, T.V. The diagnostic algorithm of assessment of EEG for estimation of brain in vestigation of central nervous system reactivity in response to photostimulation and pharmacological influence. *Neurosci. Theor. Clin. Asp.* **2005**, *1*, 12–51.
459. Koroleva, N.V.; Kolesnikov, S.I.; Dolgih, V.V. Dynamic of electroencephalogram's descriptors in children with different EEG types. *Bull. RAMH* **2007**, *2*, 49–51.
460. Gelda, A.P.; Dokukina, T.V.; Misyuk, N.N.; Cosmidiadi, A.O. Registration of electroencephalograms during psychopharmacotherapy. *Med. J.* **2008**, *4*, 16–18.
461. Misyuk, N.N.; Gelda, A.P.; Dokukina, T.V.; Cosmidiadi, A.O. Types of electroencephalograms in schizophrenia. *Med. J.* **2008**, *4*, 41–43.
462. Zvereva, Z.F.; Vanchakova, N.P.; Zolotaryova, N.N. Clinical and neurophysiological parameters in patients with discirculatory encephalopathy. *Zh. Nevrol. Psikhiatr.* **2010**, *110*, 2–15.
463. Koichubekov, B.K.; Sorokina, M.A.; Pashev, V.I.; Shaikhin, A.M. Individually-typological of CNS's regulatory processes in persons with initial signs of neurocirculatory dystonia. *Fundam. Res.* **2012**, *5*, 300–304.
464. Gibson, E.; Lobaugh, N.J.; Joordens, J.; McIntosh, A.R. EEG variability: Task-driven or subject-driven signal of interest? *NeuroImage* **2022**, *252*, 119034. [CrossRef]
465. Danilova, N.N. On individual peculiarities of the electrical activity of the cerebral cortex of humans. In *Typological Peculiarities of the Higher Nervous Activity of Humans*; Academic Psychological Science: Moscow, Russia, 1963; Volume 3, pp. 262–274.
466. Nebylitsyn, V.D. An electroencephalographic study of the properties of the strength of the nervous system and the balance of the nerve processes in humans using factorial analysis. In *Typological Peculiarities of Higher Nervous Activity in Humans*; Academic Psychological Science: Moscow, Russia, 1963; pp. 47–80.
467. Egorova, I.S. *Electroencephalography*; Meditsina: Moscow, Russia, 1973.
468. Soroko, S.I.; Bekshaev, S.S. Statistical structure of EEG rhythms and individual properties of brain self-regulatory mechanisms. *Physiol. J.* **1981**, *67*, 1765–1773.
469. Boldyreva, G.N.; Sharova, E.V.; Dobronravova, I.S. Role of cerebral regulatory structures in the formation of EEG in humans. *Hum. Physiol. (Fiziol. Cheloveka)* **2000**, *26*, 19–34.
470. Maryutina, T.M. Intermediate phenotypes of intelligence in the context of genetic psychophysiology. *Psychol. J. High. Sch. Econ.* **2007**, *4*, 22–47.
471. Clarke, A.R.; Barry, R.J.; McCarthy, R.; Selikowitz, M.; Johnstone, S.J.; Hsu, C.-I.M.; Magee, C.A.; Lawrence, C.A.; Croft, R.J. Coherence in children with attention-deficit/hyperactivity disorder and excess beta activity in their EEG. *Clin. Neurophysiol.* **2007**, *118*, 1472–1479. [CrossRef]
472. Suffin, S.C.; Emory, W.H.; Gutierrez, N.; Arora, G.S.; Schiller, M.J.; Kling, A. A QEEG database method for predicting pharmacotherapeutic outcome in refractory major depressive disorders. *J. Am. Phys. Surg.* **2007**, *12*, 104–108.
473. Almasy, L. Quantitative risk factors as indices of alcoholism susceptibility. *Ann. Med.* **2003**, *35*, 337–343. [CrossRef]
474. Porjesz, B.; Begleiter, H. Alcoholism and human electrophysiology. *Alcohol. Res. Health* **2003**, *27*, 53–160.
475. Vasilevskii, N.N.; Suvorov, N.B.; Sidorov, Y.A.; Bovtyushko, V.G. Risk factors and some features of the pathology depending on the organization type of cerebral neurodynamyics. *Vestn. Ross Akad. Med. Nauk.* **1996**, *9*, 14–18.
476. Kendler, K.; Neale, M. Endophenotype: A conceptual analysis. *Mol. Psychiatr.* **2010**, *15*, 789–797. [CrossRef]
477. Bodunov, M.V. Individual and typological characteristics of EEG structure. *J. High Nerve Act.* **1985**, *35*, 1045–1052.
478. Vasilevskii, N.N.; Soroko, S.I.; Zingerman, A.M. Psychophysiological principles of individual typological features of humans. In *Mechanisms of Brain Activity in Humans*; Bekhtereva, N.P., Ed.; Nauka: Leningrad, Russia, 1988; pp. 455–490.
479. Soroko, S.I.; Bekshaev, S.S.; Sidorov, Y.A. *The Main Types of The Brain Self-Regulation Mechanisms*; Nauka: Leningrad, Russia, 1990; p. 205.
480. Suvorov, N.B.; Zueva, N.G.; Guseva, N.L. Reflection of individual typological features in the structure of spatial interaction of EEG waves of various frequency ranges. *Hum. Physiol.* **2000**, *26*, 301–306, (Translated from Fiziologiya Cheloveka **2000**, *26*, 60–66). [CrossRef]
481. Roth, N.; Sask, G. Relations between slow (4 cps) EEG activity, sensor motor speed, and psychopathology. *Int. J. Psychophysiol.* **1990**, *9*, 121–127. [CrossRef]
482. Markina, A.V.; Pashina, A.K.; Rumanova, N.B. Correlation of the electroencephalogram rhythms with cognitive/personality-related peculiarities of the subject. *Physiol. J.* **2000**, *21*, 47–55.
483. Chorayan, O.G.; Aidarkin, E.K.; Chorayan, I.O. Individual-typological features of regulation and interaction of functional systems in different modes of activity: Review. *Valeology* **2001**, *2*, 5–15.
484. Svyatogor, I.A.; Mokhovikova, I.A.; Bekshayev, S.S.; Nozdrachev, A.D. EEG pattern as an instrument for evaluation of neurophysiological mechanisms underlying adaptation disorders. *J. High. Nerve Act.* **2005**, *55*, 178–188.
485. Konareva, I.N. Modulation of low-frequency EEG rhythms under conditions of an activation reaction: Dependence on psychological characteristics of personality. *Neurophysiology* **2011**, *42*, 42–52. [CrossRef]

486. Knott, J.R.; Gottlieb, J.S. Electroencephalographic evaluation of psychopathic personality: Correlations with age, sex, family history and antecedent illness or injury. *Arch. Neurol. Psychiatr.* **1944**, *52*, 515–519. [CrossRef]
487. Hodge, R.S. The impulsive psychopath: A clinical and electrophysiological study. *J. Ment. Sci.* **1945**, *91*, 472–476. [CrossRef]
488. Diethelm, O.; Simons, D.J. Electroencephalographic changes associated with psychopathic personalities. *Arch. Neurol. Psychiatr.* **1946**, *55*, 410–413.
489. Hill, D. EEG in episodic psychotic and psychopathic behaviour: A classification of data. *Electroencephalogr. Clin. Neurophysiol.* **1952**, *4*, 419–442. [CrossRef]
490. Mizuki, Y.; Kajimura, N.; Nishikori, S.; Imaizumi, J.; Yamada, M. Appearance of frontal midline theta rhythm and personality traits. *Folia Psychiatr. Neurol. Jpn.* **1984**, *38*, 451–458. [CrossRef]
491. Rusalov, V.M.; Rusalova, M.N.; Kalashnikova, I.G.; Stepanov, V.G.; Strel'nikova, T.N. The bioelectrical activity of the human brain in representatives of different temperamental types. *Zh. Vyss. Nerv. Deiat. Im. I P Pavlov.* **1993**, *43*, 530–542. (In Russian) [PubMed]
492. Golan, Z.; Neufield, M.Y. Individual differences in alpha rhythm as characterizing temperament related to cognitive performances. *Personal. Individ. Differ.* **1996**, *21*, 775–784. [CrossRef]
493. Sviderskaia, N.E.; Korol'kova, T.A. The effect of the properties of the nervous system and the temperament on the spatial organization of the EEG. *Zh. Vyss. Nerv. Deiat. Im. I P Pavlov.* **1996**, *46*, 849–858.
494. Tran, Y.; Craig, A.; McIsaac, P. Extraversion–introversion and 8–13 Hz waves in frontal cortical regions. *Persinal. Individ. Differ.* **2001**, *30*, 205–215. [CrossRef]
495. Knyazev, G.G.; Slobodskaya, H.R. Personality trait of behavioural inhibition is associated with oscillatory systems reciprocal relationships. *Int. J. Psychophysiol.* **2003**, *48*, 247–261. [CrossRef]
496. Chi, S.E.; Park, C.B.; Lim, S.L.; Park, E.H.; Lee, Y.H.; Lee, K.H.; Kim, E.J.; Kim, H.T. EEG and personality dimensions: A consideration based on the brain oscillatory systems. *Personal. Individ. Differ.* **2005**, *39*, 669–681. [CrossRef]
497. Knyazev, G.G. EEG correlates of personality types. *Neth. J. Psychol.* **2006**, *62*, 78–87. [CrossRef]
498. Konareva, I.N. Correlation between level of aggressiveness of personality and characteristics of EEG frequency components. *Neurophysiology* **2006**, *38*, 380–388. [CrossRef]
499. Knyazev, G.G. Antero-posterior EEG spectral power gradient as a correlate of extraversion and behavioral inhibition. *Open Neuroimag. J.* **2010**, *4*, 114–120. [CrossRef]
500. Jach, H.K.; Feuerriegel, D.; Smillie, L.D. Decoding personality trait measures from resting EEG: An exploratory report. *Cortex* **2020**, *130*, 158–171. [CrossRef]
501. Jawinski, P.; Markett, S.; Sander, C.; Huang, J.; Ulke, C.; Hegerl, U.; Hensch, T. The Big Five personality traits and brain arousal in the resting state. *Brain Sci.* **2021**, *11*, 1272. [CrossRef]
502. Gallagher, J.R.; Gibbs, E.L.; Gibbs, F.A. Relation between the electrical activity of the cortex and the personality in adolescent boys. *Psychosom. Med.* **1942**, *4*, 134–139. [CrossRef]
503. Kennard, M.K. The electroencephalogram in psychological disorders: A review. *Psychosom. Med.* **1953**, *15*, 95–115. [CrossRef]
504. Williams, D. The significance of an abnormal electroencephalogram. *J. Neurol. Psychiatr.* **1941**, *4*, 257–268. [CrossRef]
505. Hill, D.; Watterson, D. Electroencephalographic studies of psychopathic personalities. *J. Neurol. Psychiatr.* **1942**, *5*, 47–65. [CrossRef]
506. Williams, D. Neural factors related to habitual aggression: Consideration of differences between those habitual aggressives and others who have committed crimes of violence. *Brain* **1969**, *92*, 503–520. [CrossRef]
507. Akiskal, H.S.; Hirschfeld, R.M.A.; Yerevanian, B.I. The relationship of personality to affective disorders: A critical review. *Arch. Gen. Psychiatr.* **1983**, *40*, 801–810. [CrossRef]
508. Hensch, T.; Herold, U.; Brocke, B. An electrophysiological endophenotype of hypomanic and hyperthymic personality. *J. Affect. Disord.* **2007**, *101*, 13–26. [CrossRef]
509. Barnett, J.H.; Huang, J.; Perlis, R.H.; Young, M.M.; Rosenbaum, J.F.; Nierenberg, A.A.; Sachs, G.; Nimgaonkar, V.L.; Miklowitz, D.J.; Smoller, J.W. Personality and bipolar disorder: Dissecting state and trait associations between mood and personality. *Psychol. Med.* **2011**, *41*, 1593–1604. [CrossRef]
510. Klein, D.N.; Kotov, R.; Bufferd, S.J. Personality and depression: Explanatory models and review of the evidence. *Annu. Rev. Clin. Psychol.* **2011**, *7*, 269–295. [CrossRef]
511. Jeronimus, B.F.; Kotov, R.; Riese, H.; Ormel, J. Neuroticism's prospective association with mental disorders halves after adjustment for baseline symptoms and psychiatric history, but the adjusted association hardly decays with time: A meta-analysis on 59 longitudinal/prospective studies with 443 313 participants. *Psychol. Med.* **2016**, *46*, 2883–2906.
512. Widiger, T.A. Personality and psychopathology. *World Psychiatr.* **2011**, *10*, 103–106. [CrossRef]
513. Robinson, D.L. How brain arousal systems determine different temperament types and the major dimensions of personality. *Personal. Individ. Differ.* **2001**, *31*, 1233–1259. [CrossRef]
514. Knyazev, G.G.; Slobodskaya, H.R.; Safronova, M.V.; Sorokin, O.V.; Goodman, R.; Wilson, G.D. Personality, psychopathology and brain oscillations. *Personal. Individ. Differ.* **2003**, *53*, 1331–1349. [CrossRef]
515. Cuthbert, B.N.; Insel, T.R. Toward the future of psychiatric diagnosis: The seven pillars of RDoC. *BMC Med.* **2013**, *11*, 126. [CrossRef] [PubMed]

516. Trofimova, I.; Bajaj, S.; Bashkatov, S.A.; Blair, J.; Brandt, A.; Chan, R.C.K.; Clemens, B.; Corr, P.J.; Cyniak-Cieciura, M.; Demidova, L.; et al. What is next for the neurobiology of temperament, personality and psychopathology? *Curr. Opin. Behav. Sci.* **2022**, *45*, 101143. [CrossRef]
517. de Rooij, S.R. Are brain and cognitive reserve shaped by early life circumstances? *Front. Neurosci.* **2022**, *16*, 825811. [CrossRef]
518. Kent, L.; Nelson, B.; Northoff, G. Can disorders of subjective time inform the differential diagnosis of psychiatric disorders? A transdiagnostic taxonomy of time. *Early Interv. Psychiatr.* **2022**, *2022*, 1–13. [CrossRef]
519. Danilova, N.N. *Psychophysiological Diagnostics of Functional States*; Golubeva, E.A., Tushmalova, N.A., Eds.; Publishing House of Moscow State University: Moscow, Russia, 1992; p. 192.
520. Bechtereva, N.P. *Human Brain in Health and Disease*; ACT Press: St. Petersburg, Russia, 2010.
521. Kahneman, D. *Attention and Effort*; Englewood Cliffs: Prentice Hall, NJ, USA, 1973; p. 246.
522. Hegerl, U.; Wilk, K.; Olbrich, S.; Schoenknecht, P.; Sander, C. Hyperstable regulation of vigilance in patients with major depressive disorder. *World J. Biol. Psychiatr.* **2012**, *13*, 436–446. [CrossRef]
523. Wahbeh, H.; Oken, B.S. Peak high-frequency HRV and peak alpha frequency higher in PTSD. *Appl. Psychophys. Biof.* **2013**, *38*, 57–69. [CrossRef]
524. Schmidt, F.M.; Pschiebl, A.; Sander, C.; Kirkby, K.C.; Thormann, J.; Minkwitz, J.; Chittka, T.; Weschenfelder, J.; Holdt, L.M.; Teupser, D.; et al. Impact of serum cytokine levels on EEG-measured arousal regulation in patients with major depressive disorder and healthy controls. *Neuropsychobiology* **2016**, *73*, 1–9. [CrossRef]
525. Ulke, C.; Tenke, C.E.; Kayser, J.; Sander, C.; Böttger, D.; Wong, L.Y.X.; Alvarenga, J.E.; Fava, M.; McGrath, P.J.; Deldin, P.J.; et al. Resting EEG measures of brain arousal in a multisite study of major depression. *Clin. EEG Neurosci.* **2019**, *50*, 3–12. [CrossRef]
526. Hegerl, U.; Lam, R.W.; Malhi, G.S.; McIntyre, R.S.; Demyttenaere, K.; Mergl, R.; Gorwood, P. Conceptualising the neurobiology of fatigue. *Aust. N. Z. J. Psychiatr.* **2013**, *47*, 312–316. [CrossRef]
527. Hegerl, U.; Ulke, C. Fatigue with up-vs downregulated brain arousal should not be confused. *Prog. Brain Res.* **2016**, *229*, 239–254.
528. Ulke, C.; Surova, G.; Sander, C.; Engel, C.; Wirkner, K.; Jawinski, P.; Hensch, T.; Hegerl, U. Fatigue in cancer and neuroinflammatory and autoimmune disease: CNS arousal matters. *Brain Sci.* **2020**, *10*, 569. [CrossRef]
529. Okogbaa, O.G.; Shell, R.L.; Filipusic, D. On the investigation of the neurophysiological correlates of knowledge worker mental fatigue using the EEG signal. *Appl. Ergon.* **1994**, *25*, 355–365. [CrossRef]
530. Billiot, K.M.; Budzynski, T.H.; Andrasik, F. EEG patterns and chronic fatigue syndrome. *J. Neurother.* **1997**, *2*, 20–30. [CrossRef]
531. Zinn, M.A.; Zinn, M.L.; Valencia, I.; Jason, L.A.; Montoya, J.G. Cortical hypoactivation during resting EEG suggests central nervous system pathology in patients with chronic fatigue syndrome. *Biol. Psychol.* **2018**, *136*, 87–99. [CrossRef]
532. Van Luijtelaar, G.; Verbraak, M.; van den Bunt, M.; Keijsers, G.; Arns, M. EEG findings in burnout patients. *J. Neuropsychiatr. Clin. Neurosci.* **2010**, *22*, 208–217. [CrossRef]
533. Stenberg, G. Personality and the EEG: Arousal and emotional arousability. *Personal. Individ. Differ.* **1992**, *13*, 1097–1113. [CrossRef]
534. Clarke, A.R.; Barry, R.J.; McCarthy, R.; Selikowitz, M.; Brown, C. EEG evidence for a new conceptualisation of attention deficit hyperactivity disorder. *Clin. Neurophysiol.* **2002**, *113*, 1036–1044. [CrossRef]
535. Zhang, D.-W.; Johnstone, S.J.; Roodenrys, S.; Luo, X.; Li, H.; Wang, E.; Zhao, Q.; Song, Y.; Liu, L.; Qian, Q.; et al. The role of EEG localized activation and central nervous system arousal in executive function performance in children with Attention-Deficit/Hyperactivity Disorder. *Clin. Neurophysiol.* **2018**, *129*, 1192–1200. [CrossRef]
536. Ulke, C.; Sander, C.; Jawinski, P.; Mauche, N.; Huang, J.; Spada, J.; Wittekind, D.; Mergl, R.; Luck, T.; Riedel-Heller, S.; et al. Sleep disturbances and upregulation of brain arousal during daytime in depressed versus non-depressed elderly subjects. *World J. Biol. Psychiatr.* **2017**, *18*, 633–640. [CrossRef]
537. Eysenck, H.J. *The Biological Basis of Personality*; Thomas: Springfield, IL, USA, 1967; p. 420.
538. Lo, M.T.; Hinds, D.A.; Tung, J.Y.; Franz, C.; Fan, C.C.; Wang, Y.; Smeland, O.B.; Schork, A.; Holland, D.; Kauppi, K.; et al. Genome-wide analyses for personality traits identify six genomic loci and show correlations with psychiatric disorders. *Nat. Genet.* **2017**, *49*, 152–156. [CrossRef]
539. Baselmans, B.M.L.; Jansen, R.; Ip, H.F.; van Dongen, J.; Abdellaoui, A.; van de Weijer, M.P.; Bao, Y.; Smart, M.; Kumari, M.; Willemsen, G.; et al. Multivariate genome-wide analyses of the well-being spectrum. *Nat. Genet.* **2019**, *51*, 445–451. [CrossRef]
540. Moruzzi, G.; Magoun, H.W. Brain stem reticular formation and activating of the EEG. *Electroencephalogr. Clin. Neurophysiol.* **1949**, *1*, 455–473. [CrossRef]
541. Kubicki, S.; Herrmann, W.M.; Fichte, K.; Freund, G. Reflections on the topics—EEG Frequency bands and regulation of vigilance. *Pharmakopsychiatr. Neuropsychopharmakol.* **1979**, *12*, 237–245. [CrossRef]
542. Cantero, J.L.; Atienza, M.; Gómez, C.M.; Salas, R.M. Spectral structure and brain mapping of human alpha activities in different arousal states. *Neuropsychobiology* **1999**, *39*, 110–116. [CrossRef]
543. Steriade, M. Brainstem activation of thalamocortical systems. *Brain Res. Bull.* **1999**, *50*, 391–392. [CrossRef]
544. Ng, S.C.; Raveendran, P. EEG peak alpha frequency as an indicator for physical fatigue. *IFMBE Proc.* **2007**, *16*, 517–520.
545. Huang, J.; Sander, C.; Jawinski, P.; Ulke, C.; Spada, J.; Hegerl, U.; Hensch, T. Test-retest reliability of brain arousal regulation as assessed with VIGALL 2.0. *Neuropsychiatr. Electrophysiol.* **2015**, *1*, 13. [CrossRef]
546. Stoppe, M.; Meyer, K.; Schlingmann, M.; Olbrich, S.; Bergh, F.T. Hyperstable arousal regulation in multiple sclerosis. *Psychoneuroendocrinology* **2019**, *110*, 104417. [CrossRef]

547. Barry, R.J.; De Blasio, F.M.; Fogarty, J.S.; Clarke, A.R. Natural alpha frequency components in resting EEG and their relation to arousal. *Clin. Neurophysiol.* **2020**, *131*, 205–212. [CrossRef]
548. Malone, S.M.; Burwell, S.J.; Vaidyanathan, U.; Miller, M.B.; McGue, M.; Iacono, W.G. Heritability and molecular-genetic basis of resting EEG activity: A genome-wide association study. *Psychophysiology* **2014**, *51*, 1225–1245. [CrossRef]
549. Valera, F.J.; Toro, A.; John, E.R.; Schwartz, E.L. Perceptual framing and cortical alpha rhythm. *Neuropsychologia* **1981**, *19*, 675–686. [CrossRef]
550. Busch, N.A.; Dubois, J.; VanRullen, R. The phase of ongoing EEG oscillations predicts visual perception. *J. Neurosci.* **2009**, *29*, 7869–7876. [CrossRef] [PubMed]
551. Lorincz, M.L.; Kékesi, K.A.; Juhász, G.; Crunelli, V.; Hughes, S.W. Temporal framing of thalamic relay-mode firing by phasic inhibition during the alpha rhythm. *Neuron* **2009**, *63*, 683–696. [CrossRef] [PubMed]
552. Cecere, R.; Rees, G.; Romei, V. Individual differences in alpha frequency drive crossmodal illusory perception. *Curr. Biol.* **2015**, *25*, 231–235. [CrossRef]
553. Samaha, J.; Postle, B.R. The speed of alpha-band oscillations predicts the temporal resolution of visual perception. *Curr. Biol.* **2015**, *25*, 2985–2990. [CrossRef]
554. Callaway, E.; Yeager, C.L. Relationship between reaction time and electroencephalographic alpha phase. *Science* **1960**, *132*, 1765–1766. [CrossRef]
555. Surwillo, W.W. Frequency of the "alpha" rhythm, reaction time and age. *Nature* **1961**, *191*, 823–824. [CrossRef]
556. Surwillo, W.W. The relation of simple response time to brain wave frequency and the effects of age. *Electroencephalogr. Clin. Neurophysiol.* **1963**, *15*, 105–114. [CrossRef]
557. Surwillo, W.W. The relation of decision time to brain wave frequency and to age. *Electroencephalogr. Clin. Neurophysiol.* **1964**, *16*, 510–514. [CrossRef]
558. Surwillo, W.W. The electroencephalogram in the prediction of human reaction time during growth and development. *Biol. Psychol.* **1975**, *3*, 79–90. [CrossRef]
559. Woodruff, D.S. Relationships among EEG alpha frequency, reaction time, and age: A biofeedback study. *Psychophysiology* **1975**, *12*, 673–681. [CrossRef] [PubMed]
560. Klimesch, W.; Doppelmayr, M.; Schimke, H.; Pachinger, T. Alpha frequency, reaction time, and the speed of processing information. *J. Clin. Neurophysiol.* **1996**, *13*, 511–518. [CrossRef]
561. Jin, Y.; O'Halloran, J.; Plon, L.; Sandman, C.; Potkin, S. Alpha EEG predicts visual reaction time. *Int. J. Neurosci.* **2006**, *116*, 1035–1044. [CrossRef] [PubMed]
562. Kostyunina, M.B.; Kulikov, M.A. Frequency characteristics of EEG spectra in the emotions. *Neurosci. Behav. Physiol.* **1996**, *26*, 340–343. [CrossRef] [PubMed]
563. Rodriguez, G.; Copello, F.; Vitali, P.; Perego, G.; Nobili, F. EEG spectral profile to stage Alzheimer's disease. *Clin. Neurophysiol.* **1999**, *110*, 1831–1837. [CrossRef]
564. Lansbergen, M.M.; Arns, M.; van Dongen-Boomsma, M.; Spronk, D.; Buitelaar, J.K. The increase in theta/beta ratio on resting-state EEG in boys with attention deficit/hyperactivity disorder is mediated by slow alpha peak frequency. *Prog. Neuropsychopharmacol. Biol. Psychiatr.* **2011**, *35*, 47–52. [CrossRef]
565. Northoff, G.; Duncan, N.W. How do abnormalities in the brain's spontaneous activity translate into symptoms in schizophrenia? From an overview of resting state activity findings to a proposed spatiotemporal psychopathology. *Prog. Neurobiol.* **2016**, *145–146*, 26–45. [CrossRef]
566. Yeum, T.-S.; Kang, U.G. Reduction in alpha peak frequency and coherence on quantitative electroencephalography in patients with schizophrenia. *J. Korean Med. Sci.* **2018**, *33*, e179. [CrossRef]
567. Murphy, M.; Öngür, D. Decreased peak alpha frequency and impaired visual evoked potentials in first episode psychosis. *Neuroimage Clin.* **2019**, *22*, 101693. [CrossRef]
568. Fuchs, T. Temporality and psychopathology. *Phenom. Cogn. Sci.* **2013**, *12*, 75–104. [CrossRef]
569. Stanghellini, G.; Ballerini, M.; Presenza, S.; Mancini, M.; Northoff, G.; Cutting, J. Abnormal time experiences in major depression: An empirical qualitative study. *Psychopathology* **2017**, *50*, 125–140. [CrossRef] [PubMed]
570. Northoff, G.; Magioncalda, P.; Martino, M.; Lee, H.-C.; Tseng, Y.-C.; Lane, T. Too fast or too slow? Time and neuronal variability in bipolar disorder-A combined theoretical and empirical investigation. *Schizophr. Bull.* **2018**, *44*, 54–64. [CrossRef] [PubMed]
571. Ronconi, L.; Busch, N.A.; Melcher, D. Alpha-band sensory entrainment alters the duration of temporal windows in visual perception. *Sci. Rep.* **2018**, *8*, 11810. [CrossRef] [PubMed]
572. Chota, S.; VanRullen, R. Visual entrainment at 10 Hz causes periodic modulation of the flash lag illusion. *Front. Neurosci.* **2019**, *13*, 232. [CrossRef] [PubMed]
573. Surwillo, W.W. Timing of behavior in senescence and the role of the central nervous system. In *Human Aging and Behavior*; Talland, G.A., Ed.; Academic: New York, NY, USA, 1968; pp. 1–35.
574. Surwillo, W.W.; Titus, T.G. Reaction time and the psychological refractory period in children and adults. *Dev. Psychobiol.* **1976**, *9*, 517–527. [CrossRef]
575. Wedensky, N.E. Excitation, inhibition and narcosis. In *Collected Works*; Rusinov, V.S., Ed.; LGU: Leningrad, Russia, 1953; Volume 4, pp. 517–679.
576. Golikov, N.V. *Physiological Lability and Its Changes in Basic Nervous Processes*; LGU: Leningrad, Russia, 1950.

577. Posner, M. The attention system of the human brain. *Ann. Neurosci.* **1989**, *13*, 25–42. [CrossRef]
578. Cisler, J.M.; Koster, E.H.W. Mechanisms of attentional biases towards threat in the anxiety disorders: An integrative review. *Clin. Psychol. Rev.* **2010**, *30*, 203–216. [CrossRef]
579. Inanaga, K. Frontal midline theta rhythm and mental activity. *Psychiatr. Clin. Eurosc.* **1998**, *52*, 555–566. [CrossRef]
580. Martino, M.; Magioncalda, P.; Huang, Z.; Conio, B.; Piaggio, N.; Duncan, N.W.; Rocchi, G.; Escelsior, A.; Marozzi, V.; Wolff, A.; et al. Contrasting variability patterns in the default mode and sensorimotor networks balance in bipolar depression and mania. *Proc. Natl. Acad. Sci. USA* **2016**, *113*, 4824–4829. [CrossRef]
581. Northoff, G. The brain's spontaneous activity and its psychopathological symptoms—"Spatiotemporal binding and integration". *Prog. Neuropsychopharmacol. Biol. Psychiatr.* **2018**, *80 Pt B*, 81–90. [CrossRef]
582. Northoff, G.; Tumati, S. 'Average is good, Extremes are bad'-Non-linear inverted U-shaped relationship between neural mechanisms and functionality of mental features. *Neurosci. Biobehav. Rev.* **2019**, *104*, 11–25. [CrossRef] [PubMed]
583. Kao, S.-C.; Huang, C.-J.; Hung, T.-M. Frontal midline theta is a specific indicator of optimal Attentional engagement during skilled putting performance. *J. Sport Exerc. Psychol.* **2013**, *35*, 470–478. [CrossRef] [PubMed]
584. Marchand, W.R. Self-referential thinking, suicide, and function of the cortical midline structures and striatum in mood disorders: Possible implications for treatment studies of mindfulness-based interventions for bipolar depression. *Depress. Res. Treat.* **2012**, *2012*, 246725. [CrossRef]
585. Northoff, G. Spatiotemporal psychopathology I: No rest for the brain's resting state activity in depression? Spatiotemporal psychopathology of depressive symptoms. *J. Affect. Disord.* **2016**, *190*, 854–866. [CrossRef] [PubMed]
586. Aftanas, L.l.; Golocheikine, S.A. Human anterior and frontal midline theta and lower alpha refect emotionally positive state and internalized attention: High-resolution EEG investigation of meditation. *Neurosci. Lett.* **2001**, *310*, 57–60. [CrossRef]
587. Kubota, Y.; Sato, W.; Toichi, M.; Murai, T.; Okada, T.; Hayashi, A.; Sengoku, A. Frontal midline theta rhythm is correlated with cardiac autonomic activities during the performance of an attention demanding meditation procedure. *Brain Res. Cogn. Brain Res.* **2001**, *11*, 281–287. [CrossRef]
588. Cahn, B.R.; Polich, J. Meditation states and traits: EEG, ERP, and neuroimaging studies. *Psychol. Bull.* **2006**, *132*, 180–211. [CrossRef]
589. Klimesch, W.; Sauseng, P.; Hanslmayr, S. EEG alpha oscillations: The inhibition-timing hypothesis. *Brain Res. Rev.* **2007**, *53*, 63–88. [CrossRef]
590. Doppelmayr, M.; Finkenzeller, T.; Sauseng, P. Frontal midline theta in the pre shot phase of rifle shooting: Differences between experts and novice. *Neuropsychologia* **2008**, *46*, 1463–1467. [CrossRef]
591. Fox, J.J.; Snyder, A.C. The role of alpha-band brain oscillations as a sensory suppression mechanism during selective attention. *Front. Psychol.* **2011**, *2*, 154. [CrossRef] [PubMed]
592. Konareva, I.N. Locus of psychological control and characteristics of the EEG frequency components. *Neurophysiology* **2011**, *43*, 534–542. [CrossRef]
593. Arns, M.; Etkin, A.; Hegerl, U.; Williams, L.M.; DeBattista, C.; Palmer, D.M.; Fitzgerald, P.B.; Harris, A.; deBeuss, R.; Gordon, E. Frontal and rostral anterior cingulate (rACC) theta EEG in depression: Implications for treatment outcome? *Eur. Neuropsychopharmacol.* **2015**, *25*, 1190–1200. [CrossRef] [PubMed]
594. Smith, E.E.; Reznik, S.J.; Stewart, J.L.; Allen, J.J.B. Assessing and conceptualizing frontal EEG asymmetry: An updated primer on recording, processing, analyzing, and interpreting frontal alpha asymmetry. *Int. J. Psychophysiol.* **2017**, *111*, 98–114. [CrossRef]
595. Wheeler, R.E.; Davidson, R.J.; Tomarken, A.J. Frontal brain asymmetry and emotional reactivity: A biological substrate of affective style. *Psychophysiology* **1993**, *30*, 82–89. [CrossRef]
596. Davidson, R.J. Affective style and affective disorders: Perspectives from affective neuroscience. *Cogn. Emot.* **1998**, *12*, 307–330. [CrossRef]
597. Blackhart, G.C.; Minnix, J.A.; Kline, J.P. Can EEG asymmetry patterns predict future development of anxiety and depression? *Biol. Psychol.* **2006**, *72*, 46–50. [CrossRef]
598. Nusslock, R.; Shackman, A.J.; Harmon-jones, E.; Alloy, L.B.; Coan, J.A.; Abramson, L.Y. Cognitive vulnerability and frontal brain asymmetry: Common predictors of first prospective depressive episode. *J. Abnorm. Psychol.* **2011**, *120*, 497–503. [CrossRef]
599. Gollan, J.K.; Hoxha, D.; Chihade, D.; Pflieger, M.E.; Rosebrock, L.; Cacioppo, J. Frontal alpha EEG asymmetry before and after behavioral activation treatment for depression. *Biol. Psychol.* **2014**, *99*, 198–208. [CrossRef]
600. Lahey, B.B.; Krueger, R.F.; Rathouz, P.J.; Waldman, I.D.; Zald, D.H. A hierarchical causal taxonomy of psychopathology across the life span. *Psychol. Bull.* **2017**, *143*, 142–186. [CrossRef]
601. Heller, W. Neuropsychological mechanisms of individual differences in emotion, personality, and arousal. *Neuropsychology* **1993**, *7*, 476–489. [CrossRef]
602. Quaedflieg, C.W.E.M.; Meyer, T.; Smulders, F.T.Y.; Smeets, T. The functional role of individual-alpha based frontal asymmetry in stress responding. *Biol. Psychol.* **2015**, *104*, 75–81. [CrossRef] [PubMed]
603. Coan, J.A.; Allen, J.J.; Harmon-Jones, E. Voluntary facial expression and hemispheric asymmetry over the frontal cortex. *Psychophysiology* **2001**, *38*, 912–925. [CrossRef] [PubMed]
604. Thibodeau, R.; Jorgensen, R.S.; Kim, S. Depression, anxiety, and resting frontal EEG asymmetry: A meta-analytic review. *J. Abnorm. Psychol.* **2006**, *115*, 715–729. [CrossRef] [PubMed]

605. White, J.; Kivimaki, M.; Jokela, M.; Batty, G.D. Association of inflammation with specific symptoms of depression in a general population of older people: The English Longitudinal Study of Ageing. *Brain Behav. Immun.* **2017**, *61*, 27–30. [CrossRef] [PubMed]
606. Dawson, G.; Frey, K.; Panagiotides, H.; Osterling, J.; Hessl, D. Infants of depressed mothers exhibit atypical frontal brain activity: A replication and extension of previous findings. *J. Child Psychol. Psychiatr.* **1997**, *38*, 179–186. [CrossRef]
607. Robinson, R.G.; Downhill, J.E. Lateralization of psychopathology in response to focal brain injury. In *Brain Asymmetry*; Davidson, R.J., Hugdahl, K., Eds.; The MIT Press: Cambridge, MA, USA, 1995; pp. 693–711.
608. Davidson, R.J. Affective style, psychopathology, and resilience: Brain mechanisms and plasticity. *Am. Psychol.* **2000**, *55*, 1196–1214. [CrossRef]
609. Harmon-Jones, E.; Sigelman, J. State anger and prefrontal brain activity: Evidence that insult-related relative left-prefrontal activation is associated with experienced anger and aggression. *J. Personal. Soc. Psychol.* **2001**, *80*, 797–803. [CrossRef]
610. Harmon-Jones, E.; Peterson, C.K.; Harris, C.R. Jealousy: Novel methods and neural correlates. *Emotion* **2009**, *9*, 113–117. [CrossRef]
611. Drake, R.A.; Ulrich, G. Line bisecting as a predictor of personal optimism and desirability of risky behaviors. *Acta Psychol.* **1992**, *79*, 219–226. [CrossRef]
612. De Pascalis, V.; Cozzuto, G.; Caprara, G.V.; Alessandri, G. Relations among EEG alpha asymmetry, BIS/BAS, and dispositional optimism. *Biol. Psychol.* **2013**, *94*, 198–209. [CrossRef] [PubMed]
613. Terzian, H. Behavioural and EEG effects of intracarotid sodium amytal injection. *Acta Neurochir.* **1964**, *12*, 230–239. [CrossRef] [PubMed]
614. Gainotti, G. Emotional behavior and hemispheric side of the lesion. *Cortex* **1972**, *8*, 41–55. [CrossRef]
615. Henriques, J.B.; Davidson, R.J. Left Frontal Hypoactivation in Depression. *J. Abnorm. Psychol.* **1991**, *100*, 535–545. [CrossRef]
616. Davidson, R.J. Asymmetric brain function, affective style, and psychopathology: The role of early experience and plasticity. *Dev. Psychopathol.* **1994**, *6*, 741–758. [CrossRef]
617. Coan, J.A.; Allen, J.J.B. Frontal EEG asymmetry as a moderator and mediator of emotion. *Biol. Psychol.* **2004**, *67*, 7–49. [CrossRef]
618. Schutter, D.J.L.G. Antidepressant efficacy of high-frequency transcranial magnetic stimulation over the left dorsolateral prefrontal cortex in double-blind sham-controlled designs: A meta-analysis. *Psychol. Med.* **2009**, *39*, 65–75. [CrossRef]
619. Choi, S.W.; Chi, S.E.; Chung, S.Y.; Kim, J.W.; Ahn, C.Y.; Kim, H.T. Is alpha wave neurofeedback effective with randomized clinical trials in depression? A pilot study. *Neuropsychobiology* **2010**, *63*, 43–51. [CrossRef]
620. Schutter, D.J.L.G. Quantitative review of the efficacy of slow-frequency magnetic brain stimulation in major depressive disorder. *Psychol. Med.* **2010**, *40*, 1789–1795. [CrossRef]
621. Stewart, J.L.; Bismark, A.W.; Towers, D.N.; Coan, J.A.; Allen, J.J. Resting frontal EEG asymmetry as an endophenotype for depression risk: Sex-specific patterns of frontal brain asymmetry. *J. Abnorm. Psychol.* **2010**, *119*, 502–512. [CrossRef]
622. Hortensius, R.; Schutter, D.J.L.G.; Harmon-Jones, E. When anger leads to aggression: Induction of relative left frontal cortical activity with transcranial direct current stimulation increases the anger–aggression relationship. *Soc. Cogn. Affect. Neurosci.* **2012**, *7*, 342–347. [CrossRef] [PubMed]
623. Smit, D.J.A.; Posthuma, D.; Boomsma, D.I.; De Geus, E.J. The relation between frontal EEG asymmetry and the risk for anxiety and depression. *Biol. Psychol.* **2007**, *74*, 26–33. [CrossRef] [PubMed]
624. Fishman, I.; Ng, R.; Bellugi, U. Do extraverts process social stimuli differently from introverts? *Cogn. Neurosci.* **2011**, *2*, 67–73. [CrossRef] [PubMed]
625. Gray, J.A. The psychophysiological basis of introversion-extraversion. *Behav. Res. Ther.* **1970**, *8*, 249–266. [CrossRef]
626. Carver, C.S.; White, T.L. Behavioral inhibition behavioral activation and affective responses to impending reward and punishment: The BIS/BAS scales. *J. Personal. Soc. Psychol.* **1994**, *67*, 319–333. [CrossRef]
627. Matthews, G.; Gilliland, K. The personality theories of H.J. Eysenck and J.A. Gray: A comparative review. *Personal. Individ. Differ.* **1999**, *26*, 583–626. [CrossRef]
628. Brocke, B.; Battmann, W. The arousal-activation theory of extraversion and neuroticism: A systematic analysis and principal conclusions. *Adv. Behav. Res. Ther.* **1992**, *14*, 211–246. [CrossRef]
629. McAdams, D.P.; Pals, J.L. A new big five. *Am. Psychol.* **2006**, *61*, 204–217. [CrossRef]
630. Yamasue, H.; Abe, O.; Suga, M.; Yamada, H.; Inoue, H.; Tochigi, M.; Rogers, M.; Aoki, S.; Kato, N.; Kasai, K. Gender-common and –specific neuroanatomical basis of human anxiety-related personality traits. *Cereb. Cortex* **2008**, *18*, 46–52. [CrossRef]
631. Zuckerman, M. *Sensation Seeking: Beyond the Optimal Level of Arousal*; Distributed by the Halsted Press Division of Wiley; L. Erlbaum Associates: Hillsdale, NJ, USA, 1979; p. 449.
632. Samuel, D.B.; Widiger, T.A. A meta-analytic review of the relationships between the five-factor model and DSM-IV-TR personality disorders: A facet level analysis. *Clin. Psychol. Rev.* **2008**, *28*, 1326–1342. [CrossRef] [PubMed]
633. Watson, D.; Stasik, S.M.; Ellickson-Larew, S.; Stanton, K. Extraversion and psychopathology: A facet-level analysis. *J. Abnorm. Psychol.* **2015**, *124*, 432–446. [CrossRef] [PubMed]
634. Watson, D.; Ellickson-Larew, S.; Stanton, K.; Levin-Aspenson, H.F.; Khoo, S.; Stasik-O'Brien, S.M.; Clark, L.A. Aspects of extraversion and their associations with psychopathology. *J. Abnorm. Psychol.* **2019**, *128*, 777–794. [CrossRef] [PubMed]
635. Watson, D.; Stanton, K.; Khoo, S.; Ellickson-Larew, S.; Stasik-O'Brien, S.M. Extraversion and psychopathology: A multilevel hierarchical review. *J. Res. Personal.* **2019**, *81*, 1–10. [CrossRef]
636. O'Gorman, J.G.; Lloyd, J.E.M. Extraversion, impulsiveness, and EEG alpha activity. *Personal. Individ. Differ.* **1987**, *8*, 169–174. [CrossRef]

637. Wall, T.L.; Schuckit, M.A.; Mungas, D.; Ehlers, C.L. EEG alpha activity and personality traits. *Alcohol* **1990**, *7*, 461–464. [CrossRef]
638. Knyazev, G.G. Is cortical distribution of spectral power a stable individual characteristic? *Int. J. Psychophysiol.* **2009**, *72*, 123–133. [CrossRef]
639. Sampaio, A.; Soares, J.M.; Coutinho, J.; Sousa, N.; Goncalves, O.F. The Big Five default brain: Functional evidence. *Brain Struct. Funct.* **2014**, *219*, 1913–1922. [CrossRef]
640. Ikeda, S.; Takeuchi, H.; Taki, Y.; Nouchi, R.; Yokoyama, R.; Kotozaki, Y.; Nakagawa, S.; Sekiguchi, A.; Iizuka, K.; Yamamoto, Y.; et al. A Comprehensive analysis of the correlations between resting-state oscillations in multiple-frequency bands and Big Five traits. *Front. Hum. Neurosci.* **2017**, *11*, 321. [CrossRef]
641. Gianotti, L.R.R.; Dahinden, F.M.; Baumgartner, T.; Knoch, D. Understanding individual differences in domain-general prosociality: A resting EEG study. *Brain Topogr.* **2019**, *32*, 118–126. [CrossRef]
642. Roslan, N.S.; Izhar, L.I.; Faye, I.; Amin, H.U.; Saad, M.N.M.; Sivapalan, S.; Karim, S.A.A.; Rahman, M.A. Neural correlates of eye contact in face-to-face verbal interaction: An EEG-based study of the extraversion personality trait. *PLoS ONE* **2019**, *14*, e0219839. [CrossRef]
643. Martens, R.; Burton, D.; Vealey, R.; Bump, L.; Smith, D. The development of the competitive state anxiety inventory-2 (CSAI-2). In *Competitive Anxiety in Sport*; Martens, R., Vealey, R.S., Burton, D., Eds.; Human Kinetics: Chapaign, IL, USA, 1990; pp. 117–190.
644. Clark, L.A.; Watson, D. Tripartite model of anxiety and depression: Psychometric evidence and taxonomic implications. *J. Abnorm. Psychol.* **1991**, *100*, 316–336. [CrossRef] [PubMed]
645. Mathersul, D.; Williams, L.; Hopkinson, P.; Kemp, A. Investigating models of affect: Relationships among EEG alpha asymmetry, depression, and anxiety. *Emotion* **2008**, *8*, 560–572. [CrossRef] [PubMed]
646. Barlow, D.H. Disorders of emotion. *Psychol. Inq.* **1991**, *2*, 58–71. [CrossRef]
647. Watson, D. Rethinking the mood and anxiety disorders: A quantitative hierarchical model for DSM-V. *J. Abnorm. Psychol.* **2005**, *114*, 522–536. [CrossRef] [PubMed]
648. Nusslock, R.; Shackman, A.; McMenamin, B.; Greischar, L.; Davidson, R.; Kovacs, M. Comorbid anxiety moderates the relationship between depression history and prefrontal EEG asymmetry. *Psychophysiology* **2018**, *55*, 12953. [CrossRef]
649. Heller, W.; Nitschke, J.B.; Etienne, M.A.; Miller, G.A. Patterns of regional brain activity differentiate types of anxiety. *J. Abnorm. Psychol.* **1997**, *106*, 376. [CrossRef]
650. Wiedemann, G.; Pauli, P.; Dengler, W.; Lutzenberger, W.; Birbaumer, N.; Buchkremer, G. Frontal brain asymmetry as a biological substrate of emotions in patients with panic disorders. *Arch. Gen. Psychiatr.* **1999**, *56*, 78–84. [CrossRef]
651. Knyazev, G.G.; Savostyanov, A.N.; Levin, E.A. Alpha oscillations as a correlate of trait anxiety. *Int. J. Psychophysiol.* **2004**, *53*, 147–160. [CrossRef]
652. Knyazev, G.G.; Savostyanov, A.N.; Levin, E.A. Alpha synchronization and anxiety: Implications for inhibition vs. alertness hypotheses. *Int. J. Psychophysiol.* **2006**, *59*, 151–158. [CrossRef]
653. Wise, V.; McFarlane, A.C.; Clark, C.R.; Battersby, M. An integrative assessment of brain and body function "at rest" in panic disorder: A combined quantitative EEG/autonomic function study. *Int. J. Psychophysiol.* **2011**, *79*, 155–165. [CrossRef]
654. Yehuda, R.; Flory, J.D.; Pratchett, L.C.; Buxbaum, J.; Ising, M.; Holsboer, F. Putative biological mechanisms for the association between early life adversity and the subsequent development of PTSD. *Psychopharmacology* **2010**, *212*, 405–417. [CrossRef] [PubMed]
655. Mehta, D.; Klengel, T.; Conneely, K.N.; Smith, A.K.; Altmann, A.; Pace, T.W.; Rex-Haffner, M.; Loeschner, A.; Gonik, M.; Mercer, K.B.; et al. Childhood maltreatment is associated with distinct genomic and epigenetic profiles in posttraumatic stress disorder. *Proc. Natl. Acad. Sci. USA* **2013**, *110*, 8302–8307. [CrossRef] [PubMed]
656. Miller, B.F.; Seals, D.R.; Hamilton, K.L. A viewpoint on considering physiological principles to study stress resistance and resilience with aging. *Ageing Res. Rev.* **2017**, *38*, 1–5. [CrossRef] [PubMed]
657. Fleshner, M.; Maier, S.F.; Lyons, D.M.; Raskind, M.A. The neurobiology of the stress-resistant brain. *Stress* **2011**, *14*, 498–502. [CrossRef]
658. McCrory, E.J.; De Brito, S.A.; Sebastian, C.L.; Mechelli, A.; Bird, G.; Kelly, P.A.; Viding, E. Heightened neural reactivity to threat in child victims of family violence. *Curr. Biol.* **2011**, *21*, R947–R948. [CrossRef]
659. Zenkov, L.R.; Ronkin, M.A. *Functional Diagnosis of Nervous Diseases*; Medicine: Moscow, Russia, 1982; p. 432.
660. Alonso, J.F.; Romero, S.; Ballester, M.R.; Antonijoan, R.M.; Mañanas, M.A. Stress assessment based on EEG univariate features and functional connectivity measures. *Physiol. Meas.* **2015**, *36*, 1351–1365. [CrossRef]
661. Choi, Y.; Kim, M.; Chun, C. Measurement of occupants' stress based on electroencephalograms (EEG) in twelve combined environments. *Build. Environ.* **2015**, *88*, 65–72. [CrossRef]
662. Marshall, A.C.; Cooper, N.R.; Segrave, R.; Geeraert, N. The effects of long-term stress exposure on aging cognition: A behavioral and EEG investigation. *Neurobiol. Aging* **2015**, *36*, 2136–2144. [CrossRef]
663. Al-Shargie, F.; Kiguchi, M.; Badruddin, N.; Dass, S.C.; Hani, A.F.M.; Tang, T.B. Mental stress assessment using simultaneous measurement of EEG and fNIRS. *Biomed. Opt. Express* **2016**, *7*, 3882–3898. [CrossRef]
664. Stern, Y. What is cognitive reserve? Theory and research application of the reserve concept. *J. Int. Neuropsychol. Soc.* **2002**, *8*, 448–460. [CrossRef]
665. Bartrés-Faz, D.; Arenaza-Urquijo, E.M. Structural and functional imaging correlates of cognitive and brain reserve hypotheses in healthy and pathological aging. *Brain Topogr.* **2011**, *24*, 340–357. [CrossRef] [PubMed]

666. Medaglia, J.D.; Pasqualetti, F.; Hamilton, R.H.; Thompson-Schill, S.L.; Bassett, D.S. Brain and cognitive reserve: Translation via network control theory. *Neurosci. Biobehav. Rev.* **2017**, *75*, 53–64. [CrossRef] [PubMed]
667. Satz, P. Brain reserve capacity on symptom onset after brain injury: A formulation and review of evidence for threshold theory. *Neuropsychology* **1993**, *7*, 273–295. [CrossRef]
668. Perneczky, R.; Green, R.C.; Kurz, A. Head circumference, atrophy, and cognition: Implications for brain reserve in Alzheimer disease. *Neurology* **2010**, *75*, 137–142. [CrossRef]
669. López, M.E.; Aurtenetxe, S.; Pereda, E.; Cuesta, P.; Castellanos, N.P.; Bruña, R.; Niso, G.; Maestú, F.; Bajo, R. Cognitive reserve is associated with the functional organization of the brain in healthy aging: A MEG study. *Front. Aging Neurosci.* **2014**, *6*, 125.
670. Stern, Y.; Habeck, C.; Moeller, J.; Scarmeas, N.; Anderson, K.E.; Hilton, H.J.; Flynn, J.; Sackeim, H.; van Heertum, R. Brain networks associated with cognitive reserve in healthy young and old adults. *Cereb. Cortex* **2005**, *15*, 394–402. [CrossRef]
671. Fingelkurts, A.A.; Fingelkurts, A.A. Turning back the clock: A retrospective study on brain age change in response to nutraceuticals supplementation vs. lifestyle modifications. *Mediterr. J. Nutr. Metab.* **2022**. [CrossRef]
672. Fratiglioni, L.; Wang, H. Brain reserve hypothesis in dementia. *J. Alzheimers. Dis.* **2007**, *12*, 11–22. [CrossRef]
673. Stern, Y. Cognitive reserve. *Neuropsychologia* **2009**, *47*, 2015–2028. [CrossRef]
674. Cespón, J.; Miniussi, C.; Pellicciari, M.C. Interventional programs to improve cognition during healthy and pathological ageing: Cortical modulations and evidence for brain plasticity. *Ageing Res. Rev.* **2018**, *43*, 81–98. [CrossRef]
675. Nunez, P.L.; Wingeier, B.M.; Silberstein, R.B. Spatial–temporal structures of human alpha rhythms: Theory, microcurrent sources, multiscale measurements, and global binding of local networks. *Hum. Brain Mapp.* **2001**, *13*, 125–164. [CrossRef] [PubMed]
676. Jellinger, K.A. The pathology of ischemic-vascular dementia: An update. *J. Neurol. Sci.* **2002**, *203–204*, 153–157. [CrossRef]
677. Dockree, P.M.; Kelly, S.P.; Foxe, J.J.; Reilly, R.B.; Robertson, I.H. Optimal sustained attention is linked to the spectral content of background EEG activity: Greater ongoing tonic alpha (10 Hz) power supports successful phasic goal activation. *Eur. J. Neurosci.* **2007**, *25*, 900–907. [PubMed]
678. Arnau, S.; Möckel, T.; Rinkenauer, G.; Wascher, E. The interconnection of mental fatigue and aging: An EEG study. *Int. J. Psychophysiol.* **2017**, *117*, 17–25. [CrossRef]
679. Zhirmunskaya, E.A.; Makarova, G.V. Relation of separate waves' mean level of fronts' asymmetry to structure of human EEG. In *Functional States of the Brain*; Sokolov, E.N., Danilova, N.N., Khomskaya, E.D., Eds.; Moscow University Press: Moscow, Russia, 1975; pp. 113–118. (In Russian)
680. Raichle, M.E.; Snyder, A.Z. A default mode of brain function: A brief history of an evolving idea. *NeuroImage* **2007**, *37*, 1083–1090. [CrossRef]
681. Deco, G.; Jirsa, V.K.; McIntosh, A.R. Emerging concepts for the dynamical organization of resting-state activity in the brain. *Nat. Rev. Neurosci.* **2011**, *12*, 43–56. [CrossRef]
682. Allen, T.A.; DeYoung, C.G. Personality neuroscience and the five-factor model. In *The Oxford Handbook of the Five-Factor Model*; Widiger, T.A., Ed.; Oxford University Press: New York, NY, USA, 2017; pp. 319–352. [CrossRef]
683. Klimesch, W. EEG-alpha rhythms and memory processes. *Int. J. Psychophysiol.* **1997**, *26*, 319–340. [CrossRef]
684. Klimesch, W.; Freunberger, R.; Sauseng, P. Oscillatory mechanisms of process binding in memory. *Neurosci. Biobehav. Rev.* **2010**, *34*, 1002–1014. [CrossRef]
685. Müller, V.I.; Langner, R.; Cieslik, E.C.; Rottschy, C.; Eickhoff, S.B. Interindividual differences in cognitive flexibility: Influence of gray matter volume, functional connectivity and trait impulsivity. *Brain Struct. Funct.* **2015**, *220*, 2401–2414. [CrossRef]
686. Mattar, M.G.; Wymbs, N.F.; Bock, A.S.; Aguirre, G.K.; Grafton, S.T.; Bassett, D.S. Predicting future learning from baseline network architecture. *NeuroImage* **2018**, *172*, 107–117. [CrossRef]
687. Ames, A.I. CNS energy metabolism as related to function. *Brain Res. Rev.* **2000**, *34*, 42–68. [PubMed]
688. Lennie, P. The cost of cortical computation. *Curr. Biol.* **2003**, *13*, 493–497. [CrossRef]
689. Raichle, M.E.; Mintun, M.A. Brain work and brain imaging. *Annu. Rev. Neurosci.* **2006**, *29*, 449–476. [CrossRef]
690. Freeman, W.J. Origin, structure, and role of background EEG activity. Part 2. Analytic phase. *Clin. Neurophysiol.* **2004**, *115*, 2089–2107. [CrossRef]
691. Laufs, H.; Krakow, K.; Sterzer, P.; Eger, E.; Beyerle, A.; Salek-Haddadi, A.; Kleinschmidt, A. Electroencephalographic signatures of attentional and cognitive default modes in spontaneous brain activity fluctuations at rest. *Proc. Natl. Acad. Sci. USA* **2003**, *100*, 11053–11058. [CrossRef] [PubMed]
692. Fox, M.D.; Raichle, M.E. Spontaneous fluctuations in brain activity observed with functional magnetic resonance imaging. *Nat. Rev. Neurosci.* **2007**, *8*, 700–711.
693. Landers, D.M.; Boutcher, S.H. Arousal-performance relationships. In *Applied Sport Psychology: Personal Growth to Peak Performance*, 2nd ed.; Williams, J.M., Ed.; Mayfield Publishing Co.: Mayfield, CA, USA, 1993; pp. 170–184.
694. Hockey, G.R.J.; Hamilton, P. The cognitive patterning of stress states. In *Stress and Fatigue in Human Performance*; Hockey, G.R.T., Ed.; John Wiley & Sons: Chichester, UK, 1983; pp. 331–362.
695. Pariitt, C.G.; Jones, J.G.; Hardy, L. Multidimensional anxiety and performance. In *Stress and Performance in Sport*; Jones, J.G., Hardy, L., Eds.; John Wiley & Sons: Chichester, UK, 1990; pp. 43–80.
696. Could, D.; Weinberg, R.S. Sources of worry in junior elite wrestlers. *J. Sport Behav.* **1985**, *8*, 115–127.
697. Fingelkurts, A.A.; Fingelkurts, A.A.; Neves, C.F.H. Neuro-assessment of leadership training. *Coaching* **2020**, *13*, 107–145. [CrossRef]

698. Urry, H.L.; Nitschke, J.B.; Dolski, I.; Jackson, D.C.; Dalton, K.M.; Mueller, C.J.; Rosenkranz, M.A.; Ryff, C.D.; Singer, B.H.; Davidson, R.J. Making a life worth living: Neural correlates of well-being. *Psychol. Sci.* **2004**, *15*, 367–372. [CrossRef]
699. King, M.L. The neural correlates of well-being: A systematic review of the human neuroimaging and neuropsychological literature. *Cogn. Affect. Behav. Neurosci.* **2019**, *19*, 779–796. [CrossRef]
700. Cole, C.; Oetting, E.R.; Hinkle, J. Non-linearity of self-concept discrepancy: The value dimension. *Psychol. Rep.* **1967**, *21*, 58–60. [CrossRef] [PubMed]
701. Sonstroem, R.J.; Bernardo, P. Intraindividual pregame state anxiety and basketball performance: A re-examination of the inverted-U curve. *J. Sport Psychol.* **1982**, *4*, 235–245. [CrossRef]
702. Raglin, J.S.; Turner, P.E. Anxiety and performance in track and field athletes: A comparison of the inverted-U hypothesis with zone of optimal function theory. *Personal. Individ. Differ.* **1993**, *14*, 163–171. [CrossRef]
703. Murphy, P.R.; Vandekerckhove, J.; Nieuwenhuis, S. Pupil-linked arousal determines variability in perceptual decision making. *PLoS Comput. Biol.* **2014**, *10*, e1003854. [CrossRef] [PubMed]
704. McGinley, M.J.; David, S.V.; McCormick, D.A. Cortical membrane potential signature of optimal states for sensory signal detection. *Neuron* **2015**, *87*, 179–192. [CrossRef]
705. McGinley, M.J.; Vinck, M.; Reimer, J.; Batista-Brito, R.; Zagha, E.; Cadwell, C.R.; Tolias, A.S.; Cardin, J.A.; McCormick, D.A. Waking state: Rapid variations modulate neural and behavioral responses. *Neuron* **2015**, *87*, 1143–1161. [CrossRef]
706. Cavanagh, M. Mental-health issues and challenging clients in executive coaching. In *Evidence-Based Coaching: Theory, Research and Practice from the Behavioural Sciences*; Cavanagh, M., Grant, A.M., Kemp, T., Eds.; Australian Academic Press: Bowen Hills, QLD, Australia, 2005; Volume 1, pp. 21–36.
707. Sokolov, E.N.; Danilova, N.N.; Khomskaya, E.D. *Functional States of the Brain*; Moscow University Press: Moscow, Russia, 1975.
708. Cherniy, T.V. Application of method of EEG integral quantitative analysis for the estimation of zonal distinctions of electroencephalograms, laid in a concept of 'ideal norm'. In *Questions of Experimental and Clinical Medicine*; Collection of Articles; 2010; Volume 14, pp. 116–129.
709. Cona, G.; Koçillari, L.; Palombit, A.; Bertoldo, A.; Maritan, A.; Corbetta, M. Archetypes in human behavior and their brain correlates: An evolutionary trade-off approach. *bioRxiv Prepr.* **2018**; *first posted online Ma. 18*. [CrossRef]
710. Liberman, E.A.; Minina, S.V.; Shklovsky-Kordi, N.E. Quantum molecular computer model of the neuron and a pathway to the union of the sciences. *BioSystems* **1989**, *22*, 135–154. [CrossRef]
711. Igamberdiev, A.U.; Shklovskiy-Kordi, N.E. The quantum basis of spatiotemporality in perception and consciousness. *Prog. Biophys. Mol. Biol.* **2017**, *130 Pt A*, 15–25. [CrossRef]
712. Gorelick, P.B.; Furie, K.L.; Iadecola, C.; Smith, E.E.; Waddy, S.P.; Lloyd-Jones, D.M.; Bae, H.J.; Bauman, M.A.; Dichgans, M.; Duncan, P.W.; et al. Defining optimal brain health in adults: A presidential advisory from the American heart association/American stroke association. *Stroke* **2017**, *48*, e284–e303. [CrossRef]
713. Rudrauf, D. Structure-Function relationships behind the phenomenon of cognitive resilience in neurology: Insights for neuroscience and medicine. *Adv. Neurosci.* **2014**, *2014*, 462765. [CrossRef]
714. Thatcher, R.W.; Lubar, J.F. History of the scientific standards of QEEG normative databases. In *Introduction to QEEG and Neurofeedback: Advanced Theory and Applications*; Budzinsky, T., Budzinski, H., Evans, J., Abarbanel, A., Eds.; Academic Press: San Diego, CA, USA, 2008; pp. 29–62.
715. Hellhammer, D.; Meinlschmidt, G.; Pruessner, J.C. Conceptual endophenotypes: A strategy to advance the impact of psychoneuroendocrinology in precision medicine. *Psychoneuroendocrinology* **2018**, *89*, 147–160. [CrossRef]
716. McEwen, B.S. Stress, adaptation, and disease: Allostasis and allostatic load. *Ann. N. Y. Acad. Sci.* **1998**, *840*, 33–44. [CrossRef]
717. Sterling, P.; Eyer, J. Allostasis: A new paradigm to explain arousal pathology. In *Handbook of Life Stress, Cognition and Health*; Fisher, S., Reason, J., Eds.; Wiley: Chichester, UK, 1988; pp. 629–649.
718. McEwen, B.S. Allostasis and allostatic load: Implications for neuropsychopharmacology. *Neuropsychopharmacology* **2000**, *22*, 108–124. [CrossRef]
719. McEwen, B.S.; Wingfield, J.C. The concept of allostasis in biology and biomedicine. *Horm. Behav.* **2003**, *43*, 2–15. [CrossRef]
720. Anderzhanova, E.; Kirmeier, T.; Wotjak, C.T. Animal models in psychiatric research: The RDoC system as a new framework for endophenotype-oriented translational neuroscience. *Neurobiol. Stress* **2017**, *7*, 47–56. [CrossRef]
721. Kalueff, A.V.; Ren-Patterson, R.F.; LaPorte, J.L.; Murphy, D.L. Domain interplay concept in animal models of neuropsychiatric disorders: A new strategy for high-throughput neurophenotyping research. *Behav. Brain Res.* **2008**, *188*, 243–249. [CrossRef]
722. Kozel, F.A. Identifying phronotypes in psychiatry. *Front. Psychiatr.* **2010**, *1*, 141, eCollection 2010. [CrossRef]
723. Gunkelman, J. Transcend the DSM using phenotypes. *Biofeedback* **2006**, *34*, 95–98.
724. Krueger, R.F.; Markon, K.E. Reinterpreting comorbidity: A model-based approach to understanding and classifying psychopathology. *Annu. Rev. Clin. Psychol.* **2006**, *2*, 111–133. [CrossRef]
725. Insel, T.; Cuthbert, B.; Garvey, M.; Heinssen, R.; Pine, D.S.; Quinn, K.; Sanislow, C.; Wang, P. Research domain criteria (RDoC): Toward a new classification framework for research on mental disorders. *Am. J. Psychiatr.* **2010**, *167*, 748–751. [CrossRef]
726. Krueger, R.F.; Markon, K.E. A dimensional-spectrum model of psychopathology: Progress and opportunities. *Arch. Gen. Psychiatr.* **2011**, *68*, 10–11. [CrossRef]
727. Kapur, S.; Phillips, A.G.; Insel, T.R. Why has it taken so long for biological psychiatry to develop clinical tests and what to do about it? *Mol. Psychiatr.* **2012**, *17*, 1174–1179. [CrossRef]

728. Casey, B.J.; Craddock, N.; Cuthbert, B.N.; Hyman, S.E.; Lee, F.S.; Ressler, K.J. DSM-5 and RDoC: Progress in psychiatry research? *Nat. Rev. Neurosci.* **2013**, *14*, 810–814. [CrossRef]
729. Kirmayer, L.J.; Crafa, D. What kind of science for psychiatry? *Front. Hum. Neurosci.* **2014**, *8*, 435. [CrossRef]

Perspective

Decoding of Processing Preferences from Language Paradigms by Means of EEG-ERP Methodology: Risk Markers of Cognitive Vulnerability for Depression and Protective Indicators of Well-Being? Cerebral Correlates and Mechanisms

Cornelia Herbert

Applied Emotion and Motivation Psychology, Ulm University, 89081 Ulm, Germany; cornelia.herbert@uni-ulm.de

Abstract: Depression is a frequent mental affective disorder. Cognitive vulnerability models propose two major cognitive risk factors that favor the onset and severity of depressive symptoms. These include a pronounced self-focus, as well as a negative emotional processing bias. According to two-process models of cognitive vulnerability, these two risk factors are not independent from each other, but affect information processing already at an early perceptual processing level. Simultaneously, a processing advantage for self-related positive information including better memory for positive than negative information has been associated with mental health and well-being. This perspective paper introduces a research framework that discusses how EEG-ERP methodology can serve as a standardized tool for the decoding of negative and positive processing biases and their potential use as risk markers of cognitive vulnerability for depression, on the one hand, and as protective indicators of well-being, on the other hand. Previous results from EEG-ERP studies investigating the time-course of self-referential emotional processing are introduced, summarized, and discussed with respect to the specificity of depression-related processing and the importance of EEG-ERP-based experimental testing for well-being and the prevention and treatment of depressive disorders.

Keywords: cognitive vulnerability; depression; self-referential processing; emotion processing; language; event-related brain potentials

Citation: Herbert, C. Decoding of Processing Preferences from Language Paradigms by Means of EEG-ERP Methodology: Risk Markers of Cognitive Vulnerability for Depression and Protective Indicators of Well-Being? Cerebral Correlates and Mechanisms. *Appl. Sci.* **2022**, *12*, 7740. https://doi.org/10.3390/app12157740

Academic Editors: Fabio La Foresta and Serena Dattola

Received: 6 June 2022
Accepted: 22 July 2022
Published: 1 August 2022

Publisher's Note: MDPI stays neutral with regard to jurisdictional claims in published maps and institutional affiliations.

Copyright: © 2022 by the author. Licensee MDPI, Basel, Switzerland. This article is an open access article distributed under the terms and conditions of the Creative Commons Attribution (CC BY) license (https://creativecommons.org/licenses/by/4.0/).

1. Introduction

Affective disorders such as depression are among the most common mental disorders in Western industrialized societies, and are on the rise worldwide. According to the WHO (World Health Organization), already more than 300 million people worldwide currently suffer from a depressive disorder [1]. The lifetime prevalence is 16–20% for both sexes [1]. The onset of major depressive disorder is early in life, starting during adolescence, and affecting emerging adults, if untreated, during their whole lifespan [2]. Due to its high prevalence, as well as due to its multifactorial negative consequences for mental and physical health, the prevention of depressive symptoms is of outstanding clinical and health economic importance [1]. Indeed, depressive disorder affects the whole person—the body, the brain, and the mind—and related affective, motivational, and cognitive processing [3].

Regarding emotional processing, depressed individuals are found to sustain attention to negative stimuli, have difficulties in the inhibition of negative information, and, compared to non-depressed individuals, recall negative information better than positive information, e.g., [4,5]. These processing biases, described in the literature as negativity bias, have been repeatedly reported in studies investigating acute or remitted depressed individuals. Moreover, the negativity bias has been confirmed in vulnerable individuals, i.e., individuals at risk of depression e.g., see [4–9]. According to cognitive vulnerability models, two major cognitive risk factors can be identified that promote the occurrence and severity of depressive symptoms and the development of depressive symptoms prior to a clinical diagnosis of major depressive disorder [10]. These cognitive risk factors include an increased self-focus and a negative emotional processing bias [10,11]. Thus, according to cognitive vulnerability models, for the diagnostic decision of whether processing preferences for negative stimuli are specific risk markers of depressive symptoms, it is essential to determine whether these processing biases occur in reference to one's own self. Theoretically, a pronounced self-reference or self-focus, negative self-schemas, and negative attitudes towards the self are among the core characteristics of a depressive personality structure, and they are important causal disease factors characterizing depressed thinking in cognitive theories of depression [12,13]. Accordingly, cognitive vulnerability models of depression (for an overview, see [10]) assume that the aforementioned cognitive biases exist prior to the onset of depressive disorders. Moreover, it is assumed that both self-referential and emotional processing are linked in depressive disorders—specifically, it is assumed that a focus on the self, including a negative view of the self, biases information processing towards negative information [10].

In line with two-process models of information processing, cognitive vulnerability models distinguish between a fast, predominantly associative, implicit, stimulus-driven (bottom-up) processing mode and a temporally slower, reflective, and cognitively elaborate and controlled (top-down) processing mode. Both the associative and the controlled processing modes can occur during stimulus processing in a serial sequence. Any type of stimulus appraisal occurring spontaneously can be assumed to be related to the fast and associative processing level. Reflective and cognitively controlled stimulus processing, including more elaborate processing of the information, can be assumed to follow associative processing and to occur, in particular, when a task demands it, the context requires information to be elaborated and appraised in-depth, or when responses triggered by emotional stimuli have to be actively and cognitively regulated (an overview is given in [10]). As illustrated in Figure 1, according to cognitive vulnerability models, in this information processing cascade, self-reference acts as a processing filter that can bias emotion processing already at early stages of associative information processing. Moreover, it can trigger more elaborate and reflective self-related negative information processing at later stages of cognitive controlled information processing. When viewed from the perspective of information processing, according to these models, depressive vulnerability results from the interaction between self-related and emotional processing occurring at early processing stages, and the processing biases resulting from this interaction are considered the main characteristics, or risk markers, of depressive vulnerability. Therefore, individuals in whom an interaction between self-related and negative emotional processing and, consequently, a self-negativity bias can be observed should be particularly vulnerable to the experience of depressive symptoms compared to individuals not displaying self-referential negative processing. Following cognitive vulnerability models, this self-negativity bias should be best empirically observed during experimental task conditions that (a) favor self-referential associative stimulus processing (e.g., spontaneous processing or passive viewing), (b) require cognitively reflective or cognitive-controlled processing of self-related emotional content, and (c) use stimulus material that varies in emotional content and self-reference.

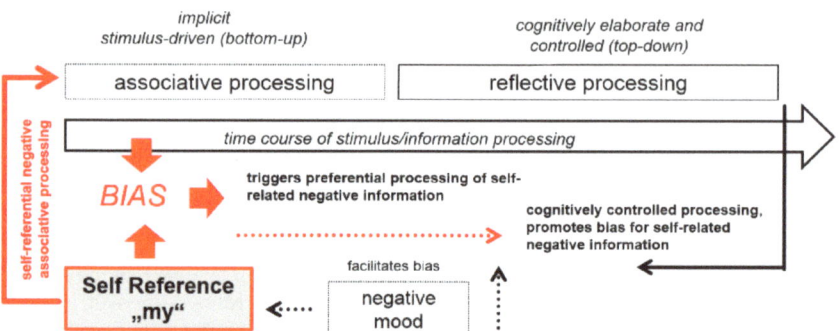

Figure 1. Information processing according to cognitive vulnerability models of depression. For further details, please see text and e.g., [10].

2. Decoding Processing Preferences by Means of EEG Methodology

Methodologically, non-invasive electroencephalography (EEG) and the analysis of event-related brain potentials (ERPs) from the electroencephalogram (EEG) are ideally suited neurophysiological techniques to unravel the processing biases of cognitive vulnerability and determine how these biases unfold over the time course of information processing. EEG methodology allows precise insights into the time course of stimulus- and information processing and their task- and mood-related changes. The latency, as well as the amplitude modulation of early and late ERPs in response to the averaged stimulus response recorded in different task contexts allow differentiated conclusions about the different stimulus and information processing stages from stimulus perception to more in-depth processing [14]. In contrast to behavioral methods, EEG methodology allows the investigation of stimulus processing and its time course even in designs in which no behavioral responses are required and no tasks are given other than to, e.g., read, watch, or listen to the stimulus presentation. EEG methods therefore might be ideally suited to study bottom-up processing without the confounding influence of task-related factors, while at the same time allowing for comparisons between spontaneous (no task) and task-related stimulus processing. Moreover, in event-related stimulus designs, a number of event-related brain potentials can be elicited that, depending on the stimulus and processing specificities, have been shown to be reliable markers of, e.g., perceptual, cognitive, semantic, sensorimotor, or affective processing [14].

Regarding emotional processing, as exemplified in Figure 2, the processing of emotional and neutral stimuli can elicit the modulation of early and late event-related brain potentials (ERPs). For example, in an EEG-ERP experiment with emotional and neutral stimuli, the amplitude amplification of early event-related potentials such as the EPN (early posterior negativity) has often been observed, irrespective of the type of stimuli presented, whether it is words or pictures of emotionally evocative scenes [15–17]. Due to its almost obligatory elicitation across many study designs (passive viewing tasks, fast stimulus presentations, or selective attentive stimulus processing), the EPN has been suggested to be a prominent neural marker of early stimulus-driven processing of high-arousing stimulus content of motivational salience [15–17]. Therefore, in the EEG, a larger EPN amplitude modulation during the processing of stimuli of negative content, compared to stimuli of positive or neutral content, would argue for a rapid, spontaneous, and bottom-up driven allocation of visual attention to negative stimulus information that, according to the two-process models mentioned above, occurs at fast—and probably still associative—levels of information processing.

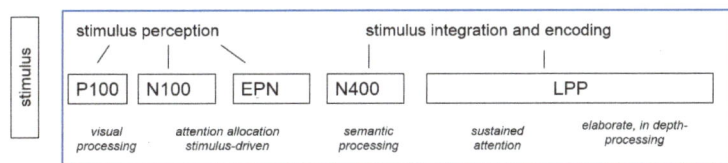

Figure 2. Temporal processing in the EEG as exemplified by the modulation of event-related brain potentials (ERPs).

In addition, the amplitude modulation of temporally later ERP components such as the N400 ERP component, typically elicited by lexical stimuli (but also found to be elicited by non-lexical stimuli such as faces), can indicate semantic processing due to facilitated lexical access [18], facilitated activation of long-term memory representation [19], or contextual integration of the stimuli [20]. Contextual integration can be established either experimentally by a sentence context [20], a specific task context, or an internal context such as the current mood and affective state of the participant [21–23]. Therefore, N400 modulation by negative or positive stimuli compared to neutral stimuli occurring approximately 400 ms after stimulus presentation during spontaneous stimulus processing tasks could suggest mood-dependent processing of negative or positive content [22], with mood acting as a context for stimulus integration and mood-incongruent stimuli being processed differently from mood-congruent stimuli under mood induction (e.g., [21]). Last, but not least, the enhanced amplitude modulations of late ERP components such as the LPP (late positive potential) have been assumed to be associated with deeper processing of the content of negative or positive stimuli (e.g., [22–24]). LPP modulation may therefore indicate a sustained attentional focus for certain information of relevance for the person. Moreover, its modulation has been shown to indicate cognitively controlled emotion processing in a number of emotion regulation studies that used implicit and explicit strategies of stimulus reappraisal or emotion suppression (e.g., [24,25]).

3. Research Gaps

Despite of the prominence of cognitive vulnerability models and the popularity of non-invasive EEG methodology, so far, the literature lacks studies that systematically investigate the time course of emotional processing and its interaction with self-referential processing in line with cognitive vulnerability models in controlled experimental laboratory settings in cognitively vulnerable individuals at risk of depression, with healthy, cognitively non-vulnerable individuals as controls. Investigating the hypothesis of cognitively vulnerable models by means of EEG-ERP methods requires appropriate experimental paradigms that allow to experimentally manipulate the self-reference and the emotionality/the emotional significance of the stimuli independently from, as well as in combination with, each other to induce and trigger self-referential as well as emotional processing at the stage of associative stimulus processing in a bottom-up fashion without any influence from higher-order, task-induced self-referential or emotional processing. This seems crucial to test the early interactions between self-reference and emotion processing postulated by cognitive vulnerability models to occur during implicit associative processing conditions compared to cognitively controlled reflective processing conditions (see Figure 1).

So far, experimental paradigms that investigated the interaction between emotion and self-referential processing in healthy controls and depressive subjects have almost exclusively induced self-referential processing explicitly via task instructions. In particular, paradigms such as the Self-Referent Encoding Task (SRET [26]) have been used, e.g., [26–29]. In this paradigm, participants are exposed to positive and negative personality traits and asked to judge which of the trait words describe them best. Afterwards, the participants are often asked to recall the words presented. The results of several studies suggest that words that have been judged to describe the self and are thus considered to be congruent with the person's own self-views are also better remembered e.g., [30], with healthy subjects differing from depressive subjects in their performance [26–29]. Studies recording EEG-ERP parameters during the SRET in healthy individuals and in individuals suffering from current or remittent depressive disorder [31–33] have found differential processing of positive and negative trait words as a function of group (healthy participants vs. participants with current or remitted depression) during later stages of stimulus processing [31–34]. Temporally earlier processing differences in the time window of early ERPs as a function of group (healthy participants vs. participants with current or remitted depression) have been reported only occasionally in the SRET [33]. Given that in the SRET, self-reference is induced via task instructions, there is a search for tasks and paradigms that allow the independent experimental manipulation of the dimensions of self-reference and emotion based on stimuli instead of task to assess not only the reflective aspects of self-referential processing, but also the associative and stimulus-driven effects of self-referential processing on emotion processing, independent from, as well as in comparison with, task-induced affordances and cognitive demands.

Experimental paradigms that allow both, the independent experimental manipulation and the joint assessment of the effects of the self-referentiality and the emotionality elicited by a stimulus are paradigms such as the so-called 'HisMine' paradigm [for an overview see below or the Supplementary Materials]. In its non-affective form, the HisMine paradigm consists of singular and plural first-person pronouns (1PP: I, my), second-person pronouns (2PP: you, yours), and third-person pronouns (3PP: he/she, hers, his), addressing the self as the subject or object of self-experience, e.g., [35–37]. In its affective form (e.g., [38,39]), self- and/or other-referential possessive pronouns ("my"/"his/her") are combined with nouns that vary in emotionality (positive, negative, and neutral) and are matched on the two emotional dimensions of valence and arousal. In addition, the stimuli are matched on several linguistic dimensions such as word length, or word frequency, or concreteness. The nouns chosen can describe positive and negative emotions (e.g., fear, anger, or happiness), emotion-inducing objects or events (e.g., gun, failure, money, or love), or neutral states or objects rated by normative study samples as neither positive nor negative in valence and low in emotional arousal (e.g., furniture, books, etc.). The nouns can be paired with 1PP, 2PP, or 3PP possessive pronouns to make reference to the reader's own emotions or to the emotions of a third person. In addition, nouns paired with articles instead of pronouns can serve as controls, and the effects triggered by self- and other-referential pronouns on emotion processing triggered by the nouns can be compared to a control condition of non-referential emotional and neutral content. Thus, in summary and in contrast to the SRET paradigm, the stimuli used in the HisMine paradigm can capture a broad range of concepts and emotional feeling states. This may—with respect to cognitive vulnerability models—allow a comprehensive insight into depression-related emotional processing biases. Instructions in the HisMine paradigm can vary from passive viewing and reading to spontaneous emotional evaluation (to trigger associative processing) to active attentive and cognitively controlled processing of the stimuli (to trigger elaborate processing). In combination with EEG methodology, this allows a reliable investigation of the time course and the neural sources of implicit stimulus-driven and task-related explicit self-referential emotional processing. Moreover, when pronouns are paired with non-verbal stimuli (e.g., faces [25]), the specificity of self-referential processing can be explored across and between emotion modalities (verbal and non-verbal). While the HisMine paradigm is conceptualized to investigate the behavioral, subjective, and neuronal correlates of self-referential emotional processing at the word–phrase level, similar approaches for the study of self-referential and emotional processing at the sentence level have been recently suggested (e.g., [40–42]).

The approaches using pronoun–noun phrases or sentences in combination with EEG methodology might be interesting for the investigation of cognitive vulnerability. At the same time, these paradigms might have the capacity to provide insight into what characterizes "normal" and healthy information processing in terms of emotion and self-referential processing. Akin to a negativity bias, a processing advantage for self-related positive information including better memory for positive than negative information has often been observed, especially in association with the above-mentioned SRET self-reference task [30]. The brain regions correlated with self-referential processing in the SRET are well known from several imaging studies (e.g., [43,44]). Nevertheless, the understanding of the time course of these biases as assessed by EEG methodology in combination with self-report methods of mental health and subjective well-being is still scarce. Assessing ERP correlates during pronoun–noun processing (e.g., HisMine paradigm) or at the sentence level might provide solutions to this: At the phrase or the sentence level, the temporal dynamics of processing preferences for self-referential negative and self-referential positive stimuli can be investigated in the same paradigm and, therefore, unravel the risk markers of cognitive vulnerability for depression and the predictive markers of well-being. In the following sections, the results from previous studies investigating the correlates and, specifically, the time course of self-referential emotional processing within the HisMine paradigm will be briefly reviewed, summarized, and discussed with respect to their importance for future studies investigating EEG-ERP-based experimental testing for the prevention, intervention, and treatment of depressive disorders and the promotion of well-being.

4. Exploring the Time Course of Stimulus-Driven, Self-Referential, and Emotional Processing by Means of EEG-ERPs

4.1. EEG Indicators of Healthy Self-Referential Emotional Processing

Using the methodological benefits of electroencephalography (EEG), a number of studies have already explored when, i.e., at which processing stages during reading, self-referential vs. other-referential pronouns are differentiated from each other [35–37], as well as at which processing stages self-referential and other referential pronoun–noun pairs referring to the reader's own emotions or to the emotions of a third person are distinguished from each other and from article–noun pairs having no personal reference or having a neutral meaning [38,39]. In addition, using functional magnetic imaging it was also explored whether brain regions belonging to the emotion, self, and reading networks are involved in these discriminatory processes [45]. In the studies with self-referential and other referential pronoun–noun pairs, the nouns (positive, negative, and neutral) were paired with possessive pronouns (my or his) or articles (the) (e.g., my fear vs. his fear vs. the fear, etc.). The possessive pronouns and articles were either shown together with the nouns without any temporal delay between pronouns and nouns (SOA = 0 ms) [39] or were presented, for example, as primes with a stimulus onset asynchrony (SOA) of 600 ms preceding the presentation of the nouns to investigate and control for expectancy effects induced by the presentation of the pronouns [38]. In neither of these studies was there an explicit self-referential processing task given to the readers; rather, the task in these studies was to silently read the presented words. This was followed by a spontaneous free recall task and a subjective evaluation task (rating task) in which the participants rated the words in terms of valence and emotion intensity (arousal). During reading and in the EEG, early and late event-related potentials (ERPs) were analyzed as cortical indicators of the time course of information processing, and ERP source estimations were used to explore the brain structures involved in the modulation of these ERPs [39]. In addition, changes in the activity of certain brain regions during reading were explored by functional magnetic resonance imaging using whole brain analysis [45]. In the EEG-ERP studies using the affective HisMine paradigm, an interaction of the two factors on which the word pairs could vary, namely, the "emotionality" of a word pair (carried by the nouns) and the "reference" (elicited by the pronouns vs. articles) occurred in the time window of the late positive potential (LPP). The LPP modulation was elicited approximately 500 ms after the stimulus onset and was most pronounced over the centro-parietal electrodes during passive reading of the words. Comparing the processing of self-referential pronoun–noun pairs against the processing of other referential pronoun–noun pairs or article–noun pairs without personal reference revealed a stronger cortical processing of self-referential positive words (self-positivity bias) in the time window of the LPP [38]. The source estimates of the

LPP modulation patterns revealed significant activation in the brain structures involved in autobiographical memory functions and in self-referential processing, which are part of the so-called self-referential processing network, i.e., the cortical midline structures (CMS) or the default mode network [46,47]. The stronger cortical processing of self-referential positive words coincided with a better recall performance in the subsequent memory test in which the participants recalled emotional words spontaneously and significantly better than neutral words, especially when they were self-referential and possessed a positive meaning [39]. In addition, analysis of the ratings obtained after the EEG session showed that participants rated emotional words higher in valence and arousal when during reading, the nouns were related to the own person (self-referential) than when the nouns were related to others (other referential), and rated self-referential positive nouns higher in arousal compared to when positive nouns had no person reference (article–noun pairs) [39]. This more in-depth processing of positive concepts, specifically when perceived as self-referential in healthy participants, accords well with the results from the more explicit self-referential processing tasks in the SRET studies (see Section 3 in this manuscript). This self-positivity has received proof from a number of behavioral studies in which healthy participants in the HisMine paradigm were asked to affectively judge the words as positive, negative, or neutral based on their gut feelings e.g., [48–50], for an overview see the Supplementary Materials. Across these behavioral valence–judgment studies, healthy participants had significantly more valence–congruent judgments for positive words than for negative or neutral words, specifically when these were linguistically made self-referential by the pronoun pairings. The results fit with theoretical assumptions of a self-positivity bias among healthy subjects. This self-positivity bias might stem from positive self-views and overall positive self-concepts [51,52], that in healthy subjects, even if sometimes too positive and optimistic, characterizes positive self-attribution in the service of well-being and mood-regulation. The experimentally observed self-positivity bias in the EEG studies thus suggests a spontaneously deeper encoding of self-referential emotional pronoun–noun stimuli during reading (LPP), as well as better memory-related recall and a faster valence-congruent appraisal, of these stimuli that is in line with these theoretical assumptions of positive self-attribution in the service of well-being.

The observations of differential processing of positive vs. negative self vs. other vs. no person-referential pronoun–noun stimuli as indicators of healthy and normal processing receive further support from recent EEG-ERP and fMRI studies from other laboratories investigating the interaction between self-referential and emotional information processing via sentences. In these studies, pronouns and emotional words are embedded in a sentence context e.g., [40–42].

When evaluated across studies, the findings from the EEG studies thus far support the notion that in healthy subjects, the impact of self-referential processing on the processing of emotional information first occurs at stages of information processing at which perceptual information is processed, lexical information is accessed, and information is elaborated in relation to subjective experiences stored in memory. Indeed, as graphically illustrated in Figure 3, in healthy participants, a cascade of processing steps seems to precede and occur in parallel prior to the self-positivity bias. Temporally earlier interaction effects of self-referential stimuli on the processing of emotional stimuli may, in healthy subjects, be restricted to a few occasions, including highly accessible information or top-down attentive processing. In support of a cascade of processing steps preceding the self-positivity bias, the results from an fMRI study [45] suggest that in healthy subjects, significant activity changes in at least three brain region of interests in trials in which the positive content of nouns and self-reference of first-person possessive pronouns are combined, including activity changes in the MPFC regions involved in self-referential processing, the left and right (anterior) insular cortex as a region of the salience network [53] and a relay or hub at the interface of cognition, emotion, and the body (interoception), and the amygdala as a significant emotion and relevance detector [54].

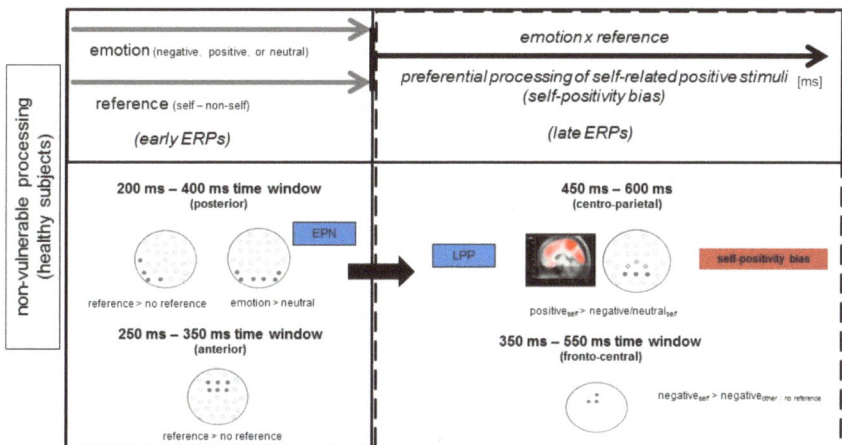

Figure 3. Cascade of processing steps preceding and underlying the self-positivity bias during reading of self-referential and other-referential pronoun-noun and non-referential article-noun pairs according to the results from [39].

4.2. EEG Indicators of Cognitive Vulnerable Self-Referential Emotional Processing

In summary, the EEG results in healthy subjects summarized in Section 4.1. provide indirect support for the assumptions that in tasks with no explicit self-referential instruction, temporally earlier interactions between self-referential and emotion processing at early stages of stimulus-driven associative processing could indeed be electrophysiological markers specific to individuals experiencing depressive symptoms or individuals prone to or at risk of or already suffering from depressive disorders. The first evidence for a self-negativity bias in the processing of pronouns and nouns as a marker of depressive symptoms was found in the EEG-ERP studies in which pronouns were used as primes for the nouns and processing of self-referential pronoun–noun pairs was compared with the processing of article noun pairs [38]. The participants' depressive symptoms were assessed via self-report standardized assessment tools, allowing assumptions about the presence of depressive symptoms and their severity of actual depressive symptoms (the BDI inventory [55]). The results showed that a deeper cortical processing (LPP) of self-referential negative rather than positive emotional words (self-negativity bias) is positively correlated with the degree of self-reported depressive symptoms. In addition, and in line with cognitive vulnerability models, this self-negativity bias started at temporally earlier processing stages—the processing of self-referential, prime- x emotional target pairs elicited a pronounced N400 potential whose amplitudes, akin to the modulation of the LPP, showed depression-congruent modulation yielding smaller N400 amplitudes for self-referential negative emotional words.

Further, though yet preliminary, evidence for even earlier temporal influences, as theoretically predicted by cognitive vulnerability models, is provided by studies investigating the EEG-ERP modulation patterns elicited during the affective HisMine paradigm in depressed and medicated individuals (with a current diagnosis of major depressive disorder) and healthy controls [56]. The HisMine paradigm was presented in multiple runs comprising passive reading/viewing conditions and conditions with instructions to pay attention to or ignore the self-referential word pairs. The first preliminary analysis of the EEG data provided in [56] suggests that depressed subjects, in contrast to the healthy control subjects, show a processing bias for negative self-referential words in the EEG which is already present during the silent reading in early time windows that started during the N100 modulation, and especially while modulating the amplitudes of the EPN Supplementary Materials [56]. In the study design, self-reference was induced by second-person possessive pronouns instead of first-person possessive pronouns. These results complement the findings from SRET studies in depressed and non-depressed participant samples, showing that besides significant group differences in the time window of elaborate processing (LPP), there are early ERP processing differences between groups. As a marker of risk of depression, more recently, early processing biases towards self-referential negative information have been shown to continue into the remission of the disorder, being still observed in remitted depressed participants [57]. The hypothesis of an early self-negativity bias in the processing of self-referential emotional content as EEG-ERP markers of depressed vulnerability is currently further evaluated and extended in a research project comprising a series of EEG-ERP studies [58]. In the studies, EEG-ERP modulation patterns are assessed in samples of healthy subjects, who, prior to study participation, were preselected into groups of depression-prone individuals vs. individuals not prone to depression, according to clinical screening and assessment tools [58].

4.3. Studying Self-Referential Emotional Processing Biases as Markers of Cognitive Vulnerability and Well-Being by Means of Language-Dependent EEG Paradigms: Potential Limitations and Advantages

The investigation of emotional and self-referential processing by means of linguistic stimuli and language paradigms might be considered a limitation or a restriction. The reservation might stem from classical theoretical definitions of emotions that define emotions from an evolutionary, biological perspective and that consider emotion processing as being largely independent from language processing; specifically, emotion perception is suggested to be primarily driven by non-linguistic physical features of stimuli that the brain, due to its biological significance, is attuned to respond to with heightened attention and the preparation of bodily preparation for fight or flight and approach or consumption (for an overview, [59,60]). While there is no doubt that emotional stimuli, be it words, pictures, or faces, can elicit similar early ERP modulation in a number of task contexts (for an overview, see [61,62]), there is also no doubt that modality- and stimulus-specific processing effects between verbal and non-verbal emotional stimuli might, nevertheless, exist. Regarding depression and its cognitive vulnerability, a particular interest in the use of language paradigms to study emotion and the self might stem from three theoretical grounds: first, to examine the theoretically assumed "errors" in cognition and self-reflection in the schemata of depressed vs. non-depressed individuals; second, to determine, in line with modern theoretical accounts of embodiment and embodied cognition, how cognitive biases are influenced by sensory and motor information among cognitively vulnerable individuals at risk of depression vs. among individuals already suffering from depressive disorders vs. healthy controls; and third, given that depression, being an affective disorder that includes mood changes, to investigate psychophysiological, neurophysiological, and neurobiological changes to determine how cognitive and bodily processes might interact while participants are construing meaning from language stimuli that might trigger feelings and subjective experience relating to one's own person.

Methodologically, an important limitation of the EEG-ERP technique with regard to its application as a diagnostic tool is its reliability and power at the individual level. However, unless the focus is on the group level, results from the EEG-ERP analysis could be used as an additional testing option, in addition to self-report diagnostic questionnaires, to distinguish between individuals at risk and those without risk of mental health conditions, specifically when EEG-ERP analysis is combined with an experimental and theoretically driven task or paradigm, such as those proposed in this perspective paper. Regarding clinical disorders, previous studies have already provided promising results for such an approach. Using EEG-ERP analysis, amongst others, in combination with more advanced preprocessing and analysis tools, or in combination with a mixed methodological analysis based on time- and frequency measures, has shown to obtain good results in, for example, attention-deficit/hyperactivity disorder (ADHD) or the prediction of Alzheimer's disease (e.g., [63,64]) or major depressive disorder, e.g., [65]. Regarding depressive disorders, EEG analyses based on time–frequency analysis and/or independent-component-based event-related synchronisation/desynchronisation analysis, as reviewed and suggested in [66], would be interesting methodological approaches to be combined in future studies with paradigms investigating alterations in self-referential emotional processing in individuals with and without the risk of depressive disorders.

5. Questions for the Future

Depressive disorder is not one disease: all disease and disorder-related changes in mood, as well as in self-reference and affective processing can, as outlined in this manuscript, theoretically manifest as biases in the processing of self-referential positive and negative information well prior to onset of the disorder. EEG and, specifically, EEG-ERP methodology are currently pursued as means that, if combined with an experimental paradigm, could help unravel the mechanisms of processing preferences and their clinical relevance as traits or state markers of the risk of mental health conditions such as depression. Although preliminary evidence has already been accumulated, future studies should principally replicate and extend the current evidence with a focus on the following guiding questions:

(1) At which stages of stimulus processing does an interaction between self-referential and emotional processing occur? Can the preliminary findings, illustrated in this manuscript, be replicated in larger cohorts of both, cognitively vulnerable and already depressed individuals vs. healthy controls?

(2) To what degree can processing preferences for self-related negative and positive stimuli, respectively, be influenced by self-related attentive and cognitively controlled processing, and which of these influences are specific for depression and its risk?

(3) Is self-negativity bias the only marker of cognitive vulnerability, or is a self-negativity bias accompanied by a reduced self-positivity bias as well (see Figure 4)?

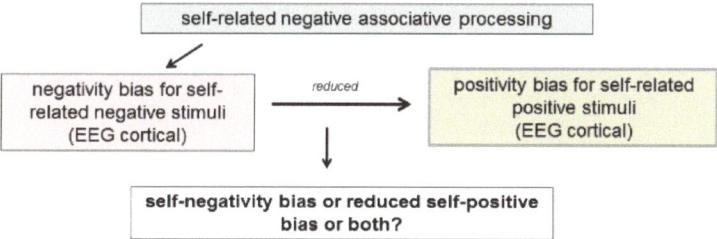

Figure 4. Extended cognitive vulnerability model (EEG cortical processing).

(4) Do the observed electrophysiological ERP correlates of the processing preferences for self-related negative or positive stimuli prove to be temporally stable markers of subjective well-being and cognitive vulnerability?

(5) Do the results vary across languages, and do they also apply to a bilingual/multilingual context?

Importantly, question 5 is receiving increasing attention in a global society in which bilingual and multilingual psychotherapy is becoming more and more prevalent. Recent research, including EEG-ERP studies, has observed a weaker emotional reactivity to negative language content in the second language than in the first, with results being further modulated by mood, proficiency, and language use, e.g., [67,68]. Future projects and studies that focus on questions (1)–(5), respectively, will fill an important gap between the understanding of processing preferences, their electrophysiological dynamics, and their psychological and clinical significance as depressive vulnerability indicators and markers of subjective well-being. Based on this evidence, experimental tests could then be developed with which depression-associated processing biases as well as indicators of well-being could be detected by means of EEG-ERP measurements and used for depression monitoring and improvement in well-being in the context of health prevention, intervention, and therapy.

Supplementary Materials: The following supporting information can be downloaded at: https://www.mdpi.com/article/10.3390/app12157740/s1.

Funding: This research was funded by the German Research Foundation (DFG), HE5880/7-1 granted to C.H.

Institutional Review Board Statement: Not applicable.

Informed Consent Statement: Not applicable.

Data Availability Statement: Not applicable.

Conflicts of Interest: The author declares no conflict of interest. The funders had no role in the design of the study; in the collection, analyses, or interpretation of data; in the writing of the manuscript; or in the decision to publish the results.

References

1. Available online: https://www.who.int/en/news-room/fact-sheets/detail/depression (accessed on 5 June 2022).
2. Solmi, M.; Radua, J.; Olivola, M.; Croce, E.; Soardo, L.; Salazar de Pablo, G.; Il Shin, J.; Kirkbride, J.B.; Jones, P.; Kim, J.H.; et al. Age at onset of mental disorders worldwide: Large-scale meta-analysis of 192 epidemiological studies. *Mol. Psychiatry* **2022**, *27*, 281–295. [CrossRef] [PubMed]
3. Kalia, M. Neurobiological basis of depression: An update. *Metabolism* **2005**, *54*, 24–27. [CrossRef] [PubMed]
4. Foland-Ross, L.C.; Gotlib, I.H. Cognitive and neural aspects of information processing in major depressive disorder: An integrative perspective. *Front. Psychol.* **2012**, *3*, 489. [CrossRef] [PubMed]
5. Gotlib, I.H.; Joormann, J. Cognition and depression: Current status and future directions. *Annu. Rev. Clin. Psychol.* **2010**, *6*, 285–312. [CrossRef]
6. Connolly, S.L.; Abramson, L.Y.; Alloy, L.B. Information processing biases concurrently and prospectively predict depressive symptoms in adolescents: Evidence from a self-referent encoding task. *Cogn. Emot.* **2016**, *30*, 550–560. [CrossRef] [PubMed]
7. Gordon, E.; Barnett, K.J.; Cooper, N.J.; Tran, N.; Williams, L.M. An "integrative neuroscience" platform: Application to profiles of negativity and positivity bias. *J. Integr. Neurosci.* **2008**, *7*, 345–366. [CrossRef] [PubMed]
8. Rude, S.S.; Durham-Fowler, J.A.; Baum, E.S.; Rooney, S.B.; Maestas, K.L. Self-report and Cognitive Processing Measures of Depressive Thinking Predict Subsequent Major Depressive Disorder. *Cogn. Ther. Res.* **2010**, *34*, 107–115. [CrossRef]
9. Watters, A.J.; Williams, L.M. Negative biases and risk for depression; integrating self-report and emotion task markers. *Depress. Anxiety* **2011**, *28*, 703–718. [CrossRef] [PubMed]
10. Beevers, C.G. Cognitive vulnerability to depression: A dual process model. *Clin. Psychol. Rev.* **2005**, *25*, 975–1002. [CrossRef]
11. Fossati, P. Is major depression a cognitive disorder? *Rev. Neurol.* **2018**, *174*, 212–215. [CrossRef]
12. Beck, A.T. *Depression: Causes and Treatment*; University of Pennsylvania Press: Philadelphia, PA, USA, 1967.
13. Beck, A.T.; Epstein, N.; Harrison, R. Cognitions, attitudes and personality dimensions in depression. *Br. J. Cogn. Psychother.* **1983**, *1*, 1–16.
14. Luck, S.J. *An Introduction to the Event-Related Potential Technique*; MIT Press: Cambridge, MA, USA, 2014.
15. Schupp, H.T.; Junghöfer, M.; Weike, A.I.; Hamm, A.O. Attention and emotion: An ERP analysis of facilitated emotional stimulus processing. *Neuroreport* **2003**, *14*, 1107–1110. [CrossRef] [PubMed]
16. Kissler, J.; Herbert, C.; Peyk, P.; Junghofer, M. Buzzwords: Early cortical responses to emotional words during reading. *Psychol. Sci.* **2007**, *18*, 475–480. [CrossRef]
17. Herbert, C.; Junghofer, M.; Kissler, J. Event related potentials to emotional adjectives during reading. *Psychophysiology* **2008**, *45*, 487–498. [CrossRef] [PubMed]
18. Lau, E.F.; Phillips, C.; Poeppel, D. A cortical network for semantics: (de)constructing the N400. *Nat. Rev. Neurosci.* **2008**, *9*, 920–933. [CrossRef] [PubMed]
19. Kutas, M.; Federmeier, K.D. Electrophysiology reveals semantic memory use in language comprehension. *Trends Cogn. Sci.* **2000**, *4*, 463–470. [CrossRef]
20. Kutas, M.; Hillyard, S.A. Reading senseless sentences: Brain potentials reflect semantic incongruity. *Science* **1980**, *207*, 203–205. [CrossRef] [PubMed]
21. Kiefer, M.; Schuch, S.; Schenck, W.; Fiedler, K. Mood states modulate activity in semantic brain areas during emotional word encoding. *Cereb. Cortex* **2007**, *17*, 1516–1530. [CrossRef]
22. Herbert, C.; Kissler, J.; Junghöfer, M.; Peyk, P.; Rockstroh, B. Processing of emotional adjectives: Evidence from startle EMG and ERPs. *Psychophysiology* **2006**, *43*, 197–206. [CrossRef] [PubMed]
23. Ferrari, V.; Codispoti, M.; Cardinale, R.; Bradley, M.M. Directed and motivated attention during processing of natural scenes. *J. Cogn. Neurosci.* **2008**, *20*, 1753–1761. [CrossRef] [PubMed]

24. Hajcak, G.; Dunning, J.P.; Foti, D. Motivated and controlled attention to emotion: Time-course of the late positive potential. *Clin. Neurophysiol.* **2009**, *120*, 505–510. [CrossRef] [PubMed]
25. Herbert, C.; Sfaerlea, A.; Blumenthal, T. Your emotion or mine: Labeling feelings alters emotional face perception—An ERP study on automatic and intentional affect labeling. *Front. Hum. Neurosci.* **2013**, *7*, 378. [CrossRef] [PubMed]
26. Kuiper, N.A.; Derry, P.A. Depressed and nondepressed content self-reference in mild depressives. *J. Pers.* **1982**, *50*, 67–80. [CrossRef]
27. Alloy, L.B.; Abramson, L.Y.; Murray, L.A.; Whitehouse, W.G.; Hogan, M.E. Self-referent information-processing in individuals at high and low cognitive risk for depression. *Cogn. Emot.* **1997**, *11*, 539–568. [CrossRef]
28. Dozois, D.J.; Dobson, K.S. Information processing and cognitive organization in unipolar depression: Specificity and comorbidity issues. *J. Abnorm. Psychol.* **2001**, *110*, 236. [CrossRef] [PubMed]
29. Dobson, K.S.; Shaw, B.F. Specificity and stability of self-referent encoding in clinical depression. *J. Abnorm. Psychol.* **1987**, *96*, 34–40. [CrossRef]
30. Matt, G.E.; Vázquez, C.; Campbell, W.K. Mood-congruent recall of affectively toned stimuli: A meta-analytic review. *Clin. Psychol. Rev.* **1992**, *12*, 227–255. [CrossRef]
31. Watson, L.A.; Dritschel, B.; Obonsawin, M.C.; Jentzsch, I. Seeing yourself in a positive light: Brain correlates of the self-positivity bias. *Brain Res.* **2007**, *1152*, 106–110. [CrossRef] [PubMed]
32. Allison, G.O.; Kamath, R.A.; Carrillo, V.; Alqueza, K.L.; Pagliaccio, D.; Slavich, G.M.; Shankman, S.A.; Auerbach, R.P. Self-referential Processing in Remitted Depression: An Event-Related Potential Study. *Biol. Psychiatry Glob. Open Sci.* **2021**. [CrossRef]
33. Shestyuk, A.Y.; Deldin, P.J. Automatic and strategic representation of the self in major depression: Trait and state abnormalities. *Am. J. Psychiatry* **2010**, *167*, 536–544. [CrossRef]
34. Dainer-Best, J.; Trujillo, L.T.; Schnyer, D.M.; Beevers, C.G. Sustained engagement of attention is associated with increased negative self-referent processing in major depressive disorder. *Biol. Psychol.* **2017**, *129*, 231–241. [CrossRef]
35. Blume, C.; Herbert, C. The HisMine-Paradigm: A new paradigm to investigate self-awareness employing pronouns. *Soc. Neurosci.* **2014**, *9*, 289–299. [CrossRef] [PubMed]
36. Herbert, C.; Blume, C.; Northoff, G. Can we distinguish an "I" and "ME" during listening?—An event-related EEG study on the processing of first and second person personal and possessive pronouns. *Self Identity* **2016**, *15*, 120–138. [CrossRef]
37. Herbert, C.; Northoff, G.; Hautzinger, M. Depressive Symptome, kardiale Regulation und kortikale Verarbeitung bei Leistungssportlern. *Dtsch. Z. Sportmed.* **2016**, *2016*, 293–300. [CrossRef]
38. Herbert, C.; Pauli, P.; Herbert, B.M. Self-reference modulates the processing of emotional stimuli in the absence of explicit self-referential appraisal instructions. *Soc. Cogn. Affect. Neurosci.* **2011**, *6*, 653–661. [CrossRef] [PubMed]
39. Herbert, C.; Herbert, B.M.; Ethofer, T.; Pauli, P. His or mine? The time course of self-other discrimination in emotion processing. *Soc. Neurosci.* **2011**, *6*, 277–288. [CrossRef] [PubMed]
40. Fields, E.C.; Kuperberg, G.R. Dynamic Effects of Self-Relevance and Task on the Neural Processing of Emotional Words in Context. *Front. Psychol.* **2015**, *6*, 2003. [CrossRef]
41. Fields, E.C.; Kuperberg, G.R. Loving yourself more than your neighbor: ERPs reveal online effects of a self-positivity bias. *Soc. Cogn. Affect. Neurosci.* **2015**, *10*, 1202–1209. [CrossRef]
42. Fields, E.C.; Weber, K.; Stillerman, B.; Delaney-Busch, N.; Kuperberg, G.R. Functional MRI reveals evidence of a self-positivity bias in the medial prefrontal cortex during the comprehension of social vignettes. *Soc. Cogn. Affect. Neurosci.* **2019**, *14*, 613–621. [CrossRef]
43. Northoff, G.; Heinzel, A.; de Greck, M.; Bermpohl, F.; Dobrowolny, H.; Panksepp, J. Self-referential processing in our brain—A meta-analysis of imaging studies on the self. *NeuroImage* **2006**, *31*, 440–457. [CrossRef]
44. Lemogne, C.; Delaveau, P.; Freton, M.; Guionnet, S.; Fossati, P. Medial prefrontal cortex and the self in major depression. *J. Affect. Disord.* **2012**, *136*, e1–e11. [CrossRef] [PubMed]
45. Herbert, C.; Herbert, B.M.; Pauli, P. Emotional self-reference: Brain structures involved in the processing of words describing one's own emotions. *Neuropsychologia* **2011**, *49*, 2947–2956. [CrossRef] [PubMed]
46. Northoff, G.; Bermpohl, F. Cortical midline structures and the self. *Trends Cogn. Sci.* **2004**, *8*, 102–107. [CrossRef] [PubMed]
47. Qin, P.; Northoff, G. How is our self related to midline regions and the default-mode network? *NeuroImage* **2011**, *57*, 1221–1233. [CrossRef] [PubMed]
48. Weis, P.P.; Herbert, C. Bodily Reactions to Emotional Words Referring to Own versus Other People's Emotions. *Front. Psychol.* **2017**, *8*, 1277. [CrossRef] [PubMed]
49. Meixner, F.; Herbert, C. Whose emotion is it? Measuring self-other discrimination in romantic relationships during an emotional evaluation paradigm. *PLoS ONE* **2018**, *13*, e0204106. [CrossRef]
50. Herbert, C.; Hesse, K.; Wildgruber, D. Emotion and self in psychotic disorders: Behavioral evidence from an emotional evaluation task using verbal stimuli varying in emotional valence and self-reference. *J. Behav. Ther. Exp. Psychiatry* **2018**, *58*, 86–96. [CrossRef]
51. Diener, E.; Napa-Scollon, C.K.; Oishi, S.; Dzokoto, V.; Suh, E.M. Positivity and the Construction of Life Satisfaction Judgments: Global Happiness is not the Sum of its Parts. *J. Happiness Stud.* **2000**, *1*, 159–176. [CrossRef]
52. Caprara, G.V.; Eisenberg, N.; Alessandri, G. Positivity: The Dispositional Basis of Happiness. *J. Happiness Stud.* **2017**, *18*, 353–371. [CrossRef]
53. Uddin, L.Q. Salience processing and insular cortical function and dysfunction. *Nat. Rev. Neurosci.* **2015**, *16*, 55–61. [CrossRef]

54. Sander, D.; Grafman, J.; Zalla, T. The human amygdala: An evolved system for relevance detection. *Rev. Neurosci.* **2003**, *14*, 303–316. [CrossRef] [PubMed]
55. BDI-II—Beck-Depressions-Inventar Revision—Hogrefe Verlag. Available online: https://www.testzentrale.de/shop/beck-depressions-inventar.html (accessed on 5 June 2022).
56. Herbert, C.; Ostermair, J.; Herbst, S.; Pauli, P.; Reif, A.; Fallgatter, A.; Herrman, M. *It's Yours! The Negativity Bias in Major Depressive Disorder is Self-Specific: Evidence from Event-Related Brain Potential Studies*; DGPPN; German Society for Psychiatry and Psychotherapy, Psychosomatics and Neurology: Berlin, Germany, 2014.
57. Auerbach, R.P.; Stanton, C.H.; Proudfit, G.H.; Pizzagalli, D.A. Self-referential processing in depressed adolescents: A high-density event-related potential study. *J. Abnorm. Psychol.* **2015**, *124*, 233–245. [CrossRef] [PubMed]
58. Herbert, C. Processing Preferences for Self-Related Emotional Words as Markers of Cognitive Vulnerability and Well-Being—Cerebral and Behavioral Correlates and Mechanisms. Project Funded by the German Research Foundation (DFG). 2019–2023. Available online: https://gepris.dfg.de/gepris/projekt/415209420 (accessed on 5 June 2022).
59. Cosmides, L.; Tooby, J. Evolutionary psychology and the emotions. In *Handbook of Emotions*, 2nd ed.; Lewis, M., Haviland-Jones, J.M., Eds.; Guilford: New York, NY, USA, 2000; pp. 91–115.
60. Lang, P.J. Emotion and motivation: Toward consensus definitions and a common research purpose. *Emot. Rev.* **2010**, *2*, 229–233. [CrossRef]
61. Kissler, J.; Assadollahi, R.; Herbert, C. Emotional and semantic networks in visual word processing: Insights from ERP studies. In *Understanding Emotions*; Elsevier: Amsterdam, The Netherlands, 2006; pp. 147–183; ISBN 0079-6123.
62. Herbert, C.; Ethofer, T.; Fallgatter, A.J.; Walla, P.; Northoff, G. The Janus face of language: Where are the emotions in words and where are the words in emotions? *Front. Psychol.* **2018**, *9*, 650. [CrossRef] [PubMed]
63. Jahanshahloo, H.R.; Shamsi, M.; Ghasemi, E.; Kouhi, A. Automated and ERP-Based Diagnosis of Attention-Deficit Hyperactivity Disorder in Children. *J. Med. Signals Sens.* **2017**, *7*, 26–32. [CrossRef]
64. Chapman, R.M.; Nowlis, G.H.; McCrary, J.W.; Chapman, J.A.; Sandoval, T.C.; Guillily, M.D.; Gardner, M.N.; Reilly, L.A. Brain event-related potentials: Diagnosing early-stage Alzheimer's disease. *Neurobiol. Aging* **2007**, *28*, 194–201. [CrossRef]
65. Mumtaz, W.; Malik, A.S.; Yasin, M.A.M.; Xia, L. Review on EEG and ERP predictive biomarkers for major depressive disorder. *Biomed. Signal Process. Control* **2015**, *22*, 85–98. [CrossRef]
66. De Aguiar Neto, F.S.; Rosa, J.L.G. Depression biomarkers using non-invasive EEG: A review. *Neurosci. Biobehav. Rev.* **2019**, *105*, 83–93. [CrossRef]
67. Wu, Y.J.; Thierry, G. How reading in a second language protects your heart. *J. Neurosci.* **2012**, *32*, 6485–6489. [CrossRef]
68. Jończyk, R.; Korolczuk, I.; Balatsou, E.; Thierry, G. Erratum to: Keep calm and carry on: Electrophysiological evaluation of emotional anticipation in the second language. *Soc. Cogn. Affect. Neurosci.* **2021**, *16*, 642. [CrossRef]

MDPI AG
Grosspeteranlage 5
4052 Basel
Switzerland
Tel.: +41 61 683 77 34

Applied Sciences Editorial Office
E-mail: applsci@mdpi.com
www.mdpi.com/journal/applsci

Disclaimer/Publisher's Note: The title and front matter of this reprint are at the discretion of the Guest Editors. The publisher is not responsible for their content or any associated concerns. The statements, opinions and data contained in all individual articles are solely those of the individual Editors and contributors and not of MDPI. MDPI disclaims responsibility for any injury to people or property resulting from any ideas, methods, instructions or products referred to in the content.

www.ingramcontent.com/pod-product-compliance
Lightning Source LLC
LaVergne TN
LVHW072321090526
838202LV00019B/2329